ELECTRONIC COMMUNICATIONS
Third Edition

DENNIS RODDY

JOHN COOLEN

Lakehead University
Thunder Bay, Ontario
Canada

Reston Publishing Company, Inc.
Reston, Virginia
A Prentice-Hall Company

Library of Congress Cataloging in Publication Data

Roddy, D.
  Electronic communications.

  Includes index.
  1. Telecommunication.   I. Coolen, John.   II. Title.
TK5101.R56   1984   621.38   84-3460
ISBN 0-8359-1598-0

© 1977, 1981, 1984
Reston Publishing Company, Inc.
A Prentice-Hall Company
Reston, Virginia

10   9   8   7   6   5   4   3   2   1

Printed in the United States of America

Contents

Preface

This book is intended for use in the final year of a technology program in telecommunications. The level of treatment presupposes a detailed knowledge of electronics, mathematics, and basic electric circuits, such as form part of most telecommunications technology programs at community colleges and technical institutes.

Some repetition of basic material is included for the sake of completeness, but the applications and illustrative examples are all directed towards the field of telecommunications. The text covers in considerable detail communications fundamentals, circuits, modulation, transmission and radiation, and systems, including data transmission and line communications.

The major additions to the third edition are new chapters on satellite communications and fiber optic communications. Also, a number of new sections have been included in the other chapters: the section on filters in Chapter 1 has been expanded to include a more detailed explanation of basic filter types, and also details of surface wave acoustic (saw) type filters; the section on the Smith chart in Chapter 12 has been expanded, and a more detailed account of microstrip and stripline circuits has been added. Some details of extremely low-frequency propagation have been included in Chapter 14, and much of Chapter 15 has been rewritten to include a more detailed explanation of antenna fundamentals. The topics of pulse and digital transmission, Chapters 11 and 17, have been expanded, particularly the sections dealing with digital filtering and pulse code modulation. An appendix, giving the answers to odd-numbered problems, has been added.

We are grateful to the many users who took the time to respond to the publisher's questionnaire regarding the text. Most of the additions and corrections to the third edition were made as a direct result of the responses received. As with the previous editions, our thanks go to the many people and companies who provided technical information. Individual acknowledgements are made at the appropriate places in the text.

Dennis Roddy
John Coolen
Lakehead University
Thunder Bay, Ontario

PART 1

COMMUNICATIONS FUNDAMENTALS

CHAPTER 1

Analysis of Passive Circuits

1.1 INTRODUCTION

Passive circuits are those which contain only resistance, inductance, and capacitance (or their equivalents, as in the case of certain types of filters described in Section 1.15). Many such circuits are encountered in electronic communication systems, and some of the more usual ones will be examined in this chapter. It is assumed that the student is familiar with the basic circuit concepts and laws (as found, for example, in B. Zeines, *Electric Circuit Analysis*, Reston Publishing Company, Inc., Reston, Va., 1972); some of the more important results are summarized in Sections 1.3 and 1.5.

1.2 SOURCE REPRESENTATION

The voltage or current source supplying a circuit may also be represented by circuit components, but should be separately identifiable from the circuit itself. The two idealized source representations most commonly used are the *constant-voltage equivalent circuit*, Fig. 1.1(a) and the *constant-current equivalent circuit*, Fig. 1.1(b). In both cases, R_s represents the internal resistance of the

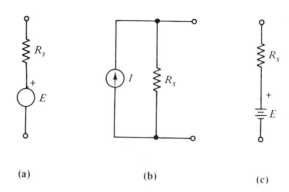

(a) (b) (c)

FIGURE 1.1. Source representations: (a) voltage; (b) current; (c) battery.

source. For the voltage circuit, E represents the generated EMF, there being zero resistance within the E circle. E is constant, i.e., independent of the loading applied to the terminals; it may, of course, be time-varying. For the current circuit, I is the generated current, the resistance of the generator branch being infinite. I is constant, i.e., independent of external loading; but also, of course, it may be time-varying.

A given source may be represented by either one of these circuits; that is, they are alternative ways of looking at the same source. The choice of which equivalent circuit to use is often a matter of personal preference, although the physical source may suggest the most appropriate one; a good example of this is the battery, represented as shown in Fig. 1.1(c).

1.3 SUMMARY OF BASIC RESULTS FOR RESISTIVE NETWORKS

A voltage V (which may be time-varying) applied to a constant resistance R results in a current I given by Ohm's law:

$$V = IR \tag{1.1}$$

Conventionally, the voltage and current are represented by arrows, as shown in Fig. 1.2(a); the current arrow shows the direction of conventional current flow (which may be real positive-charge movement in the direction of the arrow, or electron movement in the opposite direction), while the voltage arrow points to the more positive of the two terminals. Keeping in mind that conventional current flows from the more positive to the less positive terminal, it will be seen that the two arrows convey the same information, but this little bit of redundancy turns out to be useful in circuit analysis. In some texts the voltage arrow is replaced by a + sign, as shown in Fig. 1.2(b). Again it is well to keep in mind that this is not a direct-current symbol, as the voltage may be time-varying. The current may also be indicated by an arrow alongside the component symbol, Fig. 1.2(b).

The effective series resistance R_{ser} of the chain of resistors shown in Fig. 1.2(c) is given by

$$R_{ser} = R_1 + R_2 + \ldots + R_n \qquad (1.2)$$

The effective parallel resistance R_p of the network shown in Fig. 1.2(c) is obtained from

$$\frac{1}{R_p} = \frac{1}{R_1} + \frac{1}{R_2} + \ldots + \frac{1}{R_n} \qquad (1.3)$$

In terms of conductance G, this can be written as

$$G_p = G_1 + G_2 + \ldots + G_n \qquad (1.4)$$

Where only two resistors are involved, Eq. (1.3) may be rearranged to give

$$R_p = \frac{R_1 R_2}{R_1 + R_2} \qquad (1.5)$$

As a check on calculations, it is worth noting that the effective series resistive will always be greater than the greatest single resistance in the chain, while the effective parallel resistance will always be less than the least single resistance in the parallel network.

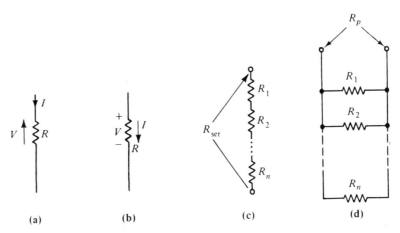

FIGURE 1.2. Symbols for voltage, current and resistance: (a) conventional symbols for voltage and current; (b) alternative symbols for voltage and current; (c) total resistance of a series group; (d) total resistance of a parallel group.

1.4 ANALYSIS OF RESISTIVE CIRCUITS

A *series-parallel circuit* is shown in Fig. 1.3(a). Note that the three 10-Ω resistors are *not* in series because they do not carry the same current. To find the effective input resistance of this circuit, i.e., the resistance between the terminals, first combine R_2 and R_3 in series to give

$$(R_2 + R_3) = 10 + 10 = 20 \ \Omega$$

Next combine R_4 in parallel with $(R_2 + R_3)$:

$$\frac{R_4(R_2 + R_3)}{R_4 + (R_2 + R_3)} = \frac{25 \times 20}{25 + 20} = 11.1 \ \Omega$$

Finally, add this value to R_1 in series, so that the effective input resistance R_{in} is

$$R_{in} = 10 + 11.1 = 21.1 \ \Omega$$

Proper application of Kirchhoff's loop (or mesh) equations can considerably simplify circuit analysis; although their use may not be justified in such a simple circuit as shown in Fig. 1.3(a), this will be used to illustrate the principle. The circuit is redrawn in Fig. 1.3(b), with a battery source E applied. The input resistance R_{in} is

$$R_{in} = \frac{E}{I_1} \tag{1.6}$$

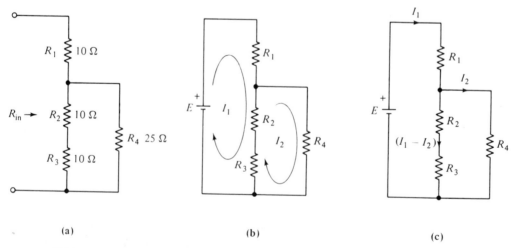

(a)   (b)   (c)

FIGURE 1.3. (a) A series-parallel circuit. The circuit currents: (b) mesh; (c) branch.

Current I_1 is imagined to circulate around the loop consisting of E, R_1, R_2, and R_3, and current I_2 around loop R_2, R_3, and R_4. The actual currents would be as shown in Fig. 1.3(c), but it will be seen that for both circuits the current carried by R_2 and R_3 is the same, $I_1 - I_2$. The advantage of the first method is that Kirchhoff's loop equations can be written down systematically; thus, from Fig. 1.3(b), following the circulating currents clockwise around each loop,

$$E = I_1(R_1 + R_2 + R_3) - I_2(R_2 + R_3)$$
$$0 = -I_1(R_2 + R_3) + I_2(R_2 + R_3 + R_4)$$

Solving these by the use of determinants,

$$I_1 = \frac{\begin{vmatrix} E & -(R_2 + R_3) \\ 0 & R_2 + R_3 + R_4 \end{vmatrix}}{\begin{vmatrix} R_1 + R_2 + R_3 & -(R_2 + R_3) \\ -(R_2 + R_3) & R_2 + R_3 + R_4 \end{vmatrix}}$$

$$= \frac{E(R_2 + R_3 + R_4)}{(R_1 + R_2 + R_3)(R_2 + R_3 + R_4) - (R_2 + R_3)^2}$$

Therefore,

$$R_{in} = \frac{E}{I_1} = \frac{(R_1 + R_2 + R_3)(R_2 + R_3 + R_4) - (R_2 + R_3)^2}{R_2 + R_3 + R_4}$$

$$= R_1 + R_2 + R_3 - \frac{(R_2 + R_3)^2}{R_2 + R_3 + R_4}$$

Using the values given in the example,

$$R_{in} = 10 + 10 + 10 - \frac{(10 + 10)^2}{10 + 10 + 25}$$

$$= 30 - \frac{400}{45}$$

$$= 21.1 \ \Omega$$

1.4.1 Simple Potentiometer

The *potentiometer* (or *potential divider*) consists of a resistor which has a variable tapping point, as shown in Fig. 1.4(a). Assume constant input voltage E, and let the potentiometer tap be set at some fraction a of the total resistance R. It is required to find the output voltage V_L in terms of the other quantities.

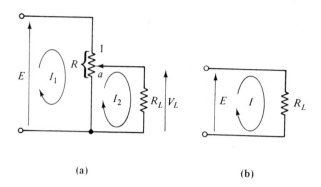

(a) (b)

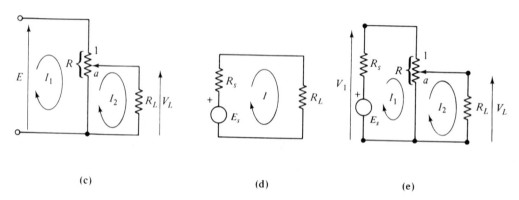

(c) (d) (e)

FIGURE 1.4. (a) Resistive potentiometer; (b) and (c): used to determine insertion loss I_2/I. (d) and (e): used to determine insertion loss I_2/I taking into account source resistance.

Applying Kirchhoff's voltage laws to the loops containing I_1 and I_2,

$$E = I_1 R - I_2 aR$$
$$0 = -I_1 aR + I_2(aR + R_L)$$

Solve for I_2:

$$I_2 = \frac{\begin{vmatrix} R & E \\ -aR & 0 \end{vmatrix}}{\begin{vmatrix} R & -aR \\ -aR & (aR + R_L) \end{vmatrix}}$$

$$= \frac{EaR}{R(aR + R_L) - (aR)^2}$$

$$= \frac{Ea}{R_L + a(1 - a)R}$$

The output voltage V_L is given by

$$V_L = I_2 R_L$$

$$= \frac{EaR_L}{R_L + a(1-a)R} \tag{1.7}$$

$$= \frac{Ea}{1 + a(1-a)(R/R_L)} \tag{1.8}$$

The potentiometer ratio is

$$\frac{V_L}{E} = \frac{a}{1 + a(1-a)(R/R_L)} \tag{1.9}$$

For $R_L \gg R$, the ratio becomes

$$\frac{V_L}{E} \approx a \tag{1.10}$$

An important network parameter is the *insertion loss*, which is defined as the ratio of load currents with, and without, the network. From Fig. 1.4(b) and (c), the insertion loss is I_2/I:

$$I = \frac{E}{R_L} \qquad\qquad\qquad \left[\text{Fig. 1.4(b)}\right]$$

$$I_2 = \frac{Ea}{R_L + a(1-a)R} \qquad (\text{as before}) \qquad \left[\text{Fig. 1.4(c)}\right]$$

Therefore,

$$\text{insertion loss} = \frac{I_2}{I}$$

$$= \frac{a}{1 + a(1-a)(R/R_L)} \tag{1.11}$$

This is seen to be the same as Eq. (1.9), but it is true only where the source resistance can be neglected. Where the source resistance has to be taken into account, e.g., in Fig. 1.4(d) and (e),

$$I = \frac{E_s}{R_s + R_L} \qquad\qquad \left[\text{Fig. 1.4(d)}\right]$$

$$E_s = I_1(R + R_s) - I_2 aR \qquad\qquad \left[\text{Fig. 1.4(e)}\right]$$

$$0 = -I_1 aR + I_2(aR + R_L)$$

Therefore,

$$I_2 = \frac{E_s aR}{(R + R_s)(aR + R_L) - (aR)^2}$$

and

$$\text{insertion loss} = \frac{I_2}{I}$$

$$= \frac{aR(R_s + R_L)}{(R + R_s)(aR + R_L) - (aR)^2} \tag{1.12}$$

The *voltage transfer function* is defined generally as

$$\frac{\text{voltage output from network}}{\text{voltage input to network}} \tag{1.13}$$

From Fig. 1.4(e), this is seen to be V_L/V_1, and a little thought will show that this is equal to the ratio V_L/E given by Eq. (1.9).

1.4.2 Attenuator Pads

An *attenuator pad* is a resistive network that is used to introduce a fixed amount of attenuation between a source and a load. One commonly used circuit is the *T* attenuator, shown in Fig. 1.5(b). The insertion loss also gives the attenuation, usually expressed in decibels.
From Fig. 1.5(a),

$$I = \frac{E}{R_s + R_L} \tag{1.14}$$

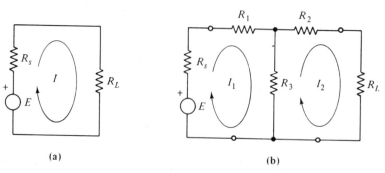

(a)   (b)

FIGURE 1.5. (a) Circuit without attenuator; (b) a T attenuator inserted in circuit (a).

From Fig. 1.5(b),

$$E = I_1(R_s + R_1 + R_3) - I_2 R_3$$
$$0 = -I_1 R_3 + I_2(R_2 + R_3 + R_L)$$

Thus,

$$I_2 = \frac{\begin{vmatrix} R_s + R_1 + R_3 & E \\ -R_3 & 0 \end{vmatrix}}{\begin{vmatrix} R_s + R_1 + R_3 & -R_3 \\ -R_3 & R_2 + R_3 + R_L \end{vmatrix}}$$

$$= \frac{ER_3}{(R_s + R_1 + R_3)(R_2 + R_3 + R_L) - R_3^2}$$

Now,

$$\text{insertion loss} = \frac{I_2}{I}$$

$$= \frac{R_3(R_s + R_L)}{(R_s + R_1 + R_3)(R_2 + R_3 + R_L) - R_3^2} \qquad (1.15)$$

and,

$$\text{insertion loss in decibels} = -20 \log_{10}\left(\frac{I_2}{I}\right) \qquad (1.16)$$

The negative sign is to show that attenuation occurs; that is, the insertion loss will come out as a positive number of decibels.

Example 1.1   Calculate the attenuation in decibels for a T pad for which $R_1 = R_2 = 31\ \Omega$, and $R_3 = 25\ \Omega$. The pad connects a 50-Ω generator to a 50-Ω load.

Solution

$$\frac{I_2}{I} = \frac{25(50 + 50)}{(50 + 31 + 25)(31 + 25 + 50) - (25)^2}$$

$$= \frac{2500}{(106)^2 - (25)^2}$$

$$= 0.236$$

$$-20 \log_{10}\frac{I_2}{I} = 12.56\ \text{dB}$$

The attenuator will usually maintain matched conditions, as described in the next paragraph, and as illustrated by Problem 4, Section 1.16.

Equation (1.15) enables the insertion loss to be calculated given the resistor values. The converse problem, that of finding resistor values that yield a specified attenuation, is more difficult. The attenuator should also provide input and output matching. Thus, three equations can be set up, one for insertion loss, one for input resistance, and one for output resistance, and in principle these can be solved for the three unknowns R_1, R_2, and R_3. In fact, it may not be possible to meet the three specified parameters, i.e., not all attenuation values are possible for all input and output resistor values. A common form of network is the symmetrical network shown in Figure 1.6 in which $R_1 = R_2 = R$, and $R_S = R_L = R_0$. For input matching $R_{\text{in}} = R_0$ and hence,

$$R_0 = R + \frac{R_3(R + R_0)}{R_3 + R + R_0} \tag{1.17}$$

From this, we have

$$\frac{R_0 - R}{R + R_0} = \frac{R_3}{R_3 + R + R_0} \tag{1.18}$$

From Eq. (1.14), $I = E/2R_0$, and because $R_{\text{in}} = R_0$, the input current for the circuit of Fig. 1.6 is also equal to I. Applying the current-divider rule to the center node gives $I_2 = IR_3/(R_3 + R + R_0)$, and therefore the insertion loss I.L. is

$$\text{I.L.} = \frac{R_3}{R_3 + R + R_0} \tag{1.19}$$

The right-hand side of this is seen to be the same as that for Eq. (1.18), and therefore,

$$\frac{R_0 - R}{R + R_0} = \text{I.L.} \tag{1.20}$$

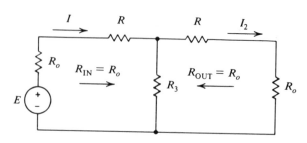

FIGURE 1.6. The symmetrical T-attenuator.

from which

$$R = R_0 \frac{1 - \text{I.L.}}{1 + \text{I.L.}} \tag{1.21}$$

This allows the resistor R to be determined for a given insertion loss. Once R is known, R_3 can be determined from Eq. (1.19), and it is left as an exercise for the student to show that Eqs. (1.21) and (1.19) give

$$R_3 = \frac{2R_0(\text{I.L.})}{1 - (\text{I.L.})^2} \tag{1.22}$$

Example 1.2   A T-type attenuator is required to provide a 6-dB insertion loss and to match 50-Ω input and output. Find the resistor values.

Solution   From Eq. (1.16), 6 dB gives an I.L. ratio of .5 : 1. Therefore, from Eq. (1.21),

$$R = 50 \frac{1 - .5}{1 + .5} = 16.67 \ \Omega$$

From Eq. (1.22)

$$R_3 = \frac{100(.5)}{1 - (.5)^2} = 66.67 \ \Omega$$

The π-network shown in Fig. 1.7 is another widely used attenuator network. The values for the π-network can be derived from those of the

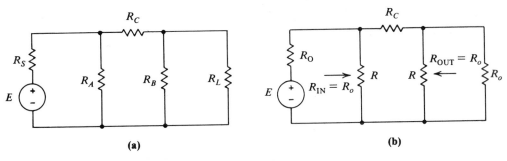

FIGURE 1.7.   (a) The π-attenuator. (b) The symmetrical π-attenuator.

T-network by use of the Y-Δ transformation equations, which are

$$R_A = \frac{R_1 R_2 + R_1 R_3 + R_2 R_3}{R_2} \qquad (1.23)$$

$$R_B = R_A(R_2/R_1) \qquad (1.24)$$

$$R_C = R_A(R_2/R_3) \qquad (1.25)$$

Alternatively of course, the π-network values may be worked out directly from the basic specifications, as was done for the T-network. For the symmetrical π-network shown in Fig. 1.7(b), it is left as an exercise for the student to show that

$$R = R_0 \frac{1 + \text{I.L.}}{1 - \text{I.L.}} \qquad (1.26)$$

and

$$R_C = \frac{R_0}{2} \frac{1 - (\text{I.L.})^2}{\text{I.L.}} \qquad (1.27)$$

A simpler type of pad is the one for which the circuit is shown in Fig. 1.8. This is frequently used to match a source to a load; that is, when it is inserted between source and load, the effective resistance seen by the source is equal in value to the source impedance, and the effective source impedance, as seen by the load, is equal in value to the load impedance. From the circuit of Fig. 1.8(a),

$$R_s = R_{\text{in}} = R_1 + \frac{R_3 R_L}{R_3 + R_L}$$

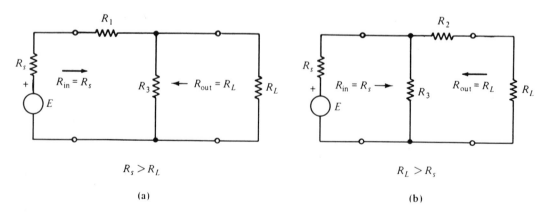

$R_s > R_L$ (a)

$R_L > R_s$ (b)

FIGURE 1.8. (a) L attenuator $R_S > R_L$; (b) L attenuator $R_S < R_L$.

So

$$R_s - R_1 = \frac{R_3 R_L}{R_3 + R_L}$$

and

$$\frac{1}{R_s - R_1} = \frac{1}{R_3} + \frac{1}{R_L} \qquad (1.28)$$

Also,

$$R_L = R_{\text{out}} = \frac{R_3(R_1 + R_s)}{R_3 + R_1 + R_s}$$

or

$$\frac{1}{R_L} = \frac{1}{R_3} + \frac{1}{R_1 + R_s} \qquad (1.29)$$

The term $1/R_3$ can be eliminated from Eqs. (1.28) and (1.29), giving

$$\frac{1}{R_s - R_1} - \frac{1}{R_L} = \frac{1}{R_L} - \frac{1}{R_s + R_1}$$

$$\therefore \quad \frac{1}{R_s - R_1} + \frac{1}{R_s + R_1} = \frac{2}{R_L}$$

or

$$\frac{2R_s}{R_s^2 - R_1^2} = \frac{2}{R_L}$$

$$\therefore \quad R_1^2 = R_s^2 - R_s R_L$$

or

$$R_1 = \sqrt{R_s(R_s - R_L)} \qquad (1.30)$$

The term R_3 can then be determined from Eq. (1.28) by substituting for R_1 from Eq. (1.30). For this result to be valid, $R_L < R_s$; otherwise, R_1 becomes an imaginary quantity. If $R_L > R_s$, the circuit of Fig. 1.8(b) must be used, and the equation becomes

$$R_2 = \sqrt{R_L(R_L - R_s)} \qquad (1.31)$$

Example 1.3   An L network is used to match a 75-Ω generator to a 50-Ω load. Determine the values of the network resistors.

Solution   The circuit of Fig. 1.8(a) must be used. Thus,

$$R_1 = \sqrt{75(75 - 50)}$$

$$= 43.3 \ \Omega$$

and

$$\frac{1}{R_s - R_1} = \frac{1}{R_3} + \frac{1}{R_L}$$

Therefore,

$$\frac{1}{75 - 43.3} = \frac{1}{R_3} + \frac{1}{50}$$

$$\frac{1}{31.7} - \frac{1}{50} = \frac{1}{R_3}$$

$$R_3 = \frac{31.7 \times 50}{50 - 31.7}$$

$$= 86.6 \ \Omega$$

Example 1.4   Repeat Example 1.3, but with $R_s = 10 \ \Omega$ and $R_L = 50 \ \Omega$.

Solution   The circuit of Fig. 1.8(b) must be used. Thus,

$$R_2 = \sqrt{R_L(R_L - R_s)}$$

$$= \sqrt{50(50 - 10)}$$

$$= 44.72 \ \Omega$$

and analogous to Eq. (1.28),

$$\frac{1}{R_L - R_2} = \frac{1}{R_3} + \frac{1}{R_s}$$

Therefore,

$$\frac{1}{50 - 44.72} = \frac{1}{R_3} + \frac{1}{10}$$

$$\frac{1}{5.28} - \frac{1}{10} = \frac{1}{R_3}$$

$$R_3 = \frac{10 \times 5.28}{10 - 5.28}$$

$$= 11.19 \ \Omega$$

1.5   IMPEDANCE AND REACTANCE

Ohm's law and Kirchhoff's laws can be extended to alternating-current circuits provided that the two additional circuit properties of inductance and capacitance are taken into account. With sinusoidal (or cosinusoidal) currents, these properties are accounted for in terms of impedance. The concepts of impedance and phasors, and the complex number representation of these, will be found in any good textbook on electric circuit analysis; see for example, B. Zeines, *Electric Circuit Analysis*, Reston Publishing Company, Reston, Virginia, (1972). A summary of the more important results is given here.

The impedance Z of a circuit is defined as the ratio of phasor voltage V across the circuit to the phasor current I through the circuit. Hence,

$$Z = \frac{V}{I} = R + jX \tag{1.32}$$

where $j = \sqrt{-1}$.

By definition, the real part R, of the impedance is the *resistance* of the circuit, and the imaginary part X, the *reactance*. The circuit may physically consist of a resistance R connected in series with a reactance X, which may be an inductance, or capacitance, or a combination of these; or, it is also possible for Eq. (1.32) to represent the equivalent impedance of a more complex network. One important equivalence representation is covered in Section 1.7.

For the circuit shown in Fig. 1.9(a),

$$
\begin{aligned}
Z &= R + jX_c \\
&= R + j(-1/\omega C) \\
&= R - j/\omega C
\end{aligned}
\tag{1.33}
$$

That is, by definition, capacitive reactance is

$$X_C = -1/\omega C \tag{1.34}$$

where ω is the angular frequency of the sinusoidal voltage or current. For the circuit of Fig. 1.7(b),

$$
\begin{aligned}
Z &= R + jX_L \\
&= R + j\omega L
\end{aligned}
\tag{1.35}
$$

That is, inductive reactance is defined as

$$X_L = \omega L \tag{1.36}$$

As shown in Fig. 1.9, the sign convention for voltage drop across a reactive element is the same as that for resistance.

Excluding for the moment mutual-inductive coupling (this is considered in Section 1.11), the effective reactance of a series network X_{ser} is

$$X_{ser} = X_1 + X_2 + \ldots + X_n \tag{1.37}$$

Where the individual reactances are all inductive, the equivalent series inductance is obtained from Eqs. (1.37) and (1.36) as

$$L_{ser} = L_1 + L_2 + \ldots + L_n \tag{1.38}$$

Where they are all capacitive, the equivalent series capacitance C_{ser} may be obtained from Eqs. (1.34) and (1.37) in the form

$$\frac{1}{C_{ser}} = \frac{1}{C_1} + \frac{1}{C_2} + \ldots + \frac{1}{C_n} \tag{1.39}$$

The effective reactance of a parallel network of reactances is given by

$$\frac{1}{X_p} = \frac{1}{X_1} + \frac{1}{X_2} + \ldots + \frac{1}{X_n} \tag{1.40}$$

Where the individual reactances are all inductive, this leads to an expression for the equivalent parallel inductance (excluding mutual induc-

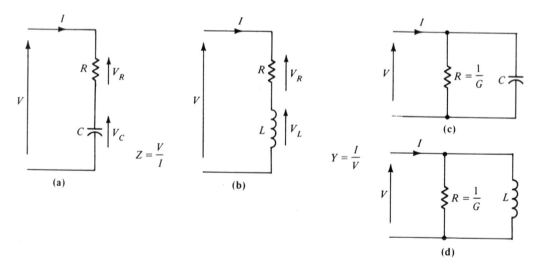

FIGURE 1.9. Impedance representation of (a) an *R-C* circuit, (b) an *R-L* circuit. Admittance representation of (c) an *R-C* circuit, (d) an *R-L* circuit.

tance) of

$$\frac{1}{L_p} = \frac{1}{L_1} + \frac{1}{L_2} + \ldots + \frac{1}{L_n} \tag{1.41}$$

Where the individual reactances are all capacitive, the equivalent parallel capacitance is obtained as

$$C_p = C_1 + C_2 + \ldots + C_n \tag{1.42}$$

As a check on calculations, it may be noted that L_{ser} should be greater than the largest single inductor in the series network, while L_p should be smaller than the smallest single inductor in the parallel network; for capacitors, C_{ser} should be smaller than the smallest capacitor in the series network, while C_p should be greater than the greatest capacitor in the parallel network.

Example 1.5   Calculate the impedance of (a) a 5-Ω resistor in a series with a 1.0-μH inductor, and (b) a 5-Ω resistor in series with a 0.02533-μF capacitor, at a frequency of 1.0 MHz.

Solution   (a) $Z = 5 + j2\pi \times 10^6 \times 10^{-6}$

$\qquad\qquad\qquad = 5 + j6.28 \ \Omega$

(b) $Z = 5 - j/2\pi \times 10^6 \times 0.02533 \times 10^{-6}$

$\qquad\qquad\qquad = 5 - j6.28 \ \Omega$

1.6   ADMITTANCE AND SUSCEPTANCE

The admittance Y of a circuit is defined as the ratio of the phasor current through the circuit to the phasor voltage across the circuit. Thus,

$$Y = \frac{I}{V} = G + jB \tag{1.43}$$

It will be seen that admittance is therefore the reciprocal of impedance, i.e.,

$$Y = 1/Z \tag{1.44}$$

By definition, the real part G of admittance is the *conductance* of the circuit, and the imaginary part B the *susceptance*. For a circuit consisting of an idealized resistance R in parallel with an idealized capacitance C, as in Fig. 1.9(c), the admittance is

$$Y = G + jB_C$$
$$= \frac{1}{R} + j\omega C \tag{1.45}$$

Thus, the conductance is

$$G = 1/R \tag{1.46}$$

and the capacitive susceptance is

$$B_C = \omega C \tag{1.47}$$

For a circuit consisting of an idealized resistor R in parallel with an idealized inductor L, as in Fig. 1.9(d), the admittance is

$$Y = G + jB_L$$
$$= G + j(-1/\omega L)$$
$$= G - j/\omega L \tag{1.48}$$

Thus, the inductive susceptance is

$$B_L = -1/\omega L \tag{1.49}$$

In practice it may not be permissible to assume idealized components. This is especially true for inductors, for which the series resistance is usually significant. The admittance of this type of component is treated in Section 1.7.

Example 1.6   Calculate, for a frequency of 1.0 MHz, the admittance of (a) a 1-kΩ resistor in parallel with a 200-pF capacitor, and (b) a 1-kΩ resistor in parallel with a 126.6-μH inductor.

Solution   (a) $$Y = \frac{1}{10^3} + j2\pi \times 10^6 \times 200 \times 10^{-12}$$

$$= 10^{-3} + j12.57 \times 10^{-4} \text{ Siemens (S)}$$

$$\underline{\underline{= 1 + j1.257 \text{ mS}}}$$

(b) $$Y = 10^{-3} - j/(2\pi \times 10^6 \times 126.6 \times 10^{-6})$$

$$\underline{\underline{= 1 - j1.257 \text{ mS}}}$$

1.7   SERIES AND PARALLEL EQUIVALENCE

Impedance is the more convenient quantity to use when dealing with series circuits, and admittance when dealing with parallel circuits, although the final value of admittance is usually converted to parallel resistance and reactance (rather than leaving it in the form of conductance and susceptance). A very useful equivalence can be established between series and parallel circuits, *at a given frequency.*

Consider first a circuit which can be represented as a resistance R in series with a reactance X.

$$Z = R + jX$$

The equivalent admittance is

$$Y = 1/Z$$

or

$$G + jB = \frac{1}{R + jX}$$

$$= \frac{R}{R^2 + X^2} - j\frac{X}{R^2 + X^2}$$

Therefore, the equivalent conductance is

$$G = \frac{R}{R^2 + X^2} \tag{1.50}$$

And the equivalent susceptance is

$$B = -\frac{X}{R^2 + X^2} \tag{1.51}$$

Example 1.7   A 100-pF capacitor has a series lead resistance of 1 Ω. Determine the equivalent parallel circuit at a frequency of 15.9 MHz.

Solution

$$X_C = \frac{-1}{2\pi \times 15.9 \times 10^6 \times 100 \times 10^{-12}}$$

$$= -100 \ \Omega$$

From Eq. (1.50),

$$G = \frac{1}{1^2 + 100^2}$$

$$\cong 10^{-4} \ \text{S}$$

Therefore the equivalent parallel resistance of the circuit is 10^4 Ω, or 10 kΩ. From Eq. (1.51),

$$B = \frac{-(-100)}{1^2 + 100^2}$$

$$\cong 0.01 \text{ S}$$

The equivalent parallel capacitance is found from application of Eq. (1.47) to be

$$B = \omega C_{eq} = 0.01$$

$$C_{eq} = 100 \text{ pF}$$

The equivalent circuits are shown in Fig. 1.10 (a) and (b). It is important to realize that the *values* shown are good only for a frequency of 15.9 MHz. Changing the frequency will change the values of the equivalent components, as illustrated in the next example.

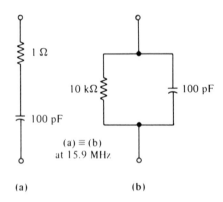

FIGURE 1.10. (a) Series and (b) parallel equivalent circuits for Example 1.6.

Example 1.8   An inductor can be represented as a 15.92-μH inductance in series with a 10-Ω resistance. Determine the equivalent parallel circuit for a frequency of (a) 10 MHz, and (b) 20 MHz, assuming the series component values remain constant.

Solution   (a) $X_L = 2\pi \times 10^7 \times 15.92 \times 10^{-6}$

$$= 1000 \text{ Ω}$$

From Eq. (1.50),

$$G = \frac{10}{10^2 + 1000^2}$$

$$\cong 10^{-5} \text{ S}$$

and

$$R_p = \frac{1}{G} = 100 \text{ k}\Omega$$

From Eq. (1.51), the equivalent susceptance is

$$B = -\frac{1000}{10^2 + 1000^2}$$

$$\cong -1.0 \text{ mS}$$

The equivalent parallel inductance is found from application of Eq. (1.49) to be

$$B_L = -1/\omega L_{eq} = -10^{-3}$$

Therefore, at a frequency of 10 MHz,

$$L_{eq} = 1/2\pi \times 10^7 \times 10^{-3}$$

$$= 15.92 \ \mu\text{H}$$

(b) Increasing the frequency to 20 MHz increases X_L to 2000 Ω. Applying Eq. (1.50) gives

$$G = \frac{10}{10^2 + 2000^2}$$

$$\cong \frac{10^{-5}}{4}$$

and therefore,

$$R_p = \frac{1}{G} = 400 \ k\Omega$$

From Eq. (1.51),

$$B = -\frac{2000}{10^2 + 2000^2}$$

$$\cong -1.0 \text{ mS}$$

Therefore the equivalent parallel inductance is 15.92 μH, as before. It will be seen in this particular example that, whereas there is negligible change in the value of equivalent parallel inductance, and indeed this is approximately equal to the series inductance value, the equivalent parallel resistance value increases approximately in proportion to the frequency squared.

Clearly a similar procedure can be followed to obtain an equivalent series circuit from a given parallel circuit. Consider a conductance G in parallel with a susceptance B, so that

$$Y = G + jB$$

The equivalent impedance is

$$Z = 1/Y$$

Hence,

$$R + jX = \frac{1}{G + jB}$$

or

$$R + jX = \frac{G}{G^2 + B^2} - j\frac{B}{G^2 + B^2}$$

Therefore the equivalent series resistance is

$$R = \frac{G}{G^2 + B^2} \tag{1.52}$$

and the equivalent series reaction is

$$X = -\frac{B}{G^2 + B^2} \tag{1.53}$$

Example 1.9 A capacitor can be represented by a capacitance of 50 pF in parallel with a dielectric loss resistance of 10 MΩ. For a frequency of 1.0 MHz, determine the equivalent series circuit.

Solution

$$G = 10^{-7}\,\text{S}$$

and

$$B = 2\pi \times 10^6 \times 50 \times 10^{-12}$$
$$= 314.2\,\mu\text{S}$$

From Eq. (1.52),

$$R = \frac{10^{-7}}{10^{-14} + (314.2)^2 \times 10^{-12}}$$
$$\cong 1.013\,\Omega$$

and from Eq. (1.53),

$$X = \frac{-314.2 \times 10^{-6}}{10^{-14} + (314.2)^2 \times 10^{-12}} \cong -\frac{1}{\underline{314.2 \times 10^{-6}}} \ \Omega$$

Thus, since $X \cong -1/B$, the equivalent series capacitance does not differ significantly from the parallel value.

1.8 SERIES RLC CIRCUIT

The total impedance for the series RLC circuit is

$$Z = R + j(X_L + X_C) \tag{1.54}$$
$$= R + j(\omega L - 1/\omega C) \tag{1.55}$$

This is the series impedance in complex form, as shown in Fig. 1.11(a). The

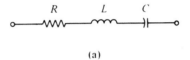

(a)

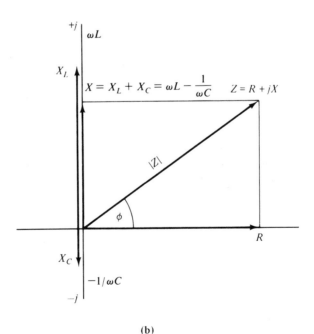

(b)

FIGURE 1.11. (a) A series *RLC* circuit; (b) its impedance diagram.

complex form contains sufficient information to determine Z completely, since

$$|Z| = \sqrt{R^2 + X^2} \qquad (1.56)$$

and

$$\phi = \tan^{-1}\left(\frac{X}{R}\right) \qquad (1.57)$$

Here,

$$X = X_L + X_C = \omega L - \frac{1}{\omega C} \qquad \text{Fig. 1.11(b)}$$

1.8.1 Series Resonance and Q Factor

The series RLC circuit is resonant when the phase angle ϕ is equal to zero. This means that X must be zero since $\phi = \tan^{-1}(X/R)$. Thus,

$$X = X_L + X_C = 0$$

or

$$X_L = -X_C \qquad (1.58)$$

Denoting this condition by the subscript $_{s0}$

$$\omega_{s0} L = -(-1/\omega_{s0} C)$$

or

$$\omega_{s0} L = \frac{1}{\omega_{s0} C}$$

Thus,

$$\omega_{s0} = \frac{1}{\sqrt{LC}} \qquad (1.59)$$

and

$$f_{s0} = \frac{1}{2\pi\sqrt{LC}} \qquad (1.60)$$

This simple but very important equation defines the *resonant frequency* of the series circuit. Clearly, at f_{s0} the impedance of the circuit is minimum and purely resistive, since

$$Z_{s0} = R + j\left(\omega_{s0} L - \frac{1}{\omega_{s0} C}\right)$$

$$= R + j0 \qquad (1.61)$$

The L or C of the circuit can be adjusted to bring the circuit into resonance with the applied frequency, a process known as *tuning*. The circuit is also referred to as a *series tuned circuit*. At frequencies below resonance the circuit impedance appears capacitive since the j coefficient is negative; at frequencies above resonance it appears inductive since the j coefficient is positive.

The widest application of the tuned circuit is as a frequency selective filter, illustrated in its simplest form in Fig. 1.12(a). The input to the system consists of two sinusoidal signals f and f_{s0} ($f \neq f_{s0}$), while the output current in the load R_L will consist almost entirely of a signal at f, providing $R_L \gg R$, the signal at f_{s0} being filtered out by the series tuned circuit across the line. The effect that the series tuned circuit has on the signal f depends on how close f is to f_{s0} and on an important characteristic of the circuit termed the *selectivity*. The selectivity, in turn, is a function of the Q factor (*quality factor*) of the circuit. The Q factor of a series tuned circuit can be defined as the ratio of inductive voltage to resistive voltage at resonance:

$$Q = \frac{V_L}{V_R}$$

$$= \frac{\omega_{s0}L}{R} \tag{1.62}$$

Note that the *resonant frequency* is used in the definition of Q for a resonant circuit. Also, at resonance, $\omega_{s0}L = 1/\omega_{s0}C$; therefore,

$$Q = \frac{1}{\omega_{s0}CR} \tag{1.63}$$

The equation for series impedance may be written in terms of Q:

$$Z = R + j\left(\omega L - \frac{1}{\omega C}\right)$$

$$= R\left[1 + j\left(\frac{\omega L}{R} - \frac{1}{\omega CR}\right)\right]$$

$$= R\left[1 + j\left(\frac{\omega}{\omega_{s0}}\frac{\omega_{s0}L}{R} - \frac{\omega_{s0}}{\omega}\frac{1}{\omega_{s0}CR}\right)\right]$$

$$= R\left[1 + j\left(\frac{\omega}{\omega_{s0}} - \frac{\omega_{s0}}{\omega}\right)Q\right] \tag{1.64}$$

$$= R(1 + jyQ) \tag{1.65}$$

where

$$y = \frac{\omega}{\omega_{s0}} - \frac{\omega_{s0}}{\omega}$$

The impedance modulus is

$$|Z| = R\sqrt{1 + y^2 Q^2} \tag{1.66}$$

The curve of $|Z|$ against frequency f/f_{s0} is sketched in Fig. 1.12(b). Obviously, the narrower the curve, the more selective the circuit is, and, as will now be shown, a high Q factor results in better selectivity.

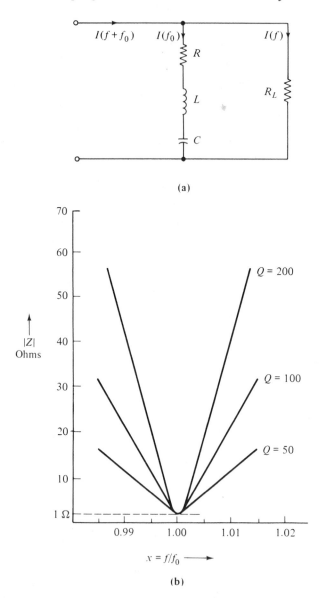

(a)

(b)

FIGURE 1.12. (a) The series tuned filter; (b) selectivity curves for various Q factors for a series RLC circuit.

The selectivity is measured in terms of the bandwidth, $B_{3\,dB}$, shown in Fig. 1.13. This is referred to as the *3-dB bandwidth* because the $\sqrt{2}$ impedance levels correspond to a 3-dB drop in current from its resonant value, for a constant voltage input. Almost any impedance level could have been chosen for defining bandwidth, but the $\sqrt{2}$ level has the advantage of simplifying the arithmetic. The specific values of y_3 corresponding to the 3-dB bandwidth points f_1 and f_2 are found by equating

$$R\sqrt{1 + y_3^2 Q^2} = R\sqrt{2}$$

$$1 + y_3^2 Q^2 = 2$$

or

$$y_3 = \frac{1}{Q} \tag{1.67}$$

Note that $y_3 = 1/Q$ must be positive.
At limit f_2,

$$y_3 = \frac{f_2}{f_{s0}} - \frac{f_{s0}}{f_2} = \frac{1}{Q}$$

therefore,

$$f_2^2 - \frac{f_{s0} f_2}{Q} - f_{s0}^2 = 0$$

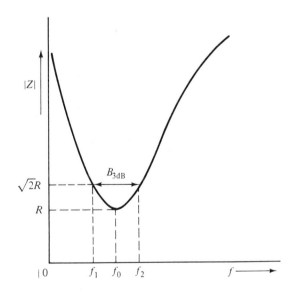

FIGURE 1.13. The 3-dB bandwidth for a series *RLC* circuit.

therefore,

$$f_2 = \frac{f_{s0}}{2Q} \pm \sqrt{\left(\frac{f_{s0}}{2Q}\right)^2 + f_{s0}^2} \qquad (1.68)$$

Note that $f_2 > f_{s0}$ and y_3 is positive.
At limit f_1,

$$y_3 = \frac{f_{s0}}{f_1} - \frac{f_1}{f_{s0}} = \frac{1}{Q}$$

therefore,

$$f_{s0}^2 - \frac{f_1 f_{s0}}{Q} - f_1^2 = 0$$

therefore,

$$f_1 = -\frac{f_{s0}}{2Q} \pm \sqrt{\left(\frac{f_{s0}}{2Q}\right)^2 + f_{s0}^2} \qquad (1.69)$$

Note that $f_1 < f_{s0}$ and y_3 is positive.

The 3-dB bandwidth is obtained from (1.68) and (1.69) as

$$B_{3dB} = f_2 - f_1 \qquad (1.70)$$

$$= \frac{f_{s0}}{Q} \qquad (1.71)$$

This shows the importance of the Q factor in determining selectivity, a high Q resulting in a narrow 3-dB bandwidth. For series tuned circuits, the Q factor typically ranges from about 10 to 300. The effect of increasing Q on selectivity is shown in Fig. 1.12(b); R is assumed to remain constant, and Q is varied by varying L/C, since

$$Q = \frac{\omega_{s0} L}{R} = \frac{1}{R}\sqrt{\frac{L}{C}}$$

1.9 PARALLEL TUNED CIRCUIT

The parallel circuit is as shown in Fig. 1.14(a), the capacitance is assumed to have negligible resistance. The admittance of the circuit can therefore be written as

$$Y = \frac{1}{R + j\omega L} + j\omega C \qquad (1.72)$$

$$= \frac{R - j\omega L}{R^2 + \omega^2 L^2} + j\omega C$$

$$= \frac{R}{R^2 + \omega^2 L^2} + j\left(\omega C - \frac{\omega L}{R^2 + \omega^2 L^2}\right) \qquad (1.73)$$

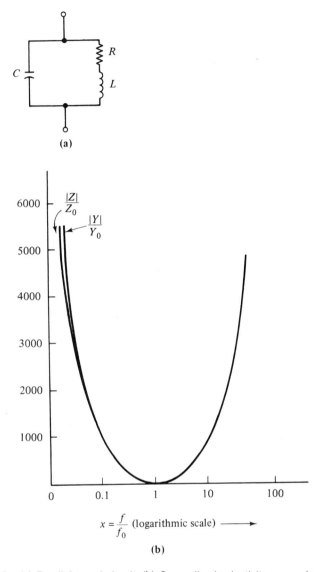

FIGURE 1.14. (a) Parallel tuned circuit; (b) Generalized selectivity curves for parallel and series tuned circuits.

At resonance the admittance is real (i.e., the j coefficients equate to zero); thus,

$$\omega_{p0}C = \frac{\omega_{p0}L}{R^2 + \omega_{p0}^2 L^2}$$

where ω_{p0} is the parallel resonant angular frequency.

Rearranging terms,

$$R^2 + \omega_{p0}^2 L^2 = \frac{L}{C} \tag{1.74}$$

Thus,

$$\omega_{p0}^2 = \frac{1}{LC} - \frac{R^2}{L^2}$$

or

$$\omega_{p0} = \sqrt{\frac{1}{LC} - \frac{R^2}{L^2}} \tag{1.75}$$

Using $\omega_{s0} = 1/\sqrt{LC}$, and $Q_s = \omega_{s0}L/R$, eq. (1.75) can be rearranged as

$$\omega_{p0} = \omega_{s0}\sqrt{1 - 1/Q_s^2} \tag{1.76}$$

Also, the Q factor for the parallel resonant circuit is

$$Q_p = \frac{\omega_{p0}L}{R}$$

Substituting for ω_{p0} from eq. (1.76) and rearranging gives

$$Q_p = \sqrt{Q_s^2 - 1} \tag{1.77}$$

For Q_s greater than about ten, it will be seen that $\omega_{p0} \cong \omega_{s0}$ and $Q_p \cong Q_s$. These approximations will be used unless otherwise stated, and resonant quantities will be indicated by means of a single "0" subscript, e.g., ω_0 for resonant angular frequency.

The admittance at resonance is purely conductive and is

$$Y_0 = \frac{R}{R^2 + \omega_0^2 L^2} \tag{1.78}$$

From Eq. (1.74), $R^2 + \omega_0^2 L^2 = L/C$; therefore,

$$Y_0 = \frac{CR}{L} \tag{1.79}$$

The impedance at resonance is purely resistive and is denoted by R_D to signify a *dynamic resistance*:

$$R_D = \frac{1}{Y_0} = \frac{L}{CR} \tag{1.80}$$

This is close to, but does not quite equal, the maximum impedance modulus obtainable. Where the approximate form of resonant frequency equation can be used, the maximum impedance coincides with the impedance at resonance.

The ratio Y/Y_0 shows how admittance varies with frequency relative to the resonant value. Usually, $\omega^2 L^2 \gg R^2$ so that Eq. (1.73) for Y can be simplified, and using Eq. (1.79) for Y_0 gives

$$\frac{Y}{Y_0} \cong \frac{L}{CR}\left[\frac{R}{\omega^2 L^2} + j(\omega C - 1/\omega L)\right]$$

$$= \frac{1}{\omega^2 LC} + j\left(\frac{\omega L}{R} - \frac{1}{\omega CR}\right)$$

$$= \frac{\omega_0^2}{\omega^2} + jyQ \qquad (1.81)$$

where yQ is as given in Eq. (1.65), and $LC = \omega_0^{-2}$. Thus,

$$|Y| = Y_0\sqrt{\left(\frac{\omega_0}{\omega}\right)^4 + (yQ)^2} \qquad (1.82)$$

The curve of $|Y|$ versus f/f_0 is similar to that for $|Z|$ versus f/f_0 for the series circuit, as shown in Fig. 1.12(b). Only at the low-frequency end do the curves differ, and the frequency has been plotted on a logarithmic scale to bring this out. It follows, therefore, that the 3-dB bandwidth for the parallel circuit is also given by

$$B_{3\text{dB}} = \frac{f_0}{Q} \qquad (1.83)$$

The parallel tuned circuit may also be used as a filter, because it presents a high impedance to signals at resonant frequency, as shown in Example 1.10. It should be noted that the dynamic impedance R_D applies only at resonant frequency. Direct currents, for example would encounter a resistance of R only, the d.c. resistance of the coil. Before working example 1.10, note that useful alternative expressions for Eq. (1.80) for R_D may be derived utilizing Eqs. (1.62) and (1.63) for the Q factor of a circuit. Thus,

$$R_D = \frac{L}{CR} \qquad (1.80)$$

$$= \omega_0 LQ \qquad (1.84)$$

$$= \frac{Q}{\omega_0 C} \qquad (1.85)$$

$$= Q^2 R \qquad (1.86)$$

Example 1.10   A parallel tuned circuit has a Q factor of 200 when tuned to resonance at 10 MHz. The value of tuning capacitance is 10 pF. Calculate (a) the dynamic impedance, and (b) the d.c. resistance, assuming this does not change with frequency.

Solution   (a) Applying Eq. (1.85),

$$R_D = \frac{200}{2\pi \times 10^7 \times 10 \times 10^{-12}}$$
$$= 318 \text{ k}\Omega$$

(b) From Eq. (1.86),

$$R = \frac{R_D}{Q^2} = \frac{318 \times 10^3}{(200)^2} = 7.96 \ \Omega$$

At frequencies below resonance, the circuit *impedance* appears inductive and above resonance, capacitive. This is most easily remembered by noting that near zero frequency, the inductance will be a near short-circuit; therefore, all the current will be inductive. And as frequency approaches infinity, the capacitive reactance approaches a short-circuit, making the total current capacitive.

1.9.1 Self-Resonance of a Coil

A real inductor has, in addition to self-inductance and resistance, a self-capacitance which is distributed throughout the coil. Because of its distributed nature, a real inductor cannot be represented exactly by any circuit, but can be approximated as shown in Fig. 1.15(a), in which the distributed properties of resistance, inductance, and capacitance are represented by lumped components R, L, and C_0. Clearly, then, the real inductor will behave like a parallel tuned circuit; that is, it will appear as an inductance below its self-resonant frequency and as a capacitance above its self-resonant frequency.

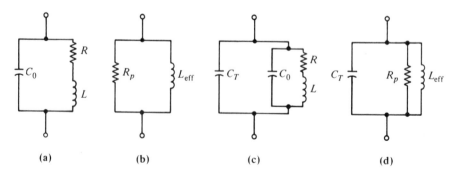

(a)   (b)   (c)   (d)

FIGURE 1.15. (a) The approximate circuit for an inductor including self-capacitance; (b) its parallel equivalence at frequencies below self-resonance; (c) a parallel tuned circuit; and (d) its equivalent circuit, taking into account C_0.

The admittance expression, Eq. (1.73), for the parallel circuit may therefore be used to represent the *coil*; making use of the approximation $\omega^2 L^2 \gg R^2$, Eq. (1.73) simplifies to

$$Y \cong \frac{R}{\omega^2 L^2} + j\left(\omega C_0 - \frac{1}{\omega L}\right)$$

$$= \frac{R}{\omega^2 L^2} - j\left(\frac{1 - \omega^2 L C_0}{\omega L}\right) \tag{1.87}$$

This allows the coil to be represented by the parallel circuit of Fig. 1.15(b), *which only applies for frequencies below the self-resonant frequency of the coil.* The effective parallel resistance R_p is seen to be obtained from

$$\frac{1}{R_p} = \frac{R}{\omega^2 L^2}$$

or

$$R_p = \frac{\omega^2 L^2}{R} \tag{1.88}$$

and the effective parallel inductance of the coil, L_{eff}, from

$$\frac{1}{\omega L_{\text{eff}}} = \frac{1 - \omega^2 L C_0}{\omega L}$$

or

$$L_{\text{eff}} = \frac{L}{1 - \omega^2 L C_0} = \frac{L}{1 - (\omega/\omega_0)^2} \tag{1.89}$$

Suppose that this coil is used in a tuned circuit (Fig. 1.15(c)) which has a resonant frequency, ω_T, well below the self-resonance, ω_0, of the coil (i.e., the external tuning capacity $C_T \gg C_0$). At circuit resonance, ω_T, the dynamic impedance of the circuit is, from Eq. (1.84)

$$R_D = Q\omega_T L \tag{1.90}$$

Now, the equivalent circuit of Fig. 1.15(d) will appear to have an effective Q factor, Q_{eff}, which can be used in Eq. (1.84) along with L_{eff} to give

$$R_D = Q_{\text{eff}} \omega_T L_{\text{eff}} \tag{1.91}$$

But since the circuits of Fig. 1.15(c) and (d) are equivalent,

$$Q\omega_T L = Q_{\text{eff}} \omega_T L_{\text{eff}}$$

so that

$$Q_{\text{eff}} = Q\frac{L}{L_{\text{eff}}} \tag{1.92}$$

Substituting from Eq. (1.89) into Eq. (1.92) gives

$$Q_{\text{eff}} = Q(1 - \omega^2 L C_0) = Q\left[1 - \left(\frac{\omega}{\omega_0}\right)^2\right] \tag{1.93}$$

where $Q = \omega_T L/R$ is the Q factor as previously defined, neglecting self-capacitance. Over a wide range of frequencies Q is approximately constant because, although ωL increases with frequency, R also increases with frequency, owing to the skin effect, and the ratio $\omega L/R$ remains approximately constant. It will be seen, however, that the effective Q factor, Q_{eff}, taking into account self-capacitance, decreases with increasing frequency.

Care must be exercised as to how Q_{eff} is used. For a parallel tuned circuit (Fig. 1.15(c) or (d)), the dynamic resistance can be calculated from either of the expressions Eqs. (1.90) or (1.91); i.e., either QL, or $Q_{\text{eff}} L_{\text{eff}}$ is used in the calculation. Alternatively, Eq. (1.85) can be used, provided that either $Q/(C_T + C_0)$ or Q_{eff}/C_T is substituted in it.

In calculating the bandwidth for a *parallel* tuned circuit from Eq. (1.83), Q, and not Q_{eff}, must be used:

$$\text{Bandwidth} = \frac{f_T}{Q} \tag{1.94}$$

This is because, for a specified resonant frequency, C_T will be adjusted to absorb C_0 (since C_T and C_o are in parallel). In the case of a series tuned circuit, C_T is no longer in parallel with C_o. Effectively, C_T resonates with L_{eff} to give a circuit Q factor of Q_{eff}, so that the bandwidth is, from Eq. (1.71),

$$\text{Bandwidth} = \frac{f_T}{Q_{\text{eff}}} \quad \text{(series circuit)} \tag{1.95}$$

It should be noted that for most Q-meters the test circuit is series tuned, so that the meter indicates Q_{eff}, and not Q.

1.10 SKIN EFFECT

The self-induced EMF in a conductor resulting from the rate of change of flux linkages opposes the current flow which gives rise to the flux (Lenz's law). Normally it is assumed that all the flux links with all the conductor.

However, the actual flux linkages increase toward the core of the conductor, since the magnetic flux within the conductor only links with the inner section; in Fig. 1.16(a), for example, flux line ϕ_1 links with the complete conductor, while flux line ϕ_2 links only with the section of radius a. The self-induced EMF is greatest at the center of the conductor, which experiences the greatest flux linkages, and becomes less toward the outer circumference. This results in the current density being least at the center and increasing toward the outer circumference, since the induced EMF opposes the current flow (Lenz's law). Of course, the lower current density at the center results in lower magnetic flux there, which tends to offset the effect producing the nonuniform distribution, and in this way, equilibrium conditions are established. The overall effect, however, is that the current tends to flow near the surface of the conductor, this being referred to as the *skin effect*. Because the current is confined to a smaller cross section of conductor, the apparent resistance of the conductor increases. The increase is more noticeable for thick conductors and at high frequencies (where the rate of change of flux linkages is high). Equally important is the fact that the resistance becomes dependent on frequency.

With coils, a special type of wire called *Litzendraht wire* (*Litz wire*, for short) is often used to reduce skin effect. Litz wire is made up of strands insulated from each other and wound in such a way that each strand changes position between center and outer edge over the length of the wire (Fig. 1.16(b)). In this manner, each strand, on average, has equal induced EMF's, so that over the complete cross section (made up of the many cross sections of the individual strands), the current density tends to be uniform.

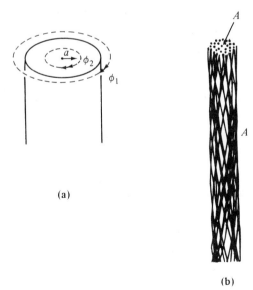

(a)

(b)

FIGURE 1.16.   (a) Magnetic flux linkage in a conductor; (b) Litzendraht wire.

1.11 MUTUAL INDUCTANCE

Reaction between inductive circuits that are physically isolated can occur as a result of common magnetic flux linkage. This effect can be taken into account by means of a mutual inductance M. For a harmonically varying current I_1, in an inductance L_1 which is magnetically coupled to an inductance L_2, the induced EMF in L_2 is given by

$$E_2 = \pm j\omega M I_1 \tag{1.96}$$

The sign to be used depends on the physical disposition of the coils and is illustrated in Fig. 1.17(a) and (b). Note that M cannot be identified with a physical winding in the sense that L_1 and L_2 can, and it has to be determined by measurement. In practice, it may prove easier to determine what is termed the coefficient of coupling k, where

$$M = k\sqrt{L_1 L_2} \tag{1.97}$$

k ranges between 0 and 1.

A useful equivalent circuit for mutual-inductive coupling is shown in Fig. 1.17(c), for the $+j\omega M$ situation; the terminals 1-1 and 2-2 are to be identified

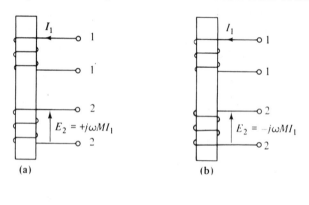

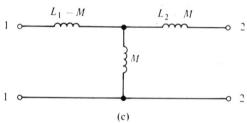

FIGURE 1.17. Mutual inductance coupling showing possible polarities (a) and (b); (c) the ac equivalent circuit for (a).

with the corresponding terminals in Fig. 1.17(a). Note that the equivalent circuit is valid only for ac conditions; it shows a dc path where none may exist in the actual circuit.

Where mutual inductance is present between two inductors connected in series, the effective series inductance is given by

$$L_s = L_1 + L_2 \pm 2M \tag{1.98}$$

Where the inductors are connected in parallel, the effective parallel inductance is given by

$$L_p = \frac{L_1 L_2 - M^2}{L_1 + L_2 \mp 2M} \tag{1.99}$$

It can be seen that Eqs. (1.98) and (1.99) reduce to Eqs. (1.38) and (1.41) when M is zero. By utilizing mutual inductance between two coils, the effective inductance can be altered in steps by making the appropriate connections, from a minimum L_{min} to a maximum L_{max}, where

$$L_{min} = \frac{L_1 L_2 - M^2}{L_1 + L_2 + 2M} \tag{1.100}$$

$$L_{max} = L_1 + L_2 + 2M \tag{1.101}$$

Furthermore, by making M variable, for example, by physically altering the spacing between the coils, a continuously variable inductance is obtained.

1.12 COUPLING CIRCUITS

Signal transfer from one circuit to another often requires some form of coupling circuit other than a direct connection. Transformer coupling, which utilizes the mutual-inductance effect, is widely used and can be considered separately in two broad areas: low frequencies (e.g., power and audio frequencies) and high frequencies (e.g., radio frequencies). Although both methods rely on mutual-inductance coupling, the practical details of the circuits are sufficiently different to warrant different methods of analysis.

1.12.1 LOW-FREQUENCY TRANSFORMERS

In the ideal low-frequency transformer, all of the magnetic flux ϕ set up by the primary ampere-turns links with the secondary winding (or secondary windings, as there may be more than one). Also in the ideal case, the voltage drops in the primary and secondary windings can be neglected, as can the power loss in the magnetic core.

Under these conditions, the applied primary voltage, V_p, is equal to the back EMF induced in the primary winding of N_p turns, which by Faraday's law of magnetic induction gives

$$V_p = N_p \frac{d\phi}{dt} \qquad (1.102)$$

Likewise, ignoring the voltage drop in the secondary winding, the EMF induced in the secondary E_s will be equal to the secondary terminal voltage, V_s, so

$$V_s \cong E_s = N_s \frac{d\phi}{dt} \qquad (1.103)$$

It follows, therefore, that

$$\frac{V_p}{V_s} = \frac{N_p}{N_s}$$

$$= n \qquad (1.104)$$

where $n = N_p/N_s$ is the turns ratio.

When the secondary is loaded such that it draws a current I_s, the secondary ampere-turns $N_s I_s$ must balance the primary ampere-turns $N_p I_p$ (otherwise, the unbalance would result in a change in induced current in such a direction as to restore balance). It follows, therefore, that

$$\frac{I_p}{I_s} = \frac{N_s}{N_p}$$

$$= \frac{1}{n} \qquad (1.105)$$

Clearly, from Eqs. (1.104) and (1.105) $V_p I_p = V_s I_s$, which is to be expected for an ideal transformer.

The load Z_L connected to the secondary may be referred to the primary Z_L' in the following way. The secondary load (Fig. 1.18(a)), is

$$Z_L = \frac{V_s}{I_s} \qquad (1.106)$$

The load, as seen from the primary terminals, is

$$Z_L' = \frac{V_p}{I_p} \qquad (1.107)$$

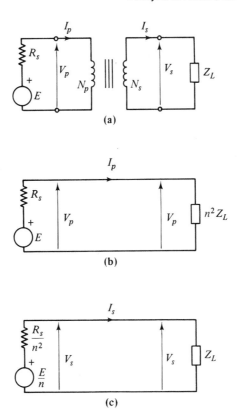

FIGURE 1.18. The ideal low-frequency transformer (a) circuit; (b) circuit referred to primary; (c) circuit referred to secondary.

Substituting for V_p and I_p from Eqs. (1.104) and (1.105), and using the relationship in Eq. (1.106), Eq. (1.107) for Z_L' may be transformed (Fig. 1.18(b)) to

$$Z_L' = n^2 Z_L \qquad (1.108)$$

Although based on the ideal transformer, this relationship is usually sufficiently accurate for, and proves to be very useful in, practical calculations.

By similar arguments, a voltage generator source of EMF E and internal resistance R_s may be transferred to the secondary so that the load appears to be fed from a source of EMF E/n and internal resistance R_s/n^2 (Fig. 1.18(c)).

The low-frequency circuit model for a practical transformer is shown in Fig. 1.19(a). In the practical transformer, a small primary current is required to set up the magnetic flux in the core; this can be accounted for by showing an inductance L_c in parallel with the ideal primary. There will also be eddy-current and hysteresis power losses in the core. These losses are dependent on the primary voltage and independent of current, and can be represented by a resistance R_c in parallel with the ideal primary. Each winding will have ohmic resistance, represented by R_p for the primary and R_s for the secondary. Each winding that carries current will produce a certain amount of magnetic flux which does not link with other windings; this is known as *leakage flux*, and its effect is represented by the inductances L_p and L_s.

The equivalent circuit of Fig. 1.19(a) may be redrawn with all components referred to either primary or secondary side, as already discussed. The circuit referred to primary is shown in Fig. 1.19(b).

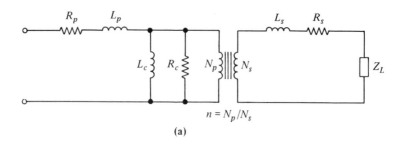

(a)

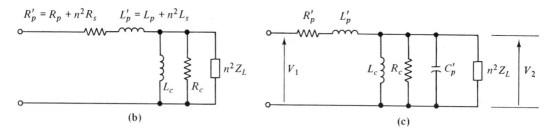

(b)   (c)

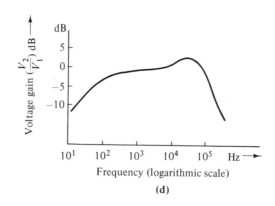

(d)

FIGURE 1.19. (a) The practical transformer equivalent circuit; (b) referred to primary; (c) referred to primary and taking capacitance into account; (d) the frequency response.

At the high-frequency end of the range it may be necessary to take into account the self-capacitances of the windings and the mutual capacitance between these. The equivalent circuit referred to the primary is shown in Fig. 1.19(c), in which C_p' takes into account all the capacitive effects. The voltage gain/frequency response for the transformer is shown in Fig. 1.19(d). At low frequencies, gain falloff occurs because of L_c. At midfrequencies, L_c and C_p' exhibit a broad parallel resonance (low-Q) response, while the effect of L_p' is negligible, so the response curve is reasonably flat in the midband range. At the higher frequencies, L_p' and C_p' exhibit series resonance, resulting in a peak in the response curve. Beyond this, the shunting effect of C_p' causes gain falloff.

It must be kept in mind that the equivalent circuits discussed are valid only for the ac signal, and cannot be used, for example, for analysis of dc conditions.

1.12.2 High-Frequency Transformers

In the radio-frequency range the parameters L_1, L_2, and k are readily measured. Therefore, it is more convenient to represent the circuit in terms of these parameters. If a magnetic core is used for tuning, for example, its losses can be taken into account by a reduced Q factor, which can also be measured.

The basic circuit is shown in Fig. 1.20(a). Z_L may be a load resistor, or it may be a tuning capacitor for the secondary. The equivalent circuit is shown in Fig. 1.20(b), in which

$$Z_p = R_p + j\omega L_p \qquad (1.109)$$

$$Z_s = R_s + j\omega L_s \qquad (1.110)$$

$$Z_m = j\omega M \qquad (1.111)$$

Analyzing the circuit from points X-X by applying Kirchhoff's voltage law to the two loops (note that C_p, the primary tuning capacitor, does not enter into

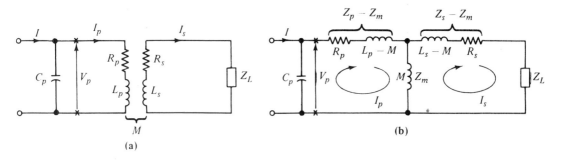

FIGURE 1.20. (a) The high-frequency transformer circuit; (b) the equivalent circuit.

the equations at this point), we have

$$V_p = I_p Z_p - I_s Z_m \tag{1.112}$$

$$0 = -I_p Z_m + I_s(Z_s + Z_L) \tag{1.113}$$

Solving these for I_p yields

$$I_p = \frac{V_p(Z_s + Z_L)}{Z_p(Z_s + Z_L) - Z_m^2} \tag{1.114}$$

The effective primary impedance, seen to the right of points X-X in Fig. 1.20(b), is

$$Z_p' = \frac{V_p}{I_p}$$

$$= Z_p - \frac{Z_m^2}{Z_s + Z_L} \tag{1.115}$$

For the case where Z_L is a load resistor R_L and Z_s is a small coupling loop such that $|Z_L| \gg |Z_s|$, the effective primary impedance becomes

$$Z_p' = Z_p - \frac{Z_m^2}{R_L}$$

$$= R_p + j\omega L_p + \frac{\omega^2 M^2}{R_L}$$

$$= R_p \left(1 + \frac{\omega^2 M^2}{R_p R_L}\right) + j\omega L_p \tag{1.116}$$

Thus the effective series resistance of the primary is seen to be increased, and is given by

$$R_p' = R_p \left(1 + \frac{\omega^2 M^2}{R_p R_L}\right) \tag{1.117}$$

Defining the Q factor of the primary circuit *by itself* as

$$Q_p = \frac{\omega_0 L_p}{R_p} \tag{1.118}$$

and of the secondary as

$$Q_s = \frac{\omega_0 L_s}{R_L} \tag{1.119}$$

and recalling that $M = k\sqrt{L_p L_s}$, Eq. (1.117) for R'_p can be written

$$R'_p = R_p\left(1 + k^2 Q_p Q_s x^2\right) \tag{1.120}$$

where $x = \omega/\omega_0$. At resonance, $x = 1$, and R'_p becomes

$$R'_{p0} = R_p\left(1 + k^2 Q_p Q_s\right) \tag{1.121}$$

The effective dynamic impedance of the circuit, from Eq. (1.80), is

$$R'_D = \frac{L_p}{C_p R'_{p0}}$$

$$= \frac{R_D}{1 + k^2 Q_p Q_s} \tag{1.122}$$

Example 1.10   A tuned primary has an undamped Q factor of 100 and is tuned to resonate at 1 MHz by a 200-pF capacitor. The primary is mutually inductive coupled by means of a secondary coil of self-inductance 0.13 mH to a load of 5 kΩ, the coefficient of coupling being $k = 0.2$. Determine the effective load referred to the primary at resonance.

Solution

$$R_D = \frac{Q_p}{\omega_0 C_p}$$

$$= \frac{100}{2\pi \times 10^6 \times 200 \times 10^{-12}}$$

$$= 79.6 \text{ k}\Omega$$

$$Q_s = \frac{\omega_0 L_s}{R_L}$$

$$= \frac{2\pi \times 10^6 \times 0.13 \times 10^{-3}}{5 \times 10^3}$$

$$= 0.163$$

Therefore,

$$R'_D = \frac{79.6}{1 + (0.2)^2 \times 100 \times 0.163}$$

$$= 48 \text{ k}\Omega$$

The effective Q factor of the primary is seen to be

$$Q'_p = \frac{Q_p}{1 + k^2 Q_p Q_s x^2}$$ (1.123)

since

$$\frac{Q'_p}{Q_p} = \frac{R_p}{R'_p}$$

Although the effective Q factor is a function of frequency (i.e., of x) at frequencies within about $\pm 10\%$ of resonance, the variation may be ignored, and the frequency response of the circuit as a whole is given by the equation for a parallel tuned circuit, Eq. (1.81). It is left as an exercise for the student to show that, for a constant-current signal source I, the primary voltage V_p is given approximately by

$$V_p = \frac{IR'_D}{\left(1 + jyQ'_p\right)}$$ (1.124)

and the load voltage by

$$V_L = V_p k \sqrt{\frac{L_s}{L_p}}$$ (1.125)

Another common arrangement is for Z_L to be a secondary tuning capacitor C_s. When the circuits are tuned to the same resonant frequency, this is known as a synchronously tuned transformer since $L_p C_p = L_s C_s = (\omega_0)^{-2}$. Analysis may be carried out in a manner similar to that for the equivalent circuit of Fig. 1.20(b), although this is much more lengthy and involved. For the case where the circuits are identical (that is, $Q_p = Q_s = Q$ and $C_p = C_s = C$), the expression for the output voltage V_2 is

$$V_2 = \frac{-jIR_D kQ}{x\left[\left(1 + (xkQ)^2 - y^2 Q^2\right) + j2yQ\right]}$$ (1.126)

At resonance, $\omega = \omega_0$, and therefore $x = 1$ and $y = 0$. Let V_{2o} represent the output voltage at resonance; then

$$V_{2o} = \frac{-jIR_D Q}{1 + k^2 Q^2}$$ (1.127)

The $-j$ shows that V_{2o} lags I by $90°$, an important result in the theory of phase discriminators.

The modulus of V_{2o} is a maximum when $kQ = 1$, as can be proved by differentiation. For $kQ = 1$, the transformer is said to be *critically coupled*; for $kQ < 1$, it is *undercoupled*; and for $kQ > 1$, it is *overcoupled*. This is illustrated in Fig. 1.21, where the modulus is plotted against x for $kQ = 0.5$, 1.0, and 1.5. A value of $IR_D = 1V$ is used as reference, and $Q = 100$.

The 3-dB bandwidth of the critically coupled circuit can be shown to be

$$B_{3dB} = \sqrt{2}\,\frac{f_0}{Q} \tag{1.128}$$

(That is, it is $\sqrt{2}$ times that of a single tuned circuit.) At the same time, it can be shown that the 60-dB bandwidth is *reduced* by a factor of about $\sqrt{10^3}$ from the single tuned circuit bandwidth, which of course is indicative of the steep sides on the response curve of the critically coupled circuit.

The double-humped curve of Fig. 1.21 is a feature of the overcoupled circuit. This is often combined with critically coupled (or slightly undercoupled) circuits to obtain a composite response which is flat along the top and which has sides that are falling away sharply. Figure 1.22 illustrates the composite response of a $kQ = 1$ and a $kQ = 1.5$ combination of transformers. The curves are plotted in decibels relative to the resonant (or mid-frequency) response, against the frequency variable $x = \omega/\omega_0$. It should be noted that the overall, or composite, curve is obtained simply by *adding* the decibel response of the individual circuit curves.

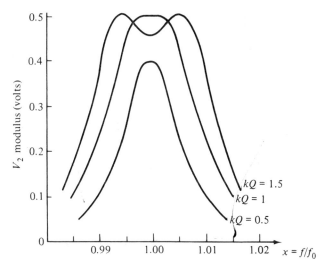

FIGURE 1.21.   Frequency response curves for coupled circuits.

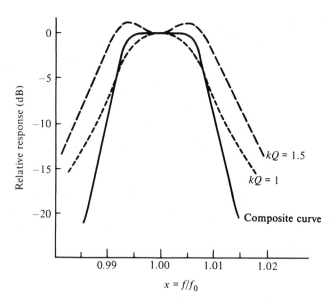

FIGURE 1.22. Composite response curve for combined critically coupled and overcoupled circuits.

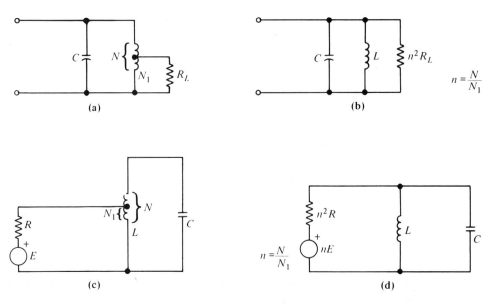

FIGURE 1.23. The tapped inductor (a) as a transformer; (b) referred to primary; (c) tapped on primary side. (d) is (c) referred to secondary.

1.12.3 Tapped Inductor

The *tapped inductor* (Fig. 1.23(a)) provides a form of mutual inductive coupling for which the coupling coefficient k (between the two sections of the coil) may be assumed to be unity. Often, the coil forms the tuning inductor of a resonant circuit as shown in Fig. 1.23(a), but this of course is not necessarily the case. The coil can be treated as a transformer of turns ratio $n = N/N_1$, and the equivalent circuit of Fig. 1.23(b) may be drawn. Note that the coil L remains in the circuit, as this corresponds to the magnetizing coil in L_c in Fig. 1.19(b).

The circuit is often used to reduce the loading effect of R_L on the Q factor of the tuned circuit. An example will illustrate this.

Example 1.12 A tuned circuit has an undamped Q factor of 200 and an undamped dynamic impedance of 1 MΩ. A load of 5 kΩ is tapped into the coil in the ratio $1:10$. Find the equivalent Q factor of the circuit.

Solution
$$n^2 R_L = (10)^2 \times 5 \times 10^3$$
$$= 0.5 \text{ M}\Omega$$

The effective dynamic impedance is the undamped R_D in parallel with the effective load $n^2 R_L$.

$$R'_D = \frac{0.5 \times 1}{0.5 + 1}$$
$$= 0.333 \text{ M}\Omega$$

Assuming that C remains unchanged, the Q factor will be reduced in the ratio R'_D/R_D:

$$Q' = 200 \times \frac{0.333}{1}$$
$$= 67$$

This is low, but had the load of 5 kΩ been connected directly across the tuned circuit, the effective Q factor would have been

$$Q' = 200 \times \frac{0.005}{1}$$
$$= 1.0$$

Note that, although the voltage gain of the circuit is reduced by a factor $1/n$, because of the effect of the step-down transformer, if the circuit forms the load of a voltage amplifier the load resistance will be approximately $n^2 R_L$, which is n^2 times what it would have been had R_L been connected directly. The overall effect on voltage gain therefore is to give a step up in voltage of approximately n times.

Where the loading effect of the source on a circuit has to be reduced, the tapping point is placed on the input side, as shown in Fig. 1.23(c). Of course, both input and load may be tapped down on the coil. From the point of view of the circuit, the source resistance R is stepped up in the ratio n^2, and the source voltage is stepped up by the voltage step-up ratio n. The equivalent circuit *referred to the secondary* is therefore as shown in Fig. 1.23(d).

1.12.4 Capacitive Tap

The arrangement of Fig. 1.24(a) is often used to couple a load resistance R_L to a tuned circuit while reducing the damping effect of R_L on the tuned circuit. The easiest method of analyzing this circuit is to find the admittance Y presented by the C_1, C_2, R_L branch, thus,

$$Y = \frac{j\omega C_1 (G_L + j\omega C_2)}{j\omega C_1 + G_L + j\omega C_2}$$

$$= \frac{G_L (\omega C_1)^2}{G_L^2 + \omega^2 (C_1 + C_2)^2} + j \frac{\omega C_1 (G_L^2 + \omega^2 C_2 (C_1 + C_2))}{G_L^2 + \omega^2 (C_1 + C_2)^2} \qquad (1.129)$$

where $G_L = 1/R_L$. Note how the admittance of G_L and C_2 in parallel (Y_p) is simply the sum of the conductance and susceptance, while the admittance of C_1 in series with the parallel circuit of C_2 and G_L requires use of the formula $Y_1 Y_p / (Y_1 + Y_p)$, where $Y_1 = j\omega C_1$.

For the tap to be effective in practice, it must be arranged that $\omega C_2 \gg G_L$, so that the expression for Y becomes

$$Y \cong \left(\frac{C_1}{C_1 + C_2} \right)^2 G_L + j\omega \frac{C_1 C_2}{C_1 + C_2} \qquad (1.130)$$

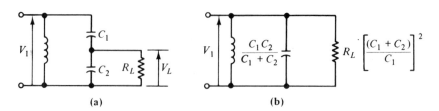

(a) (b)

FIGURE 1.24. (a) Capacitive tap; (b) referred to input.

From this, the equivalent circuit of Fig. 1.24(b) can be drawn. Clearly, the effective tuning capacity is C_1 in series with C_2, and the effective resistance across the circuit is R_L increased by the factor $[(C_1 + C_2)/C_1]^2$.

The output voltage V_L across R_L for $\omega C_2 \gg G_L$ is, from Fig. 1.24(a),

$$V_L \approx V_1 \left(\frac{C_1}{C_1 + C_2} \right) \tag{1.131}$$

This reduction in voltage is the price paid for the reduced damping on the tuned circuit.

Many other coupling arrangements are possible, and the general principles of analysis described in the previous sections can be applied. Usually there will be restrictions and approximations permissible for a given practical situation which simplify the analysis. An attempt should be made to simplify algebraic expressions and to collect terms together for comparison before these are evaluated; obviously, algebraic manipulation is useful only where it results in simplification or it brings out the significance of certain terms.

1.13   THÉVENIN'S THEOREM

Thévenin's theorem states (in part) that a linear, two-terminal network of independent sources and impedances can be replaced by a single voltage-generator equivalent circuit. The equivalent EMF E_{TH} is the open-circuit voltage appearing at the network terminals, and the equivalent internal impedance Z_{TH} is the impedance seen looking back into the network, with all the sources replaced by their internal impedances. As a specific example, consider the circuit of Fig. 1.25(a), for which it is desired to reduce the signal source and the 100-Ω input resistance to an equivalent voltage source. Looking to the left of the terminals 1-1, the open-circuit voltage (i.e., with the amplifier load removed, Fig. 1.25(b)) is

$$E_{TH} = \frac{3 \times 100}{50 + 100}$$
$$= 2\mu V$$

The internal impedance seen when the source EMF is short-circuited (Fig. 1.25(c)) is

$$R_{TH} = \frac{50 \times 100}{50 + 100}$$
$$= 33.33 \ \Omega$$

The equivalent circuit is therefore as shown in Fig. 1.25(d).

Thévenin's theorem offers a powerful means of simplifying circuits for the purpose of analysis.

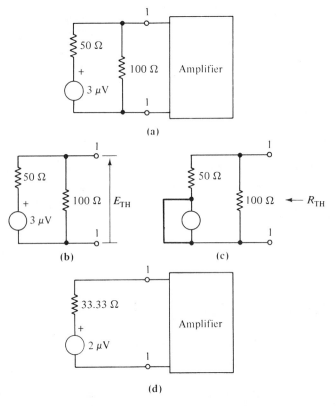

(a)

(b)   **(c)**

(d)

FIGURE 1.25. (a) Circuit example used to illustrate Thévenin's theorem; (b) load removed; (c) internal source short-circuited; (d) Thévenin equivalent circuit.

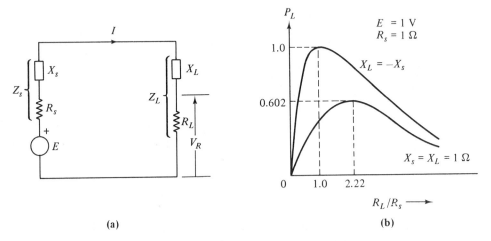

(a)   **(b)**

FIGURE 1.26. (a) Signal source delivering power to load; (b) power as a function of load impedance.

1.14   MAXIMUM POWER TRANSFER

Where a signal source is required to deliver power to a load, the power transfer should be a maximum. The average power delivered to the load Z_L (Fig. 1.26(a)) is

$$P_L = V_R I$$

$$= \left(\frac{ER_L}{\sqrt{(R_s + R_L)^2 + (X_s + X_L)^2}} \right) \left(\frac{E}{\sqrt{(R_s + R_L)^2 + (X_s + X_L)^2}} \right)$$

$$= \frac{E^2 R_L}{(R_s + R_L)^2 + (X_s + X_L)^2} \qquad (1.132)$$

Looking first at the effect of X_L on P_L, it will be seen that by making $X_L = -X_s$, P_L will be at a maximum given by

$$P_L = \frac{E^2 R_L}{(R_s + R_L)^2} \qquad (1.133)$$

R_L can now be varied to maximize this expression, and the best value is obtained by equating the differential coefficient of P_L with respect to R_L to zero:

$$\frac{dP_L}{dR_L} = \frac{E^2 [(R_s + R_L) - 2R_L]}{(R_s + R_L)^3} = 0$$

from which

$$R_s = R_L \qquad (1.134)$$

Combining this with the condition on X_L, it is seen that for maximum power transfer, the load impedance must be the complex conjugate of the source impedance, or

$$R_L + jX_L = R_s - jX_s \qquad (1.135)$$

This is known as a *conjugate match*, and although it results in maximum power transfer, it is good at only one frequency (viz. that which makes $jX_L = -jX_s$).

Substituting $R_L = R_s$ in Eq. (1.133) gives the maximum available average power:

$$P_L = \frac{E^2 R_s}{(R_s + R_s)^2}$$

$$= \frac{E^2}{4R_s} \tag{1.136}$$

Alternatively, if the constant-current generator representation of a source is used (Fig. 1.1(b)), then

$$P_L = \frac{(IR_s)^2}{4R_s}$$

$$= \frac{I^2 R_s}{4} \tag{1.137}$$

Equation (1.137) gives the maximum available power in terms of the constant current I and internal resistance R_s. This representation is used, for example, in Eq. (4.51), which is used in the determination of noise factor.

Equation (1.137) may be written in terms of source conductance $G_s = 1/R_s$ as

$$P_L = \frac{I^2}{4G_s} \tag{1.138}$$

When this is compared with Eq. (1.136), the duality of the voltage and current representations becomes apparent.

Where operation is required over a range of frequencies, the best value for Z_L is that which produces a true *reflectionless match*, which is

$$Z_L = Z_s \tag{1.139}$$

or

$$R_L + jX_L = R_s + jX_s$$

Reflectionless matching is not as efficient as conjugate matching, as illustrated in Fig. 1.26(b). However, it does give a fairly broad maximum, and there are other considerations which make it the usual choice when signal transmission over lines is required, as described in Chapter 12.

1.15 PASSIVE FILTERS

1.15.1 Filter Transfer Function

Filtering of signals in telecommunications is necessary in order to select the desired signal from the range of signals transmitted, and also to minimize the effects of noise and interference on the wanted signal. Electrical filters may be constructed using resistors and capacitors, resistors and inductors, or all three types of components, but it will be noticed that at least one reactive type of component must be present. The resonant circuits described in Sections 1.8 and 1.9, and also the tuned transformers described in Section 1.12.2, are all examples of filters. Many of the applications in telecommunications require filters with very sharply defined frequency characteristics, and the filter circuits are much more complex than simple tuned circuits. Most complex filters use all three types of components: inductors, capacitors, and resistors. The inductors tend to be large and costly, and these are now being replaced in many filter designs by electronic circuits that utilize operational amplifiers along with capacitors and resistors. Such filters are known as active filters. Active filters have many advantages over passive filters, the chief ones being that they are small in size, lightweight, less expensive, and offer more flexibility in filter design. The disadvantages are that they require external power supplies and are more sensitive to environmental changes, such as changes in temperature.

Filter design is a very extensive topic, embracing active filters, passive filters, and digital filters, and in this section only a brief introduction to passive filters will be given. In addition to passive filters designed using electrical components, various other types are available that utilize some form of electromechanical coupling. These include piezo-electric filters and electro-mechanical filters.

A filter will alter both the amplitude and the phase of the sinusoidal signal passing through it. For audio applications, the effect on phase is seldom significant. Filters are classified by the general shape of the amplitude-frequency response into *low-pass filters*, *high-pass filters*, *bandpass filters*, and *band-stop filters*. These designations are also used for digital and video filtering, but the effect on phase is also very important in these applications. A further designation is the *all-pass filter*, which affects only the phase, and not the amplitude, of the signal.

The names listed above refer to the shape of the amplitude part of the filter transfer function. The filter transfer function is defined as the ratio of output voltage to input voltage (current, but not power, could be used instead of voltage) for a sinusoidal input. Thus, if the input to the filter is a sine wave having an amplitude of x and a phase angle of θ_x, the output will also be a sine wave but with a different amplitude and phase angle in general. Let y represent the output amplitude and θ_y the output phase; then the filter transfer function is

$$H(f) = |H(f)|\underline{/\theta} = \frac{y}{x}\underline{/\theta_y - \theta_x} \tag{1.140}$$

The modulus, or amplitude-frequency part of the transfer function is $|H(f)|$, and this is sketched in Fig. 1.27(a) for the various kinds of filters listed above. The low-pass filter (LPF) is seen to be characterized by a pass-band of frequencies extending from zero up to some cut-off frequency f_c. Ideally, the response should drop to zero beyond the cut-off, but in practice there is a transition region leading to the edge of the stop band at f_s. The stop band is the region above f_s where the transmission through the filter is ideally zero. Again, in practice there will be a finite attenuation in the stop band, and also, ripple may be present in both the pass band and stop band, as shown in Fig. 1.27(a).

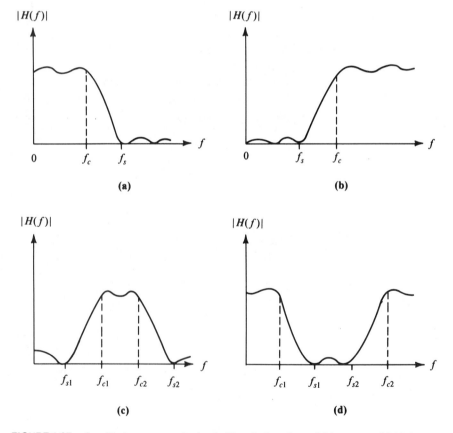

FIGURE 1.27. Amplitude response for basic filter designations: (a) low-pass, (b) high-pass, (c) band-pass, and (d) band-stop.

The high-pass filter (HPF) characteristic is shown in Fig. 1.27(b). Here, the stop band is from zero up to some frequency f_s, the transition region from f_s up to the cut-off frequency f_c, and the pass-band from f_c onwards. As with the LPF, ripple may appear in both the stop band and the pass band.

The band-pass filter (BPF) characteristic is shown in Fig. 1.27(c). The pass band is seen to be defined by two cut-off frequencies, a lower one at f_{c1}, and an upper one at f_{c2}. There is a lower transition region leading to a lower stop-band frequency limit f_{s1}. The lower stop band is from zero up to f_{s1}. At the other end, the upper transition region leads from f_{c2} to f_{s2}, and then the upper stop band extends from f_{s2} upwards. The coupled tuned circuit response shown in Fig. 1.21 is an example of a band-pass response.

The band-stop filter (BSF), or band-reject filter, response is shown in Fig. 1.27(d). This has a lower pass band extending from zero to f_{c1}, a lower transition region extending from f_{c1} to f_{s1}, a stop band extending from f_{s1} to f_{s2}, and then an upper transition region extending from f_{s2} to f_{c2} and an upper pass band extending upwards from f_{c2}.

There are a number of well-established filter designs available, each design emphasizing some particular aspect of the response characteristic. Although these designs apply to all the categories mentioned previously, they will be illustrated here only with reference to the low-pass filter. In the following sections the response curves are normalized such that the maximum value is unity.

Butterworth response. The modulus of the Butterworth response is given by

$$|H(f)| = \frac{1}{\sqrt{1 + (f/f_c)^{2m}}} \tag{1.141}$$

This gives what is termed a maximally flat response, which means that the response approaches the ideal as the frequency approaches zero. The response is sketched in Fig. 1.28(a). The order of the filter is m, an integer, and the filter response approaches more closely to the ideal as m increases. Whatever the order of the filter, it will be seen from Eq. (1.141) that at the cut-off frequency $f = f_c$, the response is reduced by $1/\sqrt{2}$, or -3dB. Thus, at the cut-off frequency the response is not abruptly "cutoff." The simple R-C filter illustrated in Fig. 2.10 is an example of a first-order Butterworth filter.

Chebyshev (or Tchebycheff) response. The Chebyshev response is given by

$$|H(f)| = \frac{1}{\sqrt{1 + \varepsilon^2 C_m^2(f/f_c)}} \tag{1.142}$$

Here, $C_m(f/f_c)$ is a function known as a Chebyshev polynomial, which for $-1 \le f/f_c \le 1$ is given by $\cos(m\cos^{-1}(f/f_c))$. This is a rather formidable expression, but it can be seen that for the range of f/f_c specified, the Chebyshev polynomial oscillates between ± 1. This produces an equiripple response in the pass band, and the coefficient ε can therefore be chosen to make the ripple as small as desired. The order of this filter response is also m, and this controls the sharpness of the transition region. The Chebyshev response is sketched in Fig. 1.28(b). It will be noticed that the cut-off frequency in this case defines the ripple pass band. For $f/f_c > 1$, the Chebyshev polynomial is given by $\cosh(m\cosh^{-1}(f/f_c))$.

Maximally flat time delay response. This type of filter is designed not for sharp cut-off, but to provide a good approximation to a constant time delay, or equivalently, a linear phase-frequency response. In other words, the

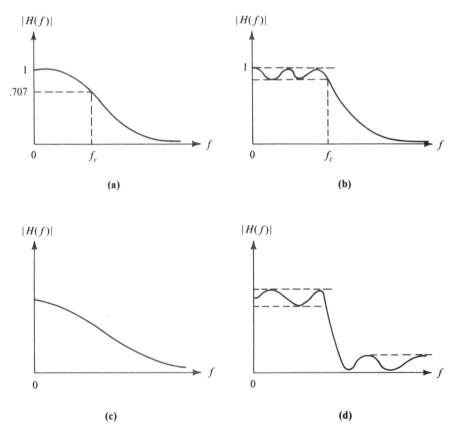

FIGURE 1.28. Sketches of the amplitude / frequency responses of several types of low-pass filters: (a) Butterworth; (b) Chebychev; (c) Maximally flat time-delay (MFTD); (d) Cauer, or Elliptic.

phase response, rather than the amplitude response, is of more importance. Such filters are required when handling video waveforms and pulses. The amplitude response is a monotonically decreasing function of frequency, meaning that it always decreases as frequency increases from zero, as sketched in Fig. 1.28(c).

The Cauer (or elliptic) filter. The filter response $H(f)$ can be written generally as the ratio of two polynomials in frequency, $N(f)/D(f)$. For the filters described so far, the numerator $N(f)$ is made constant, and the filter response is shaped by the frequency dependence of the denominator $D(f)$. In the Cauer filter, both the numerator and denominator are made to be frequency dependent, and although this leads to a more complicated filter design, the Cauer filter has the sharpest transition region from pass band to stop band. Often, in telephony applications a sharp transition band is the most important requirement. The term *elliptic filter* is also widely used for this type of filter, and comes about because the response can be expressed in terms of a mathematical function known as an elliptic function. The amplitude response of the Cauer filter has ripple in both the pass band and the stop band as sketched in Fig. 1.28(d).

1.15.2 *LC* Filters

A low-pass *LC* filter network is shown in Fig. 1.29(a). This can be considered as being made up from two end sections and an intermediate section as shown in Fig. 1.29(b). The filter can be extended by increasing the number of intermediate sections, leaving the end sections unchanged. Extending the filter in this way sharpens the cut-off or transition region of the response.

The filter can also be constructed from π-sections as shown in Fig. 1.29(c), this being similar to the T- and π-equivalences used with attenuators (Figs. 1.6 and 1.7). It is left as an exercise for the student to develop the intermediate-section and end-sections equivalent for the π-network.

A high-pass filter network is shown in Fig. 1.29(d), and this can also be made up from an intermediate T-section and two end sections. A π-type equivalent circuit can also be constructed.

A quick way of determining whether a filter is low pass or high pass is to examine the transmission paths at dc and at very high frequencies. For the circuit of Fig. 1.29(a), a dc path exists between input and output, and high frequency transmission will be blocked by the L_1 inductors. Therefore the filter is low pass. For Fig. 1.29(c), a dc path also exists between input and output, while at high frequencies the C_2 capacitors will shunt the signal to ground. Therefore this is also a low-pass filter. For Fig. 1.29(d), there is no dc path between input and output, and at high frequencies (above the series resonance of L_2C_2) the C_1 capacitors allow signal transmission, and therefore this is a high-pass filter.

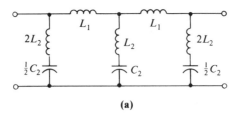

(a)

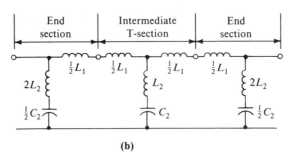

(b)

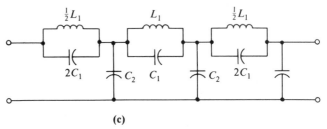

(c)

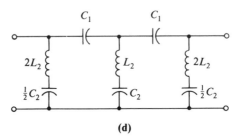

(d)

FIGURE 1.29. (a) A low-pass filter circuit, and (b) the equivalent circuit consisting of two end sections and one intermediate T-section. (c) A low-pass filter circuit with an intermediate π-section. (d) A high-pass filter circuit with an intermediate T-section.

1.15.3 Piezoelectric Crystal Filters

Piezoelectric crystals exhibit the property that when an electric potential is applied across the faces of the crystal, it physically bends or deforms. Conversely, when the same crystal is mechanically deformed by pressure, an electric potential is developed between the crystal faces. The crystal also exhibits the phenomenon of mechanical resonance when it is excited with an alternating potential of the correct frequency. The frequency of mechanical resonance is determined by the size and shape of the crystal sample in question and can be controlled over several orders of magnitude, from about 20 kHz to about 50 MHz, with considerable precision. In form, the packaged crystal is a slice of crystal cut in such a way as to give the desired mechanical resonant frequency, with isolated electrodes deposited on opposite sides so that a capacitive device is made.

Electrically, the mechanical resonance of this device makes the crystal look like a very high-Q series resonant circuit, with a capacitor in parallel with it. This capacitor causes a second parallel resonance, which occurs at a frequency that is very close to the mechanical resonant point. The reactance of a quartz crystal is plotted in Fig. 1.30, and shows that for low frequencies up to the series mechanical resonance, the crystal is capacitive. For frequencies between the series resonant and parallel resonant points, the reactance is inductive, and for frequencies above the parallel resonance, the reactance is again capacitive. At series resonance $X_{Ls} = X_{Cs}$ and the reactance is zero, and at parallel resonance $X_{Ls} = (X_{Cs} \text{ ser. } X_{Cp})$ and the reactance is infinite. The resonant frequencies of the crystal are very well defined and very stable, provided that the operating temperature is kept constant, making it very well suited as the high-Q resonant circuit which controls the operating frequency of oscillator circuits. This use of the crystal is discussed in Chapter 6.

The reactance characteristic of the quartz crystal is changed radically by placing an inductance in parallel with it. The series resonant frequency remains

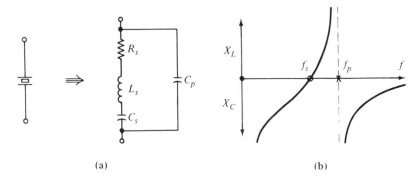

(a)   (b)

FIGURE 1.30. The quartz piezoelectric crystal: (a) the graphic symbol and the equivalent circuit of a quartz crystal; (b) variation of crystal terminal reactance with frequency.

unchanged, but the parallel resonant frequency is moved higher, so the separation between the two is greater than is the case for the crystal by itself.

Placing an inductor in series with the crystal has similar drastic effects on the reactance characteristic. In this case, however, the parallel resonant frequency remains unchanged while the series resonant frequency is caused to move lower and a second series resonant frequency is created.

The frequency separation between the series and parallel resonant frequencies of the crystal itself is small, of the order of a few hundred hertz at most for a 1-MHz crystal. Frequency spreading by means of series or parallel inductors can increase this separation to a few thousand hertz, making it possible to use the crystals as bandpass filter elements for IF amplifiers and for sideband separation.

The crystal gate shown in Fig. 1.31(a) is a narrow-band sharp-cutoff filter circuit which makes use of the reactance characteristic of the crystal itself. It has been used for separating the sidebands in SSB circuits. When the capacitance of C_2 is relatively large, a high-pass sharp-cutoff filter with the characteristics of Fig. 1.31(b) is formed. At the frequency f_∞, the reactance of the crystal is capacitive and equal in magnitude to that of the capacitor C_2, so the signal fed to the output through the crystal is equal in magnitude and opposite in phase to that fed through C_2, causing a complete cancellation at the output. At frequency f_0 the reactances are again equal in magnitude, but this time the

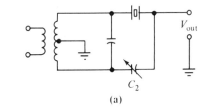

(a)

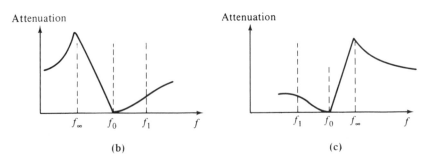

(b) (c)

FIGURE 1.31. The crystal gate: (a) the circuit; (b) a plot of attenuation versus frequency for the case where X_{C2} is small, giving a high-pass characteristic; (c) a plot of attenuation versus frequency for the case where X_{C2} is large, giving a low-pass characteristic.

crystal is inductive. The signal through the crystal is shifted by $90°$ while that through the capacitor is shifted by $-90°$, so both arrive at the output in phase with each other. For frequencies above f_0 up to f_1, just below the parallel resonant frequency of the crystal, the attenuation remains low as the signal is propagated through the capacitor C_2. Severe phase-shift distortion occurs near f_p, so the usable passband is only between f_0 and f_1. The cutoff beyond f_1 is quite gentle, but f_0 and f_∞ are only separated by a few hundred hertz, providing a sharp lower cutoff. Frequency shifting by a series inductor can be used to increase the passband to usable widths.

When the reactance of C_2 is made considerably higher, a low-pass filter with the characteristic of Fig. 1.31(c) results. Again, at f_0 the crystal reactance is equal to that of the capacitor and inductive, so the signals arrive at the output in phase with each other. At f_∞ the crystal reactance is equal to that of the capacitor and capacitive, resulting in complete signal cancellation, this time at a frequency higher than f_0 (and f_p). For frequencies below f_0 down to f_1 near f_s, the attenuation is low, determined by X_{C2}. Phase-shift distortion near f_s prevents use of frequencies below this, and again the usable bandpass can be increased by using a series inductor.

The crystal gate is inexpensive to build and uncritical in its adjustment, making it attractive, but it suffers from the disadvantage of providing a very narrow usable passband width. The crystal lattice filter is a more complicated circuit, but it provides bandwidths of a few hundred hertz to several tens of kHz and essentially flat response characteristics within the passband. Further, sharp cutoff can be provided on both the upper and lower edges of the passband.

Figure 1.32(a) shows the circuit of a full-lattice filter, and Fig. 1.32(b) shows a half-lattice filter. The attenuation characteristics of both are the same and are shown in Fig. 1.32(c). The crystals used in the lattice are matched pairs, so that crystals CR_1 and CR_2 are identical, and CR_3 and CR_4 are identical but different from CR_1 and CR_2. The crystals are chosen so that the series resonant frequency of CR_3 and CR_4 coincides with the parallel resonant frequency of CR_1 and CR_2. The inductances of the coils in the input and output circuits are effectively in parallel with the crystals and act to space out the separation between series and parallel resonance and provide the second parallel resonance of each crystal.

At $f_{\infty 1}$ the reactances of X_1 and X_3 are both inductive and equal, so the in-phase and antiphase signals fed to the output cancel, providing infinite attenuation. This is very near the parallel resonant frequency of X_1, and just past this frequency f_{01} occurs, where again the crystal reactances are equal, but X_1 is capacitive and X_3 is inductive, so the signals arrive at the output in phase with each other. $f_{\infty 1}$ and f_{01} delineate the lower transition band of the filter.

At $f_{\infty 2}$ the reactances X_1 and X_3 are equal but capacitive, so that cancellation again takes place. This occurs just above and very near the upper parallel resonant frequency of X_3. Just below $f_{\infty 2}$, the reactances are again

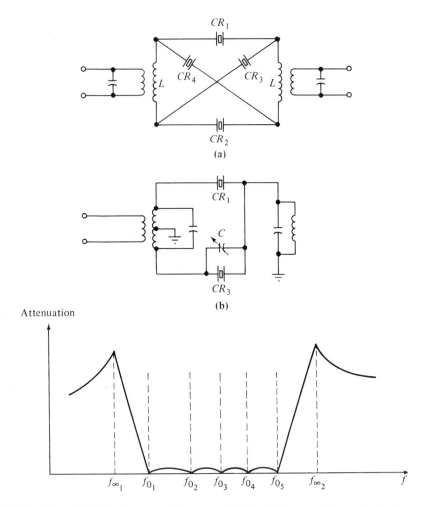

FIGURE 1.32. (a) Full-lattice crystal filter circuit; (b) half-lattice crystal circuit; (c) attenuation versus frequency for lattice filters.

equal, but opposite, so that the signals are again in phase, at f_{05}. Frequencies f_{∞_1} and f_{05} delineate the upper transition band of the filter.

At f_{03} the reactances X_1 and X_3 are again equal and opposite, so the signals arrive in phase. At frequency f_{02}, crystal 1 has zero reactance, shorting the input to the output. At frequency f_{03}, crystal 3 has zero reactance and again the output is connected directly to the input. Between these frequencies, the signal is fed to the output with very low attenuation values, providing the bandpass of the filter. Outside the band, attenuation is relatively high and can be improved by providing additional gradual cutoff filtering in tandem, such as with a normal IF amplifier filter.

1.15.4 Surface Acoustic Wave Filters

The piezoelectric crystal described in the previous section depends for its operation on *bulk acoustic waves*, i.e., mechanical vibrations that travel through the bulk of the solid. As the frequency of operation is increased, thinner crystals are required, and this sets an upper limit on frequency of about 50 MHz. It is also possible to set up *surface acoustic waves* (SAWs) on a solid, i.e., mechanical vibrations that travel across the surface of the solid. Molecules on the surface actually follow an elliptical path that penetrates a short way into the bulk. In the case of piezoelectric material, a piezoelectric EMF is generated at the surface, and this provides a means of coupling an electric signal into and out of the surface acoustic wave. The velocity of propagation of the surface acoustic wave is on the order of 3000 m/s. Since wavelength is related to frequency by $\lambda = v/f$, a frequency of, for example, 100 MHz will set up a surface wavelength of 30 μm (1 μm $= 10^{-6}$ m). The electrode structure on a SAW device requires spacings on the order of a wavelength, and thus very compact devices can be made. Since the action takes place on the surface, the bulk size can be chosen to provide mechanical strength without interfering with the surface operation. The electrodes may be deposited on the surface using one of several well-established methods in production use for silicon integrated circuit fabrication.

Fig. 1.33(a) shows the basic electrode configuration for a delay-line filter. The electrode structure consists of interdigitated metallic stripes, with the spacing between stripes that are connected together being one wavelength λ long. This is the center wavelength, and the filter has a bandpass characteristic, the response falling off as the input frequency is shifted to either side of the center frequency. The actual shape of the response curve depends on the electrode configuration, and a range of amplitude and phase combinations are possible. However, the SAW filter characteristic is always bandpass.

The input and output electrode structures are seen to be similar, and are referred to as *interdigital transducers* (IDTs). The surface acoustic wave generated by the input coupling travels out in both directions. Thus, only part of the total surface acoustic wave reaches the output IDT. The outer part is absorbed or dissipated in an electrode placed at the edge for this purpose. This prevents reflected waves from occurring. The equivalent circuit for the IDT is shown in Fig. 1.33(b). The capacitance C_0 is fixed by the geometric structure and the dielectric constant of the substrate. The susceptance component jB is zero at the center frequency of the filter and shows a periodic variation with frequency. Just around the center frequency, this susceptance is inductive for higher frequencies and capacitive for lower frequencies. The conductance G is also a function of frequency, and this in fact largely determines the filter response. The variation of G with frequency is sketched in Fig. 1.33(c).

Resonators can also be constructed from SAW devices, one arrangement being shown in Fig. 1.34(a). The end absorbers are replaced by *grating*

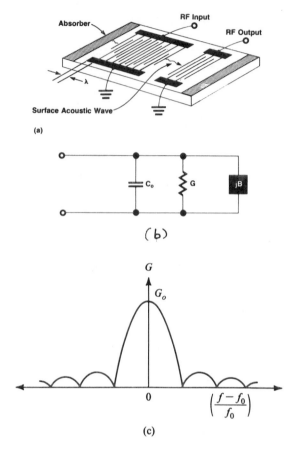

FIGURE 1.33. (a) Basic configuration for a surface acoustic wave delay line. The absorbing layer at each end reduces reflections from the edges of the substrate. (b) Equivalent circuit for an interdigital surface acoustic wave transducer (IDT). (c) Frequency response of the conductance component G. f_0 is the center frequency of the filter. ((a) and (b) are courtesy of Waguish S. Ishak, H. Edward Karrer, and William R. Shreve, Hewlett Packard Journal, December 1981.)

reflectors, which consist of an array of reflecting slots or grooves, the latter spaced $\lambda/2$ apart at the resonant frequency. Each groove reflects a small fraction of the incident surface acoustic wave, and the phasing is such that the individual reflections combine to give a peak at the output. The frequency response of a delay line filter using absorbers is shown in Fig. 1.34(b); that of one using grating reflectors is shown in Fig. 1.34 (c).

1.15.5 Electromechanical Filters

A metal disk which is mechanically driven with an axially applied oscillating force will exhibit a resonant mode which is analogous to parallel resonance in an electrical circuit. Similarly, an axially driven rod will exhibit

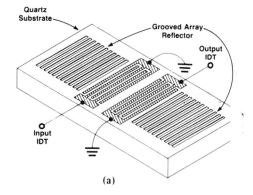

(a)

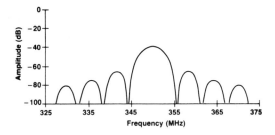

(b)

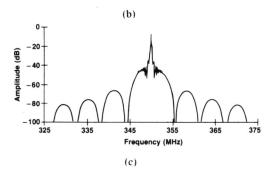

(c)

FIGURE 1.34. (a) Two-port surface-acoustic-wave resonator. The arrays of grooves at each end reflect the surface waves excited by the input IDT. The reflected waves constructively add at a frequency largely determined by the periodicity of the grooves. (b) Frequency response of a 350-MHz SAW delay line. By adding grating reflectors at each end, a resonant peak is obtained at the center frequency, shown in (c). Courtesy, Peter S. Cross and Scott S. Elliott, Hewlett Packard Journal, December 1981.

series resonance. When a series of disks and rods are interconnected to form a ladder network, and the resonances of the various components are carefully chosen, a bandpass filter with sharp cutoff characteristics results. Collins Radio has developed a mechanical filter of this type for use as sideband filters in communications receivers. These filters are electrically driven through magnetostrictive couplers and generally operate in the 100-kHz region. Bandwidths up to 5 kHz with very sharp cutoff characteristics are available. The unit is packaged in containers that are about 2 inches long and 1/2 inch square, and are arranged for printed-circuit-board mounting. The mechanical filter is pretuned at the factory, and no further adjustment is necessary or permitted, making it a very convenient component for receiver manufacture.

The ceramic filter is also a type of mechanical filter, but in this case the resonant components are in the form of disks of a piezoelectric ceramic such as barium titanate, arranged in a ladder configuration. The piezoelectric ceramic disks provide their own electromechanical conversion, and separate transducers are not necessary. In this respect the ceramic filter is more akin to the quartz crystal filter.

The ceramic filter is smaller than the mechanical filter by about half and has characteristics which are more nearly those of a complex crystal filter. They are also less complex and easier to manufacture than the mechanical filter and thus somewhat cheaper, which makes them very attractive for use in receivers. Ceramic filters are currently available in frequencies around 500 kHz and bandwidths ranging from 2 to 50 kHz.

1.16   PROBLEMS

1. A battery has an internal resistance of 0.01 Ω and an EMF of 2 V. Derive an equivalent constant-current generator circuit to represent the battery.

2. A 10-Ω and 15-Ω resistor are connected in series. Find the equivalent conductance.

3. Explain what is meant by the voltage transfer function of a circuit. Determine the voltage transfer function of the circuit shown in Fig. 1.5(b), where each resistor is 10.0 Ω.

4. For the T attenuator of Example 1.1, show that the generator sees an effective load of 50 Ω and that the load is fed by a generator of equivalent internal resistance of 50 Ω.

5. For the L network of Example 1.3, determine the insertion loss, in decibels.

6. For the L network of Example 1.4, determine the insertion loss, in decibels.

7. Design a 9-dB attenuator which has a symmetrical T network configuration and which provides input and output matching for 600 ohms.

8. Repeat problem 7 for a symmetrical π-network configuration.

9. A 1.0-mH inductor is connected in series with a resistance of 50 kΩ. Determine the impedance at a frequency of 15.9 MHz. Draw the phasor diagram for the circuit, showing the component voltages and the total voltage when a 15.9-MHz sinusoidal current of 1 mA peak flows through the inductor.

10. A 0.1-μF capacitor has an effective parallel resistance of 100 kΩ, which may be assumed constant. Determine the modulus and phase angle of the impedance at a frequency of 60 Hz.

11. A 2-μH inductor has a series resistance of 10 Ω. Determine the equivalent parallel circuit which represents the inductor at a frequency of 100 MHz. Does the inductance value change? Is the equivalent parallel resistance a function of frequency?

12. A capacitor can be represented by the parallel combination of 20 pF and 100 kΩ, these values being constant. Determine the equivalent series representation at a frequency of 100 MHz. Do the series components vary with frequency?

13. Calculate the minimum and the maximum capacitance values obtainable using three 0.01-μF capacitors.

14. Two windings are arranged such that the coefficient of coupling between them is 0.01. The self-inductances of the windings are 10 mH and 5 mH. Calculate the maximum and minimum inductance values obtainable from the combination.

15. Explain what is meant by "series resonance of a tuned circuit." A 2-μH inductor has a series resistance of 2 Ω. Calculate the capacitance value required for series resonance at 50 MHz. A 1-V rms sinusoidal signal at the resonant frequency is applied across the series circuit. Calculate the voltage appearing across the inductor and across the capacitor.

16. Show that the variable y introduced in Eq. (1.65) is given by $y \cong \pm 2\Delta f/f_0$, where $\Delta f = |f - f_0|$, at frequencies close to resonance. Using this, plot the modulus and phase angle of the series impedance of an LRC circuit as a function of $\pm 2\Delta f/B_3$, where B_3 is the -3 dB bandwidth and $R = 1$ Ω.

17. The inductor and capacitor in Problem 15 are connected in parallel. Find the parallel resonant frequency. Does this differ significantly from the series value?

18. A parallel tuned circuit is resonant at 50 MHz when the tuning capacitor is 50 pF. The Q factor of the circuit is 150. Calculate the dynamic impedance of the circuit. A sinusoidal signal of 2 V rms at the resonant frequency is applied across the circuit. Calculate (a) the total circuit current, and (b) the current in each branch. State any approximations made.

19. A coil is self-resonant at 20 MHz and has a Q factor of 200 (neglecting self-capacitance), which may be assumed constant. Calculate the effective Q factor at frequencies of 5 MHz and 10 MHz.

20. Explain briefly what is meant by skin effect and why it is undesirable. What steps may be taken to reduce skin effect in inductors?

21. To function properly, a circuit has to be connected to a 600-Ω signal source. The actual source is a 50-Ω microphone. Calculate the turns ratio of the matching transformer required. State any assumptions made.

22. A parallel tuned circuit is resonant at 10.7 MHz with a tuning capacity of 200 pF. The Q factor is 150. The circuit is loosely coupled to a 50-Ω resistive load through an untuned mutual-inductive coupling loop of self-inductance 0.1 μH and $k = 0.1$. Determine (a) the dynamic impedance of the circuit at resonance, and (b) the -3 dB bandwidth.

23. Explain briefly why an overcoupled tuned transformer should show two peaks. How can the response curves of tuned transformers be combined to produce a flat-topped response with sharply falling sides?

24. A signal source of EMF E and 1-kΩ internal resistance is tapped midway along an inductor which is tuned to parallel resonance. A 1-kΩ resistive load is also connected to the tapping point. Assuming that the resistance of the tuned circuit is negligible, show that the voltage across the tuned circuit is E.

25. Two 10-pF capacitors are connected in series to form the tuning capacitor of a 100-MHz parallel resonant circuit. The dynamic impedance of the circuit is 1 M Ω. A 10-kΩ load is connected across one of the capacitors. Determine (a) the effective dynamic impedance of the loaded circuit, and (b) the effective tuning capacity.

26. State Thévenin's theorem. Use it to show that the T attenuator and source in Example 1.1 can be replaced by an equivalent voltage generator of internal resistance 50.1 Ω and EMF $0.24E$, where E is the original source EMF.

27. Explain what is meant by a conjugate match. A signal source can be represented by an equivalent voltage generator of internal series impedance $(50 + j10)$ Ω at 200 MHz. Calculate the series RC values of load required for conjugate matching. The internal EMF of the source is 3 μV. Determine the maximum average power delivered to the load.

28. A signal source can be represented by an equivalent constant-current generator, the internal impedance of which is a 600-Ω resistance in parallel with a 10-pF capacitance. Determine the parallel load required for conjugate match at 300 MHz.

29. Explain what is meant by the transfer function of a filter. The input voltage to a filter is $5 \sin \omega t$, and the corresponding output voltage is $2 \sin(\omega t + 40°)$. What is the value of the transfer function at this frequency?

30. A square wave having the same amplitude and frequency as the sine wave input is applied to the filter of problem 29. Can the value of the transfer

function as determined in problem 29 be used to determine the output for the square wave input?

31. A second-order low-pass Butterworth filter has a -3dB frequency of 1000 Hz. Calculate the output amplitude from the filter when the input is a 1-V sinusoid at frequency (a) 100 Hz, and (b) 5000 Hz. (c) What is the transfer function modulus, in decibels, in each case?

32. On the same set of axes, plot the amplitude response for a first-order and a second-order low-pass Butterworth filter, over the range 0 to -30dB. Use a logarithmic frequency scale normalized to the -3dB value.

33. Calculate the amplitude response for a first-order Chebyshev LPF for which $\varepsilon = .25$ at frequencies (a) $f = .5f_c$, (b) $f = f_c$, and (c) $f = 5f_c$.

34. Explain how the crystal lattice filter acts to provide bandpass filtering.

CHAPTER 2

Waveform Spectra

2.1 INTRODUCTION

Signals are characterized by an amplitude/time variation of some physical quantity; for example, the ear detects the air pressure/time variation in a sound wave. When transmitted through a telecommunications channel, at some point the signal is converted to analogous voltage- and current-time variations, which for brevity will be referred to as the *signal waveform*. A fact of fundamental importance in communications is that the time waveform of a signal can be represented by a series of sine and/or cosine waves. Such a representation is termed the *spectrum* of the signal. The reason the sine (or cosine) wave is chosen as the basic waveform type is because the response of a channel to this type of wave is easily determined mathematically and by measurement, and the results can be extended to the wave represented by the series of sine or cosine waves.

A sinusoidal waveform which is described by Eq. (2.1) is shown in Fig. 2.1(a).

$$v = V_{max} \sin 2\pi f t \qquad (2.1)$$

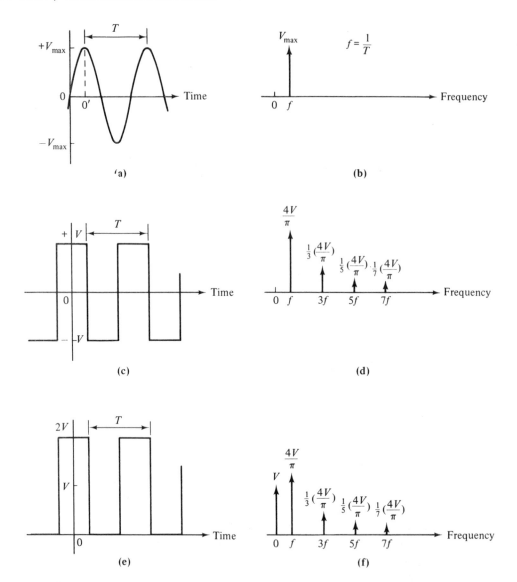

FIGURE 2.1. (a) A sinusoidal voltage waveform; (b) spectrum of a sinusoidal waveform; (c) a square voltage waveform; (d) magnitude spectrum of a square wave; (e) a square wave with a positive dc component; (f) magnitude / frequency spectrum for (e).

Alternatively, if the zero-time origin is started at $0'$ instead of 0, the wave can be described by

$$v = V_{max}\cos 2\pi ft \qquad (2.2)$$

Since the choice of time origin is arbitrary, the choice of representing the wave by a sine or cosine function will usually be unimportant. Again, it should be noted that Eq. (2.1) can be expressed as

$$v = V_{max}\cos\left(2\pi ft - \frac{\pi}{2}\right) \tag{2.3}$$

The spectrum of the sinusoidal wave is simply a straight line of height V_{max}, positioned at f on the frequency axis, as shown in Fig. 2.1(b). The amplitude spectrum takes no account of whether the wave is represented by sine or cosine. A separate phase angle/frequency graph may be shown if this information is required.

2.2   COMPLEX REPETITIVE WAVES

Any waveform, other than the sine or cosine wave, which repeats itself at regular intervals is termed a *complex repetitive wave*. The period T over which the waveform repeats is termed the *periodic time*. The spectrum for any complex repetitive wave can be found by means of a mathematical method known as *Fourier analysis*, but only the results of Fourier analysis for some common types of repetitive waveforms will be used here.

The rectangular wave shown in Fig. 2.1(c) can be represented by the Fourier series

$$v = \frac{4V}{\pi}\left(\cos \omega t - \tfrac{1}{3}\cos 3\omega t + \tfrac{1}{5}\cos 5\omega t - \tfrac{1}{7}\cos 7\omega t + \ldots\right) \tag{2.4}$$

where

$$\omega = \frac{2\pi}{T} = 2\pi f$$

Note that the symmetry of the square wave about the axes is similar to a cosine wave, and as a result the series of Eq. (2.4) contains only cosine terms.

The series has an infinite number of terms, but it can be seen that the amplitude of each term decreases as $1/n$. It will also be seen that the series contains only odd harmonics (i.e., components at frequencies f, $3f$, $5f$, etc.). The spectrum for the rectangular wave is shown in Fig. 2.1(d).

It must be clearly understood that the spectrum is more than just an alternative mathematical way of representing the wave. The component cosine waves, for example, in the series for the rectangular wave, are as physically real as the original time waveform, and could be tuned out by means of frequency-selective filters. The first three components of the rectangular wave spectrum are shown in Fig. 2.2. Adding these produces the resultant waveform shown dashed, and it will be seen that this approaches the rectangular wave shape.

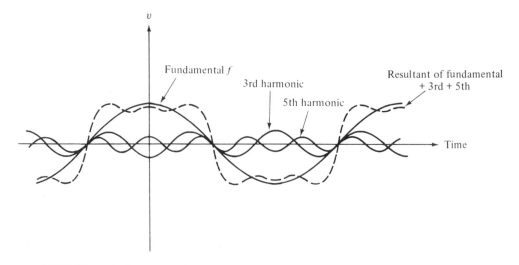

FIGURE 2.2. An illustration of how the three sinusoidal components add to approximate a square wave.

By shifting the rectangular wave up on the voltage axis (Fig. 2.1(e)), a zero-frequency component is added to the spectrum (Fig. 2.1(f)). This zero-frequency component is just the average, or "dc," value of the waveform, and is the value that would be read on an average-reading voltmeter, such as a moving-coil meter. *Any waveform that is not symmetrical in area about the time axis will have a zero-frequency component.*

The series for the square wave of Fig. 2.1(e) is

$$v = V + \frac{4V}{\pi}\left(\cos \omega t - \tfrac{1}{3}\cos 3\omega t + \tfrac{1}{5}\cos 5\omega t - \ldots\right) \qquad (2.5)$$

thus showing the zero-frequency or "dc" component to be V.

The train of rectangular pulses in Fig. 2.3(a) has the spectrum shown in Fig. 2.3(b). This has a zero-frequency component and contains both odd and even harmonics (i.e., components at frequencies f, $2f$, $3f$, $4f$, etc., where $f = 1/T$). The amplitude of the spectrum components depends on the ratio τ/T of "on" time to periodic time, the series being given by

$$v = \frac{V\tau}{T} + \frac{2V\tau}{T}\left(\frac{\sin x}{x}\cos \omega t + \frac{\sin 2x}{2x}\cos 2\omega t + \frac{\sin 3x}{3x}\cos 3\omega t + \ldots\right)$$

$$(2.6)$$

where $x = \pi\tau/T$. The series is written this way because the function $(\sin nx)/nx$ is a well-known mathematical function and is readily available in tabular form.

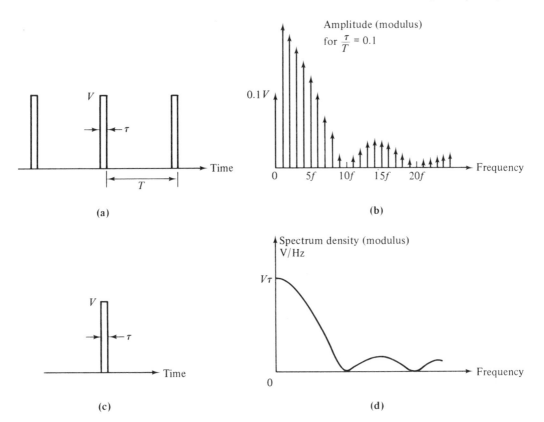

FIGURE 2.3. (a) A train of positive-going rectangular voltage pulses; (b) frequency spectrum; (c) a single positive-going rectangular voltage pulse; (d) spectrum density function of a single pulse.

The amplitude of the "nth" harmonic is seen to be

$$V_n = \frac{2V\tau}{T} \frac{\sin nx}{nx} \qquad (2.7)$$

By letting T approach infinity, the train of rectangular pulses approaches a single pulse (Fig. 2.3(c)). Also, as $T \to \infty$, $f \to 0$, and the spacing between the harmonic components shrinks to zero, so that the spectrum becomes *continuous* (Fig. 2.3(d)). By continuous, it is meant that there are no discrete harmonic frequencies, and only the envelope of the spectrum can be determined. This is given by

$$V(f) = V\tau \frac{\sin \pi f \tau}{\pi f \tau} \qquad (2.8)$$

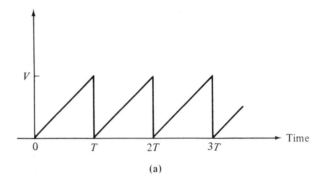

(a)

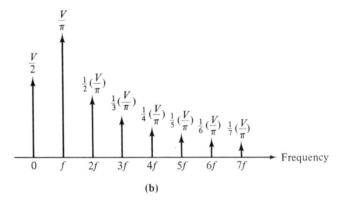

(b)

FIGURE 2.4. (a) A rising-ramp triangular voltage waveform; (b) frequency spectrum.

$V(f)$ in this case represents the *spectrum density* and has the units volts/frequency, i.e., V/Hz. The single pulse is an example of a nonrepetitive wave and is included here because its properties follow easily from those of the repetitive pulse waveform.

Another function frequently encountered in practice is the sawtooth wave, shown in Fig. 2.4(a). The Fourier series for this is

$$v = \frac{V}{2} - \frac{V}{\pi}\left(\sin \omega t + \tfrac{1}{2}\sin 2\omega t + \tfrac{1}{3}\sin 3\omega t + \ldots\right) \qquad (2.9)$$

The spectrum is shown in Fig. 2.4(b).

2.3 EFFECT OF FILTERING ON COMPLEX SIGNALS

As mentioned in Section 1.15.1, the complex transfer function of a filter can be expressed in two parts, the modulus $|H(\omega)|$ and phase shift $\theta(\omega)$. The *modulus* affects the *amplitude* of the component waves of a spectrum, appearing on the spectrum graph as a change in the height of the arrows representing the component amplitudes. The *phase shift* affects the *phase* of the component waves, and although this does not show up on the spectrum graph, it will be apparent on the output waveform. These results can be shown mathematically. Let the Fourier representation of the input wave be

$$v_i(t) = V_1 \sin \omega_0 t + V_2 \sin 2\omega_0 t + \ldots + V_n \sin n\omega_0 t \qquad (2.10)$$

Let $V_i(n\omega_0)$ represent the input spectrum, and $H(\omega)$ the transfer function of the filter. Then the output spectrum is given by

$$V_0(n\omega_0) = H(n\omega_0)V_i(n\omega_0) \qquad (2.11)$$

and the output time function is given by

$$v_o(t) = |H(\omega_0)|V_1 \sin(\omega_0 t + \theta(\omega_0)) + |H(2\omega_0)|V_2 \sin(2\omega_0 t + \theta(2\omega_0))$$
$$+ \cdots + |H(n\omega_0)|V_n \sin(n\omega_0 t + \theta(n\omega_0)) \qquad (2.12)$$

This shows that the amplitude of the nth component of the output is

$$|H(n\omega_0)|V_n \qquad (2.13)$$

where V_n is the amplitude of the nth component of input, and the phase shift of the output component relative to the input component is

$$\theta(n\omega_0) \qquad (2.14)$$

When a signal passes through a network, amplitude and phase distortion may be introduced in the transfer process. Figure 2.5 shows the transfer function of a hypothetical network and its effect on an input signal consisting of a fundamental and a second harmonic. The input spectrum is shown in Fig. 2.5(b), the modulus and phase shift of the transfer function as functions of frequency in Fig. 2.5(c) and (d). Since the modulus is constant at unity, the output spectrum is identical to the input, as shown in Fig. 2.5(e). The phase shift of this hypothetical network is shown as being zero at the fundamental frequency and π radians at the second harmonic. The second harmonic therefore undergoes a 180° phase shift. Figure 2.5(f) shows the input waveform, with the component waves shown dashed, and Fig. 2.5(g) shows the output waveform. In each case, the resultant wave can be obtained by graphically adding the component waves. The output wave is seen to differ from the input wave, and therefore distortion (in this case, phase distortion)

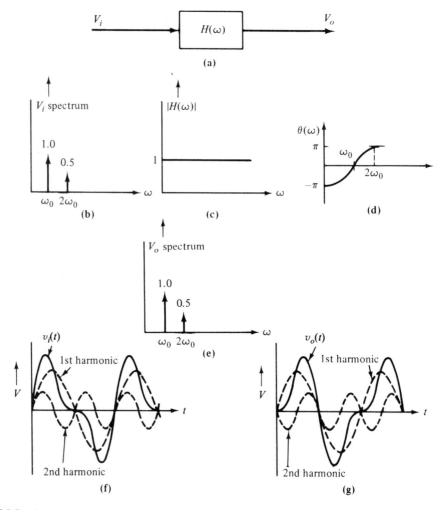

FIGURE 2.5. A network with a transfer function with constant amplitude, but with varying phase shift causing phase distortion: (a) the network; (b) input signal frequency spectrum; (c) transfer function amplitude plot; (d) transfer function phase plot; (e) output signal frequency spectrum; (f) input signal waveform; (g) output signal waveform.

has occurred. Mathematically, the result may be shown as

$$\text{Input:} \quad v_i(t) = \sin \omega_0 t + \tfrac{1}{2} \sin 2\omega_0 t \tag{2.15}$$

$$\text{Output:} \quad v_o(t) = \sin \omega_0 t + \tfrac{1}{2} \sin(2\omega_0 t + \pi)$$

$$= \sin \omega_0 t - \tfrac{1}{2} \sin 2\omega_0 t \tag{2.16}$$

This is an example of phase-shift distortion.

As a second example, consider another hypothetical network whose modulus and phase shift are shown as functions of frequency in Fig. 2.6(c) and (d). Assuming the same input as in the previous example, the output spectrum will be as shown in Fig. 2.6(e), where the fundamental component is seen to be attenuated by half. Figure 2.6(d) shows that no phase shift occurs. The resultant output wave (along with the component waves, shown dashed) will be as shown in Fig. 2.6(g). For comparison, the input wave is shown in Fig. 2.6(f).

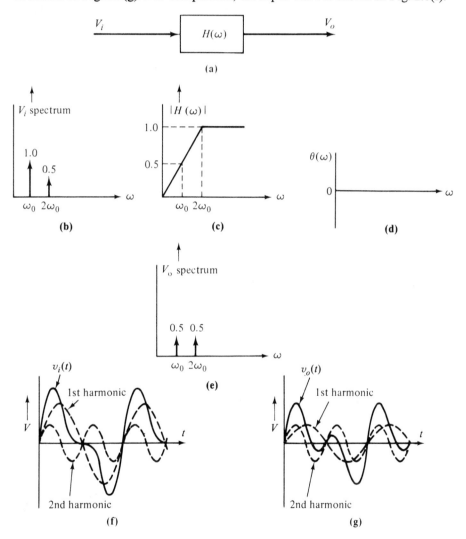

FIGURE 2.6. A network with a transfer function with a constant phase shift, but varying amplitude causing amplitude distortion: (a) the network; (b) input signal frequency spectrum; (c) transfer function amplitude plot; (d) transfer function phase plot; (e) output signal frequency spectrum; (f) input signal waveform; (g) output signal waveform.

For the input wave given by Eq. (2.15), the output wave in this case is given by

$$v_o(t) = \tfrac{1}{2} \sin \omega_0 t + \tfrac{1}{2} \sin 2\omega_0 t \tag{2.17}$$

This is an example of amplitude distortion.

The modulus and phase-shift parts of the complex transfer function are not independent; that is, given one as a function of frequency, the frequency dependence of the other is then also determined. Although in principle the output wave can be reconstructed from the spectrum and phase information, this is seldom a practical approach, and special mathematical techniques are

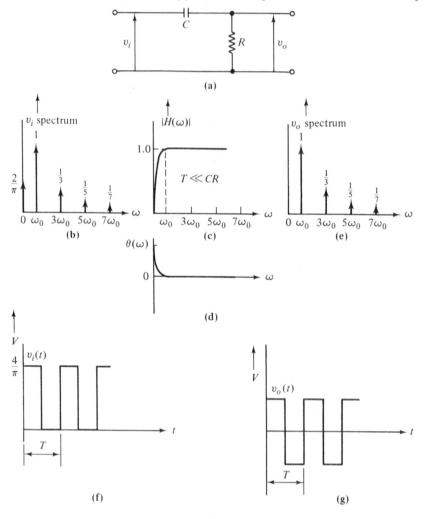

FIGURE 2.7. A high-pass filter: (a) The network; (b) Input signal frequency spectrum; (c) Transfer function amplitude plot; (e) Output signal frequency spectrum; (f) Input signal waveform; (g) Output signal waveform.

utilized to determine the shape of the output wave. However, it is important to know what components are present in the output spectrum, and a few illustrative examples follow.

2.3.1 High-Pass Filter

A *high-pass filter* (HPF) is one that passes the high-frequency components of a spectrum with negligible distortion of amplitude and phase. A simple example is the dc blocking capacitor, shown in Fig. 2.7(a). In this circuit, R represents the input resistance of the following stage. Suppose that a square wave (Fig. 2.7(f)) is applied to the input, and let the time constant CR of the filter be very much greater than the periodic time T of the wave. This is equivalent to making the resistance R very much greater than $|X_c|$ at the fundamental frequency.

The input spectrum of the square wave is shown in Fig. 2.7(b). The modulus and phase shift of the transfer function are shown as functions of frequency in Fig. 2.7(c) and (d). As can be seen, there will be negligible distortion of amplitude or phase, and so the output spectrum will be unchanged for the ac components. However, the filter does eliminate the dc component, which results in the output square wave being symmetrical about the time axis (i.e., the average value is zero).

Now, consider a similar filter, but with the time constant very much less than the periodic time of the wave (Fig. 2.8). The modulus and phase shift are shown as functions of frequency in Fig. 2.8(c) and (d), where, for definiteness, $T = 440CR$ (this makes $|X_c| = 10R$ at the seventh harmonic). In addition to eliminating the dc component, the filter reduces the other components, as far as the seventh, to approximately equal amplitudes; they are also phase-shifted by approximately $90°$. Elimination of the dc component again means that the output wave will be symmetrical about the time axis. The attenuation and phase shift of the other components results in a sharply peaked wave, as shown in Fig. 2.8(g). Although this wave could, in principle, be reconstructed from the spectrum and phase information, it is easier to consider the circuit as a differentiator in order to obtain an idea of the shape of the output wave. This is justified, since, for $T \gg CR$ (or $|X_c| \gg R$), the voltage v_c across C is very much greater than the voltage v_o across R. Hence,

$$v_c \gg v_o$$

Therefore,

$$v_i \cong v_c$$

$$\frac{dv_i}{dt} \cong \frac{dv_c}{dt}$$

$$= \frac{1}{C}\frac{dq}{dt}$$

$$= \frac{i}{C}$$

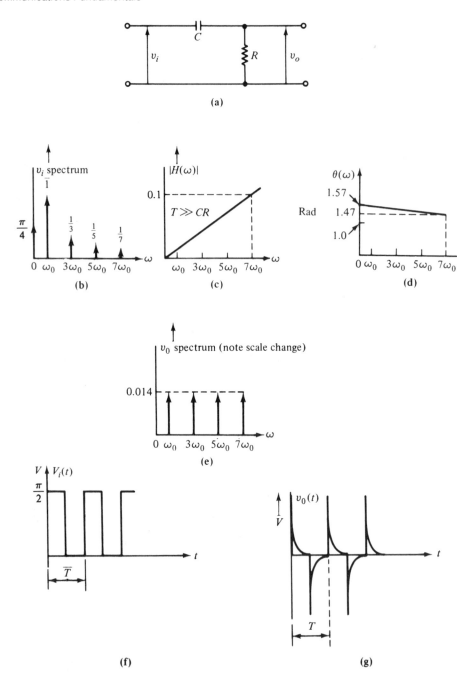

FIGURE 2.8. A differentiating network: (a) The network; (b) Input signal frequency spectrum; (c) Transfer function amplitude plot; (d) Transfer function phase plot; (e) Output signal frequency spectrum; (f) Input signal waveform; (g) Output signal waveform.

Hence,

$$i = C\frac{dv_i}{dt}$$

and

$$v_o = iR$$

$$= RC\frac{dv_i}{dt} \tag{2.18}$$

Thus the output voltage is proportional to the rate of change of the input signal, provided that v_o is small relative to v_i. The rates of change of the rise and fall sides of v_i are positive- and negative-going impulses (Fig. 2.8(g)), while the rates of change of the flat top and bottom are zero. Practical limitations will introduce an exponential approach from the impulses to the zero level as shown in Fig. 2.8(g).

2.3.2 Bandpass Filter

A simple example of a *bandpass filter* (BPF) is the parallel resonant circuit shown in Fig. 2.9(a). Here, the transfer function has to be expressed in terms of the currents, since the voltage across the circuit is common to input and output. Thus,

$$H(\omega) = \frac{I_o}{I_i} \tag{2.19}$$

The modulus and phase shift as functions of frequency are sketched in Fig. 2.9(c) and (d). For a square-wave input current (Fig. 2.9(f)), the input spectrum will be as shown in Fig. 2.9(b), and the output spectrum, when the circuit is resonant at the third harmonic, will be as shown in Fig. 2.9(e). Here again, the dc component is eliminated since the inductance L (assumed to be resistance-less) will short-circuit any dc away from the output load R_L. The phase shift is zero at the resonant frequency (the third harmonic of input). The output current and voltage (the latter of which is also the voltage across the tuned circuit) are sinusoidal at frequency $3f_0$, where $f_0 = 1/T$. This principle is used in certain types of frequency multipliers; in this example, the frequency of the signal has been tripled on passing through the filter.

2.3.3 Low-Pass Filter

As a final example, the effect of a simple *low-pass filter* (LPF) on a square-wave input will be examined. The LPF circuit is shown in Fig. 2.10(a), and the modulus and phase shift are shown as functions of frequency in Fig. 2.10(b) to (e). Two curves are shown in each case, one for $CR \ll T$ and one for $CR \gg T$. Clearly, the dc component is transmitted by this filter.

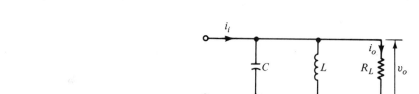

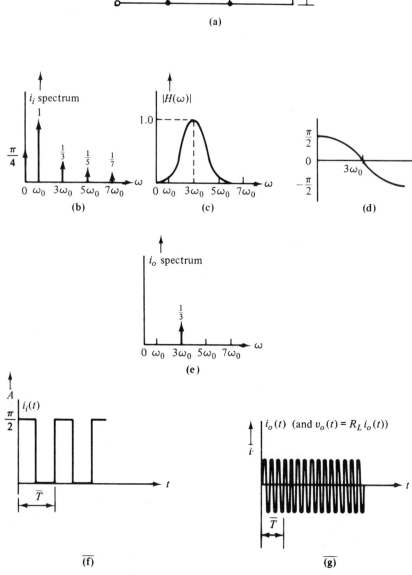

FIGURE 2.9. A bandpass filter: (a) the network; (b) input signal frequency spectrum; (c) transfer function amplitude plot; (d) transfer function phase plot; (e) output signal frequency spectrum; (f) input signal waveform; (g) output signal waveform.

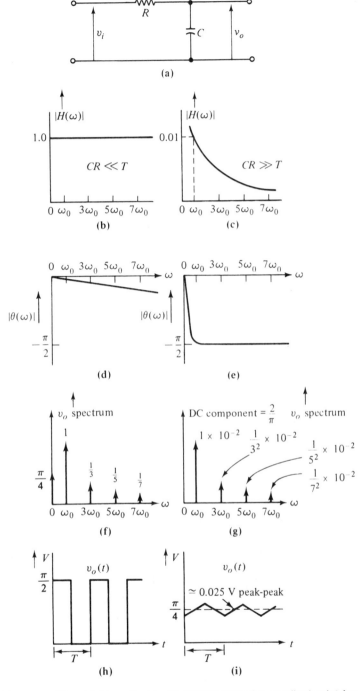

FIGURE 2.10. A low-pass filter: (a) The network; (b) Transfer function amplitude plot for CR ≪ T; (c) Transfer function amplitude plot for CR ≫ T; (d) Transfer function phase plot for CR ≪ T; (e) Transfer function phase plot for CR ≫ T; (f) Output signal frequency spectrum for CR ≪ T (same as the input signal frequency spectrum); (g) Output signal frequency spectrum for CR ≫ T; (h) Output signal waveform for CR ≪ T (same as the input signal waveform); (i) Output signal waveform for CR ≫ T.

For the condition $CR \lessdot T$, the square wave is transmitted practically unchanged, as shown by the sequence of graphs, Fig. 2.10(b), (d), (f), and (h). For $CR \gtrdot T$, the circuit integrates the square wave about the mean level. Mathematically this can be justified: for $CR \gtrdot T$, the voltage across R, v_R, is very much greater than v_o, the voltage across C, since $R \gtrdot |X_c|$. In the example shown in Fig. 2.10(c) and (e), $CR = 100T/2\pi$, which makes $R = |100 X_c|$ at the fundamental frequency. Hence,

$$v_R \gtrdot v_c$$

and therefore,

$$i \cong \frac{v_i}{R}$$

$$\therefore \qquad \frac{dq}{dt} = \frac{v_i}{R}$$

Thus,

$$C\frac{dv_c}{dt} = \frac{v_i}{R}$$

and since $v_c = v_o$, it follows that

$$v_o = \frac{1}{RC} \int v_i \, dt \tag{2.20}$$

Under these conditions the low-pass filter acts as an "integrator" circuit; its action should be compared to the high-pass filter used as a "differentiator."

Note that the capacitor charges up to the mean voltage level. The spectrum of the output in this case is shown in Fig. 2.10(g) and the output waveshape is sketched in Fig. 2.10(i). Reference to a table of Fourier series [see, for example, entries 23.7 and 23.8 in Murray R. Spiegel, *Mathematical Handbook of Formulas and Tables* (Schaum's Outline Series), McGraw-Hill Book Company, New York] shows that the triangular wave has a spectrum in which the components decrease as the square of the component number and are shifted by $\pi/2$ radians compared to the corresponding square-wave components. The low-pass filter with $CR \gtrdot T$ alters the spectrum in just this way. Note that the wave is significantly modified.

2.4 PROBLEMS

1. The Fourier series for a square wave of 1-V amplitude is

$$v(t) = \frac{4}{\pi}\left(\sin \omega t + \frac{\sin 3\omega t}{3} + \frac{\sin 5\omega t}{5} + \cdots\right)$$

By comparing this with the Fourier series for Fig. 2.1(c), sketch the square wave correctly relative to the voltage/time axes.

2. The Fourier series for a half-wave-rectified sinusoidal current wave of 1-A amplitude is

$$i(t) = \frac{1}{\pi} + \frac{1}{2}\sin \omega t - \frac{2}{\pi}\left(\frac{\cos 2\omega t}{1 \times 3} + \frac{\cos 4\omega t}{3 \times 5} + \frac{\cos 6\omega t}{5 \times 7} + \cdots\right)$$

Draw accurately to scale the spectrum up to the eighth harmonic. What is the value of the dc component?

3. The wave of Problem 1 is passed through a high-pass filter (Fig. 2.8(a)) for which $|X_c| = 10R$ at the seventh harmonic. Calculate the harmonic amplitudes up to the seventh.

4. A square-wave voltage is applied to the input of the crystal lattice filter of Fig. 1.29(a). The fundamental frequency of the square wave is at $f_{\infty 1}$ and the thirteenth harmonic at $f_{\infty 2}$. *Sketch* the output spectrum. What type of filtering action is involved?

5. The rectified current of Problem 2 is passed through a low-pass filter (Fig. 2.10(a)). Calculate the ratio of peak fundamental voltage to steady voltage at the output if the output load $R_L = 100|X_c|$ at the fundamental frequency.

CHAPTER 3

Audio Signals

3.1 INTRODUCTION

Audio waves can be analyzed by Fourier methods similar to those described in Chapter 2. For example, one complete cycle of the sound waveform *e* as in *e*at is shown in Fig. 3.1(a), and the corresponding spectrum is shown in Fig. 3.1(b). Fourier analysis for sound waves is useful in that it indicates the basic structure of the wave, but it would be impossible to carry out the analysis for every sound. For example, raising the fundamental frequency of the sound *e* increases considerably the number of harmonics in its spectrum. Also, no two voices are exactly the same, the *timbre* or quality of a speaking voice depending on the number of harmonics and their relative amplitudes. The frequency range required for speech is determined by listening tests involving large numbers of people, which also avoids the need for analyzing individual sounds.

3.2 FREQUENCY RANGE REQUIRED FOR SPEECH

Two factors determine the frequency range required for satisfactory speech transmission: the intelligibility/frequency distribution, and the energy/frequency distribution. Special listening tests called *articulation tests*

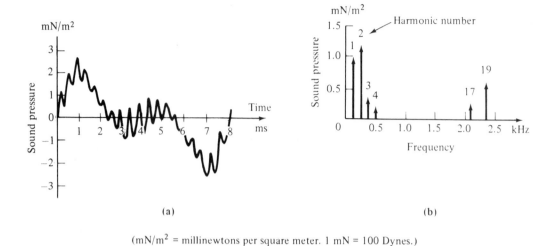

(mN/m² = millinewtons per square meter. 1 mN = 100 Dynes.)

FIGURE 3.1. (a) Pressure versus time waveform of the sound *e*; (b) frequency spectrum of the same waveform.

have been devised to measure intelligibility. In one form of articulation test, a group of listeners is asked to record syllables as they hear them announced over a loudspeaker as the frequency range of the transmission is reduced in steps. The number of correctly recorded syllables as a percentage of the total number transmitted at each step is termed the *articulation efficiency*. The syllables are unrelated to one another (i.e., they are "nonsense syllables"). In this way, results due to good guesswork are reduced. It is found in practice that an articulation efficiency of 80% corresponds to about 100% intelligibility for normal speech.

Curves of articulation efficiency against frequency are shown in Fig. 3.2(a) and (b). Figure 3.2(a) shows the effect of suppressing the high frequencies in speech. It is seen that to maintain a high articulation efficiency, frequencies up to at least 2.5 kHz must not be suppressed. Figure 3.2(b) shows the effect of suppressing the low frequencies. It is seen that all frequencies below 1.5 kHz can be suppressed without drastically reducing the intelligibility. The conclusion is, therefore, that the intelligibility is contained in the high-frequency components (1.5 to 2.5 kHz) of speech.

Looking now at the energy distribution, Fig. 3.2(c) shows the percentage of total energy transmitted to the listeners as the high frequencies are suppressed. As much as 80% is transmitted, even though all frequencies above 1 kHz are suppressed. Figure 3.2(d) shows that suppressing all frequencies below 1 kHz reduces the energy transmitted to 15%. The conclusion is, therefore, that the energy in speech sounds is contained in the low frequencies. The frequency limits adopted internationally on the basis of such tests are 300 to 3400 Hz.

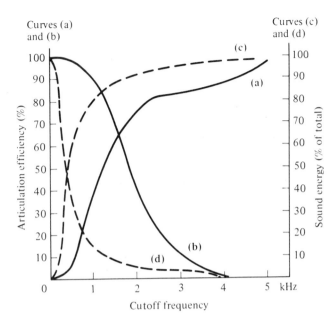

Curves (a) and (b)

Curves (c) and (d)

FIGURE 3.2. (a) Articulation efficiency as a function of upper cutoff frequency; (b) articulation efficiency as a function of lower cutoff frequency; (c) sound energy as a function of upper cutoff frequency; (d) sound energy as a function of lower cutoff frequency.

This range passes about 60% of the total energy and allows about 85% articulation efficiency. Since the type of tests involved cannot fix the frequency limits precisely, these may be different for specific applications. For example, speech transmission for entertainment purposes may require 80 to 8000 Hz in order to capture the "naturalness" of the speaker's voice.

3.3 FREQUENCY RANGE REQUIRED FOR MUSIC

Direct *frequency analysis* of the output of a large orchestra shows that the sound energy covers the frequency range 15 to 20,000 Hz. However, listening tests with a large audience, including professional musicians, have shown that a very much smaller frequency range, 160 to 8000 Hz, is preferred for *monaural* reproduction. Monaural systems do not bring out the diffuse nature of an orchestra and tend to make the performance seem unnatural. The criterion "naturalness of reproduction" or "true fidelity" therefore has little meaning for monaural systems. It should be noted that any system may be called a "high-fidelity" system provided that it meets certain minimum requirements, such as minimum bandwidth, minimum amplitude versus frequency distortion, minimum harmonic distortion, and minimum dynamic range, while still not producing "true" fidelity.

Separate tests, in which the sound passed directly from an orchestra to an audience through an "acoustical frequency filter," showed that the audience had a distinct preference for the full frequency range. In view of the conflicting evidence, and with the introduction of stereophonic systems, a frequency range of 50 to 15,000 Hz is accepted as satisfactory for reproducing music.

The various frequency-range requirements are summarized in Fig. 3.3.

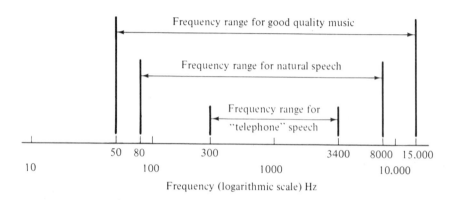

FIGURE 3.3. Frequency-range requirements for various types of audio signals.

3.4 SOUND PRESSURE LEVELS

Sound is produced by variations in air pressure around the average steady-state barometric pressure, caused by the physical movement of objects and surfaces in the air. The frequency of these variations ranges from below 1 Hz to several hundred kHz, but only those within the range from about 16 Hz to 16 kHz are audible to, or can be detected by, the human ear.

The alternating variations of pressure at a sound source cause sound waves to radiate out from the source in the same manner as waves are caused in water by a dropped stone. The waves travel with a certain velocity of propagation, which depends on the medium through which they are traveling (air, water, metal, etc.), and they transmit energy at a certain rate (expressed in watts), as discussed in Section 3.5.

Sound pressure levels may be stated in terms of peak values of pressure variations, or they may be stated in terms of average pressure variations about the barometric level. These pressure levels are stated in microbars (μbars) or in newtons per square meter, where 1 μbar $= 0.1 \ N/m^2 = 0.1$ pascal (Pa). When sound pressure levels are stated, they are usually quoted in decibels (see Appendix A) above the average threshold of hearing level, which is accepted as

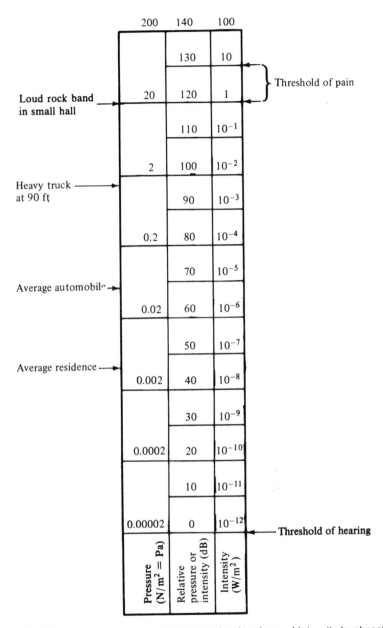

FIGURE 3.4. Comparison of sound pressure level and sound intensity level scales.

$p_0 = 2 \times 10^{-4}$ μbar of pressure variation average (0 dB), and

$$\text{SPL} = 20 \log_{10}\left(\frac{p}{p_0}\right) \qquad (3.1)$$

Sound pressure levels range from the threshold of hearing at 2×10^{-4} μbar to well above the threshold of pain at 120 dB (corresponding to 200 μbars). Since the power transmitted by a sound wave is proportional to the square of the pressure, the sound-intensity ratio in decibels, as stated in Section 3.5, coincides with the sound pressure level in decibels. A chart of sound pressure levels given in μbars, compared to the corresponding decibel level and the corresponding intensity level at the average ear, is presented in Fig. 3.4. On this chart the levels of some typical sounds are shown.

The sound power generated by a large orchestra ranges from a few microwatts at the softest passages to many watts at the loudest passages. Speech powers range from a few microwatts when whispering to a few milliwatts when shouting. However, it is the sound intensity at the listener which determines the loudness, and this depends on the acoustical properties of the room as well as on the positions of the listeners and speaker in the room. Therefore, a reproducing system does not necessarily have to produce sound powers of the same magnitude as the original sound. Usually, considerably less power is needed. For instance, in a telephone conversation, the earpiece of the handset is held close to the ear. What is important is that the reproducing system should be capable of handling the same power ratio as in the original sound. If, for example, in the original sound, power ranged from 1 mW to 10 W, the reproducing system should be capable of handling a $10^4 : 1$ power ratio, although the power output from it may be only 10 μW to 100 mW. It is also important that peak values of power are used when assessing the power output of a system.

3.5 SOUND INTENSITY

Sound intensity is defined as the average rate of transmission of sound energy in a given direction through a cross-sectional area of 1 m^2 at right angles to the direction. It is expressed in watts per square meter and is proportional to the square of the sound pressure (in newtons/square meter). Sound intensity expressed in watts/square meter is convenient for use in communications engineering, since it is expressed in the same units as electrical power.

The ear is extremely sensitive. It can detect sound intensities on the order of 10^{-13} W/m^2. This corresponds to a movement of the eardrum of 10^{-12} m, a distance equal to one hundredth the diameter of a hydrogen molecule. Hearing is not equally sensitive at all frequencies. The minimum sound intensity that can be heard is termed the *threshold of hearing*. Figure 3.5 shows how this varies with frequency. Sound intensity is usually expressed in decibels above the threshold of hearing because loudness is approximately proportional to the logarithm of intensity (see Section 3.7). A logarithmic frequency scale is also used. Decibel and logarithmic units and scales are discussed in Appendix A.

The upper limit of hearing is determined by the separation of the linkage between the outer and inner ear. This limit is felt, rather than heard, and is therefore termed the *threshold of feeling* (or threshold of pain). Figure 3.5 also shows how the threshold of feeling varies with frequency.

The power ratio between maximum and minimum intensities is 10^{12}, or 1 million million—hence the need for the decibel scale to accommodate this enormous range. The decibel zero reference for sound intensities is standardized at 10^{-12} W/m² at 1000 Hz (i.e., the threshold of hearing at 1000 Hz). The threshold levels shown in Fig. 3.5 are "averaged" results, that is, they are arrived at from the results of many tests involving many people. Although intensity can be measured, threshold level cannot. The latter depends on the opinion of the listener, or subject, and is termed a *subjective measurement*.

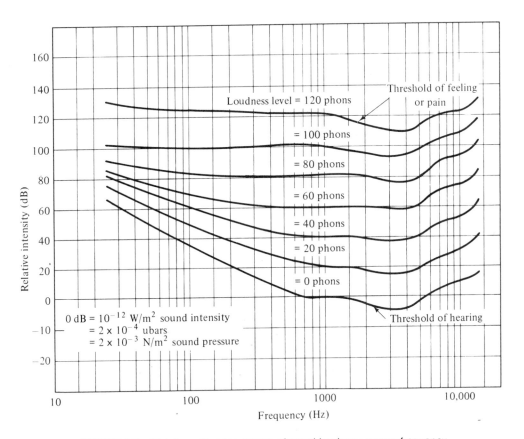

FIGURE 3.5. Fletcher–Munson curves of equal loudness versus frequency.

3.6   LOUDNESS LEVEL

Loudness is also a subjective quantity. Loudness cannot be measured in the same way that intensity can. Although two tones of different frequencies may have the same relative intensity, they will not seem equally loud. The threshold curves of Figure 3.5 represent particular *equal loudness levels*. It is possible to have varying degrees of loudness between these limits. To enable sounds at different frequencies to be compared for loudness, the intensities of tones at 1000 Hz are adopted as standard levels of loudness. To "measure" the loudness of a tone, the intensity of a 1000-Hz tone is adjusted until it is judged to have the same loudness as the unknown. The relative intensity of the 1000-Hz tone is then said to be the equivalent loudness of the unknown. A new unit, the *phon*, is introduced for loudness levels measured in this way. The relative intensity in decibels for the 1000-Hz tone is equal to the equivalent loudness level in phons.

The threshold-of-hearing curve represents a loudness level of 0 phons. The threshold-of-feeling curve represents a loudness level of 120 phons, since the relative intensity at 1000 Hz is 120 dB above the threshold of hearing. Other equal loudness curves are plotted in Fig. 3.5. These are known as *Fletcher-Munson curves*, after the engineers who first obtained them experimentally at Bell Telephone Laboratories. The curves show that at low frequencies a smaller change in relative intensity is required for a given change in loudness, compared to the high frequencies. As shown in previous sections, many sounds cover a frequency spectrum. A volume control that reduces the relative intensities by the same amount will alter the character of a sound, since it will alter the loudness of the components by unequal amounts.

The phon is a unit of *equal loudness level* but not a unit of loudness. For example, a 2 : 1 increase in loudness is not a 2 : 1 increase in the number of phons. The usefulness of the unit is that it allows sounds of the *same* loudness sensation to be compared; but it provides no connection between one loudness level and another.

3.7   LOUDNESS

The most common way of measuring loudness is to ask the listener to judge when a 2 : 1 increase in loudness sensation appears to occur when the intensity is increased. It is obvious that some initial standard is needed, and this is provided simply by defining one. The agreed standard is the loudness sensation produced by a 1000-Hz sine wave 40 dB above the *listener's* threshold of hearing. This loudness is defined as 1 *sone*. A loudness judged by the listener to be *n* times this is *n* sones. Thus 500 millisones is half the loudness sensation of the standard. At any other frequency, a loudness level of 40 phons will produce the same loudness sensation as the standard; hence the definition of the loudness standard can be extended to mean that *a pure tone of loudness 40 phons produces a loudness sensation of 1 sone.*

The relationship between loudness level and loudness is not of the straight-line form (i.e., loudness is not proportional to loudness level). The relationship has been established experimentally and is published in graphical form by the American National Standards Institute as Z24.2-1942. A useful result from this graph is that above loudness of 1 sone, a 10-phon increase is required for a 2 : 1 increase in sones.

There will also be a minimum change in intensity which is just discernible as a change in loudness. Again, this must be established experimentally, and the figure is about 1 dB for normal speech and music. For pure tones and careful experimental procedure, values as low as 0.1 dB have been recorded. These values apply over a limited range of frequencies and intensities, which include the normal speech range. Because the change is a fixed number of decibels, the response of the ear to intensity is logarithmic over the ranges involved. This means that the minimum change required is a fixed percentage of the existing intensity.

3.8 PITCH AND FREQUENCY

Pitch, like loudness, is a subjective quantity. It is determined mainly by the frequency of the sound, where this is a pure tone. Changes of pitch can occur due to changes in intensity, even though the frequency is maintained constant. In engineering practice, the definition of pitch is designed to take into account its subjective nature and also the fact that it may vary with intensity. The *mel* is the unit of pitch, and by definition a pure tone of 1000 Hz of loudness 1 sone produces a pitch of 1000 mels.

The relationship between pitch and frequency is nonlinear. By definition, a frequency of 1000 Hz establishes a pitch of 1000 mels, but a frequency of 2000 Hz corresponds to a pitch of about 1500 mels, and a frequency of 400 Hz to about 500 mels.

The ear can judge a given frequency *ratio* but not a given frequency difference. For example, a change from 1000 Hz to 3000 Hz is recognized as being the same as a change from 3000 Hz to 9000 Hz, a ratio of 3 : 1. It is not recognized as being the same as the change from 3000 Hz to 5000 Hz, although this is the same frequency difference. This property of the ear can be verified in a practical way, by playing a recording of a melody at the wrong speed. The melody will be recognizable because the frequency ratios have not altered.

3.9 INTERVALS AND OCTAVES

The ratio of two frequencies is termed their *interval*. Thus, the interval between 1000 Hz and 100 Hz is 10 : 1. An interval of 2 : 1 is termed an *octave*, and in the example above there are approximately 3.32 octaves.

Sometimes, frequency—or more loosely, pitch—is expressed in octaves. For such purposes, the reference frequency is usually 1000 Hz.

In terms of octaves, a frequency ratio f_1/f_2 may be expressed as

$$\frac{f_1}{f_2} = 2^{\text{octaves}} \tag{3.2}$$

Hence

$$\text{octaves} = \log_2\left(\frac{f_1}{f_2}\right) = \frac{\log_{10}\left(\dfrac{f_1}{f_2}\right)}{\log_{10} 2} \tag{3.3}$$

3.10 SOUND DISTORTION

If the variation of amplitude with time of a sound wave is altered in any way from the original, *distortion* is said to occur. Nearly all serious distortion is caused by the equipment in the transmission channel and is (1) nonlinear (i.e., harmonic distortion), caused primarily by tubes and transistors, or (2) linear (i.e., amplitude/frequency and phase/frequency distortion), caused by inductance and capacitance. The ear itself can introduce two effects which are similar to these forms of distortion. These are (1) generation of combination tones, and (2) masking.

Two frequencies, f_1 and f_2, can combine to produce new frequencies, $f_1 + f_2$ and $f_1 - f_2$. These are known as the *summation tone* and the *difference tone*, respectively. They may be produced outside the ear, in which case they are known as *objective* combination tones, or they may be generated inside the ear, in which case they are termed *subjective* combination tones. Since new frequencies are introduced, the effect is similar to nonlinear distortion. It does not, however, have the unpleasantness of nonlinear distortion. It can work to advantage by generating low-frequency fundamental tones which many loudspeakers and earphones cannot reproduce.

Masking is an effect that occurs in the ear where a louder sound can reduce, or even stop, the nerve voltage generated by a weaker sound. Thus, a noisy location can result in a reduction or even loss of a signal due to masking. Furthermore, high frequencies are more easily masked than low frequencies, and the effect is similar to amplitude/frequency distortion.

3.11 ROOM ACOUSTICS

The performance of an otherwise well-designed communications system can be completely ruined by a room that has poor acoustical properties. Hard, smooth surfaces reflect sound. If a room is large, echoes are created that can make speech unintelligible or result in a blurring of musical phrasing. Soft furnishings absorb sound, which results in a deadening effect. This can rob music of much of its purpose, but it is not troublesome with speech, provided sufficient energy is available to make up for that absorbed.

The best measure of the acoustic properties of a room is *reverberation time*. This is defined as the time taken for the density of sound energy in the room to drop to one millionth of its initial value. This is equivalent to the relative intensity dropping by 60 dB. The reverberation time depends on (1) the room volume, (2) the total surface area, and (3) the total absorption in the room. The materials used in the construction and the furnishings of the room will in general absorb different amounts of sound energy. The *absorption coefficient* of a material is the fraction of the total received sound energy that is absorbed by it. Thus,

$$\text{absorption coefficient} = \frac{\text{energy absorbed by surface}}{\text{total energy received by surface}}$$

The absorption coefficient not only varies with the type of material, but for a given material it varies with the frequency of the sound. Table 3.1 gives values that have been determined for various materials. It will be seen, for example, that unpainted brickwork absorbs about twice the amount of sound energy as painted brickwork. It will also be seen that special material, such as acoustic plaster, has a comparatively high absorption coefficient. This material is specially made to reduce the reverberation time of rooms. It is used in lecture theatres, telephone exchanges, and so on, where clarity of speech is essential and extraneous sounds, such as noise and echoes, must be deadened.

Consider a room that has surfaces, $S_1, S_2, S_3, \ldots$ square feet, each surface having its own absorption coefficient $a_1, a_2, a_3, \ldots$. Let the volume of the room be V cubic feet. The reverberation time T of such a room is given

TABLE 3.1. **Absorption Coefficients For Some Common Materials**

Material	500 Hz	1000 Hz
Brick, unpainted	0.03	0.04
Brick, painted	0.016	0.02
Carpet, felt, on solid floor	0.3	0.4
1/2-inch insulation board on studs	0.3	0.3
1/2-inch perforated insulation board on solid backing	0.55	0.65
Acoustic plaster	0.25	0.3

Note: The values in the table are intended to give some idea of the order of the absorption coefficient for various materials. The quality of the material will affect the absorption coefficient; for example, poor-quality carpet is likely to have a lower value than the value listed.

approximately by

$$T = \frac{0.05V}{a_1 S_1 + a_2 S_2 + a_3 S_3 + \ldots} \text{ seconds} \qquad (3.4)$$

The denominator of this expression is termed the absorption of the room. The unit of absorption is named the *sabine*, after W. C. Sabine, who laid the foundations of acoustic theory of buildings. It will be seen that each area has an individual absorption of $(a \times S)$ sabines. The advantage of the sabine unit is that it allows the absorption of objects, such as people, to be readily stated. Thus, Sabine found that the average absorption for a person was 4.7 units, or, in the present nomenclature, 4.7 sabines.

The following example will show how the reverberation time of a room is calculated.

Example 3.1   Calculate the reverberation time of a living room 8 ft high $\times$ 13 ft wide $\times$ 20 ft long = 2080 ft^3 in volume.

Solution

	Area (ft^2)	Absolute Coefficient At 1 kHz	Absorption (Sabines)
End walls, insulation board	196	0.3	58.8
Sidewalls, plaster on brick	212	0.03	6.4
Windows	36	0.075	2.7
Doors	24	0.1	2.4
Ceiling, plaster on laths	260	0.025	6.5
Carpet	200	0.4	80
Tile floor	60	0.015	0.9
Curtains			2.0
Settee and easy chairs			16
Hard furniture			5
Total absorption			180.7

$$\text{reverberation time at 1 kHz} = \frac{0.05 \times 2080}{180.7}$$
$$= 0.57 \text{ s}$$

With four persons present in the room, the total absorption would increase by $4.7 \times 4 = 18.8$ sabines. The total absorption would then be approximately 200 sabines, and

$$\text{reverberation time} \cong 0.57 \times \frac{181}{200} = 0.55$$

This reverberation time would be ideal for speech, but it is on the low side for music. By sacrificing some comfort, as for example removing the carpet, the quality of music reproduction would be improved. Preference no doubt lies with less quality and more comfort.

Generally speaking, less energy is absorbed at low frequencies, which may give rise to too great a reverberation time. Special methods are adopted to reduce this, one being the use of mechanical resonators. This may consist of a diaphragm of fibrous felt material (roofing felt has been used), the size of which is adjusted so that it is most easily set into vibration at the frequency at which it is desired to reduced the reverberation time. The energy that produces the vibration is extracted from the sound wave, and in this manner the reverberation time is reduced.

In the acoustic design of rooms the following points should be observed:

1. Concave surfaces must be avoided since they result in undesirable "focusing" of the sound.
2. Large, unbroken wall areas should be avoided; some irregularity is desirable in order to diffuse the sound.
3. Where special sound-absorbent material is used, it should be randomly distributed throughout the available wall area, not concentrated in one region.

3.12 ELECTROACOUSTIC TRANSDUCERS

Electroacoustic transducers convert sound waves to electrical signals and vice versa. At the sending end, the sound wave is converted to an electrical signal by means of a microphone, of which there is a large variety in use; only a few of these will be described here. At the receiving end, the commonly available transducers for converting the electrical signal into a sound wave are telephone receivers, headsets, and loudspeakers.

3.12.1 Microphones

Microphones are electromechanical transducers that convert changes in air pressure into corresponding changes in electrical signal. Several varieties exist, which may be classified according to five basic principles of operation. These are variable resistance, variable reluctance, moving-coil induction, variable capacitance, and piezoelectric.

Three parameters are used to describe the quality of a microphone. The first is output level, which can be described either as an absolute output level in watts, when a reference level of sound pressure signal at 1000 Hz is applied to the microphone; or in decibels referred to a standard power output level under similar input conditions. The power output level so measured gives a measure of the sensitivity of the microphone.

Frequency response is measured as a plot of the output level in decibels referred to some convenient level as in Fig. 3.6(a), and with the input signal held constant in amplitude for all frequencies, plotted against the log of frequency. The ideal microphone has a flat frequency response over the entire audio range from 16 Hz to 20 kHz, with no response at all outside this range. Practical microphones deviate widely from this ideal.

Most microphones also exhibit a directional response in their pickup characteristic. This directionality is considered in much the same manner as the directionality of antennas is treated, and polar plots of the response are provided for each unit. These polar plots are usually measured in the horizontal plane but are sometimes also provided in the vertical plane for special-purpose microphones. Figure 3.6(a) shows the frequency responses for several types of microphones, pointing up their different sensitivities, while Fig. 3.6(b) shows the polar directionality plot for a cardioid (or heart-shaped) type of microphone.

The *carbon microphone*, used as the transmitter in telephone handsets, is a variable-resistance unit. It is characterized by a limited frequency response, as shown in Fig. 3.6(a), a relatively low source resistance of about 100 Ω, and very rugged construction. Figure 3.7(a) shows the cross section of a carbon microphone. It is constructed with a metal diaphragm across one end of a metal case which is cylindrical in shape. A plungerlike metal contact is attached to the diaphragm so that movement of the diaphragm is transmitted through the plunger into the bed of carbon granules within the microphone. An insulated fixed contact is also embedded in the carbon granules to form the second electrode. When a compressing sound wave strikes the diaphragm, it pushes on the plunger and compacts the carbon granules, lowering the contact resistance between them. When the pressure is released, the resistance rises. A bias current is passed through an external load resistor from a battery to provide the electrical conversion, since varying the resistance causes similar variations in the current.

This type of microphone suffers from an inherent distortion. The variation in microphone resistance can be made to be almost directly proportional to the variation in air pressure on the diaphragm, but the current is an inverse function of the resistance, making the output signal voltage across the load resistor inversely proportional to the change of sound pressure.

The microphone resistance is made up of two components, R_0, which is the resistance of the microphone with no sound pressure signal present, and r, which is the variation of the microphone resistance about R_0 due to the varying sound pressure. Figure 3.7(b) shows the equivalent circuit of the microphone connected in its bias circuit. The output voltage is

$$v_{out} = iR_L = E \frac{R_L}{R_L + R_0 + r} = \frac{ER_L}{R_L + R_0}\left(1 + \frac{r}{R_L + R_0}\right)^{-1} \quad (3.5)$$

(a) Carbon microphone
(b) Moving-coil microphone
(c) Crystal microphone (sound cell type)

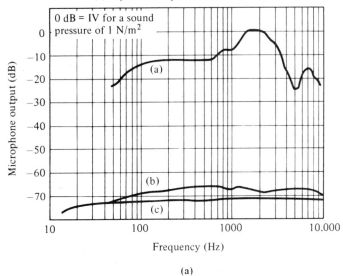

(a)

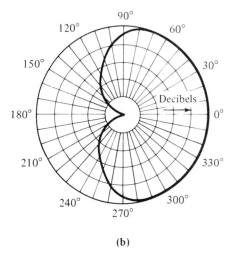

(b)

FIGURE 3.6. (a) Frequency-response characteristics of several types of microphones; (b) polar directional response of a cardioid-type microphone. (Source: Dennis Roddy, *Radio Line Transmission*, Vol. 1 (Elmsford, N.Y.: Pergamon Press, 1972).

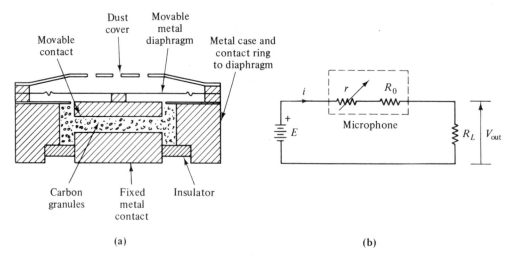

FIGURE 3.7. (a) Cross section of a carbon microphone; (b) electrical equivalent circuit of a carbon microphone.

The binomial expansion of this expression is

$$v_{out} = \frac{ER_L}{R_L + R_0} \left[1 - \frac{r}{R_L + R_0} + \left(\frac{r}{R_L + R_0} \right)^2 - \cdots \right] \qquad (3.6)$$

If r is made much smaller than $R_L + R_0$, then

$$v_{out} \cong \frac{ER_L}{R_L + R_0} \left(1 - \frac{r}{R_L + R_0} \right) \qquad (3.7)$$

If the sound wave is sinusoidal, then v_{out} is also sinusoidal:

$$v_{out} \cong \frac{ER_L}{R_L + R_0} \left(1 - \frac{r \sin \omega t}{R_L + R_0} \right) \qquad (3.8)$$

Unless the resistance variation is made to be a small percentage of the total circuit resistance, severe harmonic distortion occurs. A second disadvantage of this type of microphone is that bias current from a battery must be provided before the electrical output can be obtained.

The carbon microphone is small, cheap, very rugged, and produces a relatively high output signal level, so that it finds its chief applications in telephone systems and in portable radio systems.

The *variable-reluctance microphone* shown in Fig. 3.8(a) is one of two basic types of magnetic microphones. Both of these microphones have the advantage that they do not require bias current for their operation, but the signal output levels are correspondingly low. The variable-reluctance microphone is made with a moving diaphragm of magnetic material such as silicon steel suspended above the pole pieces of a permanent magnet. Induction coils are wound on the pole pieces and connected in series aiding. When air pressure increases on the diaphragm, the air gap in the magnetic circuit is reduced, reducing the reluctance and causing the magnetic flux to concentrate within the magnetic structure. As the flux lines move in, they cut the coil turns and induce an EMF in them. When the diaphragm moves away from the pole pieces, the air gap widens, increasing the reluctance, and the flux lines move away from the pole pieces, inducing an EMF of the opposite polarity in the coil.

This type of microphone is not often used because the diaphragm must be made of a magnetic material, and it is difficult to construct one that is not too rigid or does not have interfering mechanical resonances. The source impedance of such a microphone is high, about 5 to 10 kΩ, but its output is comparatively low, in the microvolt region.

The *moving-coil microphone* shown in Fig. 3.8(b) has basically the same structure as a moving-coil loudspeaker. An induction coil is wound on a nonmagnetic cylinder attached to the diaphragm and mounted in the cylindrical air gap of a permanent magnet. The diaphragm can be made nonmetallic and the electrical contact wires to the coil can be glued to the surface of the diaphragm. When a sound wave strikes the diaphragm, it causes the coil to move back and forth in the magnetic field, and an EMF is induced in the coil which is directly proportional to the velocity with which the coil is moving.

The light diaphragm makes it possible to construct a moving-coil microphone with very linear frequency response within the audio band, and this type

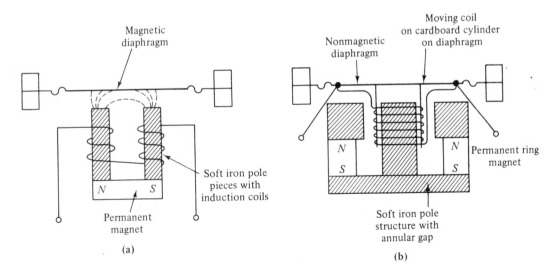

FIGURE 3.8. Magnetic microphones: (a) variable-reluctance microphone; (b) moving-coil microphone.

is often used for music recording or broadcasting applications. As mentioned, no bias is required, but the source impedance tends to be low, typically about 20 Ω, and the generated signal levels are also low, typically a few hundred microvolts. The diaphragm structure must be delicate if good frequency response is to be obtained, so this type of microphone is seldom used for telephony work.

The *capacitor microphone* shown in Figure 3.9(a) is made up of a metal diaphragm suspended above a fixed metal backplate and insulated from it. A constant dc voltage, usually several hundred volts, is applied through a load

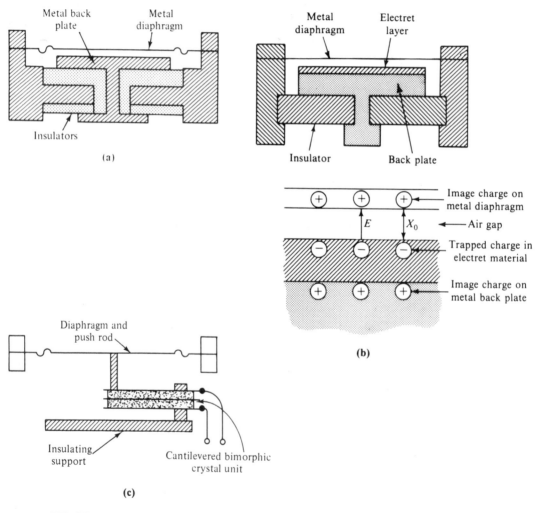

FIGURE 3.9. Other microphones: (a) capacitor microphone; (b) electret microphone; (c) piezoelectric crystal microphone.

resistor between the diaphragm and the backplate. When the diaphragm is moved by the sound wave, its capacitance varies with the changing spacing between the diaphragm and the backplate. The charge on the capacitance tends to remain constant, so the voltage across it varies directly with the sound pressure wave. If the spacing between the diaphragm and the backplate as a function of time $x(t)$ is

$$x(t) = x_0 + kP(t) \tag{3.9}$$

where x_0 is the average spacing, k is the proportionality constant relating change in distance to change in pressure, and $P(t)$ is the time-varying change of pressure from average, then the microphone capacitance is given by

$$C(t) = \frac{\varepsilon_0 A}{x(t)} \tag{3.10}$$

If the charge on the microphone capacitance is held constant, then the microphone terminal voltage is

$$V(t) = \frac{Q}{C(t)} = \left(\frac{Qx_0}{\varepsilon_0 A} \right) + \left(\frac{Qk}{\varepsilon_0 A} \right) P(t)$$

$$= V(\text{bias}) + V(\text{signal}) \tag{3.11}$$

which shows that the terminal voltage varies directly with the pressure variations in the air. Capacitances of 10 pF and voltages of 200 V are typical in capacitor microphones. Capacitor microphones with excellent frequency response have been made, and they are frequently used as "standard" microphones against which others are calibrated and in sound measuring equipment. They are delicate because of the narrow separation between the diaphragm and backplate and the thin diaphragm needed, and a high-voltage supply and signal amplifier must be located very close in use.

The "electret microphone" is a special type of capacitor microphone that has its own charge source built in, thereby eliminating the need for an external power supply. The charge source is really a charge storage device. Certain types of insulating materials will, if properly conditioned, trap large quantities of fixed charges and retain them indefinitely. One material that is frequently used for electret microphones is Teflon. Figure 3.9(b) shows the cross section of a high-quality electret microphone. In this unit, the backplate of the microphone shown in Fig. 3.9(a) is coated with a thin layer of charged Teflon, which provides the fixed charge source to condition the capacitor microphone. This thin layer contains large quantities of trapped negative charges, which induce an image charge on the back plate and on the metal diaphragm that is connected to it through an external load resistor. The trapped charge on one side and the image charge on the other establish the electric field across the gap that forms the capacitor. When air pressure causes the diaphragm to move back and forth, the capacitance changes and the terminal voltage varies according to Eq. (3.11).

A less expensive version of the electret microphone uses a plastic diaphragm for the electret material, which is coated with a thin metallic film to provide the conducting diaphragm electrode. This diaphragm is not as strong as the Teflon variety, so that supporting insulation posts are provided in the air gap to maintain separation.

The Teflon type of microphone is used for sound measuring equipment, while the metallic-film kind is used for many telecommunications applications. Care must be exercised with both electret and capacitor microphones to prevent damage due to excessive heat or humidity, or to being dropped or bumped. In the electret microphone, excessive heat or moisture may result in the trapped charge leaking off the electret material, with a resulting loss of sensitivity. Typical sensitivities are in the order of 10–50 mV/Pa.

The *piezoelectric microphone* shown in Fig. 3.9(c) is a self-generator that does not need a bias supply. Rochelle salts and certain types of piezoelectric ceramics are used as the active crystals in these microphones. A crystal of such material cut along certain planes to form a slice, and with foil electrodes attached to the two surfaces, exhibits piezoelectric characteristics. When pressure is applied to the crystal, it deforms, and a momentary displacement of charge within the structure creates a difference of potential between the two foil plates. Conversely, if an electric potential is applied between the two faces of the crystal, the crystal physically bends or deforms. This phenomenon is the same one which allows the use of quartz crystals to control oscillator frequency.

Figure 3.9(b) shows a sketch of a bimorphic crystal (really two crystals connected back to back and acting in a push–pull manner) coupled to a diaphragm. The output of such a unit is quite high, typically about 100 mV. The crystal may also be exposed directly to the sound wave instead of being coupled with a diaphragm, giving much better frequency response, but with a much lower output level, typically less than 1 mV. These microphones can be made quite rugged and provide reasonably good frequency response. They are used for a wide variety of applications from radiotelephony and broadcasting to general-purpose recording systems.

3.12.2 Telephone Receivers and Handsets

Telephone receivers are electromagnetic transducers which convert the electrical telephone signals into sound pressure waves. They are made compact and of very rugged construction to withstand the abuse of telephone usage. The original telephone receiver was a large horseshoe-shaped permanent magnet with a coil wrapped around it. A diaphragm made of magnetic material such as soft iron was suspended over the open pole pieces to complete the magnetic path and was held in place by the magnetic force of attraction. When an alternating current passed through the coil, it alternately aided or opposed the field of the permanent magnet, and the force exerted on the diaphragm was reinforced or lessened. The diaphragm vibrated in time with the varying force of the magnetic field and produced sound waves.

Figure 3.10 shows a more compact version of the same type of receiver which was in use from the early 1920s until the late 1950s in North America. Two small but very strong permanent bar magnets provide the magnetic bias field. This field is channeled to the diaphragm through the two formed soft-iron pole pieces on which the coils are mounted, as shown in Fig. 3.10(b). The whole magnetic structure is fastened to an aluminum frame which also supports the magnetic diaphragm, a perforated cover, and a dust-cover membrane held in place by a formed aluminum collar. A fiber-board cover fastened to the back of the magnetic structure carries a contact button and a contact ring through which the electrical connections are made from spring contacts in the handset.

The magnetic bias field is necessary because if it were not present, the diaphragm would vibrate at twice the frequency of the applied signal current. A magnetic diaphragm suspended above the poles of an electromagnet with no permanent magnetic field would experience no attracting force when the current through the coils was zero. The diaphragm would experience maximum attraction force when the current reached the maximum value of either polarity. Thus the diaphragm would be pulled in twice during each cycle of the applied signal current. With a permanent magnetic field component present, the diaphragm always experiences a positive attracting force. With maximum positive signal, the force would be twice the zero-signal force, and with maximum negative signal the force would approach a zero value. The diaphragm would thus oscillate around its bias position at the same frequency as the signal current.

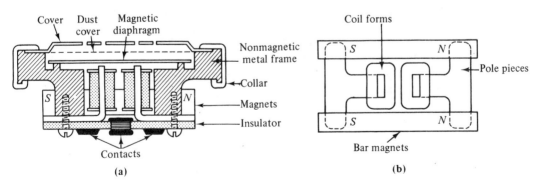

FIGURE 3.10. Magnetic diaphragm-type telephone receiver: (a) cross section; (b) magnetic structure.

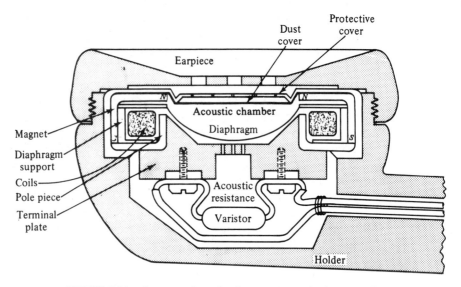

FIGURE 3.11. Cross section of a ring-armature telephone receiver.

The receiver described above has been almost entirely replaced by a different version, called a ring-armature receiver, which was introduced in the 1950s. Figure 3.11 shows a cross section of such a receiver unit. In this unit the heavy magnetic diaphragm is replaced by a very lightweight dome-shaped diaphragm made of aluminum which is fastened to a ring-shaped disk of magnetic material which forms the armature. The magnetic structure is formed by a cylindrical permanent magnet with an L-shaped cross section and an L-shaped cylindrical soft iron pole piece attached to it. The driving coil is wound inside this magnetic structure, and the ring-shaped armature is suspended in the gap between the magnet and the pole piece, supported by a nonmagnetic support ring. A Bakelite backplate provides support for the screw terminals to the coil, and also provides an acoustical resistance chamber to damp oscillations of the diaphragm. In service, a bilateral varistor diode is connected between the terminals to clip off or limit high-amplitude noise pulses which may be generated by switching equipment. The acoustical chambers in front of and behind the diaphragm are carefully designed so that the mechanical resistance encountered by the diaphragm closely matches the electrical impedance of the telephone circuit. A perforated protection cover and a dust membrane are crimped around the top of the unit, and the whole unit is mounted inside the earpiece of a handset. Typical terminal impedances range from 100 to 2000 Ω.

3.12.3 Loudspeakers

In applications where individual privacy of communication is not necessary, such as with broadcast receivers and certain types of telephone circuits, loudspeakers are used to convert the electrical voice signals back into sound waves. The most commonly used type of loudspeaker is the *cone-type moving-coil loudspeaker*. A cross section of such a speaker is shown in Fig. 3.12(a). Voice-frequency electrical currents pass through the coil of wire (voice coil), which is wound on a cylindrical form held in the annular air gap of the magnetic structure. This magnetic structure is made up of a soft iron frame, which forms one pole piece, and a very strong cylindrical permanent magnet. When a positive current is passed through the coil, a force is generated between the coil and the permanent magnet which causes the coil to move axially out of the pole structure. It is restrained lightly by the spring action of the corrugated spider mounting, which keeps the coil centered. When the current reverses, the direction of the magnetic force also reverses, and the coil is pulled back into the pole structure. When alternating current is applied to the coil, the coil moves in and out in time with the alternating current. When the coil moves, the attached paper cone also moves, and alternately compresses and expands the air in front of the speaker. The air pressure resistance

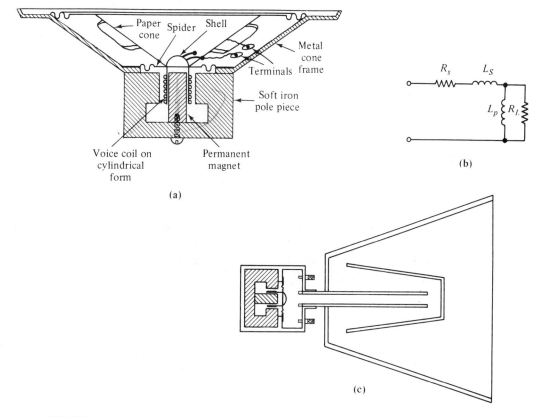

FIGURE 3.12. Loudspeakers: (a) cone-type moving-coil loudspeaker; (b) equivalent circuit of a loudspeaker; (c) horn-type loudspeaker.

presented to the cone and coil is reflected into the electrical circuit as an equivalent load resistance.

Figure 3.12(b) shows an equivalent circuit of the moving-coil loudspeaker. In this circuit, R_s represents the dc coil resistance and L_s represents the effective series inductance of the coil. R_L is the electrical equivalent of the acoustical load resistance encountered by the cone, and L_p is the electrical equivalent inductance representing the mechanical springiness of the coil structure. At very low frequencies, X_{Lp} will be very much smaller than the effective acoustical load R_L, and acoustical output will be reduced. At mid-band frequencies, X_{Lp} will be larger than R_L but X_{Ls} will be smaller than R_L, so most of the input power will be delivered to the acoustical load. At higher frequencies, X_{Ls} will predominate and reduce delivery of current to the load, again reducing the acoustical output.

The ac load impedance of the voice coil as measured at the terminals is a function of the magnetic field force, the coil size, and the efficiency of the acoustical coupling of the cone. Typical values are low, ranging from 1 Ω to about 300 Ω with 4, 8, and 16 Ω being commonly used values. Power ratings are usually stated in terms of the maximum input power that the speaker will handle without being physically damaged. Power ratings range from a few milliwatts in the case of the small 1-inch-diameter speakers used for portable receivers to several hundred watts in the case of large public-address-system speakers used in auditoriums or outdoors.

The cone of the speaker is used to improve the efficiency of coupling between the coil motion and the surrounding air. Much greater efficiency is obtained when the output from a speaker unit is fed into the throat of an acoustical horn. The horn is used as an impedance transformer, and also as a radiator in much the same manner as horn antennas are used to radiate radio waves. Figure 3.12(c) shows the structure of a typical *horn-type loudspeaker*. The driver unit is of the same type of structure as the cone-type speaker except that the cone is omitted. A shell-like dome attached to the coil couples to the air in the throat chamber of the horn. Although entirely unnecessary from an acoustical point of view, the horn structure is folded back on itself. This is done to conserve physical space, since an unfolded horn might be as much as 2 m in length and 1 m across the mouth. Such units have power-handling capabilities of more than 100 W, operating at a 16-Ω input impedance.

3.13 PROBLEMS

1. What part does resonance play in shaping speech sounds? Explain briefly why no spoken word can be represented by a single sine wave.

2. Would you expect a singing voice to generate more, less, or about the same number of harmonic waves as a speaking voice? Which voice would contain the highest frequency components? Give reasons for your answers.

3. Explain how the frequency limits are determined for speech communications circuits. How would you expect the limits to differ between commercial use and entertainment use?

4. State the upper- and lower-frequency limits generated by a large orchestra. What is the single most important factor which determines satisfactory limits for music reproduced by electrical means?

5. Write a short essay, "High-Fidelity Listening," either approving or disapproving of high-fidelity systems.

6. In reproducing sounds electrically, is it important that the original power levels be reproduced? Give reasons for your answers.

7. Define and explain the term "sound intensity," and state the units in which this is commonly measured.

8. Define and explain: (a) loudness level; (b) loudness. Explain carefully the difference between them and state the units in which each is measured.

9. Three pure tones of frequencies 100 Hz, 500 Hz, and 1000 Hz are separately adjusted to produce for each a loudness level of 100 phons. The tones are combined as a composite sound and passed through a transmission system which reduces all intensities equally by 50 dB. With reference to Figure 3.5, describe the effect to be expected on the balance of the composite sound.

10. A 1000-Hz tone has its loudness level increased from 60 phons to 80 phons. What will be the approximate increase in loudness sensation? With reference to Figure 3.5, estimate the increase in loudness sensation of a 100-Hz tone which has the same loudness level and undergoes the same change in relative intensity.

11. Explain what is meant by the pitch of a sound, and state the factors on which this depends.

12. Discuss the difference between "concert pitch" and the "mel" as a unit of pitch.

13. The pitch of a pure tone is stated to be 2000 mels. State its most likely frequency, in hertz. How many octaves is it above 1000 Hz?

14. Concert pitch is defined as a pure tone of 440 Hz. With reference to 1000 Hz, state this in (a) mels, and (b) octaves. Why can (a) be stated only as a "most likely" value?

15. Explain what is meant by "summation and difference tones." In what way can these be of advantage?

16. Describe briefly one phenomenon which makes it difficult to listen to speech in the presence of a noisy background.

17. Define and explain the term "reverberation time." Explain why this is an important factor in the overall design of a communication link.

18. It is planned to record a talk given in a theater, the reverberation time of which is 1.2 s, when the audience is present. Is the result likely to be

satisfactory? Give reasons for your answers. Suggest how this value may be modified to give better results.

19. A public telephone is to be installed in one corner of a large common room, for which it is not possible to provide a closed cubicle. Describe briefly how the effects of noise and echo would be reduced.

20. Explain how the carbon button microphone works and why it is used for telephony.

21. Explain how the ring-armature receiver works and point out the major differences between it and the magnetic diaphragm receiver.

22. In terms of power rating and physical size, what type of loudspeaker would you provide for (a) a communications receiver placed in the communications room of a police station, (b) an FM broadcast receiver placed in your living room, and (c) a public address system for an army parade square?

23. A capacitor microphone has an effective diaphragm area that is 10 mm in diameter and is separated from its backplane by an average gap of 25 μm. It uses an electret source that generates a bias voltage of 200 volts. Its diaphragm is constrained so that it moves a distance of 0.1 μm for a 1-pascal change in air pressure. Find (a) the average microphone capacity, (b) the bias charge generated by the electret, and (c) the rms variation of signal voltage for a 2-mPa rms pressure signal.

24. Express the following ratios in decibels: (a) voltage ratios: $5:1$, $10:1$, $10^6:1$; (b) current ratios: $0.1:1$, $1:1$, $10^{-10}:1$; (c) power ratios: $1:1$, $50:1$, $1:60$.

25. The voltage gain of an amplifier is given as 100 dB, this being the ratio of output to input voltage expressed in decibels. Calculate output voltage given that the input voltage is 1 μV. Can the power gain be determined from the information given?

CHAPTER 4

Noise

4.1 INTRODUCTION

Reception of a signal in a telecommunication system may be marred by noise,which can originate from a variety of sources. One obvious source, for example, could be faulty connections in equipment, a source that, in principle anyway, could be eliminated.

Noise also occurs where electrical connections that carry current are made and broken, as, for example, in the ignition system of automobiles or at the brushes of an electrical machine. Again, in principle, noise from these sources can be effectively suppressed at the source.

Natural phenomena that give rise to noise include electric storms, solar flares, and certain belts of radiation in space. The only effective means of reducing such noise is to reposition the receiving antenna where possible to minimize noise reception while ensuring that the received signal is not also seriously reduced.

There are also natural, or fundamental, sources of noise within electronic equipment; these sources are termed *fundamental* because they are an inescapable part of the physical nature of the materials used in making electronic components. Such noise is found to obey certain physical laws, and an understanding of these enables equipment to be designed in which the effects of noise are minimized.

As commonly understood, noise is audible, although in a wider sense it also covers visual disturbances such as those that occur in television reception or in chart recording of data. In the telecommunications context, *noise* will also be used to refer to the *electrical* disturbance that gives rise to the audible or visual noise, and to errors in data transmission.

4.2 THERMAL NOISE

The free electrons within a conductor are in random motion as a result of receiving thermal energy. Thus, at any given instant of time an excess of electrons may appear at one or other end of the conductor, and although the average voltage resulting from this is zero, the average power available is not zero (just as an average signal power output can be obtained from a sinusoidal voltage of zero average).

Because the noise power results from thermal energy, it is known as *thermal noise* (or sometimes *Johnson noise*, after its discoverer). It would be expected that thermal noise power would be related to the temperature of the conductor, and it is found that the average power is proportional to the absolute temperature of the conductor. It is also found that the average noise power is proportional to the frequency bandwidth, or spectrum, of the thermal noise; this will be explained in more detail later.

The law relating available average noise power to temperature and bandwidth is

$$P_n = kTB \text{ watts} \tag{4.1}$$

where

P_n = available average noise power, watts

T = temperature of conductor, kelvins

B = bandwidth of noise spectrum, hertz

k = Boltzmann's constant

    = 1.38×10^{-23} joule/kelvin

This is a particularly simple but powerful law which can be justified on physical grounds. It means that a conductor can be considered as a generator of electrical energy by virtue of its being at a finite temperature. The instantaneous voltage/time waveform would be typically as shown in Fig. 4.1(a). In Chapter 2 it is explained that every voltage/time waveform has a frequency spectrum, and in the case of thermal noise voltage, it is the power spectrum density that is significant. The power spectrum density is the average noise

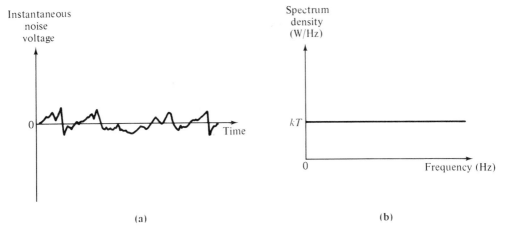

FIGURE 4.1. (a) The voltage / time variation for noise voltage; (b) the noise power spectrum density for thermal noise.

power per hertz of bandwidth, and from Eq. (4.1) this is simply

$$S_n = kT \text{ watts/hertz (W/Hz)} \qquad (4.2)$$

Thus, the spectrum density graph for thermal noise is as shown in Fig. 4.1(b). At room temperature ($T = 290\,^\circ\text{K}$), the spectrum density is

$$S_n = 1.38 \times 10^{-23} \times 290$$

$$= 4 \times 10^{-21} \text{ W/Hz}$$

For a bandwidth of 1 MHz, the available noise power is

$$P_n = S_n B$$

$$= 4 \times 10^{-21} \times 10^6$$

$$= 4 \times 10^{-15} \text{ W} \qquad (4.3)$$

This may be thought to be a small amount of power, but it has to be compared with available signal power to be seen in correct perspective. The signal may be on the order of $1.0\mu\text{V}$ EMF in a 50-Ω source, which results in an available signal power (see Eq. (1.136)) of

$$P_s = \frac{(1.0 \times 10^{-6})^2}{4 \times 50}$$

$$= 5.0 \times 10^{-15} \text{ W}$$

The signal and noise powers are seen to be of the same order of magnitude, and any subsequent amplification will not improve the signal-to-noise ratio.

Since the conductor is a generator of electrical power (albeit noise power), it has an equivalent-voltage, or an equivalent-current, generator circuit representing the noise source. Consider first the voltage equivalent circuit. The average power delivered by a voltage generator of internal rms voltage E_s and internal resistance R_s to a load R_L (Fig. 4.2(a)), when matched ($R_s = R_L$), is a maximum, given by Eq. (1.136) as

$$P_{L\,max} = \frac{E_s^2}{4R_s} \tag{4.4}$$

Applying this reasoning to a conductor of resistance R, considered as a noise generator of rms noise voltage E_n, gives

$$kTB = \frac{E_n^2}{4R}$$

$$E_n^2 = 4RkTB \tag{4.5}$$

This equation shows that the thermal noise properties of a conductor may be represented by the equivalent circuit of Fig. 4.2(b). This is one of the most useful representations of noise and is widely used in calculations of noise performance of equipment. When using it, it is best to work in terms of E_n^2 rather than E_n, for reasons that will be explained later.

The current-generator equivalent circuit for thermal noise is shown in Fig. 4.2(c), which can be derived from reasoning similar to that of the voltage circuit. The mean-square noise current (from Eq. (1.138) on substituting kTB for P_L) is

$$I_n^2 = 4GkTB \tag{4.6}$$

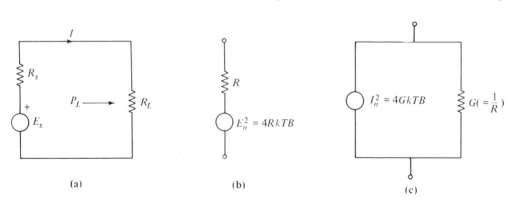

(a)　　　　　　　　　　(b)　　　　　　　　　　(c)

FIGURE 4.2. (a) A voltage generator of internal rms voltage E_s and internal resistance R_s, connected to a load resistance R_L; (b) the voltage-generator equivalent circuit for a noise source; (c) the current-generator equivalent circuit for a noise source.

where

$$G = \text{conductance} \left(= \frac{1}{R} \right)$$

If a conductor of resistance R delivers thermal noise power to a load resistance R_L, it follows that the load resistance R_L must also deliver thermal noise power to the conductor, and in thermal equilibrium the net exchange of power is zero. If, however, the load is an output indicator (i.e., a loud-speaker), the noise output will be that resulting from both R_L and R.

4.2.1 Resistors in Series

The noise-voltage equivalent circuit is shown in Fig. 4.3(a). Since two resistors in series can be replaced by a single resistor $R_s = R_1 + R_2$, the circuit can be redrawn as Fig. 4.3(b). It follows from this that

$$E_n^2 = 4R_s kTB \tag{4.7}$$

$$= 4(R_1 + R_2)kTB$$

$$= E_{n1}^2 + E_{n2}^2 \tag{4.8}$$

This shows that the resultant noise voltage *squared* is equal to the sum of the *squares* of the individual noise voltages, or, alternatively,

$$E_n = \sqrt{E_{n1}^2 + E_{n2}^2} \tag{4.9}$$

Note that just summing the noise voltages ($E_{n1} + E_{n2}$) would have given the wrong result; for this reason, it is best to work in terms of noise voltage squared.

The argument can clearly be extended to any number of resistors in series, giving for the resultant noise voltage

$$E_n^2 = 4(R_1 + R_2 + R_3 + \ldots)kTB \tag{4.10}$$

4.2.2 Resistors in Parallel

For two resistors in parallel, the equivalent noise circuit is as shown in Fig. 4.3(c). Here, it is easiest to work in terms of the conductance G and the current-generator equivalent circuit. Since two conductances in parallel can be replaced by a single conductance $G_p = G_1 + G_2$, the circuit can be redrawn as Fig. 4.3(d). From this,

$$I_n^2 = 4G_p kTB \tag{4.11}$$

$$= 4(G_1 + G_2)kTB$$

$$= I_{n1}^2 + I_{n2}^2 \tag{4.12}$$

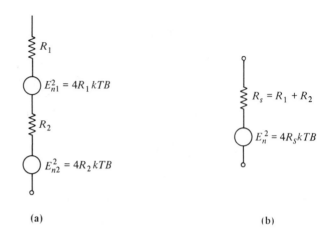

(a)

(b)

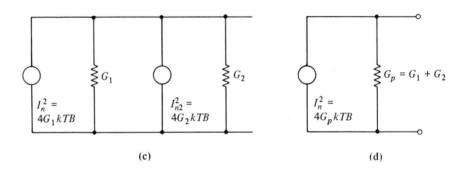

(c)

(d)

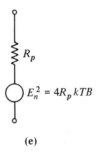

(e)

FIGURE 4.3. (a) Two noise sources connected in series; (b) the voltage-generator equivalent circuit for the series connection; (c) two noise sources connected in parallel; (d) the current-generator equivalent circuit for (c); (e) the voltage-generator equivalent circuit for (c).

The argument can be extended to any number of resistors in parallel, to give

$$I_n^2 = 4(G_1 + G_2 + G_3 + \ldots)kTB \qquad (4.13)$$

If it is desired to obtain the voltage-generator equivalent circuit for resistors in parallel, the equivalent parallel resistance R_p is used as shown in Fig. 4.3(e), where

$$\frac{1}{R_p} = \frac{1}{R_1} + \frac{1}{R_2} + \frac{1}{R_3} + \ldots$$

and

$$E_n^2 = 4R_p kTB \qquad (4.14)$$

Example 4.1   Two resistors, 20 kΩ and 50 kΩ, are at room temperature (290°K). Calculate, for a bandwidth of 100 kHz, the thermal noise voltage (a) for each resistor, (b) for the two resistors in series, and (c) for the two resistors in parallel.

Solution $kT = 4 \times 10^{-21}$ W/Hz at room temperature.

(a) For the 20-kΩ resistor:

$$E_n^2 = 4 \times (20 \times 10^3) \times (4 \times 10^{-21}) \times (100 \times 10^3)$$
$$= 32 \times 10^{-12} \text{ V}^2$$
$$E_n = 5.66 \ \mu V$$

For the 50-kΩ resistor:

$$E_n = 5.66 \times \sqrt{\frac{50}{20}}$$
$$= 8.95 \ \mu V$$

(Note the use of proportionality here.)
(b) In series,

$$R_s = 20 + 50$$
$$= 70 \text{ k}\Omega$$

Therefore,

$$E_n = 5.66 \times \sqrt{\frac{70}{20}}$$
$$= 10.59 \ \mu V$$

(c) In parallel,

$$R_p = \frac{20 \times 50}{20 + 50}$$

$$= \frac{100}{7} \text{ k}\Omega$$

Therefore,

$$E_n = 5.66 \times \sqrt{\frac{100}{7 \times 20}}$$

$$= 4.78 \text{ } \mu\text{V}$$

4.2.3 Reactance and Equivalent Noise Bandwidth

Inductive and capacitive reactances do not generate thermal noise. This follows from the fact that reactance cannot dissipate power. Consider a reactance connected to a resistor (Fig. 4.4). Assume that the reactance delivers noise power P_1 to the resistor, and that the resistor delivers noise power P_2 to the reactance. For thermal equilibrium, the power received by the reactance must equal the power delivered by it, but since the received power (i.e., the power dissipated by the reactance) is zero, the power delivered by it must also be zero.

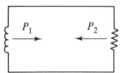

FIGURE 4.4. The power exchange between a reactance and a resistance is $P_1 = P_2 = 0$.

The effect of reactance on the noise spectrum must, however, be taken into account. It can be shown that, when noise passes through a passive filter which has a complex transfer function $H(\omega)$ (see Section 1.15.1), the noise output spectrum density S_{no}, for an input density of $S_n = kT$ (see Eq. (4.2)) is

$$S_{no} = |H(\omega)|^2 kT \tag{4.15}$$

Two important cases will be analyzed. The first is where a resistor is shunted by a capacitor (which may be the self-capacitance of R), shown in Fig. 4.5(a). The RC network forms a low-pass filter for which E_n is the input and V_n the output. The modulus of the transfer function for this filter is easily determined

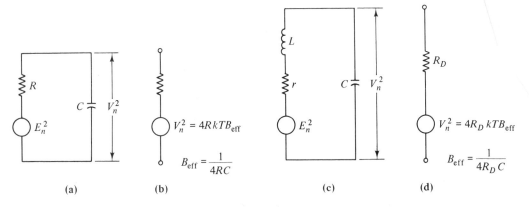

FIGURE 4.5. (a) Noise voltage acting around an RC circuit; (b) the equivalent voltage-generator circuit for (a) using the effective bandwidth; (c) noise voltage acting around a tuned circuit; (d) the equivalent voltage-generator circuit for (c).

from normal ac circuit theory to be

$$|H(\omega)| = \frac{1}{\sqrt{1 + (\omega CR)^2}} \qquad (4.16)$$

Hence,

$$S_{no} = \frac{kT}{\left(1 + (\omega CR)^2\right)} \qquad (4.17)$$

The output spectrum is seen to decrease as ω increases. The total noise power available at the output is obtained by integrating (i.e., summing) S_{no} over the complete frequency spectrum, from zero to infinity. The details of the integration are quite straightforward and will be omitted here. The result is that the total noise power available at the output, P_{no}, is

$$P_{no} = \frac{kT}{4RC} \qquad (4.18)$$

Comparing Eq. (4.18) with Eq. (4.1), an *effective* noise bandwidth B_{eff} may be defined:

$$B_{\text{eff}} = \frac{1}{4RC} \qquad (4.19)$$

This allows the equivalent circuit of Fig. 4.5(b) to be drawn. For this, the noise spectrum, instead of trailing off to infinity, is level at kT up to a bandwidth B_{eff}, at which it abruptly cuts off.

It will be seen that V_n can be simplified to

$$V_n^2 = 4RkT\frac{1}{4RC}$$

$$= \frac{kT}{C} \tag{4.20}$$

This again emphasizes the point that although C does not contribute to the noise, it limits the mean square value.

The second example is the tuned circuit of Fig. 4.5(c) in which the capacitor is assumed lossless, the only noise-generating component being the series resistance of the coil. Consider first the situation where the circuit is resonant and the noise bandwidth is restricted to a small value Δf around the resonant frequency. From the definition of Q factor, Section 1.8, the output mean square voltage is

$$V_n^2 = Q^2 E_n^2$$

$$= 4Q^2 r k T \Delta f \tag{4.21}$$

But $Q^2 r$ is equal to R_D, the dynamic impedance of the circuit [see Eq. (1.86)], and therefore

$$V_n^2 = 4R_D k T \Delta f \tag{4.22}$$

This result by itself is important, as the bandwidth is often limited physically to some small percentage of f_0 about f_0. An example will illustrate this.

Example 4.2

The parallel tuned circuit at the input of a radio receiver is tuned to resonate at 120 MHz by a capacitance of 25 pF. The Q factor of the circuit is 30. The channel bandwidth of the receiver is limited to 10 kHz by the low-frequency sections. Calculate the effective noise voltage of the tuned circuit at room temperature, as limited by the l.f. bandwidth.

Solution

$$R_d = \frac{Q}{\omega_0 C}$$

$$= \frac{30 \times 10^{12}}{2 \times \pi \times 120 \times 10^6 \times 25}$$

$$= 1.59 \text{ k}\Omega$$

$$\Delta f = 10^4 \text{ Hz}$$

$$(V_n)^2 = 4 \times 1.59 \times 10^3 \times 4 \times 10^{-21} \times 10^4$$

$$= 0.254 \times 10^{-12}$$

$$V_n = 0.50 \ \mu\text{V}$$

Where the bandwidth is unrestricted, the total noise spectrum must be taken into account using Eq. (4.15). For the circuit of Fig. 4.5(c), again considering E_n as the input and V_n as the output, the modulus of the transfer function is easily seen to be

$$|H(\omega)| = \frac{|X_c|}{|Z_s|} \tag{4.23}$$

where Z_s is the series impedance of the circuit and X_c is the reactance of C. S_{n0} is found by substituting Eq. (4.23) in Eq. (4.15), and the total available power at the output is found by integrating S_{n0} over the complete frequency spectrum as before. This is a more difficult integral than the previous one, but the result is that

$$P_{n0} = \frac{kT}{4R_DC} \tag{4.24}$$

and thus,

$$B_{\text{eff}} = \frac{1}{4R_DC} \tag{4.25}$$

The equivalent noise-generator circuit for the tuned circuit is as shown in Fig. 4.5(d).

A relationship between B_{eff} and the 3-dB bandwidth of the circuit can be found. Referring to Section 1.9, substituting Eq. (1.83) in Eq. (1.85) gives

$$R_D = \frac{1}{2\pi B_{3\,\text{dB}}C} \tag{4.26}$$

and substituting this in Eq. (4.25) gives

$$B_{\text{eff}} = \frac{\pi}{2}B_{3\,\text{dB}} \tag{4.27}$$

In the foregoing it was assumed that the Q factor remained constant, independently of frequency; this is necessary to simplify the analysis but is not fully justifiable. However, the result obtained gives a good indication of the noise expected.

The idea of an equivalent noise bandwidth B_{eff} can be extended to cover the case of an amplifier or radio receiver. Suppose that the relative power-gain, or relative voltage-gain-squared, response curve is as shown in Fig. 4.6(a). B_{eff} is given directly by the area under the response curve (since, to a different scale, the relative power-gain, or voltage-gain-squared response curve has the same shape as the power spectrum density curve). A relative response curve

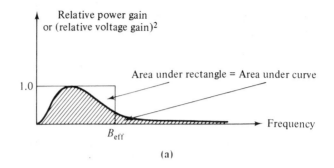

(a)

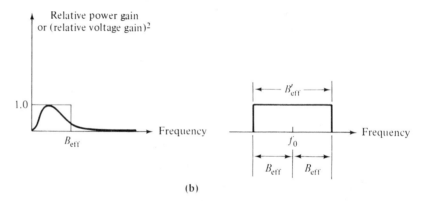

(b)

FIGURE 4.6. (a) Effective noise bandwidth as defined by the relative power-gain response curve of an amplifier; (b) effective noise bandwidth for a double-sideband receiver.

simply means that the ordinate values are scaled relative to the maximum value, and thus the maximum value of the relative curve is unity.

For a radio receiver, the noise is generated almost entirely at the front end (RF sections), while the effective bandwidth is determined by the low-frequency sections. For a single-sideband receiver (see Section 9.1), B_{eff} is determined by the area, as already described. For a double-sideband receiver (see Section 8.2.1), the effective noise bandwidth is doubled, as both sidebands contribute independently to the noise. Thus, the effective noise bandwidth B'_{eff} (Fig. 4.6(b)), is

$$B'_{\text{eff}} = 2B_{\text{eff}} \tag{4.28}$$

4.3 SHOT NOISE

A second fundamental kind of noise, *shot noise* was originally used to describe the plate current noise resulting from random fluctuations in electron emission from cathodes in vacuum tubes; the analogy is with the effect of lead shot from a gun striking a target. Shot noise also occurs in semiconductor devices where carriers are liberated into potential barrier regions, such as occur at *pn* junctions.

Shot noise is found to have a uniform spectrum density like thermal noise, and the mean-square noise current depends directly on the direct component of current. Shot noise is also a function of the operating conditions of the device, and a number of special cases have been analyzed; only two of these will be considered here, the temperature-limited diode, and the *pn*-junction diode.

A *temperature-limited diode* is a vacuum-tube diode in which the emission from the cathode is limited only by its temperature; for example, increasing the heater current will increase the temperature, and hence the shot noise current. The mean-square shot noise current is given by

$$I_n^2 = 2 I_{\text{dc}} q_e B \text{ amperes}^2 \tag{4.29}$$

where

$$I_n = \text{noise component of current, amperes}$$

$$I_{\text{dc}} = \text{direct component of current, amperes}$$

$$q_e = \text{magnitude of electron charge}$$

$$= 1.6 \times 10^{-19} C$$

$$B = \text{effective noise bandwidth, hertz}$$

The importance of Eq. (4.29) is that it shows that the noise current is known in terms of an easily measured direct current, and this is used as the basis of one type of noise-measuring method (described in Section 4.10.4).

A similar equation applies for the semiconductor *pn-junction diode*, viz.,

$$I_n^2 = 2(I + 2I_0) q_e B \text{ amperes}^2 \tag{4.30}$$

where I (in amperes) is the direct current across the junction, and I_0 (in amperes) is the reverse saturation current.

Equation (4.30) applies only at low frequencies and for low injection. (This is somewhat analogous to the temperature limitation of the vacuum-tube diode, Eq. (4.29).)

4.4 PARTITION NOISE

Partition noise occurs wherever current has to divide between two or more paths, and results from the random fluctuations in the division. It would be expected, therefore, that a diode would be less noisy than a transistor (all other factors being equal) if the third electrode draws current (i.e., the base current). It is for this reason that the inputs of microwave receivers are often taken directly to diode mixers. The spectrum for partition noise is flat.

More recently, gallium arsenide field effect transistors, which draw zero gate bias current, have been developed for low noise microwave amplification.

4.5 LOW-FREQUENCY, OR FLICKER, NOISE

Below frequencies of a few kilohertz, a component of noise appears, the spectrum density of which increases as the frequency decreases. This is known as *flicker noise* (sometimes it is referred to as $1/f$ noise). In vacuum tubes, the main causes of flicker noise are slow changes which take place in the oxide structure of oxide-coated cathodes and migration of impurity ions through the oxide.

In semiconductors, flicker noise results from fluctuations in carrier density, and it is much more troublesome in the semiconductor amplifying device than in the vacuum-tube counterpart at low frequencies. The fluctuations in carrier density cause fluctuations in the conductivity of the material; this, in turn, produces a fluctuating voltage drop when a direct current flows, which is the flicker-noise voltage. It follows that the mean-square value of this is proportional to the square of the direct current flowing.

4.6 HIGH-FREQUENCY, OR TRANSIT-TIME, NOISE

In semiconducting devices, when the transit time of carriers crossing a junction is comparable with the periodic time of the signal, some of the carriers may diffuse back to the source, or emitter. It can be shown that this gives rise to an input admittance in which the conductance component increases with frequency. This conductance has associated with it a noise current generator similar to that shown in Fig. 4.2(c), and since the conductance increases with frequency, so also does the spectrum density. A similar effect occurs in vacuum tubes when the transit time of electrons from cathode to control grid is comparable to the periodic time of the signal.

4.7 GENERATION-RECOMBINATION NOISE

In semiconductor devices, some impurity centers will be ionized on a random basis, being energized thermally; thus a random generation of carriers occurs in the device. Also, the carriers can recombine with ionized impurity centers on a random basis, either directly or through trapping centers. The overall result is that the conductivity of the semiconductor has a randomly fluctuating component which gives rise to a noise current when direct current flows through the semiconductor. The spectrum density of this type of noise has not been fully established.

4.8 EQUIVALENT NOISE RESISTANCE

It is sometimes convenient to represent the noise which originates in a device (active or passive) by means of a *fictitious* resistance R_n, assumed to generate the noise at *room temperature*, the actual device then being assumed

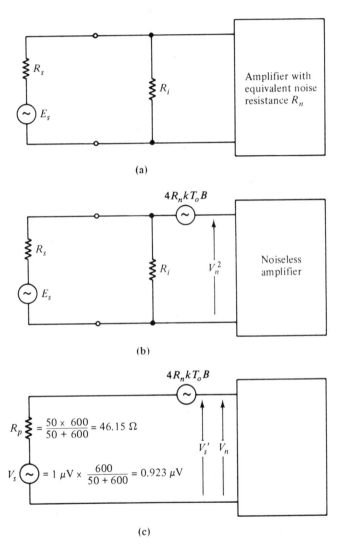

FIGURE 4.7. (a) An amplifier with input resistance R_i and equivalent noise resistance R_n; (b) the noise circuit representation for (a); (c) an amplifier of input resistance 600 Ω and noise resistance 400 Ω, fed by a source of internal resistance 50 Ω and internal EMF 1 μV—the Thévenin equivalent circuit for signal input.

noiseless. The noise resistance should be part of the manufacturer's data for the device, given for specified operating conditions. Clearly, it is important that any deviation from the specified operating conditions be taken into account in applying the concept of noise resistance.

An amplifier may have a specified noise resistance R_n and an actual input resistance R_i (Fig. 4.7(a)). The equivalent mean-square noise voltage appearing at the input terminals is

$$V_n^2 = 4(R_i + R_n)kT_0B \qquad (4.31)$$

In effect, the amplifier noise is referred to the input circuit so that the amplifier may be assumed to be noiseless (Fig. 4.7(b)). The main advantage in using R_n is that it enables the amplifier-noise contribution to be compared directly with the input-circuit contribution. It is essential to grasp the idea that R_n is a fictitious resistance which does not affect the real resistance at the input. In the previous example of Fig. 4.7(b), the input resistance remains at R_i even though the effective resistance from the point of view of noise generation is $(R_i + R_n)$.

Example 4.3

For the circuit of Fig. 4.7(a), calculate the signal voltage and the equivalent noise voltage appearing at the input terminals for an effective noise bandwidth of 10 kHz and at room temperature (290 °K), where $R_n = 400\ \Omega$, $R_i = 600\ \Omega$, $R_s = 50\ \Omega$, and $E_s = 1.0\ \mu V$.

Solution

The easiest method of solving this problem is to apply Thévenin's theorem (see Section 1.13) to reduce the input circuit to that shown in Fig. 4.7(c). Note that R_n does not affect the reduction. From Fig. 4.7(c), the terminal signal voltage V_s' is

$$V_s' = V_s = 0.923\ \mu V$$

The noise voltage V_n is obtained from

$$V_n^2 = 4(R_p + R_n)kT_0B$$

$$= 4(46.15 + 400)kT_0B$$

$$= 4 \times 446 \times 4 \times 10^{-21} \times 10^4$$

$$= 1.6 \times 4.46 \times 10^{-14}\ \text{volt}^2$$

$$V_n = 0.267\ \mu V$$

Note that this contains the noise contributions from R_s and R_i in parallel, R_n effectively being a noise source in series with R_p.

4.9 SIGNAL-TO-NOISE RATIO

In a telecommunications system the noise power in comparison with the signal power is important, and the best measure of this is the *signal-to-noise power ratio* (S/N).

Example 4.4

For the system of Fig. 4.7, the signal-to-noise power ratio is

$$S/N = \frac{V_s^2}{V_n^2} \quad \text{(since power is proportional to voltage squared)}$$

$$= \left(\frac{0.923}{0.267}\right)^2$$

$$= 11.95$$

In decibels, this is

$$(S/N)_{dB} = 10 \log_{10} 11.95$$

$$= 10.77 \text{ dB}$$

More generally, suppose that the input signal source and input resistance can be represented by the Thévenin equivalent circuit of V_s in series with R_p and that the amplifier has an equivalent noise resistance R_n (Fig. 4.7(c)); then the S/N ratio of the system is

$$S/N = \frac{V_s^2}{4(R_p + R_n)kT_0 B} \tag{4.32}$$

4.9.1 Effect of Amplification on S/N Ratio

Mixer stages in receivers have notoriously high values of equivalent noise resistance. Therefore, a large signal is required to maintain an acceptable S/N ratio, as shown by Eq. (4.32). Weak signals have to be amplified before being applied to the mixer stage, and of course the amplifier must not contribute excessive noise; otherwise, no improvement in signal-to-noise ratio results. In order to see the effect of the amplifier, it is best to transfer all noise sources to the input; that is, an equivalent source is postulated at the input which results in exactly the same output as the actual source. Figure 4.8(a) illustrates a mixer with noise resistance R_{nm} preceded by an amplifier of voltage gain A and noise resistance R_{na}. The mixer noise voltage generator V_{nm}^2 can be replaced by a noise voltage generator V_{nm}^2/A^2 at the input to the amplifier; this follows, since on going through the amplifier the mean-square noise voltage is amplified by A^2, resulting in a mixer component of noise V_{nm}^2 at the input to the mixer, as was originally the case. The equivalent circuit is shown in Fig. 4.8(b).

The total noise voltage referred to the amplifier input is then

$$V_n^2 = V_{na}^2 + \frac{V_{nm}^2}{A^2}$$

$$= 4\left(R_{na} + \frac{R_{nm}}{A^2}\right)kT_0B \qquad (4.33)$$

The S/N ratio with the amplifier connected is

$$S/N = \frac{V_s^2}{4\left(R_p + R_{na} + R_{nm}/A^2\right)kT_0B} \qquad (4.34)$$

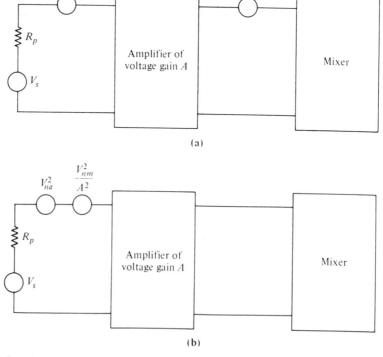

(a)

(b)

FIGURE 4.8. (a) An amplifier of voltage gain A and equivalent noise resistance R_{na}, connected between a mixer of noise resistance R_{nm} and a source of effective internal resistance R_p and EMF V_s; (b) the equivalent circuit for (a), showing all noise sources referred to the input of the amplifier.

whereas that obtainable with the signal source fed directly into the mixer would be

$$(S/N)_{\text{mixer}} = \frac{V_s^2}{4(R_p + R_{nm})kT_0 B} \tag{4.35}$$

It follows that for the amplifier to result in an improvement in S/N,

$$R_{na} + \frac{R_{nm}}{A^2} < R_{nm}$$

For really worthwhile improvement, $R_{nm}/A^2 \ll R_{na}$, so the S/N ratio becomes that of the amplifier alone.

Example 4.5   With $R_{nm} = 200$ kΩ, $R_{na} = 1$ kΩ, and $A = 100$,

$$\frac{R_{nm}}{A^2} = 20 \ \Omega$$

The mixer noise is, therefore, negligible (20 Ω compared with 1000 Ω) compared to the amplifier noise. The *improvement* in signal-to-noise ratio depends, of course, on the source resistance R_p also, and is in the ratio

$$\frac{R_p + R_{nm}}{R_p + R_{na} + R_{nm}/A^2}$$

Assuming a typical value of $R_p = 50$ Ω, the improvement ratio becomes

$$\frac{50 + 200 \times 10^3}{50 + 1000 + 20} = 187:1$$

In decibels, this is a 22.7-dB improvement in the S/N ratio. Effectively, it means that the system is 187 times more sensitive in power detection, or $\sqrt{187} = 13.67$ times more sensitive in voltage detection. Note that it is not 100 times more sensitive in voltage detection even though the voltage gain is 100. Why? For a voltage gain $A = 10$, and the other parameters as before, $R_{nm}/A^2 = 2$kΩ and the mixer noise is not negligible, although an overall improvement in S/N is still obtained. The improvement in this case is

$$\frac{50 + 200 \times 10^3}{50 + 1000 + 2000} = 65.6:1 \qquad (\text{in power})$$

This is an 18.2-dB improvement in S/N ratio, which is certainly significant.

4.9.2 Tandem Connection

Where a number of telecommunications links operate in *tandem* (i.e., one following the other), the overall S/N ratio will be less than that of any one of the links. Let M be the number of links in tandem transmitting a signal P_s. The noise powers $P_{n1}, P_{n2}, \ldots, P_{nM}$ will add so that the overall S/N ratio at the output will be

$$(S/N)_{dB} = 10 \log_{10} \left(\frac{P_s}{P_{n1} + P_{n2} + \ldots + P_{nM}} \right) \qquad (4.36)$$

Where the links are identical such that $P_{n1} = P_{n2} = \ldots = P_{nM} = P_n$, the overall S/N ratio becomes

$$(S/N)_{dB} = 10 \log_{10} \left(\frac{P_s}{MP_n} \right) \qquad (4.37)$$

$$= 10 \log_{10} \left(\frac{P_s}{P_n} \right) - 10 \log_{10} M$$

$$= (S/N)'_{dB} - M_{dB} \qquad (4.38)$$

where $(S/N)'_{dB}$ is the signal-to-noise ratio of any one link, and M_{dB} is the number of links expressed in *decilogs* (i.e., M is treated as a power ratio and the decibel value taken).

Example 4.6
Let the number of links be three, each having an S/N ratio of 60 dB. The overall S/N ratio is

$$(S/N)_{dB} = 60 - 10 \log_{10} 3$$

$$= 55.23 \text{ dB}$$

If the S/N ratio of one link is much worse than any of the others, that link will determine the overall S/N ratio. Considering the previous example again, for a given P_s let the S/N ratio for two of the links each be 60 dB, and let the third link have an S/N ratio of 40 dB. The various noise powers will be

$$P_{n1} = P_{n2} = P_s \times 10^{-6}$$

$$P_{n3} = P_s \times 10^{-4}$$

The total noise power is

$$P_{nt} = P_{n1} + P_{n2} + P_{n3}$$

$$= P_s(2 \times 10^{-6} + 10^{-4})$$

$$\cong P_s \times 10^{-4}$$

The overall S/N ratio is, therefore,

$$(S/N)_{\text{dB}} \cong 10 \log_{10} \left(\frac{P_s}{P_s \times 10^{-4}} \right)$$

$$= 40 \text{ dB}$$

The above example shows that where the S/N of one link is much worse than that of any of the others, the overall S/N is approximately that of the worst link. Signal-to-noise ratio is discussed further in relation to modulation methods in Sections 9.7 and 10.7.

4.10 NOISE FACTOR

The *noise factor* F of an amplifier (or any network) may be defined in terms of the signal-to-noise ratio as follows:

$$F = \frac{\text{available } S/N \text{ power ratio at input}}{\text{available } S/N \text{ power ratio at output}}$$

$$= \frac{P_{si}}{P_{ni}} \times \frac{P_{no}}{P_{so}} \qquad (4.39)$$

Here, P_{si} is the signal power available at the input, P_{ni} the noise power available at the input, P_{so} the signal power available at the output, and P_{no} the noise power available at the output. The temperature of the input source is defined as room temperature.

The S/N available at the output will always be less than that at the input, since any amplifier or network will add noise. Therefore, the noise factor is a measure of the amount of noise added (i.e., it is the factor by which the amplifier degrades the available signal-to-noise ratio at the input), and F will always be greater than unity.

F is frequency-dependent in many cases, and where it is determined at one frequency it is known as the *spot noise factor*. Clearly, the frequency must be stated along with the noise factor. An average value of F can also be found over the frequency range of interest, this being known as the *average noise factor*, F_{av}. In this chapter only the spot noise factor will be discussed, so it may be assumed that the frequency is known.

Available powers are used in the definition, as these can be defined unambiguously. The available power from a source is the maximum average power a source can deliver (as given by Eq. (1.136)) and is obtained under matched conditions. The available input powers (signal and noise) may not actually be delivered to the input of the amplifier because of mismatch, but the available output powers depend on the *actual* input powers, and therefore, any mismatch is taken into account.

The available power gain G is

$$G = \frac{P_{so}}{P_{si}} \tag{4.40}$$

(Do not confuse with conductance.) Therefore, from Eq. (4.39),

$$F = \frac{P_{no}}{GP_{ni}} \tag{4.41}$$

or

$$P_{no} = FGP_{ni} \tag{4.42}$$

As already stated in Eq. (4.1), $P_{ni} = kTB$, and for noise factor, temperature T is defined as T_0. Thus,

$$P_{no} = FGkT_0B \tag{4.43}$$

This shows that the output noise power is increased by a factor F over what it would have been if the amplifier had been noiseless.

4.10.1 Noise Factor F in Terms of R_n

The output S/N is given by Eq. (4.32), since both signal and noise will be amplified by the same amount and the noise of the amplifier is taken into account by R_n. The available input signal-to-noise ratio is

$$(S/N)_{\text{in}} = \frac{V_s^2}{4R_p kT_0 B} \tag{4.44}$$

Therefore,

$$F = \frac{(S/N)_{\text{in}}}{(S/N)_{\text{out}}}$$

$$= \frac{R_p + R_n}{R_p} \tag{4.45}$$

Note that the input resistance R_i of the amplifier is included in the effective source resistance R_p, and that if R_n were zero, F (noise *factor*) would be equal to unity.

Sometimes the noise factor is expressed in decibels, in which case it is customary to refer to it as the *noise figure*:

$$\text{noise figure} = F_{\text{dB}} = 10 \log_{10} F \tag{4.46}$$

Thus, in the case where $R_n = 0$, the noise *figure* would be zero dB.

4.10.2 Amplifier Input Noise in Terms of *F*

The total noise referred to the input can be represented by P_{no}/G, and it follows from Eq. (4.43) that this is

$$P_{ni(total)} = FkT_0B \qquad (4.47)$$

The source contributes kT_0B of this; therefore, the amplifier must contribute the amount P_{na}, where

$$P_{na} = FkT_0B - kT_0B$$
$$= (F - 1)kT_0B \qquad (4.48)$$

4.10.3 Amplifiers in Cascade

This is similar to the problem of an amplifier and mixer in cascade, which was solved in terms of R_n. In this case, available noise powers are used, since F is defined in terms of these. Referring to Fig. 4.9, the noise input to amplifier 1 is $F_1 kT_0B$ (from Eq. (4.47)), and therefore the noise input to amplifier 2 is G_1 times this plus the contribution from amplifier 2 itself, which is given by Eq. (4.48):

$$\text{noise input to amplifier } 2 = G_1 F_1 kT_0B + (F_2 - 1)kT_0B$$

The output noise power, which is that from amplifier 2, is G_2 times the input noise power to amplifier 2:

$$P_{no} = G_2 G_1 F_1 kT_0B + G_2(F_2 - 1)kT_0B$$

The overall gain is $G = G_1 G_2$, so the overall noise factor, from Eq. (4.41), is

$$F = \frac{P_{no}}{G_2 G_1 P_{ni}}$$
$$= \frac{G_2 G_1 F_1 kT_0B + G_2(F_2 - 1)kT_0B}{G_2 G_1 kT_0B}$$
$$= F_1 + \frac{F_2 - 1}{G_1} \qquad (4.49)$$

The argument is easily extended for additional amplifiers to give

$$F = F_1 + \frac{F_2 - 1}{G_1} + \frac{F_3 - 1}{G_2 G_1} + \cdots \qquad (4.50)$$

This is known as *Friiss' formula*.

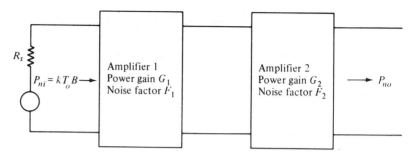

FIGURE 4.9. For two amplifiers in cascade, the overall noise factor is $F = F_1 + (F_2 - 1)/G_1$.

Example 4.7 A mixer stage has a noise figure of 20 dB, and this is preceded by an amplifier that has a noise figure of 9 dB and an available power gain of 15 dB. Calculate the overall noise figure referred to the input.

Solution
$$F_2 = 20 \text{ dB} = 100 : 1 \text{ power ratio}$$
$$F_1 = 9 \text{ dB} = 7.94 : 1 \text{ power ratio}$$
$$G_1 = 15 \text{ dB} = 31.62 : 1 \text{ power ratio}$$

Using Friiss' formula

$$F = F_1 + \frac{F_2 - 1}{G_1}$$

$$= 7.94 + \frac{100 - 1}{31.62}$$

$$= 11.07$$

This is the overall noise factor. The overall noise figure is

$$F_{dB} = 10 \log_{10} 11.07$$

$$= 10.44 \text{ dB}$$

Note that the noise figures must be converted to noise factors, and the gain converted to a power ratio, before Friiss' formula can be used.

4.10.4 Measurement of Noise Factor

The most common method of measuring the noise factor of an amplifier is to use the shot noise as produced in a diode. The diode may be a temperature-limited vacuum-tube type, or it may be a semiconductor type; in either case, the shot-noise current is given by an equation of the form

$$I_n^2 = 2I_{dc}q_eB \tag{4.29}$$

The equivalent circuit for the noise-generating diode is shown in Fig. 4.10, in which resistance R is the load resistance. This is usually chosen to match the input to the amplifier under test, and for the input to radio receivers, for example, it may be 50 Ω.

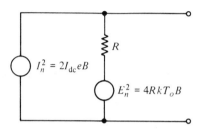

FIGURE 4.10. The equivalent circuit for a noise-generating diode.

With the noise generator connected to the input of the amplifier under test and the direct current I_{dc} set to zero, the noise output from the amplifier is

$$P_{no} = FGkT_0B \tag{4.43}$$

With the direct current now adjusted to some value I_{dc}, the available additional noise power at the input is

$$P_{n\,add} = \frac{I_n^2}{4}R \tag{4.51}$$

(This follows an application of the maximum power transfer equation, Eq. (1.138), as developed in Section 1.14.)

The additional noise power observed at the output will be

$$GP_{n\,add} = G\frac{I_n^2}{4}R$$
$$= \frac{G(2I_{dc}q_eB)R}{4}$$
$$= \tfrac{1}{2}GI_{dc}q_eBR \tag{4.52}$$

The total noise output therefore becomes

$$P'_{no} = P_{no} + \tfrac{1}{2}GI_{dc}q_eBR \tag{4.53}$$

But the new noise power output can also be considered as an increase in the original noise output by some factor X:

$$P'_{no} = XP_{no} \tag{4.54}$$

It follows, on equating Eqs. (4.53) and (4.54), that

$$XP_{no} = P_{no} + \tfrac{1}{2}GI_{dc}q_e BR$$

or

$$X(FGkT_0 B) = FGkT_0 B + \tfrac{1}{2}GI_{dc}q_e BR \tag{4.55}$$

after substituting for P_{no} from Eq. (4.43). Equation (4.55) can be rearranged to give

$$F = \frac{I_{dc}q_e R}{2(X-1)kT_0} \tag{4.56}$$

In normal measurements, the direct current is increased from zero until the noise power output is doubled (i.e., $X = 2$); the measurements are also made at room temperature, for which $kT_0 = 4 \times 10^{-21}$ J; the value for electron charge is $q_e = 1.6 \times 10^{-19}$ C. Substituting these values into Eq. (4.56) gives

$$F = 20I_{dc}R \tag{4.57}$$

Thus, the noise factor F is determined directly in terms of the direct current of the diode I_{dc}. For the particular case when $R = 50$ Ω, Eq. (4.57) becomes

$$F = I_{dc} \quad \text{(where I_{dc} is measured in mA)} \tag{4.58}$$

That is, the noise factor is *numerically* equal to the direct current in mA for the conditions stated. Keep in mind that F is a dimensionless ratio; it is *not* measured in mA.

4.11 NOISE TEMPERATURE

Yet another way of representing noise is by means of *equivalent noise temperature*. Equivalent noise temperature finds greatest use at microwave frequencies in connection with the noise at the receiver input. This noise arises from three main sources, the input amplifier, the connections to the antenna, and the antenna itself. This section deals with the first two sources, and antenna noise is discussed in Section 19.7.

4.11.1 Equivalent Noise Temperature T_e of Amplifier Input

Here it is postulated that the noise associated with the amplifier input, $(F - 1)kT_0B$, as given by Eq. (4.48), can alternatively be represented by some hypothetical temperature T_e such that

$$kT_eB = (F - 1)kT_0B \tag{4.59}$$

so that

$$T_e = (F - 1)T_0 \tag{4.60}$$

This shows that T_e is just an alternative measure for F.

Friiss' formula (Eq. (4.50)) can be rewritten in terms of an overall noise temperature. Friiss' formula is

$$F = F_1 + \frac{F_2 - 1}{G_1} + \frac{F_3 - 1}{G_2G_1} + \cdots$$

Subtracting 1 from both sides,

$$F - 1 = (F_1 - 1) + \frac{F_2 - 1}{G_1} + \frac{F_3 - 1}{G_2G_1} + \cdots \tag{4.61}$$

Substituting for the various $(F_i - 1)$ from Eq. (4.60) yields

$$\frac{T_e}{T_0} = \frac{T_{e1}}{T_0} + \frac{T_{e2}/T_0}{G_1} + \frac{T_{e3}/T_0}{G_2G_1} + \cdots$$

or

$$T_e = T_{e1} + \frac{T_{e2}}{G_1} + \frac{T_{e3}}{G_2G_1} + \cdots \tag{4.62}$$

4.11.2 Equivalent Noise Temperature of a Lossy Passive Network

A *lossy passive network* may consist of a resistive attenuator or simply a section of transmission line or waveguide. An important example is the noise contributed by the waveguide or transmission line connecting the antenna to the receiver in a satellite receiving station, described in Section 19.7. It is required to find the equivalent noise temperature of such a network, referred to the input terminals. The network is assumed matched to the source, as shown in Fig. 4.11.

Let L represent the power attenuation ratio of the network. For example, for a 3-dB attenuation, $L = 2:1$. Initially, the network is assumed to be at the same physical temperature as the source T_0. Since the network is matched to

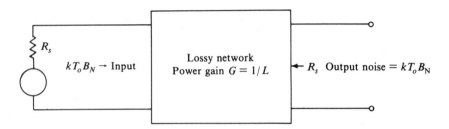

FIGURE 4.11. Noise output from a matched lossy network is equal to the source input noise.

the source, the available noise output is kT_0B_N, where B_N is the equivalent noise bandwidth. The noise power output from the network is the same as the available noise at the source under matched conditions. In effect, the network attenuates the noise from the source, but at the same time adds noise of its own. Referring the noise at the network output back to the input gives kT_0B_N/G for the network input noise, where G is the power gain of the network. The power gain is $G = 1/L$, and therefore the noise input to the network is LkT_0B_N. The source contributes kT_0B_N to this, and therefore the network contribution referred to the input is

$$P_{Nin} = LkT_0B_N - kT_0B_N$$
$$= k(L - 1)T_0B_N \qquad (4.63)$$

This may be represented by some hypothetical temperature T_e, as was done for the amplifier in the previous section. In this case,

$$T_e = (L - 1)T_0 \qquad (4.64)$$

Initially the network was assumed to be at the same temperature as the source. The noise contributed by the network depends only on *its* physical temperature (provided that matching is maintained), and therefore Eq. (4.64) may be generalized to

$$T_e = (L - 1)T_{nw} \qquad (4.65)$$

where T_{nw} is the physical temperature of the network.

Examples of the equivalent noise temperature are given in Section 19.7 in connection with satellite communications. Also, the equivalent noise temperature of an antenna is introduced in that section.

4.12 PROBLEMS

1. Explain how thermal noise power varies (a) with temperature, and (b) with frequency bandwidth. Thermal noise from a resistor is measured as 4 $\times$

10^{-17} W for a given bandwidth and at a temperature of 20 °C. What will the noise power be when the temperature is changed (a) to 50 °C?; (b) to 77 °K?

2. Calculate the rms noise voltage appearing across a 10-kΩ resistor at room temperature and for an effective noise bandwidth of 10 kHz.

3. Three 5-kΩ resistors are connected in series. For room temperature ($kT = 4 \times 10^{-21}$ J) and an effective noise bandwidth of 1 MHz, determine (a) the noise voltage appearing across each resistor, and (b) the noise voltage appearing across the series combination. (c) What is the rms noise voltage which appears across the same three resistors connected in parallel under the same conditions?

4. The noise generated by a 1000-Ω resistor can be represented by a rms current source of 4.0 nA in parallel with the 1000-Ω resistor. Show that this is equivalent to having the noise represented by a root-mean-square noise voltage source of 4 μV in series with the resistor.

5. Explain what is meant by effective noise bandwidth. A signal-measuring circuit is equivalent to a parallel combination of 1000-Ω resistance and 0.5-μF capacitance. Calculate the effective noise bandwidth.

6. Explain why inductance and capacitance do not generate noise. Explain also how they control the noise in a circuit. A parallel tuned circuit is resonant at a frequency of 1 MHz and has a Q factor of 50. The tuning capacitor is 200 pF. Calculate the effective noise bandwidth of the circuit. Also calculate, at room temperature, the rms noise voltage appearing across the circuit.

7. The tuned circuit of Problem 6 is placed at the input of a double-sideband AM receiver, and as a result the Q factor is reduced effectively to 25. The receiver has an audio bandwidth (i.e., the signal bandwidth of the audio stages) of 3 kHz. Calculate, at room temperature, the rms noise voltage at the tuned circuit that will be effective in producing an audio output.

8. Write brief notes on the sources of noise, other than thermal, which arise in electronic equipment. In particular, describe how the power spectrum density varies with frequency for the various sources.

9. An amplifier has an actual input resistance of 5000 Ω and an equivalent noise resistance of 1000 Ω referred to the input. What is the total effective input resistance of the amplifier? What is the rms noise voltage, at room temperature, at the input of the amplifier for a noise bandwidth of 1 kHz?

10. In Problem 9 a signal source of 50-Ω internal resistance is connected to the input of the amplifier via a 10 : 1 step-up matching transformer (which may be assumed lossless). What is the rms noise voltage at the input under these conditions, assuming room temperature?

11. A signal generator of internal resistance 50 Ω and internal EMF 10 μV is connected to the input of an amplifier that has an effective noise resistance

of 1200 Ω and an input resistance of 75 Ω. Calculate the S/N ratio at the input for a noise bandwidth of 1 kHz and at room temperature.

12. In Problem 10, what will be the S/N ratio at the input when the signal source has an internal EMF of 1.0 μV?

13. In a rather cheap short-wave receiver, the antenna circuit is fed directly into the mixer circuit. The input coupling is such that the antenna is matched to the 10-kΩ input resistance of the mixer. The mixer has an equivalent noise resistance of 100 kΩ. Calculate the antenna EMF necessary to produce an S/N power ratio of 20 dB if the input transformer has a step-up ratio of 10 : 1 and the effective noise bandwidth is 10 kHz. Room temperature may be assumed.   8.2 μv.

14. The short-wave receiver in Problem 13 is improved by inclusion of a voltage amplifier between antenna circuit and mixer. The amplifier has a voltage gain of 15 : 1, an equivalent noise resistance of 1000 Ω, and an input resistance of 10 kΩ, to which the antenna is matched as before. Calculate the new antenna EMF required for the 20-dB S/N ratio.

15. Three telephone circuits, each having an S/N ratio of 44 dB, are connected
   39·2dB   in tandem. What is the overall S/N ratio? A fourth circuit is now added,
   34 dB   which has an S/N ratio of 34 dB. What is the overall S/N in this case?

16. The noise factor of a radio receiver is 15 : 1. What is its noise figure? If the input S/N ratio is 35 dB, what will be the output S/N ratio?

17. Given that the noise figure of an amplifier is 11 dB, what is the fraction of the total available noise, at the input, contributed by the amplifier? 0.92

18. A mixer circuit having a noise figure of 16 dB is preceded by an amplifier having a noise figure of 9 dB and an available power gain of 25 dB. What is the overall noise figure of the combination?

19. Explain what is meant by (a) the equivalent noise temperature of an amplifier input, and (b) the equivalent noise temperature of a lossy passive network. Given that the equivalent noise temperature of an amplifier is 25 K, what is its noise factor? 1·09

20. The mixer stage of a microwave receiver has a noise figure of 11 dB and is preceded by a low-noise amplifier having an available power gain of 20 dB and an equivalent noise temperature of 33°K. Calculate the effective noise temperature of the combination referred to the input of the amplifier. 109

PART 2

ELECTRONIC COMMUNICATIONS CIRCUITS

CHAPTER 5

RF and Broadband Amplifiers

5.1 TUNED RF AMPLIFIERS

Tuned RF amplifiers are commonly used to provide the front-end selectivity and amplification in radio receivers to separate incoming signals from the antenna, to provide the precise bandpass filtering required in the intermediate frequency (IF) amplifiers of receivers, and to provide the harmonic removal filtering needed in transmitting circuits. At the higher frequencies, stability becomes a severe problem, and those types of amplifiers that have very low internal feedback, such as the common-base amplifier and the cascode amplifier, are favored. Even with this, it is often still necessary to provide special compensation or neutralization to maintain stability under operating conditions.

The bandpass filter characteristics are generated by one or more tuned circuits within the amplifier circuit. Usually voltage amplification is desired, and as a result the parallel tuned circuit is the favored one. Since the bandwidth of a tuned circuit's response is dependent on its Q, which is, in turn, dependent on the amount of resistance within or attached to the tuned circuit, special care in providing proper impedance matching to the tuned circuit is necessary to maintain the desired characteristics. This is especially true of transmitter power circuits where very low impedance circuits must be connected to the tuned circuits without lowering the Q too much.

5.1.1 Single-Tuned Class A Amplifier (Transformer Coupled with Tuned Primary)

Figure 5.1(a) shows a transformer-coupled circuit. The tuned circuit is the primary of the transformer, and the transformer acts as a load isolator. Radio-frequency transformers usually have coefficients of coupling which are much less than unity, and, as a result, the load impedance of the secondary has little effect on the primary. If the transformer were tightly coupled, the load impedance would be coupled into the primary. The real part would load the Q down and cause a widening of the bandwidth, while any reactance would detune the circuit. Similarly, the complex output admittance of the amplifier can cause detuning and loading of the primary. The detuning effects can be accounted for by making either the capacitor or the inductor adjustable over a

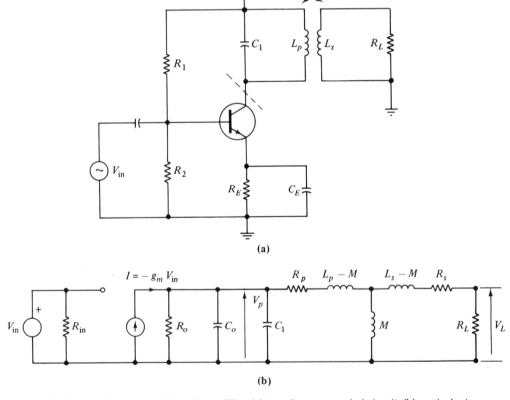

(a)

(b)

FIGURE 5.1. Single-tuned class A amplifier: (a) transformer-coupled circuit; (b) equivalent circuit.

small range. However, any loading effects must either be made negligible by efficient coupling methods or be accounted for in the desired bandwidth calculations. In some critical cases, it may be possible to adjust the shunt resistance to provide the desired bandwidth.

Impedance transformation using either tapped transformer coils or tapped capacitors, both on the primary and on the secondary of the transformer, is necessary with bipolar transistors because of their generally low impedance levels. To obtain sufficiently high Q's, especially at higher frequencies, dynamic impedance (R_D) values ranging from 10 kΩ to 1 MΩ are needed, while the base of the bipolar transistor only gives load resistances of 1 to 10 kΩ. The impedance transformation also serves to reduce the effect of parasitic reactances on the resonant frequency. The process of impedance transformation is discussed in Chapter 1.

The equivalent circuit is shown in Fig. 5.1(b), where use is made of the equivalent mutual inductance circuit of Fig. 1.18(b). Analysis follows the methods given in Section 1.12.2, as illustrated in the following example.

Example 5.1 The amplifier shown in Fig. 5.1(a) uses a transistor CE amplifier which has a transconductance of 30 mS at the resonant frequency and an output conductance comprised of 100 kΩ in parallel with 5 pF of capacity. The transformer has a primary inductance of 75 μH, secondary inductance of 100 μH, coefficient of coupling of 0.01, primary resistance of 10 Ω, and secondary resistance of 10 Ω. The primary is tuned with a 50-pF capacitor and the secondary is loaded by a 4-kΩ resistance. Find (a) the resonant frequency, (b) the effective Q of the tuned circuit, (c) the 3-dB bandwidth, (d) the small signal voltage gain at resonance, and (e) the dB rejection of a signal 150 kHz above the resonant frequency.

Solution (a) The effective tuning capacity C_p is C_0 in parallel with C_1:

$$C_p = 50 + 5 = 55 \text{ pF}$$

A calculation check will show that $1/(L_pC_p) \gg (R_p/L_p)^2$, so that the approximate form of resonant frequency, Eq. (1.59), may be used. A calculation check will also show that $R_L \gg |X_s|$; hence, from Eq. (1.116), the secondary does not detune the primary. Therefore,

$$f_0 = \frac{1}{2\pi\sqrt{L_pC_p}}$$

$$= \frac{1}{2\pi \times \sqrt{75 \times 10^{-6} \times 55 \times 10^{-12}}}$$

$$= 2.48 \text{ MHz}$$

(b) The undamped Q of the primary, Eq. (1.118), is

$$Q_p = \frac{\omega_0 L_p}{R_p}$$

$$= \frac{2\pi \times 2.48 \times 10^6 \times 75 \times 10^{-6}}{10}$$

$$= 117$$

And for the secondary, Eq. (1.119) gives

$$Q_s = \frac{\omega_0 L_s}{R_L}$$

$$= 117 \times \frac{10}{75} \times \frac{100}{4000}$$

$$= 0.39$$

Note the use of proportionality to ease the calculations. The factor $(1 + k^2 Q_p Q_s)$ for $x = 1$ (see Section 1.12.2) is equal to

$$\left(1 + k^2 Q_p Q_s\right) = (1 + 0.01^2 \times 117 \times 0.39)$$

$$= 1.005$$

Hence, Q'_p, from Eq. (1.123), is

$$Q'_p = Q_p / \left(1 + k^2 Q_p Q_s\right) = \frac{117}{1.005} = 116$$

This shows that the secondary loading is negligible; however, the loading of the transistor output resistance will be considerable and is accounted for as follows. The dynamic impedance, Eq. (1.85), including the secondary load, but not the transistor loading, is

$$R'_D = \frac{Q'_p}{\omega_0 C_p}$$

$$= \frac{116}{2\pi \times 2.48 \times 10^6 \times 55 \times 10^{-12}}$$

$$= 135 \text{ k}\Omega$$

which appears to be a resistance in parallel with L_p, and taking R_0 (Fig. 5.1(b)), into account,

$$R''_D = \frac{135 \times 100}{135 + 100}$$

$$= 57.4 \text{k}\Omega$$

The effective Q-factor, therefore, is, by Eq. (1.123),

$$Q_p'' = Q_p' \times \frac{R_D''}{R_D'}$$

$$= 116 \times \frac{57.4}{135}$$

$$= 49$$

(c)   The 3-dB bandwidth (see Eq. (1.83)) is

$$B_{3\,\text{dB}} = \frac{f_0}{Q_p''}$$

$$= \frac{2.48}{49} \text{ MHz}$$

$$= 50.6 \text{ kHz}$$

(d)   The small signal voltage gain is $A_v = (V_L/V_{\text{in}})$. Using Eqs. (1.124) and (1.125), and substituting $-g_m V_{\text{in}}$ for I,

$$A_v = -\frac{g_m R_D'' k}{\left(1 + jyQ_p''\right)} \sqrt{\frac{L_s}{L_p}}$$

Note that R_D'' and Q_p'' (not R_D' and Q_p') are used.
At resonance, $y = 0$ (see Eq. (1.65)); thus,

$$A_{v0} = -g_m R_D'' k \sqrt{\frac{L_s}{L_p}}$$

$$= 30 \times 10^{-3} \times 57.4 \times 10^3 \times 0.01 \times \sqrt{\frac{100}{75}}$$

$$= -19.9 \text{ V/V}$$

(e)   The gain modulus, relative to the resonant value, is

$$A_r = \frac{|A_v|}{|A_{v0}|}$$

$$= \frac{1}{\sqrt{1 + \left(yQ_p''\right)^2}} \quad \text{(See derivation of Eq. (7.5))}$$

At 150 kHz off-tune,

$$f = 2.48 \pm 0.15 = 2.63 \text{ (or 2.33 MHz)}$$

and by Eq. (1.65)

$$y = \left(\frac{f}{f_0} - \frac{f_0}{f} \right) = \left(\frac{2.63}{2.48} - \frac{2.48}{2.63} \right) = 0.118 \text{ (or } 0.125)$$

which gives an average of 0.121. Thus,

$$A_r = \frac{1}{\sqrt{1 + (0.121 \times 49)^2}}$$

$$= \frac{1}{6}$$

$$A_r(\text{dB}) = 20 \log A_r = -15.6 \text{ dB}$$

5.1.2 Single-Tuned Class A Amplifier (Transformer Coupled with Tuned Secondary)

Figure 5.2 shows the circuit of a single-tuned class A amplifier. The chief difference between this circuit and the one presented in the previous section is that the resonant circuit is series tuned instead of parallel tuned. Also, since the primary of the transformer is not tuned, the amplifier is feeding an untuned reactive load formed by the parallel combination of the amplifier output capacitance and the primary inductance, resulting in a net phase shift in the output voltage that is different from $180°$.

Again, analysis follows the general methods described in Chapter 1, as illustrated by the following example.

Example 5.2   The amplifier of Fig. 5.2(a) uses a transistor with a transconductance of 10 mS, an output resistance of 2 kΩ, and an output capacity of 5 pF. The other parasitic capacities are small enough to be ignored. The other circuit components have the values given below.

$$L_p = 10 \ \mu\text{H} \qquad L_s = 20 \ \mu\text{H} \qquad \text{Coupling } k = 0.01$$
$$R_p = 1 \ \Omega \qquad R_s = 0.5 \ \Omega \qquad R_L = 100 \ \text{k}\Omega \qquad C_s = 20 \ \text{pF}$$

Find the gain and bandwidth of the amplifier.

Solution   The equivalent circuit for the amplifier, using the mutual inductance equivalent circuit of Fig. 1.18(b), is shown in Fig. 5.2(b). The resonant frequency of the secondary will be used, but the effect of primary loading on the phase and gain-bandwidth will be shown. Using the methods described in Chapter 1, the resonant frequency and the impedances and admittances of Fig. 5.2(b) may be

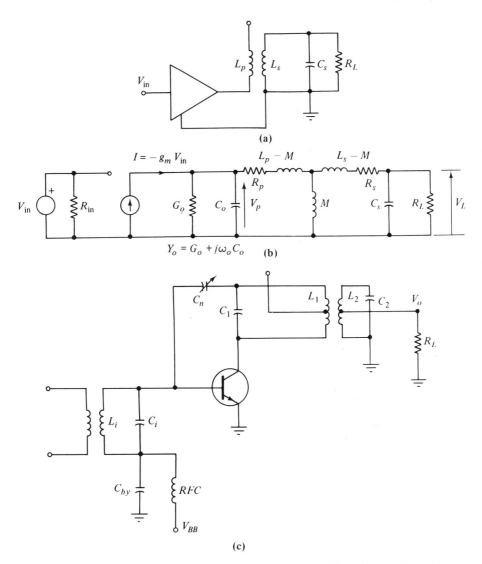

FIGURE 5.2. (a) Circuit for a transformer-coupled class A amplifier with tuned secondary; (b) equivalent circuit for (a); (c) double-tuned transformer-coupled amplifier circuit.

evaluated as

$$\omega_0 = 5 \times 10^7 \quad \text{rad/s (Eq. (1.59))}$$

$$Z_m = j7.05 \ \Omega \quad \text{(Eqs. (1.111), (1.97))}$$

$$Z_p = 1 + j500 \ \Omega \quad \text{(Eq. (1.109))}$$

$$Z_s = R_L \| C_s = 0.5 + j1000 \ \Omega \quad \text{(Eq. (1.110))}$$

$$Z_L = 10 - j1000 \ \Omega \quad \text{(Eqs. (1.52) and (1.53))}$$

$$Y_0 = (0.5 + j0.25) \times 10^{-3} \ \text{S}$$

The total impedance of the secondary at resonance is $Z_s + Z_L = 10.5 \ \Omega$ (resistive). The effective primary impedance can be evaluated, using Eq. (1.115), as

$$Z_p' = Z_p - \frac{Z_m^2}{Z_s + Z_L} = 1 + j500 + \frac{50}{10.5} \cong j500 \ \Omega$$

This shows that only the reactance of L_p is significant, the primary resistance and the resistance reflected from the secondary being negligible by comparison. The primary voltage (see Fig. 5.2(b)) is

$$V_p = \frac{-g_m V_{in}}{Y_0 + (1/Z_p')} = \frac{-g_m V_{in} Z_p'}{1 + Y_0 Z_p'} \tag{5.1}$$

This would have been $-g_m V_{in} Z_p'$ had Y_0 been zero. From Eqs. (1.112) and (1.113), the secondary current may be evaluated as

$$I_s = \frac{V_p Z_m}{Z_p (Z_s + Z_L) - Z_m^2} \tag{5.2}$$

and clearly,

$$V_L = I_s Z_L \tag{5.3}$$

Combining Eqs. (5.1), (5.2), and (5.3), and substituting numerical values gives

$$A_v = \frac{V_L}{V_{in}} = \frac{V_L}{I_s} \times \frac{I_s}{V_p} \times \frac{V_p}{V_{in}} = -7.3 \ \underline{/-15.4^\circ} \quad \text{or} \quad 16.9 \text{ dB}.$$

The bandwidth is controlled by the secondary Q factor; thus,

$$Q_s = \frac{\omega_0 L_s}{R_s'} \quad \text{(Eq. (1.62))}$$

where R'_s is the resistive component of $(Z_s + Z_L)$. And

$$BW = \frac{f_0}{Q_s} = \frac{w_0}{2\pi Q_s} = \frac{R'_s}{2\pi L_s} \qquad (\text{Eq. (1.71)})$$

$$= \frac{10.5}{2\pi \times 20 \times 10^{-6}}$$

$$= 83.6 \text{ kHz}$$

5.1.3 Double-Tuned Transformer-Coupled Amplifier

The *double-tuned transformer-coupled amplifier* circuit is formed by resonating both the primary and the secondary circuits of the output coupling transformer, as shown in Fig. 5.2(c). The transfer function of the double-tuned transformer was given in Section 1.12.2, where it is seen that if the transformer coupling is greater than a critical value, the amplitude-versus-frequency plot will exhibit a double-humped shape with steep outside skirts (Fig. 1.19). The bandwidth of the result is primarily determined by the effective Q of the primary and secondary tuned circuits, which must be the same to maintain symmetry in the response. Impedance transformation is accomplished on both sides usually by tapping the feed points down on the inductors (capacitive transformation is used in some cases).

5.2 NEUTRALIZATION

When an amplifying device is operated at frequencies near its upper cutoff frequency, a parasitic capacity between the output and input provides a path for feedback. If both the input and output circuits are tuned, then a circuit identical to the tuned input/tuned output oscillator is formed, and unless some measures are taken to stabilize the circuit, it will oscillate. Even if the input and output circuits are tuned, the feedback path may introduce sufficient phase shift to allow oscillation to occur.

Several schemes for neutralizing this feedback path are available, and three of them will be discussed here: the Hazeltine circuit, the Rice circuit, and a method for neutralizing an amplifier with common circuit feedback.

Figure 5.3(a) shows the Hazeltine neutralization scheme for tuned input/tuned output amplifiers. The output coil is center-tapped, and the tap is used as the power supply feedpoint of the amplifier. The other end of the coil, point N, is connected through a small capacitor back to the input. The portion of the circuit formed by the amplifier output voltage, the two halves of the coil, the feedback capacitance, and the neutralizing capacitance form a bridge to the

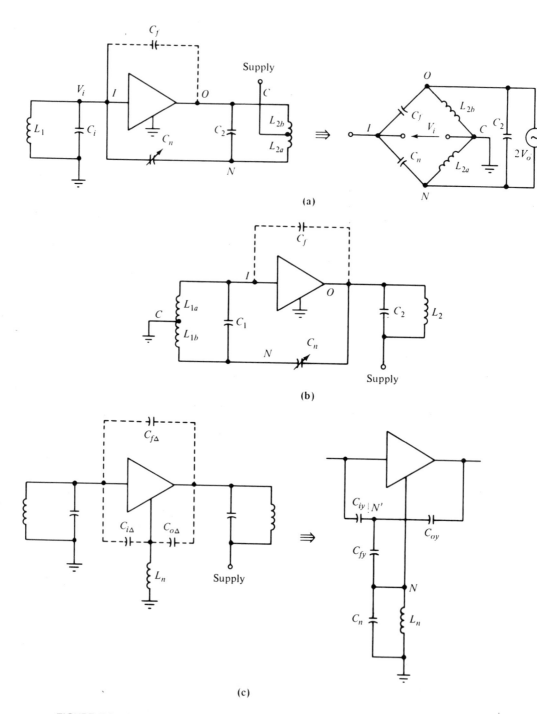

FIGURE 5.3. (a) Hazeltine neutralization system and equivalent bridge circuit; (b) Rice neutralization system; (c) Neutralization of common circuit feedback — circuit showing parasitic elements and the equivalent circuit.

input, as shown in Fig. 5.3(a). For the bridge to null, producing zero voltage across the amplifier input, it is only necessary for the impedance cross products to be equal, so that at the frequency of operation,

$$X_{Cf}X_{L_{2a}} = X_{c_n}X_{L_{2b}} \qquad (5.4)$$

and

$$C_n = C_f \frac{L_{2b}}{L_{2a}} \qquad (5.5)$$

Figure 5.3(b) shows the circuit for the Rice scheme of neutralization. In this case, the input coil is center-tapped so that $L_{1a} = L_{1b}$, and the neutralization capacitance equal to the feedback capacitance is connected between the other end of L_1 (point N) to the output of the amplifier.

A voltage appearing at the output drives a phase-shifted current through C_f and L_{1a}, producing a voltage at the amplifier input. However, the same output drives an identical-valued current in the opposite direction through C_n and L_{1b}. The currents in L_1 will be equal and opposing, so that the feedback voltage appearing at the amplifier input will be zero. Null will occur when

$$X_{cf} = X_{c_n} \qquad (5.6)$$

so that

$$C_f = C_n \qquad (5.7)$$

Both of these circuits are used for amplifiers into the VHF and UHF range, as long as no appreciable reactance appears in the amplifier common circuit. The method of adjusting both amplifiers is basically the same:

1. Turn off the amplifier power supply, but maintain input drive signal.
2. Adjust C_n.
3. Vary the output tank tuning capacitor C_2 and observe any change in input bias current.
4. Repeat steps 2 and 3 until no change occurs in the bias current when the output tank tuning is varied, indicating correct neutralization.

Figure 5.3(c) shows an amplifier that has common circuit reactance L_n, causing feedback which may be positive. This will occur at extremely high frequencies (UHF and microwave), when the inductance of the wiring must be accounted for. The parasitic capacitances of the amplifier form a delta configuration around the three terminals. Figure 5.3(c) shows the equivalent circuit resulting when the delta capacitances are replaced by their equivalent Y

capacitances. The transformation is accomplished by

$$C_{xy} = \frac{Ci \, \Delta Co \, \Delta + Co \, \Delta Cf \Delta + Cf \Delta Ci \, \Delta}{Cx \, \Delta} \tag{5.8}$$

where $x = i$, o, or f.

Now if the network formed by C_{fy}, C_n, and L_n is made series resonant, then a virtual short to ground will exist at the series resonant frequency from the point N', preventing any of the output signal from reaching the input circuit. Resonance is established when

$$X_{C_{fy}} = X_{C_n} \| X_{L_n} \tag{5.9}$$

If the input and output circuits are tuned, the adjustment procedure is the same as that stated above.

5.3 SPECIAL RF AMPLIFIERS

5.3.1 Common-Base Amplifier

The common-base amplifier is used frequently in RF amplifiers for two reasons. First, it provides usable voltage gain to a much higher upper cutoff frequency than the common-emitter stage; and second, the feedback capacity between the output and input is very much lower than that for the common-emitter stage. The same arguments are true for the field-effect transistor operating in the common gate mode.

Two things limit the frequency response of an amplifier gain. The first of these is the shunting effect, which occurs because of the parasitic capacitances between the input terminals of the amplifier, and the second is the cutoff frequency at which the amplifying device gain begins to fall. The first of these can be designed around by making the ratio of source resistance to input resistance of the amplifier sufficiently low that the break frequency formed by the net source resistance and the input capacity is considerably higher than the gain cutoff frequency. If this is done, the overall response is limited by the gain cutoff frequency of the amplifying device, and this is a function entirely of the geometry and the material used in making the device, and can only be modified by choosing a transistor with a higher cutoff frequency.

The bipolar transistor suffers from two major parasitic capacitances, the junction capacitance between base and emitter, and the junction capacitance between base and collector. The first is a function of the bias current, and the second is the reverse-biased junction capacity of the collector-base junction, which is independent of current but dependent on collector potential. (In some IC transistors a collector-substrate (ground) capacitance will also be significant.) Although these two capacitances predominate, three other very small ones exist, viz., the interelectrode capacitances between the leads of the device. At extremely high frequencies the lead inductances also have some effect, but generally these lead capacitances and inductances will not have an effect until a frequency well above that dictated by the two main capacitances is reached, and they can be ignored. In any event, their effects can be canceled by neutralization if necessary. Figure 5.4(a) shows the parasitic capacitances in the common-base model of the bipolar transistor.

When the transistor is used in the common-emitter mode, the common-base output capacitance C_{ob} is connected between the input and output. This capacitance is reflected into the input circuit multiplied by the voltage gain (the Miller effect), and reacts with the source resistance to produce a very low cutoff

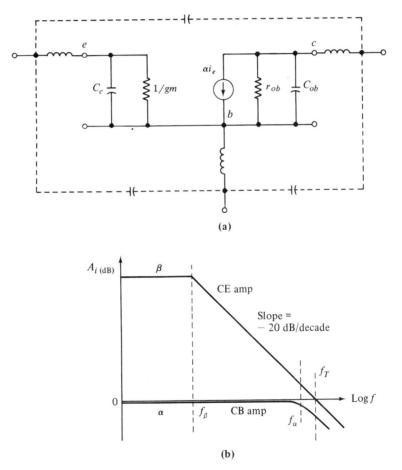

FIGURE 5.4. Common-base amplifier: (a) circuit model showing parasitic reactances; (b) plot of current gain versus log frequency for CE and CB amplifiers.

frequency, well below the gain cutoff frequency unless the source resistance is very low. This puts a limitation on the usefulness of the amplifier. C_{ob} also provides a feedback path between the collector and emitter, which at higher frequencies can make the transistor unstable (it will oscillate) and will have to be neutralized if it is to be used.

When the same transistor is used in the common-base mode, the feedback capacitance is the collector-to-emitter capacitance. But this is made up of the two capacitances C_{ob} and C_e in series, and since the intermediate base point is grounded, any feedback signal through this path is prevented from reaching the input. The two capacitances are merely connected between the input terminals and between the output terminals and can be canceled if the amplifier is tuned. The only remaining feedback capacity is that between the leads, and this usually can be kept sufficiently small as to be ignored, or can be neutralized, as mentioned in Section 5.2.1.

The current gain of the common-base amplifier is simply the α value of the transistor up to the alpha cutoff frequency. Above this value it falls off at -20 dB per decade. The frequency-dependent value of current gain is given as

$$A_{i(f)CB} = \frac{-\alpha}{1 + j(f/f_\alpha)} \qquad \text{(for CB amp)} \qquad (5.10)$$

This compares with the value for the common-emitter amplifier, which is

$$A_{i(f)CE} = \frac{-\beta}{1 + j(f/f_\beta)} \qquad \text{(for CE amp)} \qquad (5.11)$$

The two cutoff frequencies are related approximately by

$$f_\beta = \frac{f_T}{\beta} \qquad (5.12)$$

where f_T is the frequency at which the CE current gain β goes to unity. Thus,

$$f_\alpha \cong \frac{f_T}{n} \quad \text{where } n = 1 \text{ to } 2 \qquad (5.13)$$

These facts are illustrated in Fig. 5.4(b) and point out that the cutoff frequency of the CB amp is about β times the cutoff frequency of the CE amp, unless it is limited by the input-circuit time constant to less than these values.

The voltage gain of the CE amp and that of the CB amp at low frequency are virtually the same magnitude as given by

$$|A_{v_{CB}}| = |A_{v_{CE}}| = g_m R_L = A_i \frac{R_L}{R_i} \qquad (5.14)$$

where

$$g_m = I_C/V_T$$

and

$$V_T = kT/q = 0.026 \text{ V at } 300\,^{\circ}\text{K}$$

However, $A_{v_{CB}}$ is noninverting while $A_{v_{CE}}$ is inverting. Since A_i is a function of frequency in each case and the cutoff frequency is different in each case,

$$A_{v(f)} \cong -g_m R_L \frac{1}{1 + j(f/f_\beta)} \qquad \text{(for CE amp)} \qquad (5.15)$$

and

$$A_{v(f)} = +g_m R_L \frac{1}{1 + j(f/f_\alpha)} \qquad \text{(for CB amp)} \qquad (5.16)$$

It should also be noted that the input impedance is a function of frequency as well, and will begin to change radically at the frequency determined by the input circuit time constant. Since this frequency can be controlled to some extent, it is usually possible to maintain constant input impedance to frequencies beyond the gain cutoff point.

The following example will illustrate the summary presented here.

Example 5.3   A transistor has the following characteristics:

$$C_{ob} = 10 \text{ pF} \qquad I_E = 0.26 \text{ mA} \qquad V_T = 26 \text{ mV} \qquad \beta = 89$$
$$f_T = 300 \text{ MHz} = 2f_\alpha \qquad R_L = 2 \text{ k}\Omega \qquad R_s = 100 \,\Omega$$

Calculate and compare the input circuit cutoff frequency and the gain cutoff frequency for the CB and CE modes of operation of the transistor. Also, calculate and compare the current gains, voltage gains, and input resistances for the two modes.

Solution
$$R_{i_b} = \frac{V_T}{I_E} = \frac{0.026}{0.26} = 0.10 \text{ k}\Omega = \frac{1}{g_m}$$

$$R_{i_e} = (\beta + 1)R_{i_b} = 90 \times 0.1 = 9 \text{ k}\Omega$$

$$C_e = \frac{g_m}{2\pi f_T} = \frac{1}{2\pi \times 100 \times 300 \text{ MHz}} = 5.3 \text{ pF}$$

(a)   For the CE mode,

$$C_{in} = C_e + C_{ob}(1 + g_m R_L) = 5.3 + 10 \times [1 + (10 \times 2)] = 215 \text{ pF}$$
$$R_s' = R_s \| R_{i_e} = 100 \| 9000 \cong 100 \,\Omega$$

The input-circuit cutoff frequency is

$$f_{in} = \frac{1}{2\pi C_{in} R'_s} = \frac{1}{2\pi \times 215 \text{ pF} \times 100} = 7.40 \text{ MHz}$$

$$A_{v(lf)} = -g_m R_L = \frac{-2}{0.1} = -20$$

$$A_{i(lf)} \cong -\beta = -89$$

The gain cutoff is

$$f_\beta = \frac{f_T}{\beta} = \frac{300}{89} = 3.37 \text{ MHz}$$

(b) For the CB mode,

$$C_{in} = C_e = 5.3 \text{ pF}$$

$$R'_s = R_s \| R_{ib} = 100 \| 100 = 50 \text{ }\Omega$$

The cutoff frequency of the input circuit is

$$f_{in} = \frac{1}{2\pi C_{in} R'_s} = \frac{1}{2\pi \times 5.3 \text{ pF} \times 50} = 600 \text{ MHz}$$

$$A_{v(lf)} = 20 \qquad A_{i(lf)} = -\alpha \cong 1$$

The gain cutoff is

$$f_\alpha = f_T \div 2 = 300 \div 2 = 150 \text{ MHz}$$

In comparison, the CE amp is limited to an upper cutoff frequency corresponding to its β cutoff frequency. The input cutoff frequency is above that and has no immediate effect. The CB amp, similarly, is limited by its gain cutoff and not by the input circuit. The CB amp has a limit of 150 MHz as compared to the limit of 3.37 MHz for the CE amp. The voltage gains in the two cases are identical, except for the different cutoff frequencies. The input impedance of the CB amp is much lower than that of the CE amp and is more resistive than capacitive. If stray capacitances in leads are ignored, the feedback capacitance in the CB amp is negligible as compared to a value of 10 pF for the CE amp. Therefore, the CB amplifier is preferred for high-frequency operation.

5.3.2 Cascode Amplifier

The *cascode amplifier* is a composite amplifier pair which is frequently used for RF applications. It consists of a common-emitter stage followed by a common-based stage, directly coupled to each other, and combines some features of both types of amplifiers. The input resistance is essentially that of the common-emitter amplifier, the current gain essentially that of the CE amplifier, and the voltage gain essentially that of the CB amplifier, but the reverse feedback coefficient is extremely small. The result is extremely good isolation between input and output. The feedback between input and output is about the same as for the CB amplifier, so that neutralization is rarely needed.

Figure 5.5(a) shows the schematic of a tuned cascode amplifier. The input is connected between base and emitter of the first transistor operating with the correct current level. The collector of the first transistor is connected directly to the emitter of the second transistor, whose base is ac-grounded through a bypass capacitor. Another external bias supply keeps the second transistor operating at the same current level. The second collector feeds the load.

Figure 5.5(b) shows the equivalent circuit at low frequencies for the amplifier. From this circuit it can be shown that

$$R_i = R_{i1} = 1\beta_1/g_{m1} \tag{5.17}$$
$$R_o = r_{ob2} \cong \infty \tag{5.18}$$
$$A_i = A_{i1}A_{i2} = \beta_1\alpha_2 \cong \beta_1 \tag{5.19}$$
$$A_v = A_{v_1}A_{v_2} = \left(-g_{m1}R_{L1} \times g_{m2}R_{L2}\right) = -g_{m1}R_L \tag{5.20}$$

where $R_{L1} = R_{ib2} = 1/g_{m2}$ and $R_{L2} = R_L$.

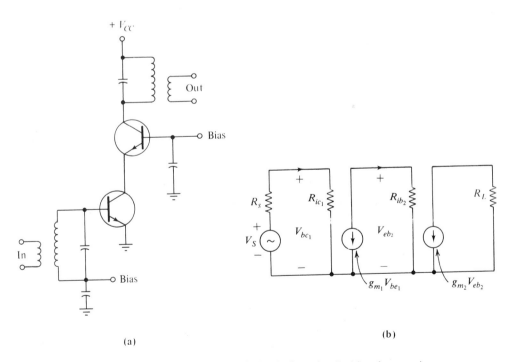

(a)

(b)

FIGURE 5.5.   Cascode amplifier: (a) circuit; (b) equivalent circuit at low frequencies.

The effective voltage gain of the first stage is approximately unity. Thus, near the cutoff frequency the collector-base capacity of the first stage C_{cb1} is multiplied only by a Miller factor of two times, to give a net input capacity of $C_i \cong C_{be1} + 2C_{cb1}$, which is much smaller than that usually associated with a CE amplifier, and its cutoff frequency is therefore much higher. The output capacity is simply the collector-base capacity of the second stage C_{cb2}.

The main advantage of this circuit is its good isolation between input and output. When it is used as a tuned amplifier, little or no neutralization is needed to maintain stability.

5.3.3 Dual-Gate MOSFETs

Compared to bipolar transistors, the insulated-gate field-effect transistor offers a number of advantages when used for RF amplification. Chief among these are higher input impedance and therefore less circuit damping; ability to handle a greater dynamic range of input signal; and less distortion, resulting in lower cross modulation (cross modulation occurs when modulation from one carrier is transferred to another as a result of nonlinearity of the amplifying device).

Insulated-gate field-effect transistors are usually referred to as MOSFETs, an acronym for metal oxide semiconductor field-effect transistors. One disadvantage of the MOSFET is the ease with which gate insulator breakdown can occur as a result of electrostatic potentials (such as can be generated by synthetic clothing). To circumvent this, most units are now fabricated with protective diodes built in as part of the gate structure, as in Fig. 5.6.

Dual-gate MOSFETs are used in RF amplifiers, as they offer certain advantages over the single-gate type. The additional gate allows automatic gain control (AGC—see Section 8.8) bias to be easily applied to the amplifier independently of the signal input, and the device is readily connected in the cascode configuration by grounding the AGC gate for signal frequencies.

Figure 5.6(a) shows in cross section the constructional features of a dual-gate MOSFET, and Fig. 5.6(b) shows the circuit symbol. The protective diodes are readily identified in these diagrams. Figure 5.6(c) shows how the unit may be connected as a cascode RF amplifier with AGC bias. The signal is applied at the 75-Ω input, the block labelled *Traps* consisting of tuned circuit filters. Fixed bias is applied to the signal gate through the 820-kΩ and 120-kΩ potentiometer network, with the input capacitor providing a dc block. The input resistance of the device and biasing network is essentially the 820 kΩ in parallel with the 120 kΩ, as the transistor itself will have an input resistance of many megohms.

The second gate is grounded for signal frequencies through the 1000-pF AGC line filter capacitor. The RF output is obtained by mutual inductive coupling from the drain circuit. The 270-Ω source resistance bypassed with the 1000-pF capacitor provides additional dc bias between source and gate.

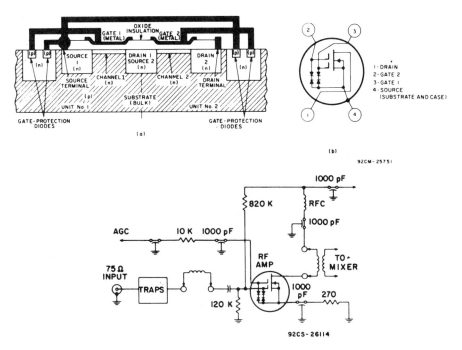

FIGURE 5.6. Dual-gate-protected *n*-channel depletion-type MOS field-effect transistor: (a) sideview cross section; (b) schematic symbol; (c) used as an RF amplifier. (Courtesy RCA Solid State Division)

The dual-gate MOSFET can also be used as a mixer, and this application is described in Section 5.4.3.

5.4 FREQUENCY CONVERSION AND MIXERS

Frequency conversion is the changing of the frequency of a carrier with its modulation from one frequency to another. This occurs when the signal is mixed with a second signal such as the output of an oscillator in such a manner that the output contains products of the two signals. One of these products will contain the sum and the difference frequencies of the two input signals. Other components are present as well, but all components except the desired one may be removed by bandpass filtering. Two general methods can be used to accomplish the mixing function, additive mixing and multiplicative mixing. The output frequency is termed the intermediate frequency (IF).

The input to a mixer is the input signal voltage, which has a magnitude of V_s at a frequency of f_s. The output is usually a current component at the IF frequency $(f_o - f_s)$, which will have a magnitude that will be proportional to V_s. A constant of proportionality can be defined which is called *conversion transconductance.* The conversion transconductance is thus given by

$$g_c = \frac{I_{IF}}{V_s} \tag{5.22}$$

where I_{IF} is the magnitude of that component of the output current at the IF, and V_s is the magnitude of the input signal voltage at f_s.

5.4.1 Additive Mixing

Additive mixing occurs when the input signal is simply added to the output of a local oscillator and then passed through a device with a nonlinear transfer function such as a diode. Diodes with essentially square-law response are usually used because this yields, under certain conditions, an almost proportional relationship between the input signal and the output signal. The output from the mixer contains many signal components, including the difference frequency and the sum frequency and several harmonics of each. Generally this output is passed directly to an IF amplifier, which acts as a bandpass filter just wide enough to pass modulation sidebands around the IF, providing whatever gain is required to boost the signal to the final detection level. Figure 5.7(a) shows the block diagram of an additive mixing system.

The input signal V_s at frequency f_s containing the modulation sidebands is added directly to the output from the local oscillator V_o at frequency f_o. A buffer amplifier is often used to provide isolation between the oscillator and the input circuit to prevent mutual detuning. The magnitude of V_o is usually made very much larger than V_s for a reason that will be discussed below. The added signals are then passed through a device such as a diode whose output contains a component that is proportional to the square of its input. The diode may be a separate device from the adder, or, as is more often the case, it may be the base-emitter junction of the amplifier transistor.

The circuit action may be analyzed in this manner. The output of the summing device produces the sum of the two signals, where

$$v_s = V_s \sin \omega_s t$$
$$v_o = V_o \sin \omega_o t$$
$$V_o \gg V_s \quad \text{and} \quad \omega_o > \omega_s$$

so that

$$v_t = v_s + v_o = \left(V_s \sin \omega_s t\right) + \left(V_o \sin \omega_o t\right) \tag{5.23}$$

If a diode with a square-law characteristic is used, then the diode characteris-

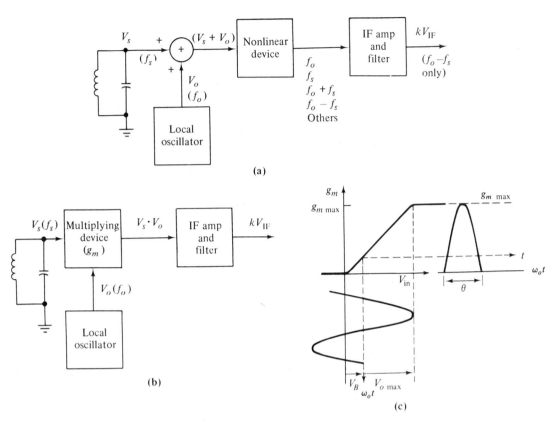

FIGURE 5.7. Superheterodyne frequency converter or mixer systems: (a) additive mixing; (b) multiplicative mixing; (c) transconductance variation in a multiplicative mixer.

tics may be stated as a power series of terms as

$$i_d = i_o + av_t + bv_t^2 + \ldots \tag{5.24}$$

The squared-term component of current is

$$i_2 = bv_t^2$$
$$= b(v_o + v_s)^2$$
$$= b(v_o^2 + v_s^2 + 2v_o v_s)$$
$$= b(V_o^2 \sin^2 \omega_o t + V_s^2 \sin^2 \omega_s t + 2V_o V_s \sin \omega_o t \sin \omega_s t) \tag{5.25}$$

The product $\sin \omega_o t \sin \omega_s t$ can be expanded by the use of trigonometric identities to give

$$\sin \omega_o t \sin \omega_s t = \tfrac{1}{2} \cos(\omega_o - \omega_s)t - \tfrac{1}{2} \cos(\omega_o + \omega_s)t$$

so that the current contains a term at the intermediate frequency, IF, which is

$$i_{(IF)} = bV_oV_s \cos(\omega_o - \omega_s)t \qquad (5.26)$$

The magnitude of $i_{(IF)}$ is

$$|i_{(IF)}| = bV_oV_s = g_cV_s \qquad (5.27)$$

from which the conversion transconductance g_c given by Eq. (5.22) is seen to be

$$g_c = bV_o \qquad (5.28)$$

The magnitude of the local oscillator voltage is made large, so that a large transconductance can be obtained. The output is then proportional to the input in amplitude but has a carrier frequency of $(f_o - f_s)$. The sidebands are also reproduced around the new carrier frequency without either amplitude or phase distortion.

5.4.2 Multiplicative Mixing

Multiplicative mixing occurs when the transconductance of the mixer circuit is caused to vary with the local oscillator voltage, so that the output current becomes a function of the product of v_o and v_s. Figure 5.7(b) shows the block diagram of a multiplicative mixer. The analysis of this circuit follows. Again, the input voltages are

$$v_s = V_s \cos \omega_s t \qquad (5.29)$$
$$v_o = V_o \cos \omega_o t \qquad \text{where } V_o \gg V_s \qquad (5.30)$$

The cosine functions are used to agree with the Fourier series given below. The mixer output current is given by

$$i_m = g_m v_s \qquad (5.31)$$

but g_m is a function of the oscillator voltage v_o. A device is chosen whose transconductance is directly proportional to the applied bias voltage magnitude, up to a saturating maximum $g_{m_{max}}$ occuring at $V_{b_{max}}$, as shown in Fig. 5.7(c). The bias is made to vary around a fixed bias point with v_o, so that the positive peaks of v_o drive the device just to saturation and the negative peaks drive it a considerable way into cutoff (class AB). The transconductance function becomes a clipped sinusoid, as shown in Fig. 5.7(c), whose value may be stated in terms of a Fourier series as

$$g_m = (a_o + a_1 \cos \omega_o t + \dots) \qquad (5.32)$$

where the Fourier coefficient a_1 is of particular interest and is also a function of the clipping angle ϕ shown in Fig. 5.7(c). Ignoring all but the fundamental component of the g_m function, the mixer output current becomes

$$i_m = g_m v_s = a_1 \cos \omega_o t (V_s \cos \omega_s t) \qquad (5.33)$$

$$= \tfrac{1}{2} a_1 V_s [\cos(\omega_o - \omega_s)t + \cos(\omega_o + \omega_s)t] \qquad (5.34)$$

Also,

$$\therefore \qquad I_{IF} = g_c V_s \qquad (5.35)$$

where

$$g_c = \tfrac{1}{2} a_1 \qquad (5.36)$$

When the clipping angle is 0.68π rad (about $122°$), g_c is found to have a maximum value of

$$g_{c_{max}} = 0.27 g_{m_{max}} \qquad (5.37)$$

5.4.3 Mixer Circuits

Many circuit combinations are possible that will work as mixers. Diodes are sometimes used as the mixing nonlinear device, but more often a transistor is used because of the additional gain that may be realized from it.

Figure 5.8(a) shows a bipolar transistor used to mix the input signal with the signal from a separate local oscillator circuit. This circuit is basically an additive circuit, because the total base-emitter signal voltage is the sum of the signal voltage and the local oscillator output voltage. The base-emitter junction serves to produce the necessary squared components. The oscillator input is made much larger than the signal input, but is kept low enough so that clipping does not occur. This produces a large value of g_m while preventing the overproduction of harmonics. The output current from the transistor contain-

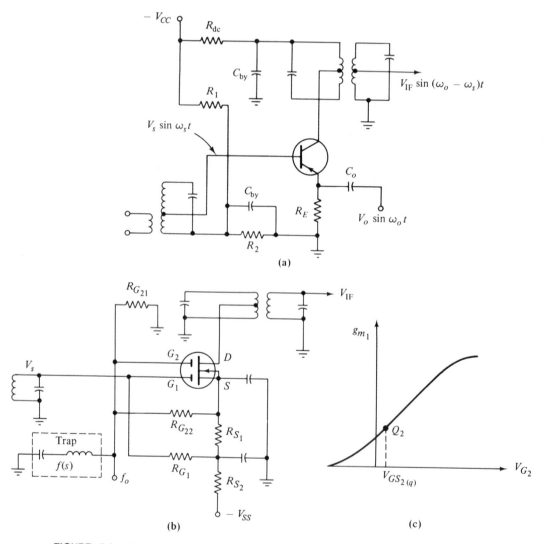

FIGURE 5.8. (a) Bipolar transistor additive mixer; (b) multiplicative mixer using a double-gate FET; (c) variation of signal gate transconductance as a function of the oscillator gate voltage.

ing the IF component drives the primary of a double-tuned IF transformer connected to the IF amplifier input stage. The signal may be coupled into the base tank circuit by mutual coupling from a previous RF stage or from an antenna, or the tank may be a ferrite rod antenna. The bias on the transistor is adjusted by the values of R_1 and R_2, so that the transistor is in conduction near the cutoff point and the maximum curvature of the base-emitter characteristic is near the operating point. The oscillator voltage drives the bias through this region of maximum curvature. Higher values of transconductance may be obtained by bypassing the emitter resistor R_E with a series resonant circuit tuned to the signal frequency, but the result will be to allow interaction between the signal tank and the local oscillator circuit. Impedance matching taps are shown on all tank circuits.

Figure 5.8(b) shows the circuit of a multiplicative mixer which uses a dual-gate field-effect transistor as the multiplying device. Figure 5.8(c) shows the effect that variation of the bias voltage on gate 2 has on the transconductance through gate 1. It is almost linear and can thus be approximated by a straight line. The input signal is fed to gate 1, which is biased into the active region just above cutoff. The oscillator output is applied to gate 2. The bias on gate 2 is such that, without the oscillator signal, a slightly positive bias is applied, keeping it up on the g_m curve near the axis. Sufficient oscillator signal is applied to drive the operating point just up to the saturation level on the positive peaks and somewhat into cutoff on the negative peaks, which should give nearly maximum conversion transconductance. A trap (series resonant circuit) that is tuned to the signal frequency is connected between G_2 and ground to shunt off any signal voltage that might be fed to this gate.

Again, the output current feeds the input of a double-tuned IF transformer. An example follows which illustrates the calculation of the conversion transconductance of such a circuit.

Example 5.4   A dual-gate FET has the following characteristics.

Gate 1: $\qquad\qquad g_{1_{max}} = 1.5 \text{ mS}$

$\qquad\qquad\qquad g_{1o} = 1.0 \text{ mS (at } V_2 = 0)$

$\qquad\qquad\qquad V_{po2} = -3.0 \text{ V}$

Gate 2: $\qquad\qquad g_{2_{max}} = 0.8 \text{ mS}$

$\qquad\qquad\qquad g_{2o} = 0.6 \text{ mS (at } V_1 = 0)$

$\qquad\qquad\qquad V_{po1} = -2.5 \text{ V}$

The circuit is arranged so that gate 2 is biased to cutoff and is driven just to saturation by the oscillator voltage. Gate 1 is biased to zero volts. It is assumed that the g–V curves are linear.

(a) Plot the g–V curves on the same axes.
(b) Find the straight-line equations for the curves.
(c) Find $V_{1\,max}$ and $V_{2\,max}$.
(d) Find the peak oscillator voltage.
(e) Find g_c.

Solution

(a) See Fig. 5.9 for the g–V curves for the FET.

(b)
$$\text{Slope of } g_1 \text{ curve} = \frac{g_{1o}}{-V_{po2}} = \frac{1.0}{3.0} = 0.333$$

$$\text{Slope of } g_2 \text{ curve} = \frac{g_{2o}}{-V_{pol}} = \frac{0.6}{2.5} = 0.24$$

The curves are straight lines with the equations

$$g_1 = 1.0 + 0.33 \, V_2$$

$$g_2 = 0.6 + 0.24 \, V_1$$

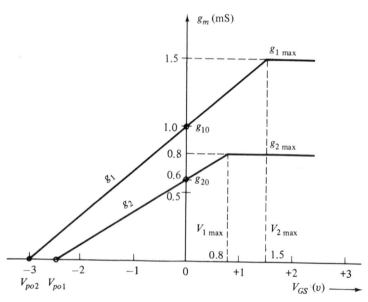

FIGURE 5.9. Transconductance versus voltage curves for a dual-gate field-effect transistor.

(c) The curves saturate at

$$V_{2_{max}} = \frac{g_{1_{max}} - 1.0}{0.333} = \frac{1.5 - 1.0}{0.333} = 1.50 \text{ V}$$

$$V_{1_{max}} = \frac{g_{2_{max}} - 0.6}{0.24} = \frac{0.8 - 0.6}{0.24} = 0.833 \text{ V}$$

(d) The peak oscillator voltage is limited to the linear section of the g_2 curve, or

$$V_{op} = V_{2_{max}} - V_{po2} = 1.5 - (-3.0) = 4.5 \text{ V}$$

(e) The mixer responds to the positive half-cycle of the oscillator voltage, so $a_1 = \frac{1}{2}g_{1_{max}}$ (the Fourier coefficient of the first harmonic of a half-wave rectified sine wave). Thus, from Eq. (5.36),

$$g_c = \tfrac{1}{4}g_{1_{max}}$$
$$= 0.375 \text{ mS}$$

5.5 IF AMPLIFIERS

The IF amplifiers used in receiver systems are special fixed-tuned units, usually with two or more cascaded stages. They incorporate a number of features: voltage gain to raise the signal level at the output of the converter to that needed by the detector; bandpass filtering action to allow only the wanted signal to pass to the detector; automatic gain control to prevent overloading of the detector; and, in the case of FM systems, signal amplitude limiting to prevent amplitude variations from interfering with detector operations.

The gain requirements of IF amplifiers are somewhat similar, regardless of the particular type of receiver. As a rule, sufficient gain is included in the RF and mixer stages ahead of the IF, so that the signal level presented to the IF input is of the order of 1 mV. Most detectors, AM, FM, or PM, require signal levels of about 1 V to perform adequately. This means that IF circuits must provide a signal voltage gain of about 1000, or 60 dB. Two or three cascaded stages usually provide the required gain. This may vary from as little as 40 dB for cheap AM broadcast receivers to as much as 80 dB for special communications receivers, but usually the 60-dB figure is adequate. This gain is the overall required, and any gain included to compensate for loss in filters and coupling circuits must be provided beyond this figure.

The filter bandpass characteristic of the IF amplifier depends on the type of signal that it is to handle. If ordinary AM broadcast signals are to be received, the ideal IF filter would have a bandpass of just less than 10 kHz. The usual AM IF frequency used is 455 kHz, so the filtering characteristic can be accomplished with tuned circuits with a Q of about 100, which is quite practical. When AM SSB signals are being received, initial filtering is still done with the 10-kHz bandwidth, but this is followed by a very sharp cutoff sideband pass filter which usually has a bandpass of about 3 kHz total. This filter is either mechanical, ceramic, or crystal lattice. Double conversion is often used in such systems, with a first IF at about 2 MHz with a 10-kHz bandwidth, followed by a second IF at either 455 kHz or perhaps a lower one between 100 and 300 kHz. The sideband filters are included in the second IF, and the total gain, including both IF strips and the second converter gain, is as before, that is, 50 to 80 dB. FM signals are usually received at VHF or higher frequencies, with a total bandwidth of about 200 kHz for broadcast and about 30 kHz for narrow-band voice communications. The IF in these cases is usually 10.7 MHz, although other frequencies have been used. Again, the Q required is about 100, which is realizable with tuned circuits. Television and microwave communications require bandwidths in the order of 6 to 12 MHz, and usually the IF for these is chosen to be in the range of 40 to 100 MHz. At these frequencies discrete-component tuned circuits are still usable, along with transistor amplifiers.

Stagger tuning is a technique often used to obtain the bandpass filter characteristic of the IF amplifier. Figure 5.10(a) shows the circuit layout for a stagger tuned IF amplifier that might be used for an AM receiver. It consists of two amplifier stages which provide enough gain to make up for the losses in the transformers and which supply the overall net gain requirement, and three interstage transformers. The amplifiers are basically two class A small-signal amplifiers with tuned input and tuned output. As such, either individual-stage neutralization or overall neutralization must be provided to maintain stability.

The three transformers are all undercritically coupled, so that no double humping occurs in their response. This means that each of the tuned circuits is effectively isolated from all the others and acts independently. Each has a Q which results in a much narrower bandwidth than is required overall. The first and last circuit are tuned to the center IF frequency, circuits 2 and 3 are tuned to about 2 kHz on either side of the center frequency, and circuits 4 and 5 are tuned to about 4 kHz on either side of center. The individual tuned circuit responses are sketched in Fig. 5.10(b). Because the tuned circuits are isolated and do not react with one another, their responses (in dB) add directly, to give an overall response similar to Fig. 5.10(c). The ripple across the top can usually be kept within 1 dB, and skirt steepness can cause the response to drop as much as 60 dB between 4 and 6 kHz off center. The form of this response is similar to a Chebyshev filter, and quite often such filters are used in place of tuned transformers.

It has already been noted that in the case of SSB systems it is necessary to include a special sideband filter with about a 3-kHz bandwidth and extremely steep skirts. This usually takes the form of a mechanical, crystal, or

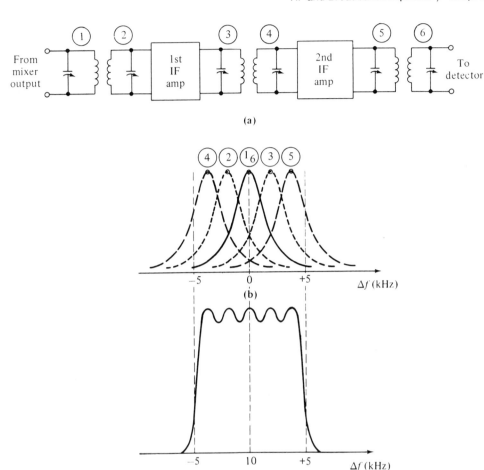

FIGURE 5.10. A stagger tuned IF amplifier: (a) Circuit; (b) frequency responses of each tank taken independently; (c) the overall IF circuit response.

ceramic lattice filter. With the increasing availability of operational amplifiers designed to operate at IF frequencies and to provide as much as 100 dB of gain in a single unit, the use of such filters as the only IF filtering is becoming commonplace. Circuit design becomes very much simpler if one can simply pick an amplifier with sufficient gain and a filter with the required bandpass off the shelf.

Automatic gain control (AGC) is usually included in the first stage of the IF amplifier if it is to be used for AM signals. This is especially necessary if there is no RF stage to prevent strong signals from overloading the detector circuit and causing distortion.

In the FM systems, amplitude variations in the signal level presented to the detector are undesirable, because they cause distortions in the output.

Amplitude limiting is provided in one stage of the IF amplifier of such receivers. These circuits provide maximum gain until the output signal level reaches a predetermined threshold. For any signal level above this, the output remains constant at the value of the threshold. One scheme for accomplishing this limiting function is to make one stage (usually the last one) of the IF amplifier a class B or class C amplifier with fixed bias, followed by a tuned circuit. As the input signal amplitude increases, the tube or transistor is driven farther and farther into conduction until saturation occurs. For signal levels higher than that necessary to saturate the amplifier, no further increase in the current amplitude can occur, and the output signal remains at constant amplitude. Another method, used with IC amplifiers, is to use a delayed AGC-type negative feedback system. The amplifier is left with maximum gain for signal levels lower than threshold, but lowered for higher levels. The response of this limiting AGC must be fast enough to correct for changes at frequencies up to the maximum modulation frequency.

Figure 5.11(a) shows the schematic for a typical AM transistor IF amplifier. Two PNP transistors provide the gain, both of them operating in the class A mode. Three stagger tuned transformers provide the coupling. The input and output of each transformer is tapped down to provide the necessary impedance matching because the input and output resistances of the amplifiers are too low to provide the required Q. Bias for the second stage is fixed, while that for the first stage is derived from an AGC signal from the detector. Neutralizing capacitors are provided on each stage to stabilize them. The tapped transformer coils make neutralization very easy. Decoupling of the power supply leads to each stage is necessary if stability is to be maintained. The transistors must have a beta cutoff frequency at least as high as the IF if their full gain is to be realized.

Figure 5.11(b) shows an FM IF strip made with integrated circuit amplifiers. The first IC is an MFC 4010, which is a limiting amplifier designed specifically for 10.7-MHz IF circuits. It provides up to 60 dB of gain and limiting at an output of about 200 mV. The second IC is an MC 1357 circuit which includes further limiting gain and a complete quadrature detector circuit. Selectivity is provided by the tuned input transformer and a ceramic IF filter. The only other tuned circuit required is the simple parallel tuned network in the detector to provide the necessary phase shift. All the other components are capacitive bypass or coupling, or resistive bias networks. These chips were specifically designed for use in inexpensive FM receivers, and their cost is very low. The whole strip can be assembled on a printed circuit card about 5 × 10 cm in size.

5.6 BROADBAND VIDEO AMPLIFIERS

Amplifiers used to amplify the video signals in television systems require flat amplitude response and linear phase response from 10 Hz to about 6 MHz. To allow such operation, several things must be accomplished in the design of

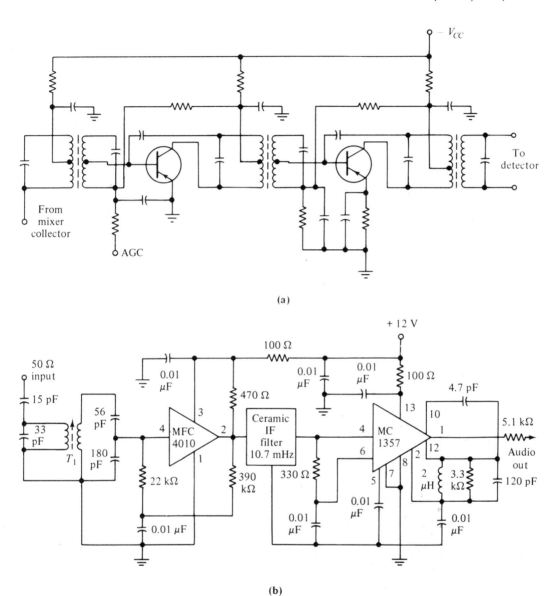

FIGURE 5.11. (a) A transistorized AM IF strip; (b) an integrated circuit FM IF amplifier including detector. (Extract from Motorola Application Note AN543). (Courtesy of Motorola, Inc.)

the amplifier. The first of these is to choose an amplifying device with a sufficient gain-bandwidth product so that the transistor limitation on frequency response falls above the required upper cutoff frequency. For instance, if a bipolar transistor were chosen, and if it were to be operated without feedback, its beta cutoff frequency would have to be above the upper cutoff frequency. If the midband gain were reduced, either by loading or applying negative feedback, the gain cutoff frequency would move upward and would be roughly given by the gain-bandwidth equation

$$(G)(BW) = A_v f = A_{v_{max}} f_u \tag{5.38}$$

where $A_{v_{max}}$ is the voltage gain for frequencies below the upper cutoff frequency f_u, and A_v is the voltage gain at some frequency f above f_u. RC coupling is used in such amplifiers, and the coupling circuits used also limit the frequency response. The size of the series coupling capacitors used between stages limits the lower cutoff frequency. There are three ways to handle this problem: use a larger capacitor, provide a low-frequency compensation network, or dc couple the amplifier.

Increasing the size of the capacitor is easier as far as low frequencies are concerned, but if the capacitor is increased too much, its physical size will make the shunt stray capacitance too large, spoiling the high-frequency response. It may be necessary to use the more complicated compensation networks shown in Fig. 5.12(b). Figure 5.12(a) shows the uncompensated amplifier and its approximate equivalent circuit. In this circuit, if the frequency is lower than that given by the RC time constant and the corner frequency is

$$\omega = \frac{1}{C_2(R_i + R_L)} \tag{5.39}$$

the gain drops off at the rate of -20 dB/decade of frequency. Increasing the value of C_2 lowers the value of the corner frequency by the inverse ratio. The responses of both amplitude and phase vs. frequency are sketched as the solid lines in Fig. 5.12(c).

The compensation network of Fig. 5.12(b) uses a split load resistor. An additional resistor R_2 shunted by the capacitor C_1 is added in series. At midband frequencies, C_1 shunts out R_2, and only the normal load resistance R_L has effect. For frequencies lower than the original corner frequency, the capacitance C_2 allows more and more of the resistor R_2 to have effect, causing the gain of the amplifier to rise, thereby canceling out the loss due to C_2. The overall gain function is given approximately by

$$A_v(\omega) \cong A_{v_{mid}} \frac{\left(1 - j\frac{1}{\omega C_1 R_L}\right)}{\left(1 - j\frac{1}{\omega C_1 R_2}\right)\left(1 - j\frac{1}{\omega \frac{C_1 C_2}{C_1 + C_2}(R_i + R_L)}\right)}$$

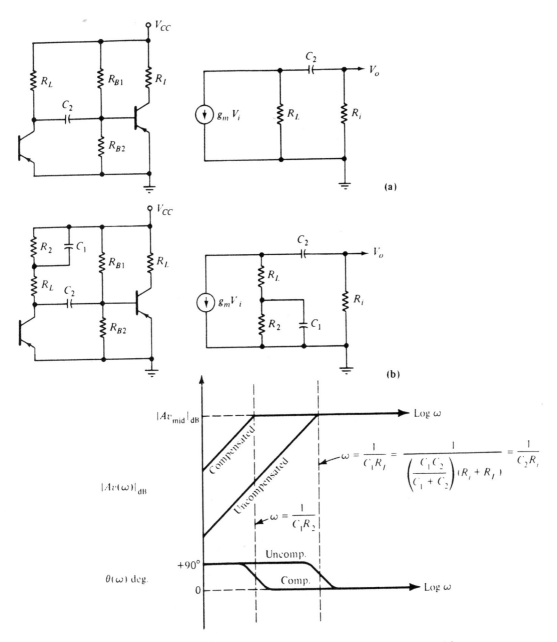

FIGURE 5.12. Low-frequency compensation of *RC*-coupled amplifiers: (a) uncompensated amplifier with its equivalent circuit; (b) compensated amplifier with its equivalent circuit; (c) Bode plot approximations for both the uncompensated and the compensated amplifiers.

where $R_2 \gg 1/\omega C_1$. By making the time constant $C_1 R_L$ equal to $C_2 R_i$, the gain function reduces to

$$A_v(\omega) = \frac{A_{v\,\text{mid}}}{\left(1 - j\dfrac{1}{\omega C_1 R_2}\right)} \tag{5.41}$$

and the lower cutoff frequency is now controlled by the time constant $C_1 R_2$ which is much lower than that of the uncompensated amplifier. The results of the compensation are illustrated in Fig. 5.12(c): both the amplitude- and phase-cutoff frequencies are lowered.

At the high-frequency end, cutoff is usually caused by the effects of shunt capacity between the signal circuit and ground. This is due to input capacity in the following amplifier, to output capacity in the amplifier in question, and to wire stray capacitance. The upper cutoff frequency of this amplifier can be raised by including a small inductor in series with the load resistor, as shown in Fig. 5.13(a). Assuming that the natural cutoff frequency of the amplifying device is much higher than that caused by the shunt capacity and that both the amplifier output resistance and the following-amplifier input resistance are much larger than the load resistance, the equivalent circuit of the uncompensated amplifier becomes that of Fig. 5.13(b), where C_t is the sum of the output capacitance C_o, the input capacitance C_i, and the stray capacitances C_s. The cutoff frequency is given by

$$\omega = \frac{1}{C_t R_L} \qquad \text{where } C_t = C_o + C_i + C_s \tag{5.42}$$

When the inductor is included with the load resistor, the equivalent circuit becomes a very heavily damped parallel-resonant circuit, as shown in Fig. 5.13(c). The resonant frequency of this circuit is given by

$$\omega = \frac{1}{\sqrt{LC_t}} \tag{5.43}$$

It can be proved that, for this circuit to provide "maximally flat" frequency response, we must have

$$L = 0.360 R_L^2 C_t \tag{5.44}$$

Since the values of C_t and R_L are already fixed, it is only necessary to calculate the required value of L from Eq. (5.44). Generally, this will give an improvement in the cutoff frequency of about two times, as illustrated by the following example.

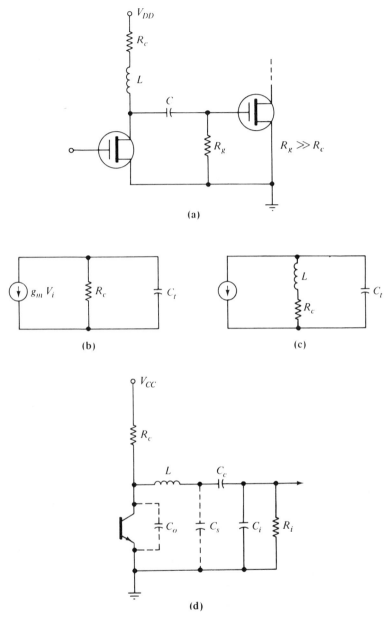

FIGURE 5.13. The shunt peaked compensated FET amplifier: (a) circuit showing compensating inductor; (b) equivalent circuit without compensation; (d) series-compensated amplifier.

Example 5.5   An *RC*-coupled amplifier has an output capacitance of 10 pF and a load resistance of 2 kΩ. It is coupled to another amplifier with an input capacitance of 25 pF, and the net stray-wiring capacitance adds another 5 pF. Both the output and input resistances are much larger than R_L and can be ignored. Find the uncompensated upper cutoff frequency, the value of inductance necessary to compensate for maximally flat response, and the new upper cutoff frequency.

Solution   Uncompensated,

$$C_t = C_o + C_s + C_i = 10 + 25 + 5 = 40 \text{ pF}$$

$$\omega = \frac{1}{C_t R_L} = \frac{1}{40 \text{ pF} \times 2 \text{ k}\Omega} = 12.5 \text{ Mrad/s}$$

$$f = \frac{\omega}{2\pi} = 2.0 \text{ MHz}$$

Compensated maximally flat,

$$L = 0.36 \times 2^2 \times 10^6 \times 40 \times 10^{-12} = 57.6 \ \mu\text{H}$$

$$f_{comp} = \frac{1}{2\pi\sqrt{LC_t}} = 3.32 \text{ MHz}$$

This is the frequency at which the peaking circuit resonates, and the actual 3-dB cutoff frequency will be slightly higher.

Figure 5.13(d) shows a different way of compensating such an amplifier. In this case, a small inductor is included in series with the coupling capacitor, and the process is known as *series compensation*. Although the calculations involved in such a compensation are quite complex, somewhat better results can be obtained than with the shunt-compensation technique. However, unless the application is critical, the extra effort involved is not worth it, and the shunt-peaking circuit is more often used. Extra care in choosing the amplifying devices and operating modes to minimize input and output capacitances is more profitable. Stray capacitances can be reduced by careful layout, keeping all signal lines as short as possible.

5.7   CLASS C AND LINEAR AMPLIFIERS

5.7.1   Class C Amplifier

A class C amplifier is any amplifier that conducts over less than one-half of its input drive cycle. With tuned class C amplifiers, the output is forced to be a sine wave because of the resonant action of the output tank circuit. If the output tank is tuned to the fundamental of the driving frequency, the amplifier current pulses the tank once every cycle. If it is tuned to the second harmonic, the tank is pulsed every other cycle. Since current is flowing in the amplifying device only during short current pulses, very high efficiencies can be realized. Theoretical maximum efficiencies for fundamental operation range to over 90%, and for this reason class C amplifiers are used as power output stages and modulators of transmitters.

Very high power transmitters in all frequency ranges still make use of vacuum power tubes operating in the class C mode, but for the frequencies up to 1 GHz at moderate power levels, transistors are becoming popular. A good analysis of a tube-type class C amplifier has been given by B. Zeines in *Electronic Communication Systems* (Prentice-Hall, Inc., Englewood Cliffs, N.J., 1970).

Figure 5.14(a) shows the circuit of a transistorized fixed-bias class C amplifier. The input signal is inductively coupled to the input tank circuit, which is tuned to the fundamental frequency. Fixed bias is provided through a decoupling network consisting of an RF choke and a bypass capacitor, through the input tank to the base, from a dc source. This dc bias keeps the base circuit biased well beyond cutoff, and no base current can be drawn until the input signal voltage exceeds the cutoff threshold voltage. Base current then flows only during the positive peaks of the input voltage and is limited in magnitude by the transistor input resistance.

Figure 5.14(b) shows the approximate base voltage and current waveform that result. Conduction begins when the input voltage v_s exceeds the threshold voltage, which is the sum of the bias voltage and the base-emitter cut-in voltage:

$$|V_{s\,\text{thresh}}| = |V_{BB}| + |V_{BE}| \qquad (5.45)$$

The *conduction angle* is the portion of the input angle during which conduction takes place (Fig. 5.14(b)). This given approximately by

$$\Theta_b = 2\cos^{-1}\left(\frac{|V_{s\,\text{thresh}}|}{V_{sp}}\right) \qquad (5.46)$$

The collector current will be approximately the same shape as the base current pulse *provided* that the transistor does not saturate. Maximum drive conditions are those which provide just enough base current so that the collector voltage is driven just to zero at its negative peak. Under these conditions, maximum power output will occur. If the collector is driven into saturation, harmonic distortion (clipping) in the load voltage will result, and signals will be transmitted at one or more of the higher harmonics, a situation that is usually undesirable.

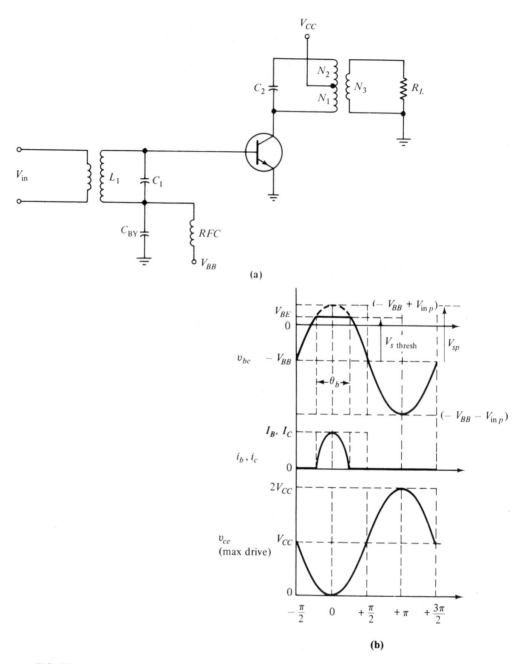

FIGURE 5.14. Class C amplifier: (a) a typical circuit using fixed bias; (b) waveforms for full-drive conditions.

It must be emphasized that the collector current waveform will be a clipped sinusoid only if the transistor is perfect. However, it is not, since the base-emitter junction does not behave as a perfect rectifier at high frequencies due to its nonlinear reactive component, and the current gain of the transistor varies with the current level. As a result, the actual current pulse in the collector will be narrower and sharper than the expected sinusoid and will contain more harmonics.

An analysis of the transistor gain functions will yield only grossly approximate results unless complicated nonlinear models are used, so this analysis is not attempted here.

A Fourier analysis of the ideal current pulse in the collector circuit, however, indicates how much of the current pulse contributes to the signal. The ideal current pulse is assumed to be a truncated cosine pulse (i.e., a cosine wave that has had the bottom sliced off it). The Fourier analysis of the ideal pulse shows that for a given conduction angle and peak pulse amplitude I_c, the total current is given by

$$i_c = I_{dc} + i_{fund} + i_{2nd} + i_{3rd} + i_{4th} + \cdots$$

$$= I_c n_{dc} + I_c n_1 \cos \omega t + I_c n_2 \cos 2\omega t + \cdots \qquad (5.47)$$

Figure 5.15 shows plots of n_{dc}, n_1, n_2, n_3, and n_4 versus the conduction angle Θ_b for $0° < \Theta_b < 180°$. These plots show that the fundamental and dc components rise steadily with increasing conduction angle, but that the second, third, and fourth harmonics peak between $60°$ and $120°$. Thus, this is the best range of operation for frequency multiplication operation. The ratio of the fundamental component to the dc component, which determines the collector efficiency, peaks in the same range. In practice, the bias or drive is adjusted to yield peak output for minimum dc current.

When the output tank circuit is tuned to the fundamental frequency, its parallel reactance becomes infinite, leaving only the dynamic load resistance R''_L in the collector circuit, at fundamental frequency. At higher frequencies, the tank circuit is capacitive and shorts most of the higher harmonic current to ground. The collector signal voltage is

$$v_{ce} = V_{CC} - i_c Z_{tank} \cong V_{CC} - n_1 I_c R''_L \sin \omega_o t \qquad (5.48)$$

where n_1 in this case is the value of the coefficient of the fundamental for the real collector current waveform under operating conditions.

From Eq. (5.48), the peak fundamental voltage is $n_1 I_c R''_L$. At full drive, the magnitude of the peak fundamental voltage is equal to the supply voltage V_{CC}.

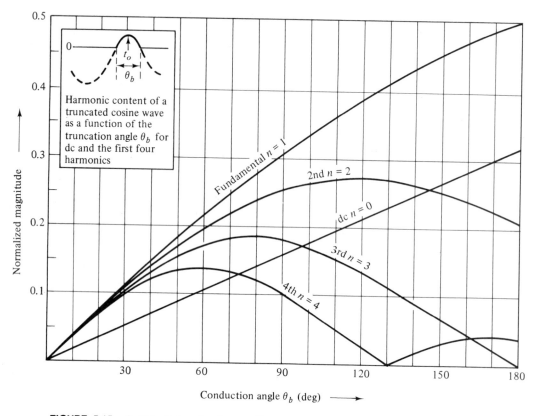

FIGURE 5.15. A plot of the Fourier coefficients of a class C current waveform as a function of the conduction angle θ_b.

The signal power output at maximum drive may now be calculated as

$$P_{o_{max}} = \frac{(I_c n_1)^2 R_L''}{2} = \frac{V_{CC}^2}{2R_L''} = \frac{V_{CC} I_c n_1}{2} \tag{5.49}$$

The dc power input to the collector circuit is

$$P_{dc} = V_{CC} I_{dc} = V_{CC} n_{dc} I_C \tag{5.50}$$

and the collector efficiency is calculated from these powers as

$$\eta_c = \frac{P_o}{P_{in_{dc}}} = \frac{n_1}{2n_o} \times 100\% \tag{5.51}$$

This shows that the circuit need only be adjusted to given maximum RF power output at a minimum dc input current to ensure best efficiency.

If the output tank circuit is tuned to a higher harmonic n_x, all the other harmonics are shorted, and wherever n_1 appears in the power equations, n_x is substituted. Lower efficiencies result, but the circuit will act as a frequency multiplier.

Power is also dissipated in the base circuit, and a certain amount of signal power must be applied to drive the amplifier to maximum output. Under maximum drive conditions the circuit power gain can be calculated as

$$A_p = \frac{P_o}{P_{in}} \qquad (5.52)$$

Drive signal conditions are usually monitored in operation by providing a dc milliammeter to measure the dc base current and a peak-reading RF voltmeter to measure the base signal voltage. A second dc ammeter which is used to indicate correct output tank tuning is provided in the collector circuit. When there is sufficient drive on the base circuit, and the output tank is tuned through resonance, the collector current will dip to a minimum value when the tank is tuned to the drive frequency. At other frequencies a higher value of current will flow because of the shunting effect of the tank reactances. The circuit is adjusted to give maximum power output (just at saturation) without producing undue harmonic output.

In the above analysis the base bias was provided by an external source. This is often inconvenient, and self-bias is used instead. The preferred form of self-bias is to place a parallel RC combination in the emitter lead of the transistor. The capacitor is then charged by the dc component of the collector current to provide a dc voltage for biasing the base which is proportional to the input signal level. As a result, the circuit will automatically adjust itself to small variations in input signal level. The time constant RC of the emitter circuit is made relatively much longer than the fundamental frequency period.

5.7.2 Push–Pull Class C Amplifier

The class C amplifier injects energy into the output tuned circuit only during the short time at the peak of the input cycle during which the amplifier goes into conduction. It relies on storing sufficient energy in the tuned circuit during this short period to maintain the output oscillation during the remainder of the cycle, similar to a flywheel or pendulum effect in mechanical systems. A large improvement in the efficiency of energy transfer and output waveform linearity can be obtained by operating two class C amplifiers in push–pull into the same tuned output circuit, with one amplifier supplying the positive peaks and the other supplying the negative peaks.

A push–pull class C amplifier is shown in Fig. 5.16. It uses two common base transistor amplifiers, and their operation in each case is similar to that discussed in the previous section.

The input circuit is a transformer whose secondary is tuned by a split-stator capacitor with the rotor grounded. (The net tuning capacitance is the series equivalent of the capacitances of the two sections.) The emitter of the first transistor is fed from the top of the input tank circuit, and the emitter of the second transistor is fed 180° out of phase with the first from the bottom of the input tank circuit. The bias circuit acts in conjunction with the emitter-base junctions to provide a bias voltage V_B which is proportional to the average of the driving signal amplitude, with the transistors being driven farther into cutoff as the signal increases. A small fixed bias is supplied from V_{EE} as well, which serves to hold the no-signal bias just below the cutoff level, which allows a maximum range of modulation without modulation clipping at minimum.

The collectors of the two transistors are connected to opposite ends of the output tank circuit, which is also tuned with a split-stator capacitor that has a grounded rotor. The collector supply current is introduced through a center tap on the tank transformer primary, by means of a radio frequency choke (RFC) or isolating inductor and a dc ammeter which monitors the power input level. The load is coupled through the secondary of the transformer, which provides the proper impedance matching.

With an unmodulated carrier signal applied to the input, some of the signal is rectified by the base-emitter junctions to develop a bias voltage V_B across the capacitor C_B. As shown in Fig. 5.16(b), under these conditions collector current only flows for a short period of time when the input signal is near its peak value, producing a train of short collector current pulses whose amplitude is adjusted to one-half of the allowed maximum. As the modulation signal increases to a positive maximum, the input carrier amplitude increases, driving the transistors harder into conduction and increasing the peak current value to maximum. At the same time, more bias voltage is generated across C_B, and the average emitter-base voltage rises. When the signal modulation decreases to a negative maximum, the input carrier decreases to minimum and the collector current peaks reduce to minimum. At the same time, less bias voltage is developed and V_B decreases toward V_{EE}.

Because of the push–pull arrangement, while transistor Q1 is conducting on a positive carrier peak, Q2 is cut off, and while transistor Q2 is conducting on a negative carrier peak, Q1 is cut off. This "pushes" the output voltage one way during the positive half-cycle and "pulls" it the other way during the negative half-cycle.

The common base mode of operation reduces parasitic coupling between input and output to a minimum, so that neutralization is not usually required to maintain stable operation. A metal shield is usually placed so that it separates the collector and emitter circuits and improves isolation of the input and output.

Push–pull amplifiers help to reduce harmonic distortion in the output waveform by cancellation. It can be shown that, provided that the circuit is exactly balanced, all the even-numbered harmonics are canceled out when the

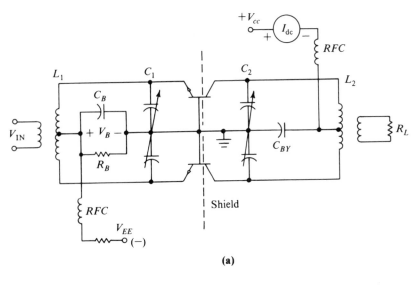

(a)

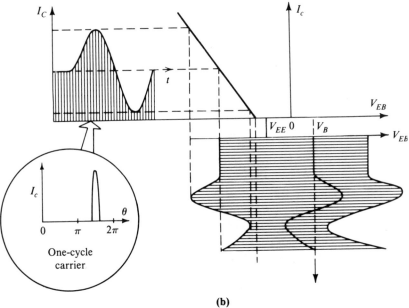

(b)

FIGURE 5.16. Push-pull class C power amplifier: (a) circuit; (b) transfer curve for one transistor showing relationship between collector current and emitter-base voltage.

two opposing signals are added in the output transformer primary. The second harmonic is usually the largest of these and the hardest to suppress, so that class C push–pull aids in the suppression of the second harmonic.

5.7.3 Linear Power Amplifier

The class C power amplifier provides the best amplifying efficiency, and its use in the output stages of transmitters is desirable from that point of view. However, its transfer function is not linear. That is, its voltage gain is not the same at all input signal amplitudes. This occurs because the bias is derived from the input signal and is made proportional to it. As a result, as signal increases, bias voltage increases and the gain drops. This is not serious if the waveform being amplified has a constant amplitude, as is the case for angle modulation or for interrupted carrier telegraph.

However, when an amplitude-modulated signal is passed through a class C amplifier, the envelope waveform of the modulation becomes distorted because of the changing bias at different signal levels. This distortion in the modulation is unacceptable, especially in single sideband systems, and as a result the class C amplifier cannot be used for this application and a linear amplifier must be used. The most commonly used linear amplifier is the class B amplifier. Figure 5.17(a) shows the circuit of a push–pull class B amplifier. The circuit uses a pair of common base transistors, and except for the way in which the bias is applied, it is identical to the class C amplifier in Fig. 5.16(a). Although bipolar transistors are used here, high-power VFETs are commonly used as well, and at very high power levels vacuum tetrode tubes are still used.

Bias for the transistors is provided from a fixed supply, and no base leak is provided. The input tank circuit is coupled directly to the emitters. The bias level is set so that with no input signal present, both transmitters are just cut off; or they may be operated with a very small amount of collector current to counteract crossover distortion in the same way as with the class B audio amplifier.

The transfer curve for one of the transistors is shown in Fig. 5.17(b). The bias voltage is fixed at V_{EE}, which is made equal to the cut-in voltage of the emitter-base junction of the transistor. Thus, with no carrier applied, both transistors will be just cut off. With carrier present, but with no modulation, the input signal amplitude is adjusted so that collector current half-cycle peaks reach one-half of the allowed maximum. When modulation reaches a positive maximum, the transistors are driven harder into conduction on the carrier peaks, and the collector current peaks rise to a maximum. At negative modulation, the input carrier is reduced to minimum and the output current peaks are reduced to minimum. In this case, however, with the fixed bias at cutoff, each collector current pulse lasts for a full half-cycle of the input, and each transistor carries what looks like a half-wave rectified signal current. The two transistor collector currents are antiphase, and when they add in the output transformer they result in an undistorted signal.

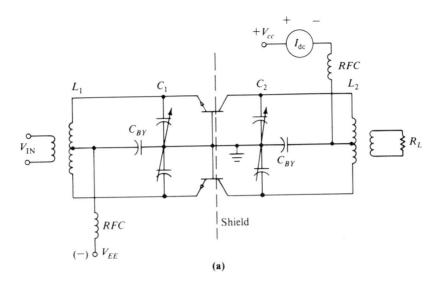

(a)

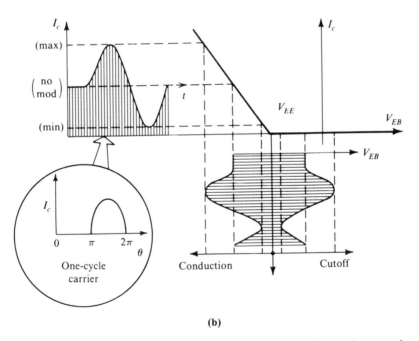

(b)

FIGURE 5.17. Class B linear push-pull amplifier: (a) circuit; (b) transfer curve for one transistor showing the relationship between collector current and emitter-base voltage.

Power-conversion efficiency in the class B amplifier is much lower than with the corresponding class C amplifier. The ideal class B push–pull collector circuit efficiency is 78%, while that for the class C amplifier can be as high as 90%. Linearity is gained at the expense of efficiency and the fact that larger transistors are needed for the same signal power output because of the greater heat dissipation. Parallel-connected transistors are often used in push–pull circuits in order to obtain higher output power levels.

One way around the problem of using class C for amplitude-modulated signals is to use a class C final amplifier and modulate it at high level, using collector modulation. With careful design and the use of negative feedback, it is possible to modulate the class C amplifier in a linear manner. It is done at the expense of very high power modulating amplifiers and large modulation transformers. At microwave frequencies, amplifying devices with sufficient gain for the linear mode of operation are difficult to obtain, and one reason why frequency modulation is preferred to amplitude modulation at microwaves is that class C amplifiers can be used.

5.8 TRANSMITTER AND AMPLIFIER MATCHING

5.8.1 Transmitter Matching Networks

Transmitter matching networks have two basic functions. First, they form the final tuned circuit in the transmitter chain and must provide sufficient Q so that the harmonics produced by the final amplifier are filtered out before reaching the antenna. Second, they must provide the impedance matching that is necessary between the final amplifier and the load if the desired power transfer is to be obtained. Inherent in the first function is that if the load has any reactive component at the point of connection, this reactive component must be canceled by the matching network *before* the impedance matching function can be performed. Usually the transmitter will feed the antenna through a transmission line, which requires a second matching network at the antenna end so that no standing waves appear on the line itself. The impedance of the terminated line at the feed end is purely resistive if this match is properly made.

Several types of LC circuits can be used for the impedance-matching function. These include the single-tuned transformer, the L-match and the Π-match. At microwave and VHF frequencies, stub matching and quarterwave transformer sections of line are used to accomplish the same functions. An analysis of these three circuits follows.

Transformer match. The circuit shown in Fig. 5.18(a) is seen to be similar to that of Fig. 5.1(a). Applying the analysis of Section 1.12.2 (and bearing in mind the approximations involved), the load impedance at resonance, as seen by the transmitter, is R'_D as given by Eq. 1.122, and when the transformer is matched for maximum power transfer, $R_s = R'_D$. The circuit Q is Q'_p as given by Eq. 1.123, but matching reduces this by a factor of two, so that the overall Q factor, including the effect of R_s, when matched, is

$$Q_{eff} = \tfrac{1}{2} Q'_p \tag{5.53}$$

L-match. This circuit is often used for transmitter output matching because of its simplicity, and either of the forms shown in Fig. 5.18(b) and (c) may be used. For either network, let the input branch be represented by $Z_1 = jX_1$, and the series branch containing the load R_L by $Z_2 = R_L + jX_2$ (i.e., the network elements are assumed lossless). Using the results of Section 1.7, the admittance of either network, as seen by the transmitter, is

$$Y = -j\frac{1}{X_1} + \frac{R_L}{R_L^2 + X_2^2} - j\frac{X_2}{R_L^2 + X_2^2} \tag{5.54}$$

At resonance, the j-coefficients sum to zero to give

$$-\frac{1}{X_1} = \frac{X_2}{R_L^2 + X_2^2} \tag{5.55}$$

or

$$R_L^2 + X_2^2 = -X_1 X_2 \tag{5.56}$$

The admittance at resonance then becomes

$$Y_{res} = \frac{R_L}{R_L^2 + X_2^2} \tag{5.57}$$

$$= -\frac{R_L}{X_1 X_2} \tag{5.58}$$

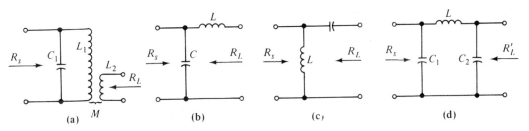

FIGURE 5.18. Transmitter matching networks: (a) transformer match; (b) L-match with series L; (c) L-match with series C; (d) Π-match.

Hence, the dynamic impedance (see Eq. (1.80)) is

$$R_D = -\frac{X_1 X_2}{R_L} \tag{5.59}$$

For either network, this reduces to

$$R_D = \frac{L}{CR_L} \tag{5.60}$$

The effective Q factor of the network is simply the Q of the branch containing R_L:

$$Q = \frac{X_2}{R_L} \tag{5.61}$$

When the circuit is matched, $R_s = R_D$, and this reduces Q by a factor of two, so that the overall Q factor, including the effect of R_s, is

$$Q_{\text{eff}} = \tfrac{1}{2}Q \tag{5.62}$$

The resonance condition, Eq. (5.56), can be expressed as

$$R_L^2 + X_2^2 = \frac{L}{C}$$

Therefore,

$$X_2^2 = \frac{L}{C} - R_L^2$$

For $X_2 = \omega L$, the resonant frequency may be evaluated as

$$\omega_o^2 = \frac{1}{LC} - \frac{R_L^2}{L^2} \tag{5.63}$$

which is the condition expressed by Eq. (1.75). Equation (5.63) can be rearranged as with Eq. (5.60),

$$\omega_o^2 = \frac{1}{LC}\left(1 - \frac{CR_L^2}{L}\right)$$

$$= \frac{1}{LC}\left[1 - \left(\frac{R_L}{R_s}\right)^2\right] \tag{5.64}$$

and when the impedance is matched, R_s can be substituted for R_D. It also follows from Eq. (5.64) that the equivalent inductance is

$$L_{\text{eff}} = \frac{L}{1 - \left(\dfrac{R_L}{R_s} \right)^2} \tag{5.65}$$

For $X_2 = -1/\omega C$, a similar analysis yields

$$\omega_o^{-2} = LC \left[1 - \left(\frac{R_L}{R_s} \right)^2 \right] \tag{5.66}$$

and

$$C_{\text{eff}} = C \left[1 - \left(\frac{R_L}{R_s} \right)^2 \right] \tag{5.67}$$

In practice, both X_2 and X_1 are made adjustable. X_2 is adjusted to give the necessary value of Q to provide the proper impedance ratio, and then X_1 is adjusted to tune out the reactance. The tuning of the transmitter proceeds by observing (for a class C amplifier) the plate (or collector) current meter. X_2 is progressively adjusted to increase the value of plate current, and after each change, X_1 is adjusted to the point of minimum plate current, or the bottom of the "dip," indicating resonance. Adjustment continues until the plate (collector) current at the bottom of the resonant dip is a maximum, indicating that a power match has been obtained and the circuit is in resonance.

Π-match. The circuit for a π-match is shown in Fig. 5.18(d). The parallel combination of R_L' and C_2 can be converted, by the method of Section 1.7, to an equivalent series impedance

$$Z_L = \frac{R_L'}{1 + Q_L^2} - j \frac{R_L' Q_L}{1 + Q_L^2} \tag{5.68}$$

$$= R_L - jR_L Q_L \tag{5.69}$$

where

$$Q_L = \omega_o C_2 R_L' \tag{5.70}$$

The circuit can now be reduced to that of Fig. 5.18(b), and the previous analysis can be applied, but with

$$R_L = \frac{R_L'}{1 + Q_L^2} \qquad X_2 = \omega L - R_L Q_L$$

Defining $Q_1 = (R_s/X_1)$, then since $R_s = R_D$ when matched, Eq. (5.59) now yields $X_2 = -Q_1 R_L$. Substituting this in the resonant condition (Eq. (5.56)) gives, after some manipulation,

$$\frac{R'_L}{R_s} = \frac{1 + Q_L^2}{1 + Q_1^2} \tag{5.73}$$

The effective dynamic impedance of the circuit when matched is $R_s \| R_D = \frac{1}{2} R_s$. The effective circuit Q factor when the impedance is matched is

$$Q_{\text{eff}} = \frac{1}{2} Q_1 = \frac{R_s \omega_o C_1}{2} \tag{5.74}$$

The resonant frequency will be determined to a good approximation by L, C_1, C'_2 in series, where $C'_2 = C_2/(1 + Q_L^{-2})$.

Circuit adjustment is accomplished by changing C_2 to adjust the impedance ratio and changing C_1 to maintain resonance, with basically the same "dipping" method already outlined. An example follows to illustrate the calculations.

Example 5.6 A Π-match is to be used to match a transmitter with a 4-kΩ output impedance into a 75-Ω coaxial cable to a matched antenna. The frequency of operation is 12 MHz and the desired effective Q is 20. Find the values of C_1, C_2, and L when the transmitter is tuned and matched for maximum power transfer.

Solution Find C_1 from Eq. (5.74):

$$-X_{C1} = \frac{R_s}{2Q_{\text{eff}}} = \frac{4000}{2 \times 20} = 100 \ \Omega$$

$$\omega_o = 2\pi f_o = 2\pi \times 12 \ \text{MHz} = 75.4 \ \text{Mrad/s}$$

$$C_1 = -\frac{1}{\omega_o X_{C_1}} = \frac{-1}{-100 \times 75.4 \ \text{Mrad}} = 133 \ \text{pF}$$

To find the value of C_2, begin with

$$Q_1 = 2Q_{\text{eff}} = 40$$

From Eq. (5.73),

$$Q_L^2 = \left[\frac{R'_L}{R_s}(1 + Q_1^2) \right] - 1 = \left[\frac{75}{4000}(1 + 40^2) \right] - 1 = (5.39)^2$$

From Eq. (5.70),

$$\omega_o C_2 = \frac{Q_L}{R_L} = \frac{5.39}{75} = 0.0072 \text{ S}$$

$$C_2 = \frac{0.072}{75.4 \times 10^6} = 955 \text{ pF}$$

Next, find the equivalent series capacitances:

$$C_1' \cong C_1 = 133 \text{ pF}$$

$$C_2' = C \frac{Q_L^2}{Q_L^2 + 1} = 955 \times \frac{5.39^2}{5.39^2 + 1} = 923 \text{ pF}$$

Finally, find the inductance:

$$C_{\text{eff}} = C_1' \text{ ser. } C_2' = \frac{133 \times 923}{133 + 923} = 116 \text{ pF}$$

$$L = \frac{1}{\omega^2 C_{\text{eff}}} = \frac{1}{(75.4 \text{ Mrad})^2 \times 116 \text{ pF}} = 1.52 \text{ } \mu\text{H}$$

5.8.2 Line Matching of Amplifiers

At UHF and higher frequencies it is impossible to build discrete components of the small size required to form the impedance matching and tuning networks needed for amplifiers. Resonant cavities and shorted sections of transmission line are used to form these networks at higher frequencies, above about 300 MHz, and occasionally at lower VHF frequencies as well. Wavelengths for this band of frequencies range from 10 cm to 1 cm, so that shorted stubs can be made easily to act as inductors, capacitors, and tuned circuits, and short sections of line can be used as transformers for impedance matching.

At VHF frequencies, sections of coaxial line and occasionally resonant cavities are used for the purpose. For the higher frequencies, waveguide sections are used instead. Microstrip and stripline printed-circuit versions of wave guides are popular for low-power UHF circuits, where all the components necessary for an amplifier except the transistor can be realized in printed-circuit form. Figure 5.19 illustrates the method. Suppose the output admittance at the collector is $G_o + jB_o$ (assumed capacitive for the purposes of illustration). Short-circuited stub 1 is inductive in accordance with Section 12.11(i), and parallel-resonates with jB_o to leave G_o. Length l of the transforming section is selected in accordance with the general impedance equation (Eq. (12.82), the reciprocal relationship actually being used for admittance) to

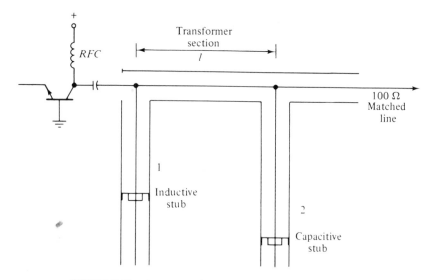

FIGURE 5.19. A common base amplifier with line matching.

transfer G_o to an admittance $(1/100) + jB$ (assumed inductive for illustration). Stub 2 is also selected in accordance with Section 12.11(i) to parallel-resonate with $+jB$, leaving an output admittance of $1/100$, or a resistance of 100 Ω. In practice, the calculations would most likely be carried out with the aid of a Smith chart. Note that stub 1 will be less than $\frac{1}{4}\lambda$ long, whereas stub 2 will be greater than $\frac{1}{4}\lambda$ long, since both are short-circuited stubs. Stub 2 could have been made an open-circuited stub with a length less than $\frac{1}{4}\lambda$, but generally open-circuited stubs should be avoided (see Section 12.11(ii)).

5.9 PROBLEMS

1. The circuit of Fig. 5.1(a) uses a transistor with an output resistance of 2000 Ω and an output capacitance of 10 pF. It is to amplify signals of 25 MHz. The transformer has a primary inductance of 1 μH and a series resistance of 1 Ω. If the secondary is loosely coupled so that the secondary load has negligible effect on the primary Q, find (a) the size of capacitor needed to tune the amplifier, (b) the effective Q, and (c) the bandwidth of the circuit.

2. The circuit of Fig. 5.2(a) uses a transformer with $L_s = L_p = 500$ μH and $k = 0.01$. The coil resistances are small enough to be ignored. If the amplifier has an $r_o = 100$ kΩ and a $g_m = 20$ mS and the circuit is to have a bandwidth of 10 kHz at an operating frequency of 250 kHz, find (a) the secondary tuning capacitance; (b) the secondary parallel load resistance if no appreciable coupling from the primary exists, and (c) the voltage gain at resonant frequency.

3. An amplifier has an internal feedback capacitance of 10 pF, which has to be neutralized. It operates at 5 MHz, and the output transformer is connected with a tapped primary at $N_a : N_b = 1 : 3$. Its primary inductance is 5 μH. If Hazeltine neutralization is to be used, calculate the size of neutralizing capacitor needed.

4. If the input transformer is tapped with the same ratio as in Problem 4, calculate the size of neutralizing capacitor for Rice neutralization of the same amplifier.

5. An amplifier has an input capacity of 50 pF, an output capacity of 5 pF, and a feedback capacity of 10 pF. Calculate the size of inductor needed to common-circuit-neutralize this amplifier at 50 MHz.

6. Calculate the conversion transconductance of a multiplicative mixer circuit for which g_{max} is 2 mS at $V_{B_{max}} = 2$ V. The bias level is adjusted so that the Fourier coefficient a_1 is 0.25 and the local oscillator generates a peak-to-peak voltage of 1 V.

7. An additive mixer uses a diode whose characteristic can be approximated by the equation $I_d = 0.002V_d + 0.0005V_d^2$. Find the conversion transconductance if the oscillator voltage produced is 0.5 V peak to peak.

8. Explain the principle of stagger tuning of transformer-coupled IF amplifiers.

9. Why is stagger tuning preferable to in-line tuning?

10. Calculate the lengths of tuning stubs and transformer sections needed if an amplifier with an output resistance of 1500 Ω is to be matched to a 600-Ω line at 50 MHz. The line has a velocity factor $(1/\sqrt{\varepsilon_r})$ of 0.8, and the same type of line is to be used for the sections.

11. Why is the common base amplifier preferred for many RF amplifier applications?

12. The uncompensated amplifier of Fig. 5.1(a) has a first-stage load resistance of 5 kΩ, a coupling capacitance of 1.0 μF and an input resistance to the second stage of 10 kΩ. (a) Find the uncompensated lower cutoff frequency. (b) Find the values of C_1 and R_2 needed to extend the lower cutoff frequency to 1 Hz.

13. The FET amplifier of Fig. 5.13(a) has $R_c = 5$ kΩ, $C = 10$ μF, $C_o = 2$ pF, $C_i = 15$ pF, $C_s = 1$ pF. (a) Find the uncompensated upper cutoff frequency. (b) Find the value of the compensating inductor that gives maximally flat compensation. (c) Find the compensated value of the upper cutoff frequency.

14. A transmitter output amplifier is matched to the load by a tuned transformer. The output resistance of the amplifier is 10 kΩ, the load resistance is 600 Ω, and the frequency is 1 MHz. (a) What turns ratio is required for the transformer, assuming a coupling coefficient of unity. (b) If the effective Q is to be 20, determine the values of L_1 and C_1.

15. A 10-MHz transmitter with an output resistance of 5 Ω is to be matched to a 50-Ω transmission line using an L-match network. (a) Which circuit configuration (Fig. 5.18(b) or (c)) should be used? Why? (b) Find the values of L and C required. (c) What is the effective Q of the network?

16. The transmitter of Problem 15 is to be matched using a Π-network with an effective Q of 25. Find the values of C_1, C_2, and L required.

17. The amplifier in Fig. 5.2(c) uses a transistor with $g_m = 20$ mS, $r_o = 200$ kΩ, $C_o = 5$ pF. The transformer $L_1 = L_2 = 150$ μH, $k = 0.02$, $R_p = R_s = 5$ Ω, and both coils are tapped at $N/4$. The load $R_L = 20$ kΩ. Find (a) values of C_1, C_2 to tune for 455 kHz, (b) the gain at resonance. (c) Is the circuit under- or overcoupled? (d) What size resistors must be placed in parallel with L_1 and L_2 to balance and critically couple the circuit ($kQ = 1$; see Section 1.12.2). (e) What is the 3 dB bandwidth? (Note: This is a more difficult problem.)

18. Two bipolar transistors with the characteristics given in Example 5.3 are connected in a Cascode amplifier circuit and biased for 0.26 mA of collector current. The circuit drives an 18-kΩ resistive load and the source resistance is 100 Ω. Find (a) the mid-frequency voltage gain, and (b) the expected upper cutoff frequency for the amplifier.

19. List the tuning procedure for a class C power amplifier.

20. Explain why push–pull operation is preferred for RF power amplifiers.

CHAPTER 6

Oscillators

6.1 INTRODUCTION

Electronic communications systems could not operate without sources of sinusoidal electrical waves. Many types of oscillator circuits are used to produce these sinusoids, and a few of the more commonly used ones are presented on the following pages.

Feedback oscillators, *RC* and tuned *LC* types, are discussed first. These circuits are applicable to frequencies from the audio range up to the VHF range, and may use any convenient three-terminal amplifying device. Discussion here is limited to the bipolar transistor and the field-effect transistor. The crystal-controlled oscillator is included as a tuned *LC* type, and a discussion of the factors affecting the frequency stability of oscillators is included.

The principles of negative-resistance oscillators using two-terminal negative resistance devices are presented.

Although it is not an oscillator in its own right, the frequency synthesizer is included here because it is revolutionizing the frequency-control scene in communications, making possible complex systems which until recently have just not been economically feasible.

6.2 POSITIVE-FEEDBACK OSCILLATOR

Positive feedback forms the basis for many of the more commonly used oscillator circuits. Figure 6.1 shows the general form of the feedback amplifier circuit. Under certain conditions, positive feedback can be created, and oscillations will build up in the circuit.

Barkhausen stated the conditions necessary for a feedback circuit to oscillate. Simply stated, the *Barkhausen criterion* says that, if a feedback circuit is to *maintain* oscillation, (1) the net gain around the feedback loop must be no less than 1, and (2) the net phase shift around the loop must be a positive integer multiple of 2π radians or $360°$ with both conditions existing together only at the frequency of oscillation.

$$\text{Loop Gain} = |GH| \underline{/\Theta} \tag{6.1}$$

where $n = 0, 1, 2, \ldots, \Theta = n360°$, and $|GH| \geq 1$.

Most electronic amplifiers are of the sort which create a phase inversion or a phase shift of $180°$ and some value of gain greater than unity. In oscillator circuits, an amplifier of this sort is included in the forward part of the loop (G in Fig. 6.1), and a passive phase-shifting network of some type is used in the feedback part of the loop (H in Fig. 6.1). From Fig. 6.1, the total loop gain can be stated as the product of the two complex numbers G and H. The overall gain with feedback is expressed as

$$A_v = \frac{G}{1 - GH} \tag{6.2}$$

where

A_v = overall circuit complex gain (magnitude and phase angle)

G = forward-portion complex gain

H = feedback-portion complex gain

GH = loop complex gain

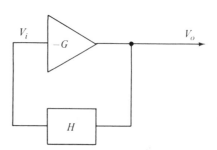

FIGURE 6.1. The oscillator as a positive-feedback amplifier.

In many oscillators, G and H each have a phase shift of $180°$ or are negative. If the magnitude $|GH|$ of the loop gain is unity, the overall gain with feedback is ∞, indicating that oscillations will take place. If the magnitude $|GH|$ is greater than unity, oscillations will grow. In practical oscillators, these oscillations continue to grow until amplifier nonlinearities such as saturation limit $|GH|$ to unity. After that, the magnitude adjusts itself automatically to maintain unity gain, or a "stable" condition of oscillations.

The frequency of oscillations is determined by components in H which create the additional $180°$ phase shift. These components are chosen carefully so that $180°$ phase shift occurs at only one frequency, and the circuit oscillates at that frequency.

6.3 RC PHASE-SHIFT OSCILLATOR

The RC phase-shift oscillator is a direct application of the theory of feedback as expressed by the Barkhausen criterion. Figure 6.2 shows the arrangement of the circuit. Virtually any RC network that will provide the required phase shift can be used. The one shown uses three RC lead sections in cascade, each with the same time constant (i.e., equal R's and equal C's). At the frequency of oscillation, each section provides about one-third of the total required phase shift of $180°$. The amplifier provides a further $180°$ phase shift and the necessary gain to compensate for the losses of the shifting network.

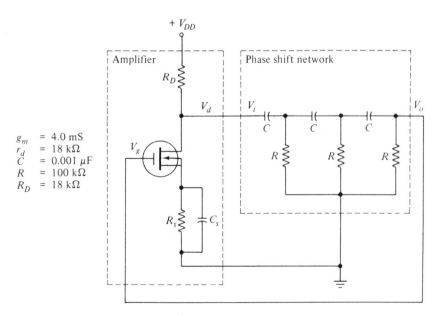

$$g_m = 4.0 \text{ mS}$$
$$r_d = 18 \text{ k}\Omega$$
$$C = 0.001 \text{ }\mu\text{F}$$
$$R = 100 \text{ k}\Omega$$
$$R_D = 18 \text{ k}\Omega$$

FIGURE 6.2. The RC phase-shift oscillator.

The analysis of the phase-shift network shows that at the frequency for which the phase shift of the network is $180°$, the losses of the network reduce the output voltage V_o to $1/29$ of the input voltage V_i, and the frequency at which this occurs is given by

$$f_{180} = \frac{1}{2\pi\sqrt{6}\,RC} \quad \text{and} \quad \left|\frac{V_o}{V_i}\right|_{180} = \frac{1}{29} \tag{6.3}$$

The amplifier must provide a minimum gain of 29 in order to maintain oscillations. In practice, a gain of about 10 dB larger than this would be provided to guarantee that oscillations will start.

Since the output of the phase-shift network is connected to the input of the amplifier, the input resistance of the amplifier will load the phase-shift network sufficiently to change is characteristics. It is therefore desirable for the amplifier to have an input resistance much higher than the value of R used in the phase-shift network, or some isolating network must be provided so that the network is undisturbed by the connection of the amplifier.

The amplitude of oscillations is controlled by the fact that as amplitude increases, the amplifier drives into a nonlinear portion of its characteristics, and its effective gain is reduced to the value 29. The output can be maintained more nearly sinusoidal if the gain of the amplifier is maintained nearer to the value 29, so that the peaks do not drive so far into the nonlinear region. An amplifier with automatic gain control can be used to provide an almost pure sinusoidal output.

The phase-shift oscillator is most often used to generate low-frequency sinusoidal signals for testing purposes. The variation of the values of C or R or both allow adjustment of the frequency of oscillation over very wide ranges, from a few hertz to several megahertz. Ganged capacitors are used to allow for frequency adjustment.

Example 6.1   The oscillator of Fig. 6.2 uses a FET which is biased to give a transconductance $g_m = +4.0$ mS with a dynamic drain resistance $r_d = 18$ kΩ. Its interelectrode capacities can be ignored, as can the loading of the amplifier output by the shifting network. Find

(a)   The frequency of oscillation.
(b)   Whether or not the circuit will oscillate.

Solution   (a)   The frequency of oscillation is

$$f_o = \frac{1}{2\pi\sqrt{6}\,RC} = \frac{1}{6.28 \times 2.45 \times 0.001\ \mu\text{F} \times 0.1\ \text{M}\Omega} = 650\ \text{Hz}$$

(b)   The complex input impedance of the phase-shift network works out to about $83 - j270$ kΩ; in parallel with the 18-kΩ drain resistor, its loading effect is negligible. (It causes about a 3° error in the phase-shift calculation.) Thus the load of the amplifier may be taken as 18 kΩ. The voltage gain is calculated as

$$A_v = -g_m(r_d \| R_D) = -4.0 \times (18 \| 18) = -36$$

This gain is sufficient to maintain oscillations.

6.4 TUNED LC OSCILLATOR

The tuned LC oscillator is another example of the feedback type of oscillator. This equivalent circuit is valid only in determining the startup gain of the circuit. The circuit is shown in Fig. 6.3(a), with its equivalent circuit in Fig. 6.3(b). The gain in this case is provided by a common emitter amplifier, although any amplifier that provides sufficient inverting gain will suffice. The load of the amplifier is comprised of the parallel tuned LC tank circuit, in parallel with the amplifier output resistance, and the amplifier input resistance reflected back through the transformer. The transformer provides a convenient coupling medium to the input and serves primarily to provide the 180° phase shift necessary for positive feedback. The coupling capacitor, base resistor, and the base-emitter junction provide a clamping circuit which biases the transistor toward cutoff as the signal level rises. This causes a reduction in the amplifier gain, until the loop gain falls to unity, just sufficient to maintain oscillations.

At the parallel resonant frequency of the tank circuit made up of the transformer primary inductance and the capacitor, the net susceptance of the tank is zero, and the tank circuit looks like a pure resistance. Under these conditions, the phase shift in the amplifier is 180° and the transformer provides a further 180° shift, yielding a net shift of 0° (or 360°) for the loop. If the frequency should drop below the resonant value, the tank looks like an inductance and causes the voltage gain to lead the 180° value by a small amount. If the frequency is high, the tank is capacitive, and the output voltage lags. Oscillation can take place only at the frequency for which the loop phase shift is zero, that is, at the parallel resonant frequency of the tank circuit.

The amplifier must only make up for the circuit losses and the turns ratio of the transformer. Since the transformer can provide a turns ratio greater than unity, it is possible to make an oscillator of this type using an amplifier with a voltage gain less than unity!

Resistive loading of the tank circuit does two things. It causes a slight shift of the resonant frequency, and it causes the frequency determination to be less stable. Thus it is desirable to use a tank circuit with a high Q, especially at the higher frequencies. The resonant frequency of the oscillator, along with the

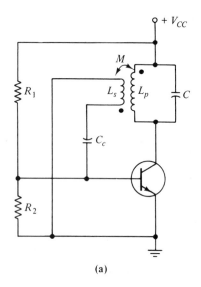

(a)

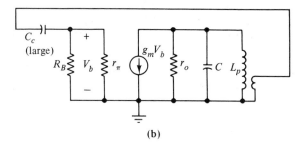

(b)

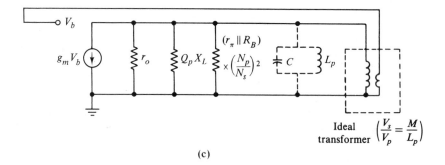

(c)

FIGURE 6.3. The tuned LC oscillator: (a) circuit; (b) ac equivalent circuit; (c) simplified ac equivalent circuit.

gain required of the amplifier, is given by

$$f_o = \frac{1}{2\pi\sqrt{LC}} \qquad |A_v|_{(amp)} \geq \frac{L_p}{M} \qquad (6.4)$$

assuming amplifier output resistance has negligible effect. The transformer voltage transfer ratio V_s/V_p is M/L_p, as found from Eqs. (1.84) and (1.112).

The base resistors provide the initial bias conditions to allow oscillations to start. The transistor is biased to the middle of its active region, where it provides the greatest gain. After the oscillations start to build up, each successive positive peak drives the base into conduction and charges up the coupling capacitor more negatively. This puts a bias on the base of the transistor during the rest of each cycle, which cuts the transistor off. The mode of operation of the transistor after this is as a class C amplifier. The magnitude of the oscillations will continue to build up until the circuit nonlinearities or self-bias reduces the loop gain to unity.

Output can be obtained from the circuit either by capacitive coupling from the collector or by another secondary winding on the transformer. Usually, a buffer amplifier with a constant high value of input impedance will follow to prevent load changes from affecting the oscillator stability.

Example 6.2

The circuit of Fig. 6.3 uses a transistor with $\beta = 100$, $r\pi = 5$ kΩ, $gm = 20$ mS, $ro = 50$ kΩ. The transformer has a mutual inductance of 25 μH, a primary inductance of 100 μH, and a Q of 100 at the resonant frequency. The tuning capacitor is 100 pF. Secondary circuit loading is negligible. Find

(a) The frequency of oscillation.
(b) Whether the circuit will oscillate or not.

Solution

(a) The frequency of oscillation is

$$f_o = \frac{1}{2\pi\sqrt{LC}} = \frac{1}{6.28 \times \sqrt{200 \ \mu\text{H} \times 100 \ \text{pF}}} = 1.59 \ \text{MHz}$$

(b) If parallel Q_p is defined as $Q_p = R_p/X_L$, and $R_p = R_D$, then the amplifier load resistance is

$$R'_L = ro\|R_D = ro\|Q_p X_L$$

$$= 50 \ \text{k}\Omega\|(100 \times 2\pi \times 1.59 \ \text{MHz} \times 100 \ \mu\text{H})$$

$$= 33.3 \ \text{k}\Omega$$

The amplifier gain is

$$A_v = -gmR'_L$$
$$= -20 \times 33.3$$
$$= -676$$

Hence

$$\frac{L_p}{M} = \frac{100}{25} = 4 \lessdot |A_v|$$

This is more than sufficient to guarantee oscillations.

6.5 TUNED LC OSCILLATORS

6.5.1 General LC Oscillator Form

The general form of a tuned circuit oscillator is shown in Fig. 6.4. An inverting amplifier provides the necessary gain, and a delta network of three complex impedances forms the phase-shifting network between the input and output of the amplifier. The circuit oscillates at the frequency for which the loop phase shift is $0°$, and the loop gain is equal to or greater than unity. The amplitude of oscillations is limited by the amplifier being driven into its nonlinear region, where its gain is reduced.

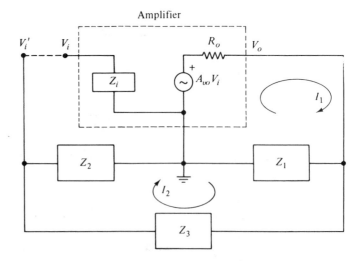

FIGURE 6.4. The general oscillator circuit. Z_1, Z_2, and Z_3 form a series resonant circuit.

The circuit can be analyzed by breaking the loop at the amplifier input and calculating the loop gain from V_i to V_i'. If the amplifier input impedance is large enough to be ignored, two loop equations can be written as follows:

$$A_{vo}V_i = +I_1(R_o + Z_1) - I_2(Z_1)$$
$$0 = -I_1(Z_1) \qquad + I_2(Z_1 + Z_2 + Z_3)$$

Also,

$$I_2 = \frac{V_i'}{Z_2}$$

and

$$Z_i \cong \infty$$

Solving these equations gives

$$A_{v(\text{loop})} = \frac{V_i'}{V_i}$$

$$= -A_{vo}\frac{Z_1Z_2}{Z_1^2 - (Z_1 + Z_2 + Z_3)(R_o + Z_2)} \qquad (6.5)$$

But at resonance $(Z_1 + Z_2 + Z_3)$ forms a high-Q series resonant circuit, the loop resistances of which are small enough to be ignored. Also, the sum of the three reactances is zero, so that

$$Z_1 + Z_2 + Z_3 = X_1 + X_2 + X_3 = 0 \qquad (6.6)$$

and the loop gain reduces to

$$A_{v(\text{loop})} = -A_{vo}\frac{X_2}{X_1} \geq 1 \qquad (6.7)$$

A number of commonly used resonant circuit oscillators can be analyzed using this procedure. Some of these are summarized in Table 6.1 according to

TABLE 6.1. Resonant-Circuit Feedback Oscillators

Circuit	Z_1	Z_2	Z_3
Hartley oscillator	L	L	C
Colpitts oscillator	C	C	L
Clapp oscillator	C	C	Ser. LC (net L)
Tuned in/tuned out oscillator	Par. LC (net L)	Par. LC (net L)	C
Pierce crystal oscillator	C	C	Crystal (net L)

the reactance types used for the three impedances, and include the familiar Hartley and Colpitts circuits.

6.5.2 Colpitts Oscillator

A field-effect-transistor Colpitts oscillator is shown in Fig. 6.5(a). Z_1 and Z_2 are capacitors, and Z_3 is an inductor. The RF choke provides a low-resistance dc path for the collector current while blocking the signal. The coupling capacitor C_c and the base resistor R_g act to bias the transistor into the class C mode once oscillations have started.

The resonant frequency, as determined from Eq. (6.6), is

$$f_o = \frac{1}{2\pi\sqrt{L_3 C_{eq}}} \qquad (6.8)$$

where $C_{eq} = C_1 C_2 / (C_1 + C_2)$, the value of C_1 in series with C_2. The minimum gain required of the amplifier to maintain oscillation is derived from Eq. (6.7),

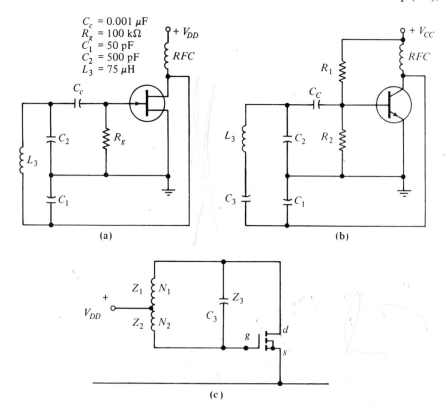

FIGURE 6.5. (a) Colpitts oscillator; (b) Clapp oscillator; (c) Hartley oscillator.

for $A_{v(\text{loop})} \geq 1$:

$$|A_{vo}| \geq \frac{X_1}{X_2} = \frac{C_2}{C_1} \tag{6.9}$$

Example 6.3

The Colpitts oscillator of Fig. 6.5(a) uses an *n*-channel junction FET biased to give an initial $g_m = 2$ mS and an $r_d = 10$ kΩ. Find the frequency of oscillation and whether or not oscillations will start.

Solution

From Eq. (6.8), the frequency of oscillation will be

$$C_{\text{eq}} = \frac{C_1 C_2}{C_1 + C_2} = \frac{500 \times 50}{500 + 50} = 45.5 \text{ pF}$$

$$f_o = \frac{1}{2\pi \sqrt{L_3 C_{\text{eq}}}}$$

$$= \frac{1}{2\pi \sqrt{75 \times 10^{-6} \times 45.5 \times 10^{-12}}}$$

$$= 2.72 \text{ MHz}$$

The open-circuit gain of the amplifier is

$$A_{vo} = -g_m r_d = -2 \text{ mS} \times 10 \text{ kΩ} = -20$$

From Eq. (6.7), the total loop gain is

$$A_{v(\text{loop})} = -A_{vo}\left(\frac{C_1}{C_2}\right) = +20 \times \frac{50}{500} = +2 \, (> 1)$$

Therefore, oscillations will start.

6.5.3 Clapp Oscillator

The *Clapp oscillator* is a particular version of the Colpitts circuit. The difference between the two is that X_3 is comprised of an inductor and capacitor in series, such that the net reactance at resonance is inductive. The capacitors used for X_1 and X_2 are much larger than the one used in X_3, so the series resonant frequency is controlled mainly by C_3. As a result, changes in the input impedance of the amplifier, which appear in parallel with C_1, do not have as much effect on the resonant frequency as they do in the Colpitts circuit, and the result is a much more stable oscillator circuit. The Clapp oscillator is shown in Fig. 6.5(b).

The resonant frequency is determined by L_3 in series with an equivalent capacitance made up of the series combination of C_1, C_2, and C_3, as given by Eq. (6.8), where

$$C_{eq} = C_1 \text{ ser. } C_2 \text{ ser. } C_3 = \frac{1}{\dfrac{1}{C_1} + \dfrac{1}{C_2} + \dfrac{1}{C_3}} \qquad (6.10)$$

Equation (6.9) gives the gain requirement for self-starting oscillations.

6.5.4 Hartley Oscillator

The *Hartley circuit* is one of the most frequently used of the resonant oscillators. It fits the general form of the oscillator presented in Eq. (6.4) but uses two inductances for X_1 and X_2 and a capacitance for X_3. Three distinct types arise, depending on whether the two coils are uncoupled, are two independent coils with mutual coupling, or are two parts of a single tapped coil. The latter configuration is nearly always used in practice and is analyzed here.

The circuit utilizing an *n*-channel enhancement-mode insulated-gate FET is shown in Fig. 6.5(c). The biasing for this takes advantage of the special characteristics of the transistor and is particularly simple, no resistors being required. The tapped coil forms a mutually inductive coupled circuit for which the coefficient of coupling k is unity (see Section 1.12.3), and the impedances Z_1 and Z_2 must take account of this. Assuming coil resistance is zero, detailed analysis gives

$$\frac{Z_2}{Z_1} = \frac{X_2}{X_1} = \frac{L_2 + M}{L_1 + M} = \frac{N_2}{N_1} \qquad (6.11)$$

(*Note*: this is not $(N_2/N_1)^2$, as might have been expected with $L \propto N^2$.)
The resonant frequency is given by

$$f_o = \frac{1}{2\pi\sqrt{L_t C_3}} \qquad (6.12)$$

where $L_t = L_1 + L_2 + 2M$, L_1 and L_2 are the inductances of sections N_1 and N_2 excluding mutual inductance, and $M = \sqrt{L_1 L_2}$. Substituting Eq. (6.11) in Eq. (6.7), the maintenance condition becomes

$$A_{v(\text{loop})} = -A_{vo}\frac{N_2}{N_1} \geq 1 \qquad (6.13)$$

Example 6.4 The Hartley oscillator shown in Fig. 6.5(c) is biased to give $g_m = 1.5$ mS and a drain resistance of 10 kΩ. The coil is tapped to give a turns ratio $(N_1/N_2) = 3$, with $L_t = 64$ μH and $C_3 = 39$ pF. Calculate the frequency of oscillation, and determine whether oscillation will self-start.

Solution   From Eq. (6.12),

$$f_o = \frac{1}{2\pi\sqrt{64\ \mu\text{H} \times 39\ \text{pF}}} = 3.19\ \text{MHz}$$

From Eq. (6.13),

$$A_{v(\text{loop})} = -(-1.5 \times 10) \times \tfrac{1}{3} = 5\ (> 1)$$

Therefore oscillation will be self-starting.

6.6 CRYSTAL OSCILLATORS

The characteristics of quartz peizoelectric crystals were discussed in Section 1.15.2 on filters. Their use as the frequency control elements in oscillator circuits is discussed here. Crystals can be used to control any of the tuned LC oscillators which were discussed previously by appropriate connections. The crystal may be used to replace an entire LC tank circuit, or it may be used to replace one of the reactances in a tank circuit. The Pierce crystal oscillator circuit illustrates the method.

The Pierce oscillator, like the Clapp oscillator, is basically a Colpitts oscillator in which the inductor is replaced by the crystal. The circuit is shown in Fig. 6.6(a), with an equivalent circuit shown in Fig. 6.6(b) in which the crystal has been replaced by its equivalent circuit. The resonant frequency of the circuit is determined by the series resonance of the circuit made up of C_1, C_2, C_s, and L_s. C_1 and C_2 are both very much larger than C_s, so the resonant

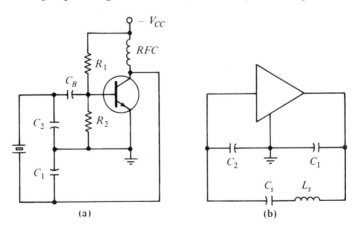

FIGURE 6.6.   The Pierce crystal oscillator: (a) circuit; (b) equivalent circuit.

frequency is almost entirely dependent on the value of C_s. The input capacity of the amplifier, C_1, and C_2 are effectively swamped out by C_s, and the resonant frequency is almost the series value of the crystal itself. The feedback ratio is the ratio of X_{C2}/X_{C1} (from Eq. (6.7)), and the required loop gain is determined in the same manner.

Any energy withdrawn from the circuit to drive successive amplifier stages is the equivalent of spoiling the Q of the crystal, and special coupling circuits must be used to minimize the loading effect. In practice, a parallel tuned tank circuit tuned to the desired frequency is placed in the collector circuit, and the next stage is transformer coupled through this tank. The load impedance presented to the amplifier can thus be increased to the point where it does not greatly affect the crystal Q.

Example 6.5   The circuit of Fig. 6.6 uses a 5-MHz quartz crystal and a transistor with $gm = 10$ mS, $\beta = 100$, $ro = 100$ kΩ, $C_2 = 5000$ pF; the circuit is coupled to a 12.5 kΩ load resistance. Find the value of C_1 necessary to give a 20-dB guarantee that oscillations will be self-starting.

Solution

$$A_{vL} = -gm(ro)\|R_L)$$
$$= -10(50k\|12.5k)$$
$$= -100 \ (40 \text{ dB})$$

Note that this is the voltage gain which will be obtained if the reactive network is disconnected from the transistor collector, or A_{vo}, in Eq. (6.7). Next,

$$A_{v(\text{loop})} = \frac{-A_{vo}C_1}{C_2} = +10 \qquad \text{(from the 20-dB requirement)}$$

$$C_1 = \frac{10C_2}{|A_{vo}|} = \frac{10 \times 5000}{100} = 500 \text{ pF}$$

6.7   STABILITY

The stability of oscillator operation can be discussed from a number of points of view. First, will the oscillator produce the desired frequency? Will the oscillator start up by itself and maintain oscillations under all normal load conditions? Will it maintain oscillations at the desired level under all load conditions? Will it produce a sinusoidal output, or will it contain harmonics? These topics will each be discussed in turn.

6.7.1 Frequency Stability

Practically all the components in an oscillator will have some effect in determining the frequency of oscillation. The frequency of the tuned oscillator, for example, is dependent on the series *impedance* of $Z_1 + Z_2 + Z_3$. The analysis given above assumed that these were all ideal isolated reactances (of negligible resistance), which is not quite the truth. Amplifier output and input resistances and circuit resistances spoil the Q and broaden out the impedance curve near resonance. Also, the parasitic capacities in the amplifier appear directly in parallel with Z_1 and Z_2. These parasitics can vary over a wide range, depending on the amplifier supply voltage, temperature, and loading, and unless some effort is made to isolate these changing elements from the tank circuit, they will cause changes to occur in the resonant frequency.

The amount of change occurring in the amplifier parameters can be reduced by providing power to the amplifier from a voltage-regulated source and by using a buffer amplifier with a high input impedance to isolate the oscillator from successive stages. Temperature effects can be minimized by operating the oscillator at low power levels and providing thermal isolation against ambient temperature variations. (The whole oscillator circuit may be mounted in a temperature-regulating crystal oven, for instance.)

Isolation of the tank circuit from the amplifier parasitics can be improved by means of impedance transformation. In the Hartley oscillator this may be accomplished by providing a second tap point on the coil L_2 so that the amplifier input capacity only appears in parallel with a fraction of L_2. Further, the ratio L/C can be increased (within limits) so that the reactance of the coil L_2 is very much larger than the parasitic reactance and its detuning effect will not be so pronounced. For the Colpitts circuit, the ratio C/L can be increased so as to swamp the parasitic capacity, but the degree of improvement is limited by the small size of coil that may result.

The Clapp circuit provides a greater degree of stability, because the oscillating frequency is almost entirely due to the series reactance of Z_3, and any convenient values of L_3 and C_3 may be used to yield the correct frequency. Isolation from the amplifier may be improved by several orders of magnitude. The price that must be paid for this improvement, however, is that more gain is required from the amplifier to maintain oscillations.

The tuned circuit itself may experience changes of parameters which will cause a change of oscillator frequency. Temperature changes can cause significant changes of parameters. These temperature changes may be partially compensated by choosing elements whose temperature coefficients cancel out in the circuit, and the whole circuit may be placed in a temperature-controlled environment. Each component in the tank circuit must be a rigid entity and must be protected from mechanical vibrations. Even very small distortions of a coil will cause a shift of the oscillating frequency. Microphonics, or changes of the tank-circuit parameters by sound waves or mechanical vibrations, cause frequency modulation of the oscillator and are a common and serious problem. Acoustic shielding and shock mounting are a must for the oscillator.

A low Q in the tank circuit (because of coil resistance or loading) results in a broadening of the notch which appears in the impedance-versus-frequency curve for the tank circuit. The result is that the frequency of oscillation can shift a considerable amount near the true oscillation frequency. The higher Q becomes, the less this shift is allowed to become. Circuit Q's from 10 to 1000 are practical in ordinary LC circuits and, with careful design, yield frequency stabilities to about 1 part in 10^4. For stabilities greater than this, crystals must be used. A quartz crystal run at ambient temperature can give stabilities of about 1 in 10^5, while a temperature-controlled oscillator can give about 1 in 10^6.

6.7.2 Self-starting

Self-starting is ensured by guaranteeing that the amplifier is biased to a point before oscillation begins where its gain provides several dB more than the minimum according to the Barkhausen criterion. If self-starting is to be guaranteed under all operating conditions, it is necessary to make sure that the amplifier is unaffected by temperature and supply variations, as mentioned above.

6.7.3 Amplitude Stability

The amplitude of oscillations is determined by factors which reduce the amplifier gain to the point where the loop gain is unity. These factors include an inherent nonlinearity of the amplifying device, such as the onset of saturation or self-bias, and an externally provided amplitude clipper. In any case, the amplitude is subject to variation with load, operating conditions, and supply variations.

A more stable mode of operation can be obtained by using an amplifier whose gain can be controlled by the application of an external bias voltage. An amplitude detector is used to provide a bias voltage which increases with increasing signal amplitude, and this increasing bias is used to decrease the gain of the amplifier (automatic gain control). Very good amplitude stability can be obtained by this means, even under varying load conditions.

6.7.4 Linearity

Some harmonic distortion will result in the output waveform of any oscillator which relies on the amplifier nonlinearity to limit oscillation amplitude. The larger the signal amplitude is allowed to become, the more pronounced is the nonlinearity of operation, and the more harmonic content that will be produced.

Linear operation with a minimum of harmonic generation can be obtained by operating the oscillator at signal levels well below the total range of the amplifier used. The degree to which this can be done is limited by the amount of gain margin necessary to start and maintain oscillations, because the reduced amplitude is obtained by reducing the loop gain. A more reasonable way to do this is to use an amplifier that does operate linearly over a wide range of amplitudes, and to use an error feedback system to limit the amplitude of oscillations to well within the linear region of operation. Practically pure sinusoidal output can be obtained by this latter method.

6.8 NEGATIVE RESISTANCE OSCILLATORS

Certain two-terminal devices exhibit the property of negative resistance over some portion of their characteristics; that is, over some portion of their operating range, an increase in terminal voltage causes a decrease in terminal current. The dynamic resistance which is the inverse slope of the I-V curve of the device is a negative quantity in this range. If such a device is connected in series or in parallel with a tuned LC tank circuit and biased to a stable operating point within the negative resistance portion of its characteristics, so that the magnitude of its negative dynamic resistance is greater than the damping or loading resistance in the tuned circuit, the circuit will oscillate. In other words, the negative resistance device acts as an energy source which replaces the tank energy dissipated in its internal resistance and the load.

Two basic configurations are possible. The tank circuit may be series-resonant, with the negative resistance made part of the series loop; or it may be parallel-resonant, with the negative resistance placed in shunt with it. The parallel-resonant circuit is more properly called a negative conductance oscillator. Figure 6.7 illustrates these two circuits. The series impedance of the series-resonant circuit is given by

$$Z_s = R_s + j\left(\omega L_s - \frac{1}{\omega C_s}\right) + r \tag{6.14}$$

If the resistance r has a value equal in magnitude to R_s, and is negative, the real portion of the impedance expression disappears, and the circuit will maintain oscillations. If the negative impedance has a larger magnitude than R_s, then oscillations will grow in magnitude until some nonlinearity decreases the negative resistance to the critical value. The circuit will oscillate at the frequency for which the series reactance is zero. The condition guaranteeing oscillation is

$$|r| \geq |R_s| \tag{6.15}$$

with the frequency of oscillation given by

$$f = \frac{1}{2\pi\sqrt{L_s C_s}} \tag{6.16}$$

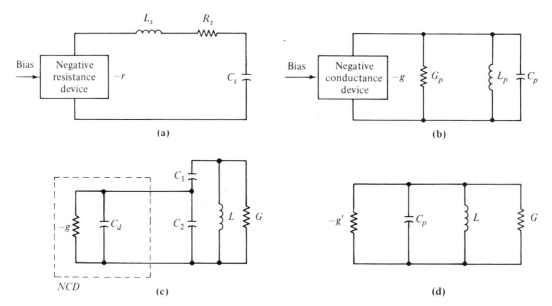

FIGURE 6.7. (a) Negative resistance series-resonant oscillator; (b) negative conductance parallel-resonant oscillator; (c) loosely coupled negative conductance oscillator circuit; (d) equivalent circuit.

The parallel circuit shown in Fig. 6.7(b) is the dual of the series circuit. The admittance expression for the parallel circuit is

$$Y_p = G_p + j\left(\omega C_p - \frac{1}{\omega L_p}\right) + g \qquad (6.17)$$

where the condition guaranteeing oscillation is

$$|g| \geq |G_p| \qquad (6.18)$$

with g negative, and

$$f = \frac{1}{2\pi\sqrt{L_p C_p}} \qquad (6.19)$$

gives the frequency of oscillation. Again, as the amplitude of oscillation increases, the nonlinearity of the negative conductance limits the amplitude by reducing the magnitude of g at the extreme excursions.

Practical negative resistance devices such as tunnel diodes also have a reactive component present in their characteristics. For the tunnel diode, this reactive component is the junction capacity of the diode, and since its value is dependent on diode voltage and temperature, it can cause undesired shifts in the resonant frequency of the oscillator, especially at higher frequencies, where the tank capacity is quite small. Greater stability is usually accomplished by using an impedance transformation between the negative resistance device and the tank circuit. Capacitive coupling is preferred because the resulting circuit is less prone to parasitic oscillation. Figure 6.7(c) shows how the coupling is accomplished for a parallel resonant circuit. The effective negative conductance g' felt by the tank circuit is obtained from Eq. (1.117) as

$$|g'| \cong |g|\left(\frac{C_1}{C_1 + C_2}\right)^2 \tag{6.20}$$

The reactive component of the negative resistance device is in parallel with C_2, and if C_2 is very much larger than C_1, the detuning effect will be small. However, the ratio C_2/C_1 cannot be made too large, or it will become impossible to maintain oscillations because the condition

$$|g'| \cong |G_p| \tag{6.21}$$

cannot be met.

6.9 FREQUENCY SYNTHESIZERS

The frequency synthesizer is not a frequency generator in the same sense as an oscillator, but is a frequency converter, which uses a phase-locked loop and digital counters in a phase-error feedback system to keep the output running in a fixed phase relation to the reference signal. Output frequency stabilities are determined by the stability of the reference oscillator, which is typically a crystal-controlled oscillator circuit.

The principles of the frequency synthesizer were developed about 1930 but only found application in very sophisticated equipment because of the cost of the components. Microcircuit chips designed especially for this application are available now at very low cost, and frequency synthesizers are finding increasing application for channel selection in communication equipment.

6.9.1 Phase-Locked Loop

The heart of the frequency synthesizer is the phase-locked loop. A simple phase-locked loop is illustrated in Fig. 6.8, and its operation may be described as follows. A stable oscillator produces a square-wave reference frequency f_r which provides one of the inputs to the phase-detector circuit. This reference frequency may be any convenient value, but is usually chosen so that a crystal oscillator circuit may be used. A voltage-controlled oscillator (VCO) generates the final output frequency f_o and is designed so that it will tune over the whole range from the minimum frequency to the maximum frequency desired. Its output is fed directly to the load and also is used to drive a programmable binary counter which provides the function of frequency division ($\div N$, where N is the number programmed into the counter). The output of the counter is a square wave at the reference frequency which provides the second input to the phase-comparator circuit.

The phase comparator is a circuit which produces a dc signal whose amplitude is proportional to the phase difference between the reference signal f_r and the counter output f_o/N. This dc signal is filtered to smooth out noise and slow the response of the circuit to prevent overshoot or oscillations and is applied as the control input to the VCO. When the phase difference between the two signals f_r and f_o/N is zero, the dc output from the phase comparator is just exactly that needed to tune the VCO to the frequency Nf_r. If a phase difference exists between the two, the bias applied to the VCO will change in a direction to raise or lower the frequency f_o just sufficiently so that the phase difference will disappear. Once the VCO output reaches the value Nf_r, it will "lock onto" that frequency, and the feedback loop will prevent it from drifting.

The output frequency f_o is adjusted to a new value by changing the number by which the counter divides. This is accomplished by means of thumbwheel switches or by means of a register into which a new number for N can be entered to control the set point of the counter. The number N is the number of pulses which the counter will count before it recycles, coded in binary.

6.9.2 Prescaling

The simple frequency synthesizer described above will only produce output frequencies which are integer multiples of the reference frequency f_r. If other frequencies intermediate between these values are desired, prescaling must be used. Another reason for the use of prescaling is because at high frequencies (above 100 MHz) programmable counters are not available. Fixed-modulus prescale counters are used to count down to a frequency below

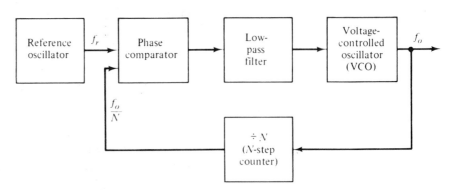

FIGURE 6.8. Basic phase-locked loop frequency synthesizer.

the 100-MHz limit, and then the prescaler output can drive a low-frequency programmable counter, which is readily available.

Figure 6.9 shows how a prescaler circuit can be used to allow division by a noninteger number (a number that contains a fractional part). The prescaler circuit is a two-modulus counter; that is, in one mode it produces an output for every P input pulses, and in the other mode an output for every $P + 1$ pulses. Two low-frequency programmable counters count the output pulses from the prescaler circuit, the main counter counting B pulses and the second counter counting A pulses.

At the beginning of a cycle, both counters are set to their programmed numbers (i.e., B and A). As long as the A counter countains a nonzero number, the prescaler will be conditioned to count in the $P + 1$ mode, so the counter chain will count down for $(P + 1)A$ pulses until the A counter goes to zero. At this point, the prescaler circuit will be forced to count in the P mode, and also, the input to the A counter will be turned off so that the A counter will remain in the zero state until the B counter completes its count. At the point where the A counter has reached the zero state, the B counter will contain the number $(B - A)$ and will then proceed to count down from $(B - A)$ on every Pth pulse from the output. When the B counter reaches zero, both counters reset to their programmed numbers, and the cycle begins again.

The result of this prescaling procedure is shown in Eq. (6.22), which relates the output frequency to the reference frequency in terms of the three

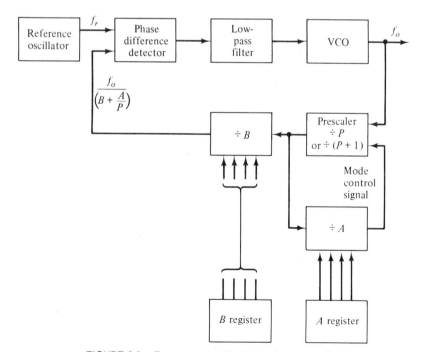

FIGURE 6.9. Frequency synthesizer using prescaling.

counter moduli.

$$f_o = Nf_r$$

$$= \left(B + \frac{A}{P}\right)Pf_r$$

$$= [(B - A)(P) + (A)(P + 1)]f_r \qquad (6.22)$$

Since any fractional number may be stated to a very near approximation as the ratio of two integers, the number of precise frequencies which may be "dialed up" on this frequency synthesizer is very much expanded. A further advantage is that only the prescaler circuit need operate at very high frequencies, and the programmable counters can be made from readily available low-frequency components.

6.9.3 Applications

The most obvious application of the frequency synthesizer is as a digitally programmed (manual or remote) signal generator which may be used for testing purposes, frequency measuring at radio-monitoring stations, or in laboratory frequency-measuring apparatus. Because of the ease of remote setting of the digital numbers to select the output frequency, the synthesizer is increasingly being used in computer-controlled testing stations and other computer-controlled apparatus.

The frequency synthesizer is also used in multichannel communication link transceivers, where it is necessary to frequently switch from one channel to another. The frequency synthesizer generates the local oscillator frequency for the receiver mixer and also the primary frequency source for the transmitter. The switching is usually done manually, but may also be done automatically, as would be the case in a frequency-diversion system. Since the output frequency will have a stability comparable to that of the reference, which may be a crystal oscillator, the flexibility of a variable-frequency oscillator is obtained with the stability of the crystal oscillator. Since most of the components used are inexpensive microcircuits, the result is high-quality but inexpensive communication equipment.

6.10 PROBLEMS

1. A phase-shift oscillator is to be designed for a frequency of 10 kHz, using a special transistor (Darlington pair) with $ri = 100$ kΩ, $\beta = 2000$, $gm = 20$ mS. Base-bias resistors give a parallel combination of 100 kΩ. The third resistor in the phase-shift network is replaced by the amplifier input resistance, and the other two resistors are to have the same value. Calculate the values of R and C, and the value of R_c if the amplifier is to have 6 dB more gain than the minimum needed to maintain oscillations. Check that the input impedance of the phase-shift network does not load the amplifier.

2. For the *LC* tuned oscillator of Fig. 6.3, calculate the frequency of oscillation and the decibel amount by which the gain exceeds the minimum required to maintain oscillations, when the following circuit conditions exist:

turns ratio	= 3 : 1	R_1	= 22 KΩ
L_p	= 10 mH	R_2	= 1 kΩ
R_p (coil resistance)	= 20 Ω	$r\pi$	= 8 kΩ
C	= 1 μF	β	= 64
C_c	= 1 μF	gm	= 8 mS

3. A Colpitts oscillator has a coil with an inductance of 50 μH and is tuned by a capacitor of 300 pF across the amplifier input and 100 pF across the output. Find the frequency of oscillation and the minimum gain which the amplifier must provide to maintain oscillations.

4. A Clapp oscillator is used to tune a radio receiver. For the circuit of Fig. 6.5(b), C_1 = 1000 pF, C_2 = 10,000 pF, L_3 = 50 μH, and C_3 is a 5- to 150-pF variable capacitor. (a) Find the tuning frequency range of the oscillator. (b) If the amplifier uses a bipolar transistor with an $r\pi$ of 2 kΩ and an ro = 10 kΩ, find the minimum β the transistor must have to guarantee oscillations.

5. The Hartley oscillator of Fig. 6.5(c) uses a FET with a g_m of 2 mS and an r_d of 20 kΩ. The total coil inductance is 20 μH with a turns ratio N_1/N_2 of 10. It is tuned with a 10-pF capacitor. Find the frequency of oscillation and the amplifier gain margin (amount by which the gain exceeds the minimum required to maintain oscillations) in decibels.

6. A crystal oscillator has the crystal connected between the amplifier input terminals and a parallel-resonant tank circuit connected between the amplifier output terminals. A small capacitor provides the feedback path between the input and output. The crystal is series-resonant at 1.50000 and parallel-resonant at 1.50001 MHz. (a) To what frequency must the output tank circuit be tuned to guarantee oscillations? What will the frequency of oscillation be? How stable is this in parts per million? (b) If the tank circuit were tuned to the second harmonic, would the circuit still oscillate? (c) At what frequency would the crystal vibrate under these conditions?

7. A negative conductance oscillator uses a coil with L = 10 μH and Q_o = 100, in parallel with a capacitor of 50 pF. Find the minimum value of conductance that the negative conductance device must generate, and the frequency of oscillation.

8. The frequency synthesizer in Fig. 6.9 is run from a 1.0000-MHz crystal reference oscillator to produce an output of 146 MHz. The main counter will only work up to 80 MHz, and a prescaler *P* which divides by 10 is used. Find the values of *B* and *A* which are appropriate to these conditions.

CHAPTER 7

Receivers

7.1 INTRODUCTION

Radio receivers perform a number of functions. They serve to separate a wanted radio signal from all other radio signals which may be picked up by the antenna, and to reject all the others. They amplify the separated signal to a usable level. Finally, they separate the intelligence signal from the radio carrier and pass it to the user.

This chapter will examine the operating principles of some of the more commonly used radio receivers, treating them mostly on a block-diagram basis and referring to previous chapters for the individual detailed circuits.

7.2 SUPERHETERODYNE RECEIVERS

Early receivers used for the reception of amplitude-modulated signals or interrupted-carrier telegraph signals used the tuned radio frequency, or TRF, principle. They were simply a chain of amplifiers, each tuned to the same frequency, followed by a detector circuit. These receivers suffered from poor adjacent signal selectivity, especially when required to tune over wide frequency ranges, because the Q of tuned circuits changes with frequency.

The superheterodyne receiver was developed to improve this adjacent channel selectivity by placing most of the frequency selectivity in the *intermediate frequency* (IF) stages after the first frequency conversion. Obtaining this selectivity at the IF is much easier, because the circuits remain fixed-tuned at the IF and do not vary as different stations are selected. The *superheterodyne principle* is such that when two sinusoidal signals of different frequencies are

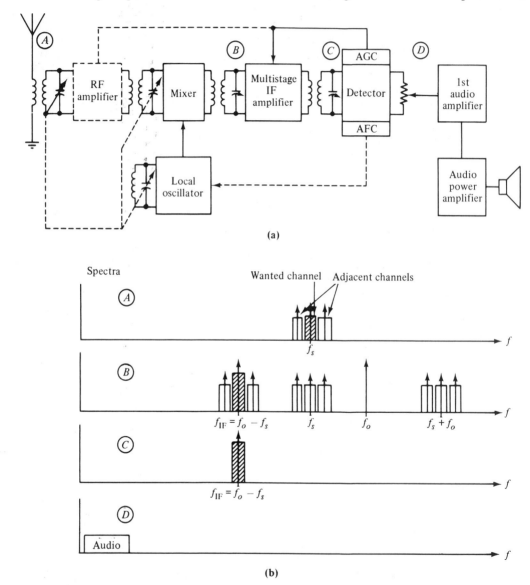

(a)

(b)

FIGURE 7.1. (a) A superheterodyne receiver circuit; (b) signal spectra in the superheterodyne receiver.

mixed together so that they multiply (or add and are passed through a circuit with a nonlinear transfer function), the output signal will contain signal components at the sum, the difference, and each of the two original frequencies. There will also be mixtures of the harmonics of these signals, but if the two base frequencies are carefully chosen, these will not interfere.

This principle forms the basis of amplitude modulation and all frequency-conversion operations—for example, the frequency multiplexing of telephone channels. The superheterodyne broadcast receiver was the original application of this principle, and is still one of the largest. The name "superheterodyne" is a contraction of "supersonic heterodyne," or the production of beat frequencies above the range of hearing.

The basic superheterodyne receiver is illustrated in Fig. 7.1. The first stage is a tuned RF amplifier, the main purpose of which is to improve signal-to-noise ratio (see Section 4.9.1). It also gives some improvement in RF selectivity and a reduction of oscillator reradiation. However, it is usually omitted in cheaper receivers. The output from the tuned RF stage is fed to the signal input of an oscillator-mixer circuit in which the heterodyning takes place. The oscillator circuit is usually tuned by capacitance tuning, and the three tuning capacitors are mechanically ganged to a single control knob. The oscillator and mixer may be separate circuits, or they may be combined as in the autodyne mixer circuit.

The mixer output (the difference frequency for down-conversion in the receiver) is fed to two tuned IF amplifiers, which are fixed-tuned and provided with sufficient selectivity to reject channel signals. The output from the IF amplifier is fed to the detector, where the audio signal is extracted, or demodulated. The detector also provides signals for automatic gain control (AGC) in AM receivers or automatic frequency control (AFC) in FM receivers. The automatic gain control signal is applied to one or more of the IF and RF amplifiers, while the AFC signal is used to correct the local oscillator frequency. The audio output is passed through a volume control to an audio amplifier, which usually consists of one low-level voltage amplifier followed by a power amplifier, and finally is coupled to a speaker.

7.3 CHOICE OF INTERMEDIATE AND OSCILLATOR FREQUENCIES

The intermediate frequency (IF) is generally chosen to be lower than the lowest signal frequency to be converted, although in some special cases an up-conversion may be used. The actual frequency chosen is a compromise. The lower the IF, the easier it is to obtain the selectivity needed to reject adjacent channel signals. However, the rejection of image signals (see Section 7.4) is improved by going higher in frequency. If the IF is too high, falling inside the signal frequency range, direct feed-through may occur.

The choice of oscillator frequency range is mostly determined by the tuning capacitor tuning range, stated as a ratio, thus:

$$\text{range} = \frac{C_{max}}{C_{min}} = \left(\frac{f_{max}}{f_{min}}\right)^2 \tag{7.1}$$

The oscillator frequency f_o may be made either lower or higher than the signal frequency f_s. If it is made lower, then

$$\text{IF} = f_s - f_o \tag{7.2a}$$

If it is made higher, then

$$\text{IF} = f_o - f_s \tag{7.2b}$$

If the oscillator frequency is made lower than the signal frequency, the range over which the oscillator must tune will be larger than if the oscillator frequency is chosen to be higher. If a high value of IF is also chosen, the tuning range of the oscillator capacitor will become excessive. This problem is less critical at higher signal frequencies but is severe in the MF band, as illustrated in the following example.

Example 7.1   Find the tuning range necessary for the oscillator capacitor in an MF superheterodyne receiver which tunes over the range of signals from 500 kHz to 1600 kHz and uses an IF of 465 kHz if the oscillator is

(a)   Higher than the signal frequency.
(b)   Lower than the signal frequency.

Solution   The frequency range of the signal frequency is 500 to 1600 kHz, which gives a frequency ratio of

$$\frac{f_{s_{max}}}{f_{s_{min}}} = \frac{1600}{500} = 3.2$$

Since the capacitance necessary to tune the circuit varies as the inverse square of the frequency (see Eq. (1.60)), the ratio of capacitance needed becomes

$$\frac{C_{s_{max}}}{C_{s_{min}}} = \left(\frac{f_{s_{max}}}{f_{s_{min}}}\right)^2 = 3.2^2 = 10.24$$

(a)   For $f_o > f_s$, by Eq. (7.2b), the range of f_o is

$$465 + (500 \text{ to } 1600) = (965 \text{ to } 2065) \text{ kHz}$$

This gives a frequency ratio for the oscillator of

$$\frac{f_{o_{max}}}{f_{o_{min}}} = \frac{2065}{965} = 2.14$$

and a capacitance ratio for the oscillator capacitor of

$$\frac{C_{o_{max}}}{C_{o_{min}}} = 2.14^2 = 4.58$$

(b) For $f_o < f_s$, by Eq. (7.2a), the range of f_o is

$$-465 + (500 \text{ to } 1600) = 35 \text{ to } 1135 \text{ kHz}$$

The corresponding ratios are

$$\frac{f_{o_{max}}}{f_{o_{min}}} = \frac{1135}{35} = 32.4 \quad \text{and,} \quad \frac{C_{o_{max}}}{C_{o_{min}}} = 32.4^2 = 1052 \qquad \text{(Impractical!)}$$

The upper oscillator range must be used. This combination of IF and oscillator frequencies is typical of nearly all AM broadcast receivers for the MF band intended for the domestic market.

7.4 IMAGE REJECTION

The superheterodyne mixer circuit produces signal components at the IF frequency which are the difference between the oscillator frequency and the signal frequency. Equation (7.2) shows that the signal can be either above or below the oscillator frequency and still produce an IF signal. If the front end (RF stage) of the receiver does not have very selective tuning, which is the case if only a single tuned circuit is used, the RF stage, when tuned to a signal frequency at $f_o - $ IF, will also respond to a signal at $f_o + $ IF. This other signal is referred to as the image frequency f_i, which is given by

$$f_i = f_s \pm 2\text{IF} \tag{7.3}$$

where the plus sign is used for the case when $f_o > f_s$ and the minus sign when $f_o < f_s$. The image frequency can only be rejected by the selectivity of tuned circuits placed *before* the mixer. Once the image frequency has been converted to the IF, it cannot be separated from the desired signal.

The front end of a receiver is one or more parallel tuned resonant circuits which act as filters. One such circuit is illustrated in Fig. 7.2(a), where a

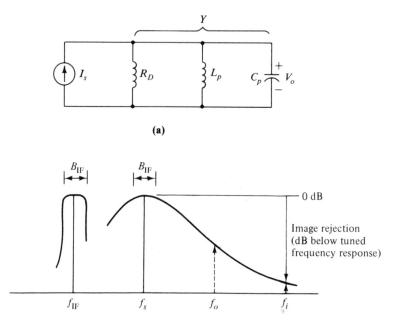

FIGURE 7.2. (a) Parallel tuned circuit. (b) Image frequency rejection.

parallel tuned circuit is driven by a constant current source I_s (an amplifier) and produces an output voltage V_o. Its response at a frequency near but not at resonance can be found as follows. The parallel admittance magnitude of the resonant circuit is given by

$$|Y| = Y_o\sqrt{\left(\frac{\omega}{\omega_o}\right)^4 + (yQ)^2} \cong Y_o\sqrt{1 + (yQ)^2} \qquad (1.82)$$

where

$$y = \frac{\omega}{\omega_o} - \frac{\omega_o}{\omega} \qquad (1.65)$$

and where ω is less than one decade above or below ω_o. The output voltage in this frequency range is given by

$$|V_o| = \frac{|I_s|}{|Y|} = \frac{I_s}{Y_o\sqrt{1 + (yQ)^2}} \qquad (7.4)$$

which reduces to $V_o(\text{res}) = I_s/Y_o$ at resonance ($\omega = \omega_o$). The relative response at a frequency off resonance is given by

$$A_r \triangleq \frac{|V_o|}{V_o(\text{res})} = \frac{1}{\sqrt{1+(yQ)^2}} \tag{7.5}$$

for a single tuned circuit. If several tuned circuits are included and isolated by amplifiers, then the overall response is given by the product

$$A_r(\text{tot}) = A_r(1)A_r(2) \tag{7.6}$$

The ratios A_r are usually stated in decibels (dB) using Eq. (A4).

Figure 7.2(b) shows the response of a single tuned circuit plotted against frequency. Peak response occurs at the signal f_s, with only a slight drop in response at the limits of the IF bandpass, which is also shown. The oscillator f_o is shown at $f_s + \text{IF}$, and the image is shown at $f_s + 2\text{IF}$. The image-rejection ability is the amount in decibels by which the tuned circuit response is down from its resonant response, and can be readily calculated for the single-circuit case. The following example illustrates the problem.

Example 7.2   An AM broadcast receiver has an IF of 465 kHz and tunes to 500 kHz with a Q of 50 at that frequency. Calculate the image rejection in dB.

Solution   From Eq. 7.5, the relative response at the image frequency is

$$A_r = \frac{1}{\sqrt{1 + y^2 Q^2}}$$

By Eq. (7.3),

$$f_i = f_s + 2\text{IF} = 500 + (2 \times 465) = 1430 \text{ kHz}$$

By Eq. (1.65),

$$yQ = \left(\frac{1430}{500} - \frac{500}{1430} \right)50$$

$$= 126$$

$$A_r \cong \frac{1}{126} \quad (\text{or} -42 \text{ dB})$$

Thus, if a station with the same signal strength as the desired signal is received at the image of 1430 kHz, it will appear in the output attenuated by -42 dB. If a tuned RF amplifier with a second tuned circuit were included, then the selectivity would be doubled, and the image rejection would become -84 dB —more than enough.

A phenomenon similar to image reception which occurs at higher frequencies is that of *double spotting*. This occurs for exactly the same reason as does image response, and happens in the following way. As the local oscillator is tuned down the spectrum, a signal f_s is received at the frequency $f_o - \text{IF}$. As the local oscillator frequency is reduced further, the same signal is again received, this time with the local oscillator tuned below the signal so that f_s (now the image frequency) is now $f_o + \text{IF}$. It will be reduced in amplitude, depending on how good the image rejection of the RF amplifier is. The effect of double spotting is that the same station appears to be received for two different settings of the tuning dial.

One solution to both image response and double spotting is to improve the sharpness of selectivity of the RF amplifier or to increase the IF so that the image lies outside the response of the amplifier, or a combination of both.

Another solution to the problem of image response is to choose a first IF frequency which lies above the maximum frequency to be tuned, with the local oscillator tuning the range above that frequency. The image frequencies for such a system lie even higher, above the oscillator frequencies and can easily be removed in the RF stage by a low-pass filter with a cutoff below the first IF. The system has been somewhat impractical until recently because IF filters with a narrow enough bandpass at the higher frequencies were not available. Recent advances in crystal and ceramic filters have provided a range of narrow band IF filters for frequencies in the VHF range, and the system is now quite practical.

7.5 ADJACENT CHANNEL SELECTIVITY

The selectivity of the RF stages of a receiver are a function of frequency, being best at low frequencies and getting poorer as frequency increases. Also, it is difficult to get several high-Q tuned circuits to track properly when they are tuned. For this reason, the selectivity of the RF stages of most receivers is purposely left much wider than is necessary for single-channel operation, and the final selectivity is obtained in the IF amplifier.

Channel assignments in an increasingly crowded spectrum are made as close together as possible, with 10-kHz spacing being typical of AM stations in the MF and HF bands. Wider spacing is used for FM stations, and even wider spacing for television and microwave systems. It should be possible for two stations to occupy adjacent channels with minimum spacing between them, and the receiver should be able to separate them. The ideal IF channel bandpass characteristic is illustrated in Fig. 7.3(a). It has a flat-topped response within the band, is centered on the channel frequency f_1, has vertically straight skirts, and has uniform rejection, ideally infinite, outside of the band.

Under these ideal conditions, no signal from the band around the adjacent channel frequency f_2 will interfere with the signal of f_1. Practical filters are far from ideal, however, and more or less of the adjacent channel signal will be allowed through. A good receiver should provide 60- to 80-dB rejection of adjacent channels, perhaps more for high-quality broadcast. Typically, IF selectivity has in the past been obtained by using several cascaded high-Q tuned circuits in different combinations. The most straightforward system simply uses an IF amplifier consisting of two or more amplifiers and

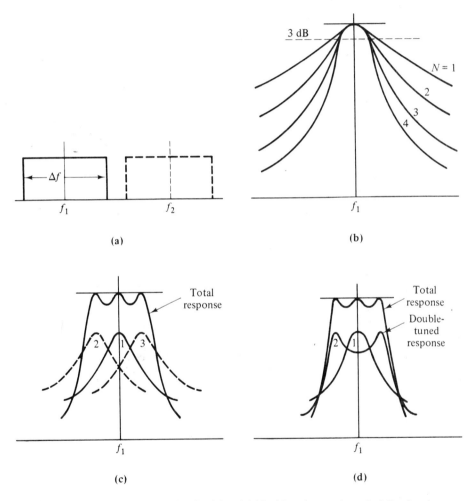

FIGURE 7.3. Adjacent channel selectivity: (a) ideal bandpass characteristic showing an adjacent channel; (b) response of cascaded single-tuned circuits; (c) response of stagger-tuned circuits; (d) response of overcoupled double-tuned circuits.

several undercoupled transformers—all tuned to the IF. Each contributes a single resonance curve, and the total IF response is simply the product of all the IF curves. The Q's are chosen so that the overall response has a 3-dB bandwidth of the required bandpass, and steeper skirts are obtained for each additional tuned circuit in the chain. This system suffers somewhat from both amplitude and phase distortion over the channel bandwidth, although reasonable adjacent channel selectivity can be obtained with five circuits. The response of this system is illustrated in Fig. 7.3(b).

Better in-band distortion characteristics can be obtained with the single-tuned system without undue sacrifice of selectivity by stagger tuning. An odd number of tuned circuits is required, and these are tuned so that one is on the center frequency and each successive pair is tuned to a pair of frequencies equidistant and successively further from the center frequency, as illustrated in Fig. 7.3(c). The overall response is again the product of the individual responses, but this time it has several peaks, forming a ripple across the top of the bandpass. This ripple can be "smoothed" by adding more tuned circuits tuned closer together. The steepness of the skirts of the response is dependent on the number of tuned circuits as well.

The overcoupled double-tuned transformer offers a unique variation to the stagger-tuning problem. When two circuits are tuned to the same frequency and then tightly coupled, the overall response will have a double hump, as shown in Fig. 1.19. The frequency deviation of the humps from center is dependent on the degree of overcoupling, and the steepness of the skirts is dependent on the circuit Q. Two or more of these transformers are cascaded to make the total IF filter. This system has the advantage over the stagger-tuned system that isolating amplifiers need only be placed between pairs of tuned circuits instead of between individual circuits, thus reducing the total number of amplifiers needed. A typical IF amplifier using the overcoupled transformers would consist of three transformers and two amplifiers. The first transformer would be undercoupled, and each of the others double-tuned and overcoupled, providing a bandpass with five-hump ripple. The ripple amplitude is easily kept to less than 1 dB, and phase distortion is not a serious problem. The analysis of amplifiers of this type is presented in Chapter 5.

IF bandpass in more expensive communications receivers and in special-purpose receivers of recent vintage commonly use one or more integrated circuit amplifiers and special filters such as the crystal lattice filters or mechanical resonator filters discussed in Section 1.15.

High-frequency operational amplifiers have become available in integrated-circuit form, making it possible to build active filters with a wide variety of bandpass characteristics without using inductors. This allows a large reduction in the physical size of receivers. Entire IF amplifiers can now be built in single chips, and entire receivers can be built with the use of two or three such chips. This topic is discussed in more detail in Section 5.5.1.

7.6 SPURIOUS RESPONSES

The section on image rejection pointed out that a careful choice of IF and good tracking of high-Q coils in the RF stage would effectively suppress the image frequency. Spurious responses are another matter. These will result when a second received signal near the desired signal in frequency passes through the RF tuned circuits and enters the mixer. Since the mixer is inherently a nonlinear device, harmonics of the oscillator frequency and of the desired signal, together with the undesired signal, will all be produced and mixed in the mixer. When the sum or difference of any of these products falls within the passband of the IF, it will interfere with the desired IF component as a spurious response.

Generally, an IF response will be produced if either of the equations

$$\text{IF} = \pm nf_o \mp mf_u \tag{7.7}$$

where

f_o = oscillator frequency

f_u = undesired signal

$m = 1, 2, 3, \ldots$ (harmonic number of the signal)

$n = 1, 2, 3, \ldots$ (harmonic number of the oscillator)

is satisfied. Rearranging these two equations gives equations which yield all the possible undesired frequencies that could cause interference in terms of the oscillator frequency:

$$f_u = \frac{n}{m} f_o \pm \frac{\text{IF}}{m} \tag{7.8}$$

Any signal that falls within $\pm \frac{1}{2}$ of the IF bandwidth (± 5 kHz for normal AM) of the unwanted signal frequency f_u will cause interference with the desired signal, if the unwanted signal is strong enough after passing through the receiver front end.

The reduction of the spurious responses requires a number of things. First, the front end must be carefully designed so that no more bandwidth than is absolutely necessary is provided, to reduce the strength of the interfering signals. Next, the oscillator must be carefully designed so that it puts out as little harmonic content as possible. And finally, the mixer should be designed and adjusted so as to produce as few harmonics and at as low an amplitude as possible. Any effort to reduce the spurious responses must, of course, take place before the mixing process if it is to be successful. The undesired signals cannot be removed once mixing has taken place.

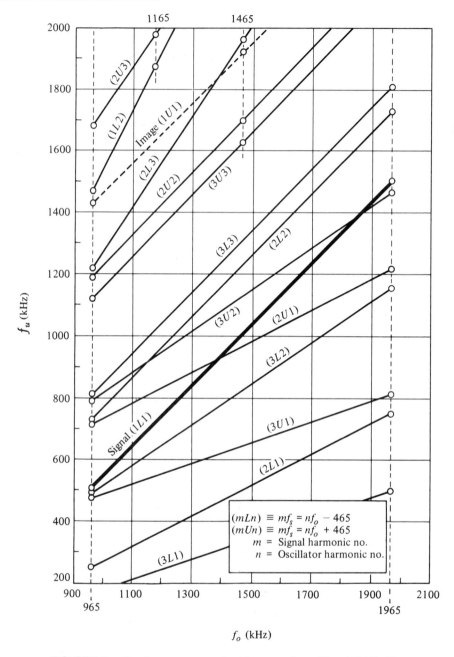

FIGURE 7.4. Spurious response chart for a receiver with a 465-kHz IF.

The spurious responses of a receiver may be plotted against oscillator frequency in the manner of Figure 7.4, over the tuning range of the receiver. This spurious response chart can then be used to identify graphically spurious signals that occur.

Example 7.3   An AM broadcast receiver operates with a 465-kHz IF and has its oscillator tuned above the signal frequency. Its front end has a -60 dB bandwidth of 500 kHz.

(a)   Find all frequencies which can cause an IF response for values of m, n up to 3 (18 in all).
(b)   Identify the signal and image frequencies.
(c)   Which of the unwanted signal frequencies may cause an interfering response in this receiver?
(d)   Will a strong signal at 1200 kHz cause an interference tone, and if so at what tone frequency?

Solution   (a)

n	m	$\left(\dfrac{n}{m}\right)f_o$	$\pm\left(\dfrac{1}{m}\right)$IF	f_u(diff)	f_u(sum)
1	1	1565	465	1100	2030
1	2	782.5	232.5	550	1015
1	3	521.7	155	366.7	676.7
2	1	3130	465	2665	3595
2	2	1565	232.5	1332.5	1797.5
2	3	1043.3	155	888.3	1198.3
3	1	4695	465	4230	5160
3	2	2347.5	232.5	2115	2580
3	3	1565	155	1410	1720

where $f_o = f_s + \text{IF} = 1100 + 465 = 1565$ kHz.
(b) $f_s = 1100$ kHz. $f_i = 2030$ kHz.
(c)   Signals falling within $f_s \pm \frac{1}{2}\text{BW} = 1100 \pm 250$ kHz may cause interference:
  (i) $n = 1$, $m = 2$ (signal with oscillator 2nd harmonic), at 1015 kHz.
  (ii) $n = 2$, $m = 3$ (2nd harmonic of signal with 3rd harmonic of oscillator), at 888 kHz.
  (iii) $n = 2$, $m = 3$, at 1198.3 kHz.
(d)   A signal at 1200 kHz will cause an interference tone because it falls within 5 kHz of the f_u value at 1198.3 kHz. The tone produced will be at $1200 - 1198.3 = 1.7$ kHz.

7.7   TRACKING

The oscillator tuning of a superheterodyne receiver should follow or track the signal circuit tuning so that for all settings of the dial, the difference between the two is precisely the IF. It was shown in Example 7.1 that for the

MF band, a capacitance tuning range of 10.24 : 1 is required in the signal circuit and a range of 4.58 is required in the oscillator circuit. A capacitance considerably smaller than that for the RF stage is required for the oscillator section if proper tracking is to be obtained.

This different capacitance may be obtained in a number of ways. For single-band receivers, a special tuning capacitor can be used which is designed to track over the proper range for that band. In this case the rotor of the oscillator section can be made with a smaller number of plates, giving a lower total capacitance, and one of the plates can be segmented so that small adjustments can be made throughout the tuning range, giving almost perfect tracking.

When a receiver is made to tune more than one band, however, specially made capacitors cannot be used, and ganged capacitors with identical sections must be used. A different value of inductance and special extra capacitors called *trimmers* and *padders* are used to adjust the capacitance of the oscillator to the proper range. The circuit may be made to adjust with only a small padder capacitor in series with the tuning capacitor. Or the circuit may be adjusted with a single trimmer capacitor in parallel with the tuning capacitor, or a combination of both may be used. Figure 7.5 illustrates the methods of connecting the tuned circuits when equal-ganged capacitors are used. Either trimmer or padder adjustment allows the oscillator to match the desired frequency at either end of the band, but not in the middle, while more or less tracking error is created in midband. For the padder circuit, the oscillator tunes below the frequency it should in midband, so the IF created is higher than it should be, and a positive error is created. The trimmer circuit causes the oscillator to tune high, and a negative error is produced. Also, a much larger error is produced by the padder circuit. The combination circuit can be adjusted to give zero error at three points across the band, at each end, and at the middle.

The value of padder capacitor required for the circuit of Fig. 7.5(a) is found as follows:

1. Find the minimum and maximum oscillator frequencies and the required oscillator capacitance ratio.
2. Obtain the capacitance ratio and maximum value of the signal circuit tuning capacitance.
3. The oscillator tuning capacitance is given by

$$C_o = C_s \text{ ser. } C_p = \frac{C_s C_p}{C_s + C_p} \tag{7.9}$$

and will have exactly the correct values at $f_{o_{\max}}$ and $f_{o_{\min}}$. Therefore, using ratios,

$$\frac{C_{o_{max}}}{C_{o_{min}}} = \frac{C_{s_{max}} \text{ ser. } C_p}{C_{s_{min}} \text{ ser. } C_p} = \frac{C_{s_{max}} \left(C_{s_{min}} + C_p \right)}{C_{s_{min}} \left(C_{s_{max}} + C_p \right)} \qquad (7.10)$$

and this equation can be solved directly for C_p.

4. The oscillator coil value is then found as

$$L_o = \frac{1}{\left(2\pi f_{o_{min}} \right)^2 C_{o_{max}}} = \frac{1}{\left(2\pi f_{o_{max}} \right)^2 C_{o_{min}}} \qquad (7.11)$$

The value of trimmer capacitor required in Figure 7.5(b) is found in the same way except that in step 3,

$$C_o = C_s \text{ par. } C_t = C_s + C_T \qquad (7.12)$$

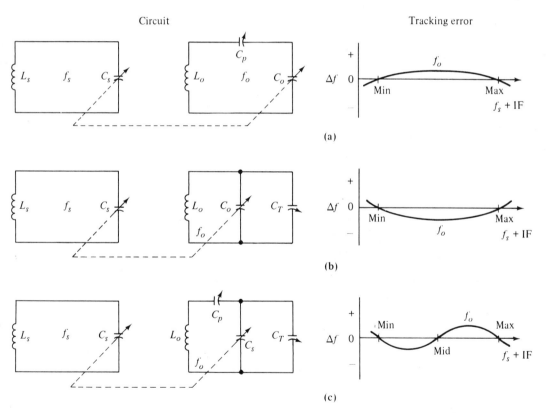

FIGURE 7.5. Superheterodyne receiver tracking methods: (a) padder tracking ; (b) trimmer tracking; (c) combination or three-point tracking. In each case both sections of the tuning capacitor have the same value, and the resulting tracking error is shown.

For three-point tracking, Figure 7.5(c), two equations in terms of C_p and C_T must be set up.

Example 7.4   Find the value of padder capacitor and oscillator inductor to give two-point tracking for the receiver of Example 7.1, assuming the value of $C_{s_{max}}$ to be 350 pF. Also, find the error in oscillator tracking frequency for a signal frequency of 1 MHz.

Solution   (a)   From Example 7.1(a),

$$\frac{C_{o_{max}}}{C_{o_{min}}} = 4.58 \qquad \frac{C_{s_{max}}}{C_{s_{min}}} = 10.24 \qquad \begin{array}{l} f_{o_{max}} = 2065 \text{ kHz} \\ f_{o_{min}} = 965 \text{ kHz} \end{array}$$

Now,

$$C_{s_{min}} = \frac{350}{10.24} = 34.2 \text{ pF}$$

and from equation (7.10)

$$4.58 = (10.24)\frac{(34.2 + C_p)}{(350 + C_p)}$$

giving

$$C_p = 221.3 \, pF$$

(b) $$C_{o_{max}} = \frac{350 \times 221.3}{350 + 221.3} = 135.6 \text{ pF}$$

$$C_{o_{min}} = \frac{34.2 \times 221.3}{34.2 + 221.3} = 29.6 \text{ pF}$$

giving

$$\frac{135.6}{29.6} = 4.58$$

as required. Then

$$L_o = \frac{1}{(2\pi \times .965 \text{ MHz})^2 \times 135.6 \text{ pF}} = 201.5 \, \mu\text{H}$$

(c)   At 1 MHz,

$$\frac{C_{s_{max}}}{C_{s_{mid}}} = \left(\frac{f_{mid}}{f_{min}}\right)^2 = \left(\frac{1000}{500}\right)^2 = 4$$

Then

$$C_{s_{mid}} = \frac{350}{4} = 87.5 \text{ pF}$$

and

$$C_{o_{mid}} = \frac{87.5 \times 221.3}{87.5 + 221.3} = 62.7 \text{ pF}$$

giving an actual value of

$$f'_{o_{mid}} = \frac{1}{2\pi\sqrt{L_o C_{o_{mid}}}} = \frac{1}{2\pi \times \sqrt{201.5 \ \mu H \times 62.7 \ pF}}$$

$$= 1418 \text{ kHz}$$

The desired value is

$$f_o = 1000 + 465 = 1465 \text{ kHz}$$

so that the tracking error is

$$err = 1418 - 1465 = -47 \text{ kHz}$$

The oscillator frequency is lower than it should be, so the station will appear at a point higher on the dial than marked, and the signal tuning circuit will be tuned 47 kHz high.

7.8   AUTOMATIC GAIN CONTROL

When a receiver without automatic gain control (AGC) is tuned to a strong station, the signal may overload the later IF and AF stages, causing severe distortion and a disturbing blast of sound. This can be prevented by using a manual gain control on the first RF stage, but usually some form of AGC is provided. The AGC derives a varying bias signal which is proportional to the average received signal strength and uses this bias to vary the gain of one or more IF and/or RF stages. When the average signal level increases, the size of the AGC bias increases, and the gain of the controlled stages decreases. When there is no signal, there is a minimum AGC bias, and the amplifiers produce maximum gain.

Simple AGC is used in most domestic receivers and many cheaper communications receivers. In simple AGC receivers the AGC bias starts to increase as soon as the received signal level exceeds the background noise level, and the receiver immediately becomes less sensitive. The AM detector used in these receivers is a simple half-wave rectifier which produces a dc level that is proportional to the average signal level. This dc level is put through an *RC* lowpass filter to remove the audio signal and then applied to bias the base of the RF and/or IF stage amplifier transistor. The time constant of the filter must be such that it is at least 10 times longer than the period of the lowest modulation frequency received, which is usually around 50 Hz, or about 0.2 s. If the time constant is made longer, it will give better filtering, but it will cause an annoying delay in the application of the AGC control when tuning from one signal to another. The circuit of Fig. 7.6(a) uses a time constant of about $\frac{1}{4}$ s.

The circuit shown uses the main signal detector diode for two purposes, detection and the provision of AGC bias. Some compromise in the performance of each is necessary, and in better receivers a second detector is used, especially for the AGC. Also, signals may be picked off earlier in the IF and fed to a separate IF amplifier to supply the AGC, thus reducing loading on the IF circuits.

Better response can be obtained by including more gain in the feedback loop. This is accomplished by providing a dc amplifier stage after the filter section. This amplifier produces a low enough source resistance so that several RF/IF stages may be easily driven from the same AGC line.

Connection to the RF stage in transformer-coupled circuits is easily accomplished. The lower end of the input transformer secondary is isolated from ground with a bypass capacitor and connected directly to the AGC line. A resistance back to the collector supply provides the bias current to the base necessary to maintain full class A gain when a very low signal or no signal is present. When several stages operating at the same frequency are connected to the same AGC line, decoupling between them must be used to prevent instability. This can easily be done by feeding the AGC to the earlier stage through a second filter section with the same time constant and providing good local bypassing at each stage input point.

Delayed AGC is used in most of the better communications receivers. Delayed AGC is obtained when the generation of the AGC bias is prevented until the signal level exceeds a preset threshold, and then increases proportionally after that. The threshold may be fixed by the circuit design or may be adjustable, and is usually adjusted to start taking effect when the signal has risen nearly to the level which produces the receiver maximum output under maximum sensitivity conditions (full gain). A delayed AGC response characteristic is illustrated in Fig. 7.6(b), where it is compared with responses for no AGC and simple AGC.

Delayed AGC is easiest to achieve when an AGC amplifier is included in the circuit. In this case, the amplifier is biased well beyond cutoff by a fixed bias source from which the AGC bias is subtracted. The AGC level then must overcome the fixed bias before any bias is passed on to the controlled amplifiers.

AGC is not usually supplied in the less expensive FM receivers, because these have sufficient amplifier gain so that the last stage operates in saturation for most signals anyway to obtain the necessary amplitude limiting for good detection. AGC may be provided on some FM receivers to prevent overloading the RF stage, and in this case it is used to control the RF stage so that saturation of the earlier IF stages does not occur on very strong signals.

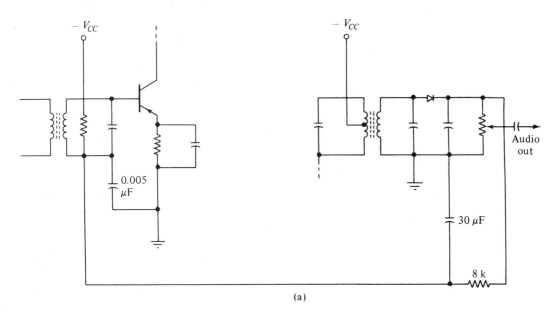

(a)

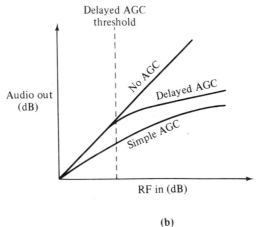

(b)

FIGURE 7.6. Automatic gain control: (a) simple AGC applied to an IF amplifier; (b) response of a receiver with either simple or delayed AGC compared to one without AGC.

Figure 10.21 shows the block diagram of an FM receiver which uses AGC. In this case a sample of the IF signal is extracted just before the input to the limiting IF amplifier which would remove variations in signal level. This sample is applied to a special detector used only to obtain the AGC signal, which is a peak amplitude detector circuit similar to that in Fig. 7.6. The derived AGC signal is then applied to control the RF preamplifier and the first IF amplifier. Its time constant would be similar to that used in an AM receiver.

In SSB receivers the size of the RF signal is zero at zero modulation, rising to a maximum during peak signal periods. There is no average carrier level to use for an AGC reference. AGC in this case is derived by taking a sample of the IF signal just before the sideband filters and SSB detector and applying that to a peak detector specially included for the AGC. An SSB receiver which uses this scheme is shown in Fig. 9.8. The AGC signal derived is used to control the RF and both IF amplifiers and also provides the gating signal to turn on the squelch, or muting, circuit. Because the signal has large time gaps between syllables, it is necessary that the AGC circuit provide fast attack, but very slow decay. Otherwise there will be annoying dropouts between syllables. Attack times of a few milliseconds with decay times of a few seconds are reasonable. Usually a means will be included to let the operator adjust the decay time to suit his preference.

In television receivers a form of *keyed AGC* is used. The waveform of the horizontal sync pulse shown in Fig. 18.19(c) is comprised of a black level which is used as an AGC threshold, with a high sync pulse superimposed on top of it. This pulse triggers the horizontal oscillator and starts the positive flyback pulse in the horizontal output. If there is no signal, the positive pulse from the flyback is used to charge the AGC line to a value near zero volts (average after filtering). When a strong signal is encountered, a negative-going pulse whose amplitude is proportional to the height of the received sync pulse above the black level is subtracted from the AGC charging pulse, and a negative AGC control voltage is generated to reduce the gain of the RF amplifiers. During that portion of the horizontal period outside of the horizontal sync pulse, the AGC circuit is allowed to "coast," discharging slowly through a resistive load. As shown in Fig. 18.20, the derived AGC signal is then applied to the RF preamplifier and two or more stages of the IF amplifier.

7.9 DOUBLE-CONVERSION RECEIVER

The front-end selectivity of any receiver must reject the first image frequency located at 2IF above the desired signal frequency, and for this to occur, the IF should be high. However, as the IF gets higher, its bandpass gets larger. Beyond the HF band (about 30 MHz), it becomes impossible to obtain the required bandpass with good image rejection using ordinary circuits. Thus,

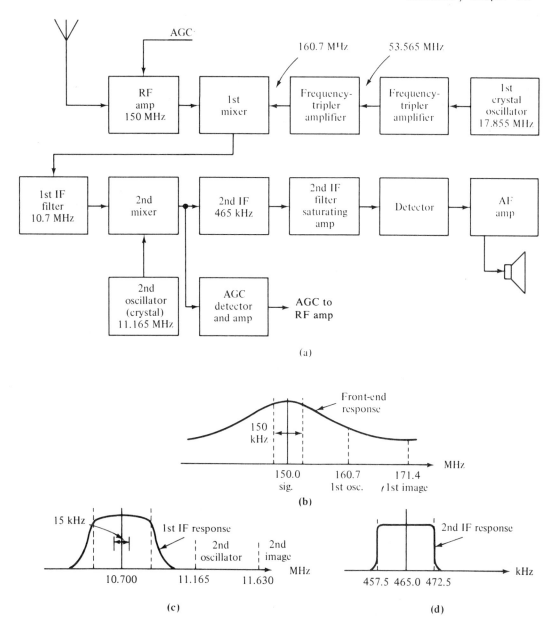

FIGURE 7.7. A double-conversion superheterodyne receiver: (a) block schematic; (b) RF stage response; (c) first IF stage response; (d) second IF stage response.

single-conversion superheterodyne receivers using only one IF are seldom used beyond 20 or 30 MHz.

The *double-conversion receiver* offers a way out of the dilemma, allowing the receiver to get good image rejection and narrow IF bandpass without compromise. Fig. 7.7(a) shows the block diagram of a double-conversion receiver which might be used for the 150 MHz FM mobile band. A high first IF of 10.7 MHz is chosen, which allows sufficient selectivity to be obtained in the RF stage tuning to reject the image of the first IF at 171.4 MHz. However, with 15-kHz spacing between channels, several adjacent channels are passed to the first mixer. Variable tuning is seldom used, although more than one channel may be used by switching, so the first oscillator is usually a set of crystals selected by a switch to operate an oscillator in the 10-MHz region, followed by a series of harmonic multiplier stages to obtain the desired oscillator signal in the 160-MHz region. The RF stage is fixed-tuned also, and usually is broad enough to pass several channels near the 150-MHz frequency, spaced 15 kHz apart. Figure 7.7(b) illustrates this RF-stage selectivity.

The first IF filter is a filter block designed with a bandpass of about 150 kHz (sufficient for FM broadcast signals) centered on 10.7 MHz. Usually the mixer stages will supply sufficient gain, and a separate amplifier is not required. The filters used are usually of the crystal-lattice type. Piezoelectric ceramic filters have also been developed which are somewhat cheaper than quartz crystals, but these do not provide as high Qs and have somewhat poorer selectivity characteristics. Figure 7.7(c) shows the selectivity of the first IF, which is sharp enough to eliminate the second image at 11.63 MHz but not sharp enough to eliminate the adjacent channels. Figure 7.7(d) shows the selectivity of the second IF filter centered on 465 kHz and 15 kHz wide, rejecting the adjacent channels at 480 kHz and 450 kHz.

Most of the communications in the VHF band above 30 MHz use FM, so the IF stages will be designed to saturate and provide amplitude limiting. An AGC signal is derived from the unlimited IF signal separately and is used to control the gain of the IF and RF amplifiers.

7.10 HF COMMUNICATIONS RECEIVERS

Communications receivers for use in the HF bands from 2 to 30 MHz are usually multipurpose receivers, built for several modes of communication. These modes of communication include normal AM broadcast signals; SSB voice communications, usually of the reduced or pilot-carrier type; interrupted carrier telegraph, including Morse, teletype, and digital data; and audio subcarrier telegraph (the subcarrier modulation may be interrupted tone, amplitude-modulated tone, FSK, or PSK, and there may be several channels on one voice carrier). These receivers are made continuously tunable over each of several frequency ranges covering the total spectrum from 1 to 30 MHz. Extra bands to cover the LF and MF bands from 100 kHz to 1.6 MHz may also be included.

Many different features may be included in a given communications receiver, and the block schematic of Fig. 7.8 shows an arrangement which illustrates several of these. The circuit is basically a double-conversion super-heterodyne receiver designed for reception of amplitude-modulated signals. The RF amplifier is double-tuned and designed to track the local oscillator. A multisection switch allows connection of one of four sets of coils and trimmers to a three-gang mechanical tuning capacitor, so that the range from 500 kHz to 30 MHz may be tuned in four overlapping ranges:

1. The MF broadcast range, from 500 to 1600 kHz;
2. 1.5 to 4.5 MHz, covering mostly aircraft and ship communications;
3. 4.5 to 10 MHz, covering general communication channels and two international broadcast bands;
4. 10 to 30 MHz, covering some international broadcasts and general communications on a sporadic basis, depending on ionospheric conditions.

Bandspread is provided by a separate variable capacitor in parallel with the main oscillator tuning capacitor, which allows a calibrated small variation of tuning around the main-dial indicated frequency. It is usually arranged to allow tuning above and below the indicated frequency.

The first IF is chosen in this case to be 465 kHz. The bandpass on this first IF can be fairly broad, say ± 15 kHz. Most of the adjacent channel selectivity results in the second IF, which is here chosen to be 150 kHz. A fixed local oscillator at 615 kHz provides the down-conversion.

The second IF is fed directly to an ordinary AM detector and the AGC circuit. The AGC in this circuit serves several functions. On fading signals, it performs its usual function of controlling the gain of the RF/IF amplifiers. The signal from the AGC detector drives a panel meter, which gives an indication of signal strength, calibrated in decibels. This signal-strength meter is useful in comparing the signal quality of received signals. The AGC signal also drives a muting or "squelch" circuit, which disconnects the audio output when signal strength falls below an adjustable threshold level, so disturbing noise during no-signal intervals is eliminated. A switch is provided to allow disabling of this circuit while tuning to different channels. Another switch allows disabling of the entire AGC circuit during reception of Morse telegraph.

A complete SSB detection circuit, including a carrier amplifier, switchable upper and lower sideband filters, and a balanced modulator, are provided. The sideband filters can be crystal, ceramic, or mechanical. Provision may also be made for ISB signals by including a local oscillator. Automatic frequency control may be used to stabilize the first oscillator when receiving reduced carrier SSB signals as well.

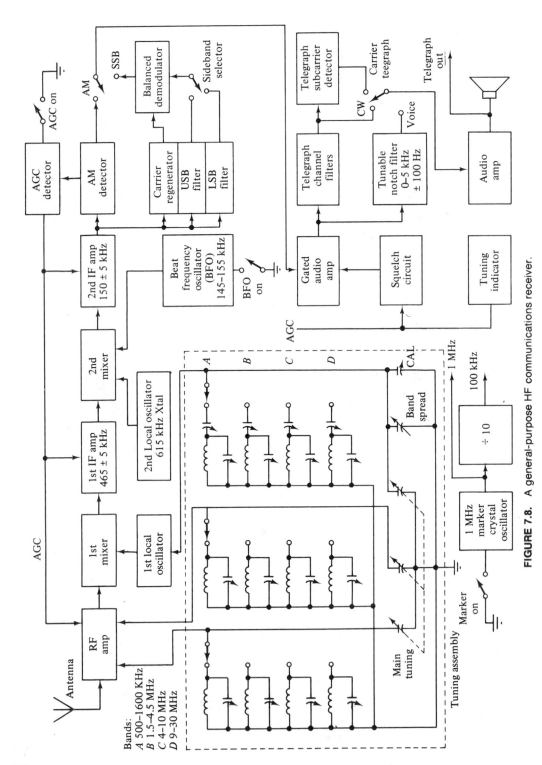

FIGURE 7.8. A general-purpose HF communications receiver.

Bands:
A 500–1600 KHz
B 1.5–4.5 MHz
C 4–10 MHz
D 9–30 MHz

250

The reception of interrupted carrier telegraph signals poses a special problem. An ordinary detector will not put out an audible signal. This is corrected by including a beat-frequency oscillator (BFO), which can be tuned to any position within the passband of the second IF. If this BFO is tuned 1 kHz to one side of center frequency, and a telegraph signal is being received at center frequency, the two will mix and produce a 1-kHz modulation as long as the carrier is on. The result is demodulated to produce the resulting telegraph signals as bursts of 1-kHz tone which are audible. If an external relay is to be driven from the telegraph signal, either the tone version produced by beating or the AGC signal may be used to drive it. Since received signals may be located within a few hundred hertz of each other in the spectrum, separation may be difficult. Sometimes a separate tunable notch filter which may be tuned to any position in the audio passband is used to "notch out" unwanted signals. Narrow-band IF filters are sometimes provided as well, which allow selection of passbands down to as narrow as 400 Hz.

When FSK telegraph is to be received, special telegraph channel filters are provided in the audio circuit. For FSK, "mark" and "space" frequencies combine with the BFO signal to produce mark and space tones. A separate filter is used to separate each of these tones and pass them on to a digital circuit, which reconstructs the telegraph signal in pulsed dc form. Since these filters are narrow-band, they can also be used to separate the beat frequency of an interrupted carrier signal from background noise.

Tuning calibration is sometimes a problem in multiband receivers, and a special "marker" oscillator is usually provided to facilitate the recalibration. The marker oscillator is usually a 100-kHz (or a 1-MHz) crystal oscillator which may be switched on when needed. It has a trimmer that allows a few hertz variation of the marker frequency. The output signal drives a saturating amplifier to produce harmonics at 100-kHz intervals across the entire reception range. Dial-calibration adjustment is allowed either by another panel-mounted trimmer on the oscillator or by a mechanical offset device on the tuning scale. Calibration is carried out by first tuning the receiver to a frequency standard station (such as WWV on 10.0000 MHz) and zero-beating the marker oscillator. Zero-beating is achieved by simply adjusting the oscillator frequency until the audible beat-tone frequency drops to zero. This will be accurate within the low-frequency cutoff limit of the receiver audio circuits, usually about 100 Hz. Then the receiver is tuned to the portion of the dial in which calibration is desired. The bandspread trimmer is adjusted to zero, and the main dial is adjusted by using the BFO set to the center IF frequency and zero-beating with the main tuning. Then the calibration adjustment is made to bring the scale calibration marks into line. The BFO and marker may then be turned off, and the dial will give a true indication as long as the bandspread tuner is kept at zero.

Many variations of this type of receiver have been produced, ranging in price from less than $100 for inexpensive kit models to several thousand dollars for the more sophisticated models. The advent of LSI, of course, will make a distinct improvement in the level of sophistication which can be obtained for a given price. Until recently, vacuum tubes provided the best characteristics from the point of view of noise, linearity, and spurious response, but they tended to be bulky. Entirely transistorized circuits are most commonly available now, and soon integrated-circuit versions should be available.

7.11 PROBLEMS

1. A receiver tunes the 3–30 MHz HF band in one range, using an IF of 40.525 MHz. Calculate the range of oscillator frequencies, the range of image frequencies, and the types of filters needed to make the receiver function properly.

2. A superheterodyne receiver is to tune the range from 4–10 MHz, with an IF of 1.8 MHz. A triple-gang capacitor with a maximum capacity per section of 325 pF is to be used. Find (a) the RF-circuit coil inductance; (b) the RF-circuit frequency tuning ratio; (c) the RF-circuit capacitance tuning ratio; (d) the required minimum capacity per section; (e) the oscillator frequency maximum and minimum and the frequency ratio; and (f) the capacitance ratio for the oscillator.

3. (a) Calculate the image frequency range for the receiver of Problem 2. Do any image frequencies fall in the receiver passband? (b) If the front-end circuits have combined effective Q of 50 at the top end of the band, calculate the image rejection ratio in decibels at that frequency.

4. (a) Calculate all the possible spurious response frequencies for the receiver of Problem 2, assuming that it is tuned to a signal at 7.3 MHz and that only harmonics up to the third are significant. (b) Which ones fall within ± 2IF of the signal frequency at 7.3 MHz?

5. The oscillator section of the receiver in Problem 2 is to be made, using a padder capacitor. Determine the value of padder capacitor and oscillator inductor needed.

6. Repeat Example 7.4 for the trimmer circuit of Fig. 7.5(b).

7. Find the values of C_T and C_p required for the circuit of Fig. 7.5(c), assuming the receiver of Example 7.1, a 350-pF-per-section tuning capacitor, and a tracking crossover frequency of 1000 kHz. (Note: This is a more difficult problem.)

PART 3

MODULATION OF SIGNALS

CHAPTER 8

Amplitude Modulation

8.1 INTRODUCTION

To modulate means to regulate or adjust, and specifically in the case of telecommunications it means to regulate some parameter of a high-frequency *carrier* wave by means of the lower-frequency information signal. The need for modulation first arose in the radio transmission of low-frequency (e.g., audio-frequency) signals. It was found that for efficient radiation, antenna dimensions had to be of the same order as the wavelength of the signal being radiated. As shown in Appendix B, Eq. (B.4), the frequency f and wavelength λ of an electromagnetic wave are related through the phase velocity v_p by

$$f\lambda = v_p \tag{B.4}$$

Most low-frequency information signals have frequencies on the order of 1 kHz, and since electromagnetic waves in space travel at the speed of light, the wavelength would be

$$\lambda = \frac{300 \times 10^6}{1000} \text{ m} = 300 \text{ km}$$

which is about 188 miles. It is obviously impractical to build antennas of this size.

The problem is overcome by using the low-frequency signal to modulate a higher frequency (shorter wavelength) signal called a *carrier wave*, which is then radiated. The carrier wave is always sinusoidal, and its voltage-time variation is represented by the equation

$$e_c = E_{c_{max}} \sin(\omega_c t + \theta) \tag{8.1}$$

The parameters of this wave that may be modulated are (1) $E_{c_{max}}$ for amplitude modulation; (2) f_c (or $\omega_c = 2\pi f_c$) for frequency modulation; (3) θ for phase modulation. Frequency and phase modulation both come under the general heading of angle modulation, which is covered in Chapter 10.

Modulation also led to the development of a form of transmission known as *frequency-division multiplexing*, which is discussed in Chapter 9.

Most of the important properties of amplitude modulation can be studied using the assumption that the modulating (low-frequency) signal is a sine or cosine wave (it having been shown in Chapters 2 and 3 that real basic signals are composed of series of sine/cosine waves). Unless otherwise stated, the modulating signal will be assumed to be a single sinusoid represented by

$$e_m = E_{m_{max}} \sin \omega_m t \tag{8.2}$$

where

$$\omega_m = 2\pi f_c$$

8.2 AMPLITUDE MODULATION

When a carrier wave is amplitude modulated, the amplitude of the carrier voltage waveform is caused to vary directly with the modulating voltage, so that

$$e = \left(E_{c_{max}} + e_m \right) \sin \omega_c t \tag{8.3}$$

where e is the instantaneous voltage of the modulated signal, $E_{c_{max}}$ is the peak carrier voltage without modulation, and e_m is the instantaneous modulating voltage.

Figure 8.1 shows the time variation of the modulated signal over one cycle, assuming that both the carrier and the modulating signal are sinusoids. The peaks of the carrier cycles may be joined to form an *envelope wave*, which is given by

$$e_{env} = E_{c_{max}} + e_m \tag{8.4}$$

where e_{env} is the instantaneous value of the envelope waveform.

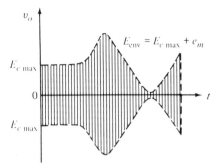

FIGURE 8.1. Waveform of an amplitude-modulated signal.

Substituting for e_m from Eq. (8.2) into Eq. (8.4) and for e_{env} from Eq. (8.4) into Eq. (8.3) gives the modulated signal voltage as

$$e = e_{env} \sin \omega_c t$$
$$= \left(E_{c_{max}} + E_{m_{max}} \sin \omega_m t \right) \sin \omega_c t \qquad (8.5)$$

A useful measure of the degree of modulation is the *modulation index m*, defined as

$$m = \frac{E_{m_{max}}}{E_{c_{max}}} \qquad (8.6)$$

In terms of m, Eq. (8.5) may be written as

$$e = E_{c_{max}}(1 + m \sin \omega_m t) \sin \omega_c t \qquad (8.7)$$

Without loss of generality, the carrier amplitude may be assumed to be 1 V, so that Eq. (8.7) becomes

$$e = (1 + m \sin \omega_m t) \sin \omega_c t \qquad (8.8)$$

Equation (8.8) is sketched in Fig. 8.2 for three different values of m. It will be seen that, for m greater than unity, the inward peaks of the envelope are clipped, as the carrier completely disappears when the modulator circuit is driven into cutoff. This condition must be avoided, as it results in distortion of the modulating signal; such distortion also produces a form of interference known as sideband splatter, described in the next section.

The smallest value of m is clearly zero (corresponding to $E_{m_{max}} = 0$), so the practical limitations on m may be conveniently expressed by $0 \le m \le 1$.

8.2.1 Frequency Spectrum

The idea of a spectrum has already been introduced in Chapter 2, where it is seen that the spectrum shows the amplitude and frequency of the component sine and cosine waves making up a complex wave. The amplitude-modulated wave of Eq. (8.8) is complex, and therefore it also can be analyzed

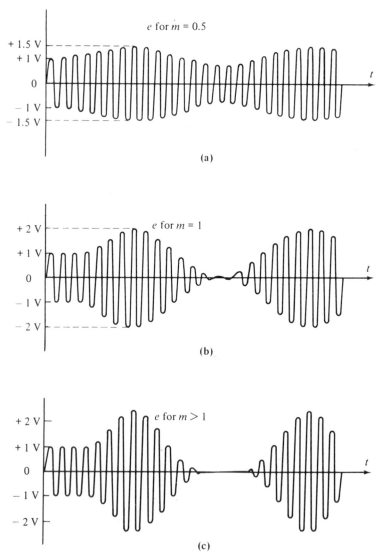

FIGURE 8.2. Modulated output voltage waveform for various values of modulation index m: (a) for $m = 0.5$ (undermodulated); (b) for $m = 1.0$ (fully modulated); (c) for $m > 1.0$ (overmodulated).

into sine and cosine components:

$$e = (1 + m \sin \omega_m t) \sin \omega_c t$$

$$= \sin \omega_c t + m \sin \omega_m t \sin \omega_c t$$

$$= \sin \omega_c t + \frac{m}{2} \left[\cos(\omega_c - \omega_m)t - \cos(\omega_c + \omega_m)t \right] \qquad (8.9)$$

The student should be able to derive this result making use of the trigonometric identity

$$\cos(A \pm B) = \cos A \cos B \mp \sin A \sin B \qquad (8.10)$$

Equation (8.9) consists of three separate components which may be considered to be three separate sinusoidal generators in series, as shown in Fig. 8.3(a). The first term on the right-hand side is clearly the carrier wave of amplitude 1 V and frequency f_c, since $\omega_c = 2\pi f_c$. The second term is a cosine wave of amplitude $\frac{1}{2}m$ and frequency $f_c - f_m$, since $\omega_c - \omega_m = 2\pi(f_c - f_m)$. This component is known as the *lower side frequency*. The third term is also a cosine wave of amplitude $\frac{1}{2}m$ and frequency $f_c + f_m$. This is known as the *upper side frequency* component. The spectrum representing the amplitude modulated wave is shown in Fig. 8.3(b). It will be recalled that $E_{c_{max}}$ was set equal to 1 V. For other values of $E_{c_{max}}$, the heights of the spectrum arrows would be correspondingly scaled.

The frequency analysis expressed by Eq. (8.9) is of more than mathematical interest, and, in fact, one of the most important practical means of signal transmission, known as *single-sideband transmission*, is based on it. (Single-sideband methods are described in Chapter 9.)

Example 8.1   A carrier wave of frequency 10 MHz and peak value 10 V is amplitude-modulated by a 5-kHz sine wave of amplitude 6 V. Determine the modulation index and draw the spectrum.

Solution   By equation (8.6),

$$m = \frac{6}{10} = 0.6$$

The side frequencies are $10 \pm 0.005 = 10.005$ and 9.995 MHz. The amplitude of each side frequency, from Eq. (8.9), is $(m/2)E_{c_{max}}$ (note that this is the same as $(\frac{1}{2})E_{m_{max}}$), which is 3 V. The spectrum is shown in Fig. 8.3(c).

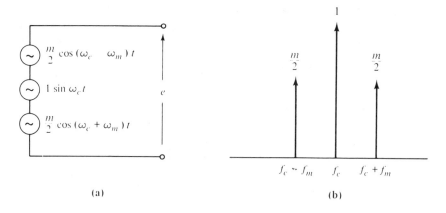

(a)   (b)

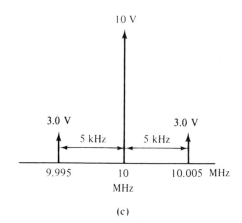

(c)

FIGURE 8.3. (a) Generator representation of a modulated wave; (b) spectrum of a 100%-modulated waveform; (c) spectrum of a 10-MHz carrier modulated to 50% with a 5-kHz signal.

8.2.2 Average Power

Equation (8.9), expressed slightly differently, is

$$e = \sin \omega_c t + \frac{m}{2} \cos \omega_L t - \frac{m}{2} \cos \omega_U t$$

which, for brevity, may be written as

$$e = e_C + e_L - e_U \qquad (8.11)$$

where

e_C = instantaneous carrier voltage

e_L = instantaneous lower side frequency voltage

e_U = instantaneous upper side frequency voltage

The instantaneous power developed by the modulated wave across a resistor R is

$$p = \frac{e^2}{R}$$

$$= \frac{(e_C + e_L - e_U)^2}{R}$$

$$= \frac{e_C^2}{R} + \frac{e_L^2}{R} + \frac{e_U^2}{R} + \frac{2}{R}(e_C e_L - e_L e_U - e_C e_U) \qquad (8.12)$$

The squared terms represent the instantaneous power from each of the component waves. The average power in a sine or cosine wave is $P_{av} = E^2/R$, where E is the rms voltage. The rms carrier voltage is $E_c = 0.707 E_{c_{max}}$, and the average power associated with the first squared term (the carrier) on the right-hand side of Eq. (8.12) may be written as

$$P_c = \frac{E_c^2}{R} \qquad (8.13)$$

The rms side frequency voltage is $E_s = 0.707(m/2)E_{c_{max}} = (m/2)E_c$, and therefore the average power in each side frequency component (corresponding to the instantaneous power terms, second and third on the right-hand side of Eq. (8.12) is

$$P_{sf} = \frac{E_s^2}{R}$$

$$= \frac{m^2}{4}P_c \qquad (8.14)$$

Considering now the cross-product terms of Eq. (8.12) (the terms within the parentheses), it is easily shown, in a manner similar to that of Eq. (8.9), that these cross-product terms may be expressed as the sum and difference of sine and cosine waves. This means that the instantaneous power represented by each cross-product term is also sinusoidal (or cosinusoidal), and therefore the mean, or average, value is zero. The total average power P_T in the modulated wave therefore consists of the sum of the separate component averages:

$$P_T = P_c + \frac{m^2}{4}P_c + \frac{m^2}{4}P_c \qquad (8.15)$$

Therefore,

$$P_T = P_c\left(1 + \frac{m^2}{2}\right) \qquad (8.16)$$

For 100% modulation, $m = 1$, so that the average power in either side frequency is, from Eq. (8.14),

$$P_{sf} = \tfrac{1}{4}P_c \tag{8.17}$$

Also, from Eq. (8.16), the total average power is

$$P_T = 1.5P_c \tag{8.18}$$

Therefore, the ratio of single side frequency power to total power is $1/6$. Since the side frequency contains all the information (amplitude and frequency) of the modulating signal, it follows that only one side frequency need be transmitted, which results in a much more efficient use of transmitted power. In practice, of course, the modulating signal will be complex, so the power ratio of $1/6$ will not necessarily apply; also, a sideband as discussed in Section 8.2.4, rather than just a side frequency, will have to be transmitted. It is still generally true, however, that much more efficient use of transmitted power is made with single-sideband transmission.

8.2.3 Effective Voltage and Current

Let the *effective*, or rms, *voltage* of the modulated wave be E; then the total average power P_T can be expressed as

$$P_T = \frac{E^2}{R} \tag{8.19}$$

Comparing this with Eq. (8.16),

$$\frac{E^2}{R} = P_c\left(1 + \frac{m^2}{2}\right)$$

$$= \frac{E_c^2}{R}\left(1 + \frac{m^2}{2}\right) \tag{8.20}$$

where E_c is the rms voltage of the unmodulated carrier. It follows from Eq. (8.20) that

$$E = E_c\sqrt{1 + \frac{m^2}{2}} \tag{8.21}$$

A similar argument can be applied to currents, resulting in

$$I = I_c\sqrt{1 + \frac{m^2}{2}} \tag{8.22}$$

where I is the rms current of the modulated carrier and I_c is the rms current of the unmodulated carrier. Equation (8.20) is used as the basis for one method of measuring modulation index, by measuring antenna current, although the method is not very sensitive. Let the ratio of modulated to unmodulated rms currents be

$$I_r = \frac{I}{I_c} \tag{8.23}$$

Then in terms of I_r, Eq. (8.20) becomes

$$m = \sqrt{2(I_r^2 - 1)} \tag{8.24}$$

By monitoring antenna current and setting the reading corresponding to I_c equal to unity, the ammeter may be calibrated directly in terms of m. It is essential that a true rms-reading ammeter be used.

Example 8.2   The rms antenna current of a radio transmitter is 10A when unmodulated, rising to 12 A when the carrier is sinusoidally modulated. Calculate the modulation index.

Solution   By Eq. (8.23),

$$I_r = \frac{12}{10} = 1.2$$

By Eq. (8.24),

$$m = \sqrt{2(1.2^2 - 1)} = 0.94$$

8.2.4 Modulating Signal for a Complex Wave

Suppose, as shown in Chapter 2, that the modulating signal is of the form

$$e_m = E_1 \sin \omega_m t + E_2 \sin 2\omega_m t + E_3 \sin 3\omega_m t + \ldots \tag{8.25}$$

The modulated carrier wave will be given by Eq. (8.3):

$$e = \left(E_{c_{max}} + e_m \right) \sin \omega_c t$$

It follows that an expansion similar to that used to obtain Eq. (8.9) may be applied, resulting in a pair of side frequencies for each harmonic component of the modulating wave. A modulation index for each harmonic component may be defined:

$$m_1 = \frac{E_1}{E_{c_{max}}}$$

$$m_2 = \frac{E_2}{E_{c_{max}}} \quad \text{etc.}$$

The spectrum is shown in Fig. 8.4(a). It is seen from this that *sidebands* occur, consisting of the various upper and lower side frequencies.

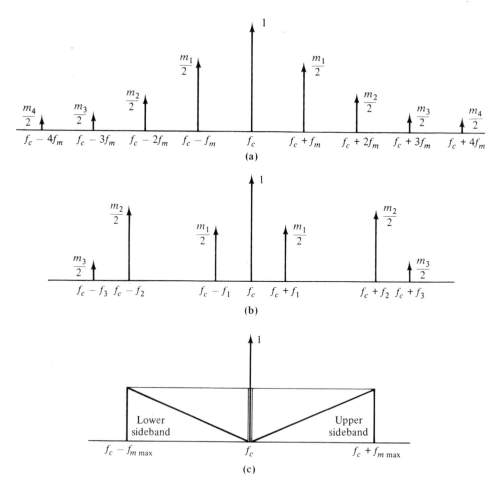

FIGURE 8.4. Frequency spectrum of an AM carrier signal modulated with a signal containing several frequencies: (a) spectrum containing harmonically related modulating frequencies; (b) spectrum containing several unrelated modulating frequencies; (c) modulation band representation for a complex modulating signal such as music or speech.

Similar reasoning may be applied when the modulating signal consists of sine waves which are not necessarily harmonically related, e.g.,

$$e_m = E_1 \sin \omega_1 t + E_2 \sin \omega_2 t + E_3 \sin \omega_3 t$$

the spectrum then being as shown in Fig. 8.4(b).

In general, the spectrum for the modulated wave is usually shown as in Fig. 8.4(c), where the modulating signal consists of a band of frequencies such as speech or music. Modulating signal frequencies for AM broadcast signals in the MF band (550 to 1600 kHz) are limited to a maximum of 5 kHz, and station allocations are spaced 10 kHz apart, requiring a transmission bandwidth of 10 kHz.

An expression for the average power in a complex-modulated wave can be derived in exactly the same manner as Eq. (8.16), giving

$$P_T = P_c\left(1 + \left(\tfrac{1}{2}\right)\left(m_1^2 + m_2^2 + m_3^2 + \ldots\right)\right) \tag{8.26}$$

It follows that the rms values for complex wave modulation are given by

$$E = E_c\sqrt{1 + \left(\tfrac{1}{2}\right)\left(m_1^2 + m_2^2 + m_3^2 + \ldots\right)} \tag{8.27}$$

$$I = I_c\sqrt{1 + \left(\tfrac{1}{2}\right)\left(m_1^2 + m_2^2 + m_3^2 + \ldots\right)} \tag{8.28}$$

Measurement of rms voltage or current to determine modulation index will yield an effective value m_{eff}, where

$$P_T = P_c\left(1 + \frac{m_{\text{eff}}^2}{2}\right) \tag{8.29}$$

Comparison of Eqs. (8.26) and (8.29) gives

$$m_{\text{eff}} = \sqrt{m_1^2 + m_2^2 + m_3^2 + \ldots} \tag{8.30}$$

It must be noted that in order to avoid overmodulation, the *modulation depth* must not exceed 100%, and this is generally more restrictive than the condition $m_{\text{eff}} \leq 1$. Modulation depth is defined as the ratio of peak downward change in amplitude to carrier amplitude, expressed as a percentage. This is illustrated in Fig. 8.5, thus:

$$\text{modulation depth} = \frac{E_m}{E_{c_{\max}}} \times 100\% \tag{8.31}$$

Only for sinusoidal modulation does $E_m = E_{m_{\max}}$, and then

$$\text{modulation depth} = m \times 100\% \tag{8.32}$$

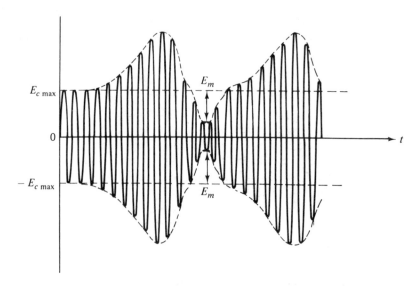

FIGURE 8.5. Voltage waveform of an AM carrier modulated with a complex signal. The values of $E_{c_{max}}$ and E_m are used in Eq. (8.31) for calculation of modulation depth.

8.2.5 Oscilloscope Display of Modulation

A pattern such as the one illustrated in Fig. 8.5 may be obtained directly on an oscilloscope, and modulation depth measured directly from this. A better method, the *trapezoidal method*, is to apply the modulated wave to the vertical deflection circuit of the oscilloscope and the modulating signal to the horizontal deflection circuit. The resulting display will be as shown in Fig. 8.6(c), where for simplicity a sinusoidal modulating voltage is assumed. Until point A on the modulating wave is reached, the display traces out a vertical line between A and A' on the oscilloscope. As the modulating voltage moves on to B, the vertical display gradually increases in length until it reaches a maximum at B-B'. As the modulating voltage decreases to C, the display reduces in length, passing through A-A' again, and reaching a minimum at C-C'.

Comparing Fig. 8.6(c) with 8.6(b), we have

$$L_1 = 2\left(E_{c_{max}} + E_{m_{max}}\right)$$
$$L_2 = 2\left(E_{c_{max}} - E_{m_{max}}\right)$$

Therefore,

$$\frac{L_1}{L_2} = \frac{E_{c_{max}} + E_{m_{max}}}{E_{c_{max}} - E_{m_{max}}}$$

$$\frac{L_1}{L_2} = \frac{1 + m}{1 - m}$$

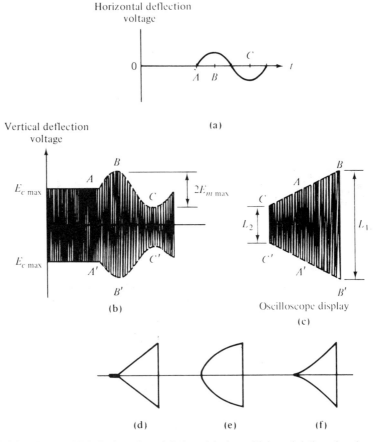

FIGURE 8.6. Trapezoidal display of modulation: (a) sinusoidal modulating signal applied to the horizontal input; (b) modulated carrier waveform with sinusoidal modulation signal applied directly to the vertical input of the oscilloscope; (c) resulting trapezoidal display; (d) pattern showing overmodulation; (e) and (f) patterns showing envelope distortion.

from which

$$m = \frac{L_1 - L_2}{L_1 + L_2} \qquad (8.33)$$

It can be seen that the trapezoidal pattern will be obtained on the oscilloscope for modulating voltages other than sinusoidal, since the horizontal deflection is synchronized with the peaks of the vertical deflection (the modulation envelope). If overmodulation occurs, the trapezoid will become a triangle with a tail as shown in Fig. 8.6(d); the onset of this condition is sharply pronounced, so it is a good indication of modulation conditions. Also, distortion in the modulation process shows up as distortion on the diagonal sides, as indicated in Fig. 8.6(e) and (f).

8.2.6 Amplitude-Modulator Circuits

Several methods of accomplishing amplitude modulation of an RF carrier exist, but the most commonly used one is to modulate the output electrode DC supply voltage to a class C tuned power amplifier. This method will be discussed here. Although vacuum tubes will continue to be used for very high power transmitters, transistors are being used at increasingly higher power levels. Even in the higher power units modulation is initially performed at a low power level and then the modulated signal is boosted by another power amplifier after the modulation stage. As a result, only transistor modulator circuits will be considered here.

Figure 8.7(a) shows the circuit of a bipolar transistor collector modulator. The amplifier is a common emitter class C amplifier similar to the circuit discussed in Section 5.7.1. Tuned transformer T1 couples the unmodulated carrier wave into the base circuit of the amplifier transistor. Bias components R_B, C_B act with the base-emitter junction to create the self-bias voltage required to maintain class C operation, so that the transistor only conducts for a very brief period near the positive peak of each input cycle.

These collector current pulses drive the output tuned tank circuit formed by C_2 and the primary inductance L_2 of the output transformer T2. The modulating signal v_m is added to the dc supply voltage V_{CC}(dc) in the

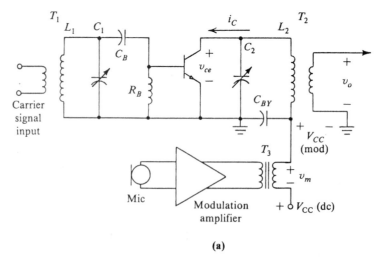

(a)

FIGURE 8.7. A bipolar transistor collector modulator: (a) circuit; (b) collector supply voltage variation with modulation; (c) collector voltage variation with modulation; (d) output voltage variation with modulation; (e) collector current variation with modulation.

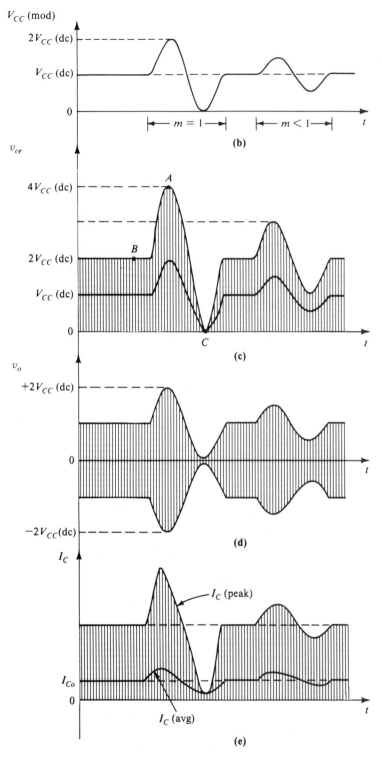

FIGURE 8.7. (Continued).

secondary of the modulation amplifier output transformer T3, so that the effective dc supply provided to the modulator circuit $V_{CC}(\text{mod})$ is not constant, but varies around the dc level with the modulation signal. This voltage is given by

$$V_{CC}(\text{mod}) = V_{CC}(\text{dc}) + v_m$$
$$= V_{CC}(\text{dc}) + E_{m_{\max}} \sin \omega_m t \qquad (8.34)$$

where the modulating voltage v_m is sinusoidal as given in Eq. (8.2). If the modulator is driven to its maximum modulation level allowed without creating distortion, then $V_{CC}(\text{mod})$ will swing between zero and $2V_{CC}(\text{dc})$, at which point $E_{m_{\max}}$ will be equal to $V_{CC}(\text{dc})$ and m will be 1. Substituting this into Eq. (8.6) and substituting from Eq. (8.6) for $E_{m_{\max}}$ into equation (8.34) gives

$$V_{CC}(\text{mod}) = V_{CC}(\text{dc})(1 + m \sin \omega_m t) \qquad (8.35)$$

The time waveforms of $V_{CC}(\text{mod})$ for $m = 1$ and for $m < 1$ are shown in Fig. 8.7(b).

The carrier input drive to the amplifier is adjusted so that at all levels of $V_{CC}(\text{mod})$ the voltage on the collector swings between zero and $2V_{CC}(\text{mod})$ around an average value of $V_{CC}(\text{mod})$. This happens only if the transistor is driven just to, but not beyond, saturation at the positive peak of each carrier cycle. Figure 8.7(c) shows the collector voltage waveform for one cycle of modulation at $m = 1$ and another at $m < 1$. At point A, modulation voltage is at a positive maximum, so that the carrier voltage is oscillating between zero and $V_{CE}(\text{max}) = 4V_{CC}(\text{dc})$, around an average value of $V_{CC}(\text{mod})\text{max} = 2V_{CC}(\text{dc})$. At point B, the modulation voltage is zero, and the carrier voltage on the collector swings between zero and $2V_{CC}(\text{dc})$ around the average value of $V_{CC}(\text{dc})$. At point C, the modulation voltage reaches a negative peak, and both the average and peak values of collector voltage are reduced to zero. These conditions at points A and C represent the maximum modulation levels which can be allowed without causing envelope distortion in the output.

When the signal produced at the collector is passed through the output transformer T2, the "dc," or average, level corresponding to $V_{CC}(\text{mod})$ is subtracted from the collector voltage, leaving only the RF carrier oscillating about zero as shown in Fig. 8.7(d), which is plotted for the conditions in Fig. 8.7(c). This voltage can easily be seen to be a proper AM signal by comparing its waveform to that in Fig. 8.1.

The equation for the collector voltage at any value of supply voltage $V_{CC}(\text{mod})$ is given by the equation

$$v_{ce} = V_{CC}(\text{mod})(1 + \sin \omega_c t) \qquad (8.36)$$

as long as the carrier is sinusoidal and the collector voltage swings symmetrically to zero on each cycle, without being clipped. Substituting from Eq. (8.36) for $V_{CC}(\text{mod})$ gives

$$
\begin{aligned}
v_{ce} &= V_{CC}(\text{dc})(1 + m \sin \omega_m t)(1 + \sin \omega_c t) \\
&= V_{CC}(\text{dc})(1 + m \sin \omega_m t) \\
&\quad + V_{CC}(\text{dc})(1 + m \sin \omega_m t)\sin \omega_c t
\end{aligned}
\tag{8.37}
$$

Subtracting out the average level given by Eq. (8.35) leaves the output voltage v_o as

$$
v_o = V_{CC}(\text{dc})(1 + m \sin \omega_m t)\sin \omega_c t
\tag{8.38}
$$

which can be seen to be the same as the general amplitude-modulated voltage given by Eq. (8.7).

If the modulator operates in a linear manner, then the current flowing in the collector (which is a train of positive pulses) will have an "average" value which will vary in a sinusoidal manner around the value which occurs for zero modulation signal, as shown in Fig. 8.7(e). This average value can be expressed as

$$
I_c(\text{avg}) = I_{co}(1 + m \sin \omega_m t)
\tag{8.39}
$$

The direct current flowing from the collector power supply is found by taking the average of this current, which, since the modulation component is alternating and has a zero average, becomes a constant value

$$
I_{dc} = \text{avg}(I_c(\text{avg})) = I_{co}
\tag{8.40}
$$

for all values of modulation. The power drawn from the collector supply is given by

$$
P_{dc} = V_{CC}(\text{dc})I_{dc} = V_{CC}(\text{dc})I_{co}
\tag{8.41}
$$

which also remains constant for all values of modulation. It must be noted that this is the power input required to the modulator when the modulating voltage is zero.

With modulation, the total power input to the collector is, with substitutions from Eqs. (8.35), (8.39), and (8.41) and some manipulation,

$$
\begin{aligned}
P_{coll} &= V_{CC}(\text{mod})I_c(\text{avg}) \\
&= V_{CC}(\text{dc})I_{co}(1 + m \sin \omega_m t)^2 \\
&= P_{dc}\left(\left(1 + \frac{m^2}{2}\right) + \left(2m \sin \omega_m t - \frac{m^2}{2}\cos 2\omega_m t\right)\right)
\end{aligned}
\tag{8.42}
$$

which is varying with modulation. It has an average value which, since the sinusoidal components have zero average value, is just

$$P_{coll}(\text{avg}) = P_{dc}\left(1 + \frac{m^2}{2}\right) \tag{8.43}$$

Subtracting out the power delivered by the dc supply gives, from Eqs. (8.41) and (8.43), the modulator output power:

$$P_o(\text{mod}) = P_{coll}(\text{avg}) - P_{dc} = \frac{m^2}{2}P_{dc} \tag{8.44}$$

The carrier voltage input or drive to the modulator must be adjusted as the modulation voltage varies, so that at all times the current pulses are just the right amplitude to cause the collector voltage to just bottom out at zero on each carrier negative peak. This condition is shown on the transistor collector characteristics for the load line bb′ with the collector supply voltage set at $V_{CC}(\text{dc})$, Fig. 8.8(a). It is assumed that the input carrier level is just right so that at the positive peak the collector voltage reaches almost to zero and i_c reaches its maximum at point b′. The current pulse is shown in Fig. 8.8(b) and the collector voltage swing is shown in time synchronism in Fig. 8.8(c).

If the modulation voltage is now increased so that the load line rises as shown by line aa′, but the drive level is kept the same, then the collector current will rise only to the level at point a″, at which point the collector voltage is still considerably above zero. This represents a condition of under-drive, with the current pulse limited to a smaller amplitude than desired and the collector voltage being reduced in amplitude with its negative peak lifted off the axis. More base drive would have to be applied in order to raise the peak current up to the desired level at a′ and increase the output voltage to its proper magnitude.

If, now, the modulation voltage is decreased so that the load line moves down to line cc′, the current pulse will be clipped off square at c′ as the transistor is driven into saturation. There would still be sufficient drive to force i_c to point c″ if it were possible for the transistor to conduct with negative voltage applied. The output voltage amplitude is reduced to its desired level, but it is deformed because the saturation of the transistor clamps the output voltage to zero for the duration of the conduction angle. The result is the creation of excess carrier harmonics in the output voltage, although no serious change in the modulation envelope occurs.

Figures 8.8(d) and (e) show the effect of insufficient drive on the collector and output voltage waveforms, resulting in envelope clipping. While this clipping may be avoided simply by increasing the drive to the maximum value needed to prevent clipping at the highest positive peak of the modulation voltage, this solution results in excessive harmonic content in the output which

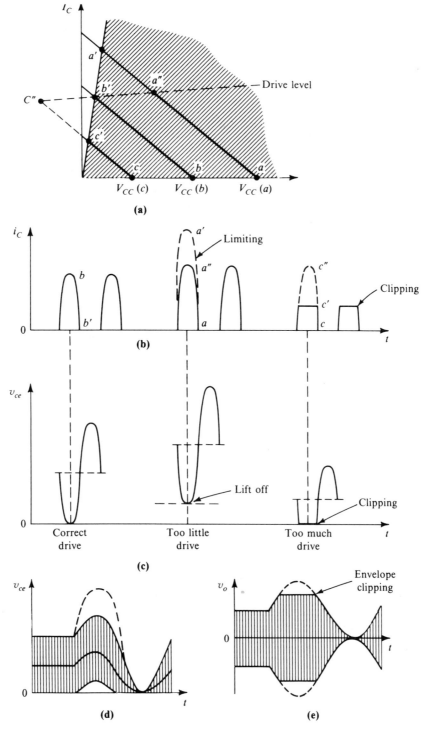

FIGURE 8.8. Effects of incorrect carrier level drive on a collector modulator: (a) collector characteristic showing variation of load line with modulation; (b) effects on collector current pulses; (c) effects on collector carrier waveform; (d) effects on collector voltage over modulation cycle; (e) envelope clipping as a result of insufficient drive.

it may not be possible to filter out. A better solution is obtained either by supplying the class C modulator amplifier with an extra dc bias level which increases and decreases with the modulation voltage or by partially modulating a previous stage so that the input signal level rises and falls with the modulation voltage. The latter solution is used in the circuit of Fig. 8.9.

In this circuit, the transistor Q_3 stage is fully modulated, while diodes MSD6100 (dual-package diodes) allow stage Q_2 to be modulated on the upward modulation swing; when the modulating voltage swings below 13.6 V, the diode connected to the modulated supply ceases to conduct, thus cutting Q_2 off from the modulation while the other diode conducts, connecting Q_2 to the unmodulated 13.6-V supply. This means that the RF drive to Q_3 is increased at the same time as the collector voltage increases because of modulation, thus increasing the drive to the Q_3 output stage and preventing clipping.

The tuned output stage for Q_3 is the series circuit of 470 pF and L_3, while the various radio-frequency chokes and capacitors in the collector line are for the purposes of filtering. The modulating amplifier is not shown in the figure.

8.2.7 Demodulation

It is, of course, necessary to provide a circuit at the receiver which enables the information signal—the envelope—to be recovered from the modulated wave. The most common circuit in use is the envelope detector, which produces an output voltage proportional to the envelope of the input wave. The basic circuit is shown in Fig. 8.10(a). The diode acts as a rectifier and can be considered an "on" switch when the input voltage is positive, allowing the capacitor C to charge up to the peak of the RF input. During the negative half of the RF cycle, the diode is "off," but the capacitor holds the positive charge previously received, so the output voltage remains at the peak positive value of RF. There will, in fact, be some discharge of C, producing an RF ripple on the output waveform, which must be filtered out.

As the input voltage rises with the modulation cycle, the capacitor voltage has no difficulty in following this, but during the downward swing in modulation the capacitor may not discharge fast enough, unless a discharge path is provided by the resistor R. The time constant of the CR load has to be short enough to allow the output voltage to follow the modulation cycle, and yet long enough to maintain a relatively high output voltage. The constraints on the time constant are determined more precisely in the next section.

Applying Kirchhoff's voltage law to the circuit, the diode voltage v_d is found to be

$$v_d = e - v \tag{8.45}$$

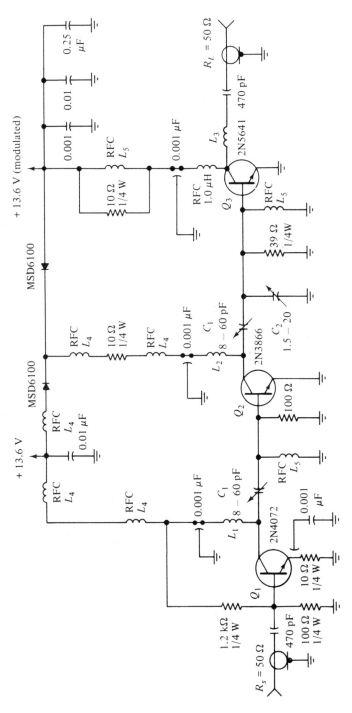

FIGURE 8.9. A transistorized collector modulator circuit. Extracted from *Motorola Application Note AN507*. (Courtesy of Motorola, Inc.)

L_1 = 6T #26 wire wound on toroid (micrometals T30-13) with $\frac{3}{32}$ spacing

L_2 = 2T #26 wire wound on toroid (see L_1) with $\frac{1}{8}$ spacing

L_3 = 2T #26 wire wound on toroid (see L_1) with $\frac{5}{16}$ spacing

L_4 = RF bead (one hole), $\frac{1}{8}$''

L_5 = Ferrite Choke (Ferroxcube VK 200)

C_1 = 8 – 60 pF (Arco 404)

C_2 = 1.5-20 pF (Arco 402)

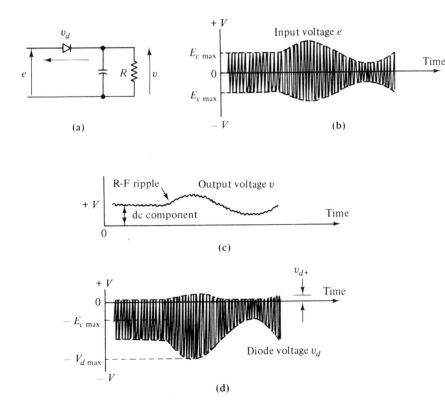

FIGURE 8.10. The diode envelope detector: (a) the basic detector circuit; (b) the modulated input waveform; (c) the output voltage waveform; (d) the voltage across the diode.

where e is the input voltage and v the output voltage. Figure 8.10(b) shows e, and Figure 8.10(c) shows v, both for sinusoidal modulation. By graphically subtracting v from e, the graph of v_d in Figure 8.10(d) is obtained. It is interesting to see that v_d is positive only for very short periods, as indicated by the peaks $v_d +$, and it is during these peaks that the capacitor is charged to make up for discharge losses. Also, the peak voltage across the diode is twice the output voltage, which can rise to $4E_{c_{max}}$ at 100% sinusoidal modulation. This should be compared with the conditions at the collector of the modulator described in the previous section.

Diagonal peak clipping. This is a form of distortion which occurs when the time constant of the CR load circuit is too long, thus preventing the output voltage from following the modulation envelope. Figure 8.11(b) shows how the CR discharge curve may control the shape of the output voltage. At some time t_A, the modulation envelope starts to decrease more rapidly than the capacitor discharges, so the output voltage follows the discharge law of the CR circuit until it once again meets up with the modulation envelope on the rise, at t_B.

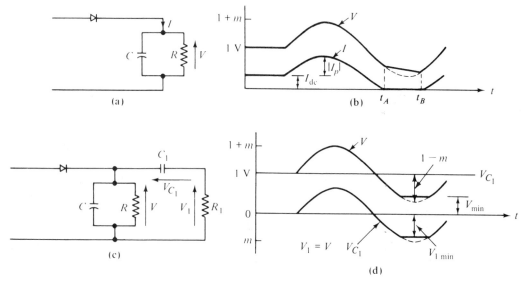

FIGURE 8.11. (a) Basic detector circuit; (b) output waveform showing diagonal peak clipping; (c) detector circuit showing the next stage input resistance; (d) detector output voltage showing negative peak clipping.

For sinusoidal modulation, the condition necessary for avoiding distortion of this type is found as follows. The diode (Fig. 8.11(a)) appears to supply the RC load with a current I, the modulation component of which has a peak value

$$|I_p| = \frac{m}{|Z_p|} \tag{8.46}$$

(where $|Z_p|$ is the modulus of $R\|C$, at modulating frequency) and a dc component

$$I_{dc} = \frac{1}{R} \tag{8.47}$$

(Recall that $E_{c_{max}} = 1$ V.)
    If the envelope falls faster than the capacitor discharges, the diode ceases to conduct (since the capacitor voltage biases it off), and the apparent supply current I goes to zero. During the period the current is zero, the load voltage follows the discharge law of the CR network, resulting in the diagonally clipped peak as shown in Fig. 8.11(b). Clearly, for the avoidance of diagonal peak clipping,

$$I_{dc} \geq |I_p|$$

Thus,

$$\frac{1}{R} \geq \frac{m}{|Z_p|}$$

or

$$\frac{|Z_p|}{R} \geq m_{max} \qquad (8.48)$$

where the maximum modulation index m_{max} is typically about 0.9.

Negative peak clipping. This is similar to diagonal peak clipping and results from the loading effect of the network following the diode load, C_1 and R_1 (Figure 8.11(c)). C_1 is a dc blocking capacitor, and R_1 represents the input resistance of the following stage. C_1 is large ($\omega_m C_1 R_1 \gg 1$), so that the modulation component of voltage is passed unattenuated to R_1 and, as a result, C_1 maintains a constant charge at the mean load voltage ($= 1$ V). V, and hence V_1, follow the modulation envelope, maintaining the relationship $(V - V_1) = V_{C_1} =$ constant. However, V cannot drop below the minimum level set by

$$V_{min} = V_{C_1}\frac{R}{R + R_1} = \frac{R}{R + R_1} \qquad (\text{for } V_{C_1} = 1 \text{ V}) \qquad (8.49)$$

Below this minimum level, capacitor C can no longer discharge to follow the modulation envelope because it is now being charged from C_1 through the voltage divider formed by R and R_1. Instead it is held at the voltage V_{min} which is the output voltage of the voltage divider from C_1, whose voltage does not change appreciably over the modulation cycle. Eq. (8.49) is derived by applying Kirchhoff's voltage law to the loop formed by R_1, C_1, R. From Fig. 8.11(d), it can be seen that the minimum level of V, which is $(1 - m_{max})$, must be kept greater than V_{min} in order to avoid negative peak clipping:

$$(1 - m_{max}) \geq \frac{R}{R + R_1}$$

Hence,

$$\frac{R_1}{R + R_1} \geq m_{max} \qquad (8.50)$$

Eq. (8.47) can be rearranged as

$$\frac{R_p}{R} \geq m_{max} \qquad (8.51)$$

where $R_p = (R \cdot R_1)/(R + R_1)$. Equation (8.51) should be compared with Eq. (8.48).

8.3 AMPLITUDE-MODULATED TRANSMITTERS

The *amplitude-modulated (AM) transmitter* is basically the same as the keyed CW transmitter described in Section 17.13, except that the final power-amplifier stage is replaced by a modulator amplifier. Figure 8.12(a) shows the block diagram of a typical AM transmitter. The carrier source is a crystal-controlled oscillator at the carrier frequency or a submultiple of it. This is followed by a tuned buffer amplifier and a tuned driver, and if necessary frequency multiplication is provided in one or more of these stages.

The modulator circuit used is generally a class C power amplifier that is collector modulated as described in Section 8.2.6. The audio signal is amplified by a chain of low-level audio amplifiers and a power amplifier. Since this amplifier is controlling the power being delivered to the final RF amplifier, it must have a power driving capability which is one-half the maximum power the collector supply must deliver to the RF amplifier under 100% modulation conditions. A transformer-coupled class B push-pull amplifier is usually used for this purpose.

Low-power transmitters with output powers up to 1 kW or so may be transistorized, but as a rule the higher-power transmitters use vacuum tubes in the final amplifier stage, even though the low-level stages may be transistorized. In some cases where the reliability and high overall efficiency of the transistor is mandatory, higher powers can be obtained by using several lower-power transistorized amplifiers in parallel. The system is complicated, and usually the vacuum-tube version will do the same job at lower capital cost.

Sometimes the modulation function is done in one of the low-level stages. This allows low-power modulation and audio amplifiers, but it complicates the RF final amplifier. Class C amplifiers cannot be used to amplify an already modulated (AM) carrier, because the transfer function of the class C amplifier is not linear. The result of using a following class C amplifier would be an undesirable distortion of the modulation envelope that contains the signal. A linear power amplifier, such as the push-pull class B amplifier (Section 5.7.3), must be used to overcome this problem (Fig. 8.12(b)). Unfortunately, the efficiency of this type of amplifier is lower than that of the comparable class C amplifier, resulting in more costly equipment. Larger tubes or transistors must be used which are capable of dissipating the additional heat generated.

The output of the final amplifier is passed through an impedance-matching network which includes the tank circuit of the final amplifier. The Q of this circuit must be low enough so that all the sidebands of the signal are passed without amplitude/frequency distortion, but at the same time must present an appreciable attenuation at the second harmonic of the carrier frequency. The

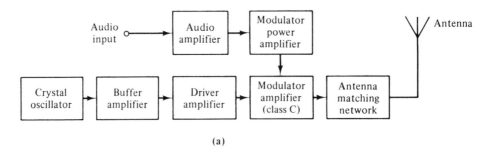

(a)

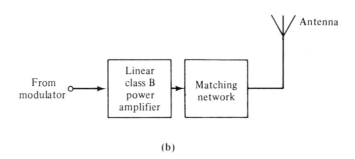

(b)

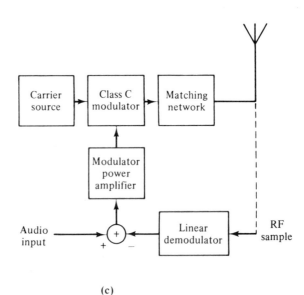

(c)

FIGURE 8.12. Amplitude-modulated transmitters: (a) transmitter with a modulated class C final power amplifier; (b) linear class B push-pull power amplifier used when modulation takes place in a low-level stage; (c) negative feedback applied to linearize a class C modulator.

bandwidth required in most cases is a standard 3 dB at ±5 kHz around the carrier. For amplitude-modulation broadcast transmitters, this response may be broadened so that the sidebands will be down less than 1 dB at 5 kHz, where music programs are being broadcast and very low distortion levels are desired, or special sharp-cutoff filters may be used. Because of the high power levels present in the output, this is not usually an attractive solution.

Negative feedback is quite often used to reduce distortion in a class C modulator system. The feedback is accomplished in the manner shown in Fig. 8.12(c), where a sample of the RF signal sent to the antenna is extracted and demodulated to produce the feedback signal. The demodulator is designed to be as linear in its response as possible and to feed back an audio signal that is proportional to the modulation envelope. The negative feedback loop functions to reduce the distortion in the modulation.

8.3.1 AM Broadcast Transmitters

Most of the domestic AM broadcast services use the medium-wave band from 550 to 1600 kHz. International AM broadcasts take place in several of the HF bands scattered from 1600 kHz up to about 15 MHz. The mode of transmission in all cases is double-sideband full carrier, with an audio baseband range of 5 kHz. Station frequency assignments are spaced at 10 kHz intervals, and power outputs range from a few hundred watts for small local stations to as much as 100 kW in the MW band and even higher for international HF transmitters.

A main requirement of an AM broadcast transmitter is to produce, within the limits of the 5-kHz bandwidth available, the highest possible fidelity. The modulator circuits in the transmitter must produce a linear modulation function, and every trick available is used to accomplish this. A typical AM broadcast transmitter is shown in Fig. 8.13. The crystal main oscillator is temperature-controlled to provide frequency stability. It is followed by a buffer amplifier, and then by tuned class C amplifiers which provide the necessary power gain to drive the final power amplifier. For high power output, vacuum tubes would be used as described next. The modulator system is the "triple equilibrium" system, in which the main part of the modulation is performed by plate-modulating the final class C power amplifier. Secondary modulation of both the final grid and the plate of the driver stage is also included to compensate for bias shift in the final amplifier that is due to the nonlinear characteristic of the amplifier.

The final power amplifier is a push-pull parallel stage in which each side of the push-pull stage is comprised of several vacuum tubes operating in parallel, to obtain the power required. A further advantage of this system is that if one or more of the tubes in the system should fail, the remaining tubes will provide partial output until repairs can be made, thus making a more

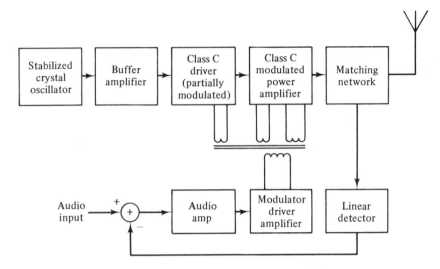

FIGURE 8.13. An AM broadcast transmitter.

secure system. Power dissipation in these final tubes can be as high as 50 kW, in addition to several kilowatts of heater power. Water cooling systems are used to dissipate the large quantities of heat produced.

The modulator amplifier is an audio frequency push-pull parallel amplifier, which is transformer-coupled to the modulator. The audio preamplifier stage includes a difference amplifier and an envelope detector which demodulates a sample of the transmitter output and uses the signal to provide negative feedback. This feedback further linearizes the modulation characteristic of the system.

Antenna systems for AM transmitters are large and usually must be located at some point remote from the studio operations. All the studio signal operations are performed at relatively low levels and transmitted to the main transmitter location, either over telephone wire lines or a radio link such as a microwave system. Quite often the transmitters are unattended and are remotely controlled from the studio location. Service personnel make periodic visits to do routine maintenance.

8.4   AM RECEIVERS

The general principles of the superheterodyne receiver are described in Chapter 7, and specific operating details of the AM envelope detector are discussed in Section 8.2.7. Most receivers in use today are assembled from discrete components, although there is a trend toward the use of integrated circuits for subsections in the receiver. Therefore, in this section, a very commonly encountered transistorized receiver will be described, followed by the description of two integrated circuit type receivers.

8.4.1   Discrete Component AM Receiver

The circuit for a standard broadcast receiver using discrete components is shown in Fig. 8.14. This is a superheterodyne receiver, transistor Q1 functioning as both a mixer and an oscillator in what is known as an autodyne mixer. The oscillator feedback is through mutual inductive coupling from collector to emitter, the base of Q1 being effectively grounded at oscillator frequency.

The AM signal is coupled into the base of Q1 via coil L1. Thus, it is seen that Q1 operates in grounded base mode for the oscillator while simultaneously operating in grounded emitter mode for the signal input.

Tuned IF transformer T1 couples the IF output from Q1 to the first IF amplifier Q2. The output from Q2 is also tuned-transformer-coupled through T2 to the second IF amplifier Q3. The output from Q3, at IF, is tuned-transformer-coupled to the envelope detector D2, which has an RC load consisting of a 0.01 μF capacitor in parallel with a 25-kΩ potentiometer. This potentiometer is the manual gain control, the output from which is fed to the audio preamplifier Q4. The audio power output stage consists of the push-pull pair Q5, Q6.

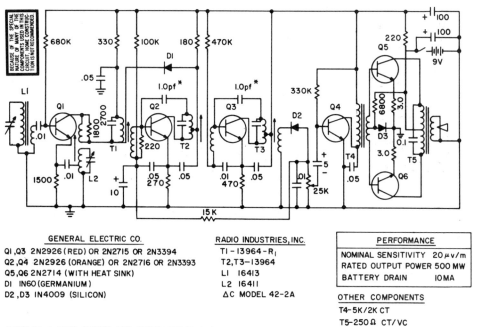

GENERAL ELECTRIC CO.

Q1,Q3 2N2926 (RED) OR 2N2715 OR 2N3394
Q2,Q4 2N2926 (ORANGE) OR 2N2716 OR 2N3393
Q5,Q6 2N2714 (WITH HEAT SINK)
D1 IN60 (GERMANIUM)
D2,D3 IN4009 (SILICON)

RADIO INDUSTRIES, INC.

T1 - 13964 - R₁
T2,T3 - 13964
L1 16413
L2 16411
ΔC MODEL 42-2A

PERFORMANCE	
NOMINAL SENSITIVITY	20 μv/m
RATED OUTPUT POWER	500 MW
BATTERY DRAIN	10 MA

OTHER COMPONENTS

T4-5K/2K CT
T5-250 Ω CT/VC

✳USE 1.0pf WITH 2N2926 AND 2N3391 SERIES TRANSISTORS, 0.5pf WITH 2N2715 SERIES.

FIGURE 8.14.   A six-transistor 9-V receiver. (Courtesy General Electric Company.)

Automatic gain control (AGC) is also obtained from the diode detector D2, the AGC filter network being the 15-kΩ resistor and 10-μF capacitor (Fig. 8.14). The AGC bias is fed to the Q2 base.

Diode D1 provides auxiliary AGC action. At low signal levels D1 is reverse biased, the circuit being arranged such that the collector of Q1 is more positive than the collector of Q2. As the signal level increases, the normal AGC bias to Q2 reduces Q2 collector current, resulting in an increase in Q2 collector voltage. A point is reached where this forward-biases D1, the conduction of D1 then damping the T1 primary and so reducing the mixer gain.

8.4.2 AM Receiver Using Integrated Circuit Subsystem

Figure 8.15 shows the details of the integrated subsystem, and Figure 8.16 shows how this is connected as an AM receiver. The package itself, illustrated in Fig. 8.15(a), is a 14-pin dual-in-line package, nominally measuring $19 \times 6.3 \times 3.2$ mm, and yet it contains six major circuit blocks, as shown in Figure 8.15(b). These blocks are integrated on one chip, the complete subsystem circuit being shown in Fig. 8.15(c). The circuit action is better understood in relation to the AM receiver application shown in Fig. 8.16. It will be immediately seen that certain bulky components, notably the tuned circuits, have to be added externally.

The AM signal input is fed via a tuned coupled circuit to pins 11 and 12, which are the RF inputs. Pin 11 is grounded at RF, while pin 12 is connected to the base of transistor Q2 (Fig. 8.15(c)). Transistors Q2 and Q3 form a cascode amplifier (see section 5.3.2). The RF output at pin 13 is tuned-circuit-coupled to the mixer input, pin 1. The mixer transistors are Q6, Q7, connected as an emitter-coupled pair with the base of Q7 being grounded for RF and IF signals. The local oscillator signal is generated by another emitter-coupled pair Q4, Q5 (requiring an external tuned circuit at pin 2). The oscillator is seen to control the current to the mixer pair (Fig. 8.15(c)), and hence multiplicative mixing occurs between the AM input and the oscillator signals. An IF output is obtained from the mixer output, pin 14, which is tuned-transformer-coupled back into the integrated circuit at pin 7. This goes to the input of a cascode IF amplifier pair Q9, Q10, and the amplified IF output, at pin 6, is fed via a tuned transformer to an envelope detector.

The AGC in this case employs a diode system separate from the detector diode. The IF signal is coupled via the external 5-pF capacitor connected between pins 5 and 6 to the AGC diode D2 (see Fig. 8.15(c)). The rectifying action of D2, along with the filtering action of the externally connected capacitor at pin 10 (Fig. 8.16), produces an AGC bias which is fed through the coil between pins 11 and 12 (Fig. 8.16) to the base of the RF amplifier Q2. Transistor Q1, along with associated circuitry, provides the reference bias for the AGC operation, and diode D1 provides auxiliary AGC action for large signals through its damping action on the IF_2 transformer in a manner similar to that described for diode D1 in Fig. 8.14.

PIN CONFIGURATION

N PACKAGE

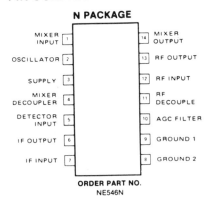

ORDER PART NO.
NE546N

(a)

BLOCK DIAGRAM

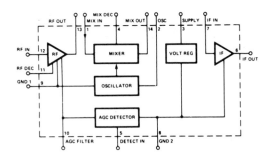

(b)

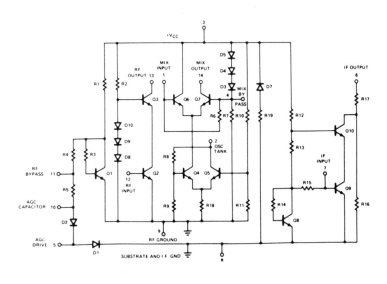

(c)

FIGURE 8.15. AM radio receiver subsystem available in integrated circuit form: (a) the package details, nominally measuring 19 × 6.3 × 3.2 mm; (b) block diagram of the subsystem; (c) equivalent schematic of the subsystem. "Permission to reprint granted by Signetics Corporation, a subsidiary of U.S. Philips Corp., 811 E. Arques Avenue, Sunnyvale, CA 94086."

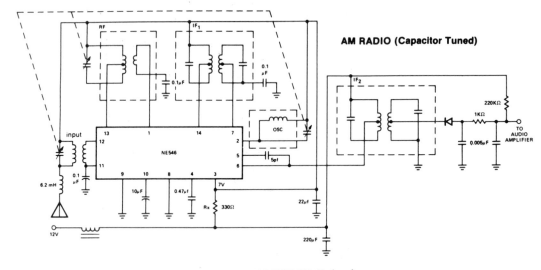

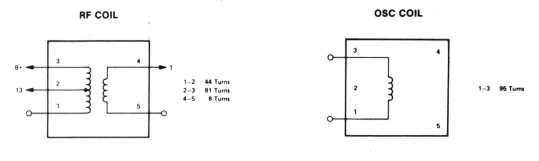

AM RADIO (Capacitor Tuned)

VARIABLE CAPACITOR (Air Varicon)
ANT & RF   13pF ~ 190pF
OSC   12pF ~ 80pF

ANTENNA COIL
    10mm φ < 120mm Ferrite Antenna

RF COIL

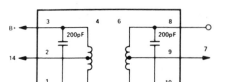

    1–2   44 Turns
    2–3   81 Turns
    4–5   8 Turns

OSC COIL

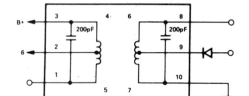

    1–3   95 Turns

1st. IF COIL

2nd. IF COIL

FIGURE 8.16. Details of an AM radio receiver utilizing the subsystem of Fig. 8.15. The receiver is for the standard broadcast band. "Permission to reprint granted by Signetics Corporation, a subsidiary of U.S. Philips Corp., 811 E. Arques Avenue, Sunnyvale, CA 94086."

Diode D7 and resistor R19 form the voltage regulator, and the diode chains D10, D9, D8 and D3, D4, D5 provide bias levels for Q3 and Q7, respectively. Transistor Q8 sets the bias level for Q9.

The active circuitry is easily integrated on a single chip, and because transistors are much more economical to fabricate than resistors and capacitors in integrated circuits, the circuit design philosophy is to use transistors and diodes wherever possible for bias control.

8.4.3 AM Receiver Using a Phase-Locked Loop (PLL)

Figure 8.17(a) shows the basic circuit blocks in a phase-locked loop (PLL), yet another approach to the use of integrated circuits in receivers. The input signal is compared in phase with a locally generated oscillation. The

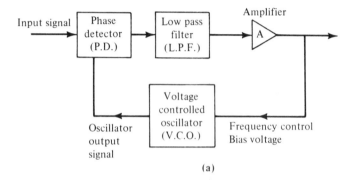

(a)

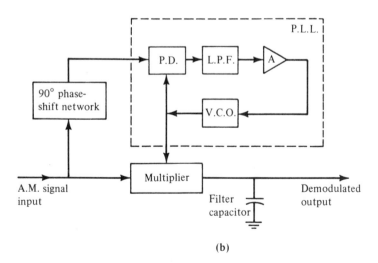

(b)

FIGURE 8.17.   (a) The basic phase-locked loop (PLL). (b) AM receiver utilizing the PLL.

output of the phase detector contains a low-frequency component which reduces to dc when both signals are of the same frequency. The low-pass filter (LPF) selects this component, which is then amplified and fed back to the oscillator. This bias alters the frequency of the oscillator in such a way as to reduce the output of the phase detector. Since the oscillator frequency is controlled by bias voltage, it is termed a voltage controlled oscillator (VCO). In practice, the oscillator is often a multivibrator type circuit producing a square wave output. For a sinusoidal input signal, the circuit, when properly adjusted, will lock the oscillator fundamental frequency to the input frequency, and the dc output of the phase detector is then proportional to $V_i \cos \theta_i$, where V_i is the amplitude of the input and θ_i is the phase angle between input and oscillator fundamental. This dc bias automatically alters the frequency of the oscillator such that the bias itself reduces to zero, a condition which is met when the phase angle θ_i is $90°$.

When applied to detection of AM signals, the VCO is locked exactly in frequency and phase to the carrier, and hence to compensate for the $90°$ phase difference which occurs within the PLL, the carrier is externally shifted by a further $90°$, as shown in Fig. 8.17(b). It does not matter whether the total phase difference is zero or $180°$, the fundamental component of the VCO output will be proportional to $\sin \omega_c t$, where ω_c is the angular frequency of the carrier.

Let the modulating signal be represented by $m(t)$ and the AM signal, therefore, by $(1 + m(t))\sin \omega_c t$. The output from the multiplier, Fig. 8.17(b) is proportional to

$$(1 + m(t))\sin \omega_c t \sin \omega_c t = \frac{(1 + m(t))}{2}(1 - \cos 2\omega_c t) \qquad (8.52)$$

PHASE LOCKED AM RECEIVER

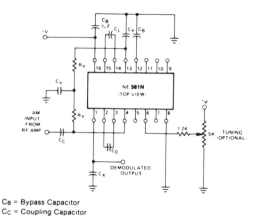

C_B = Bypass Capacitor
C_C = Coupling Capacitor

(a)

FIGURE 8.18. (a) Phase-locked AM receiver.

SCHEMATIC DIAGRAM OF 561N

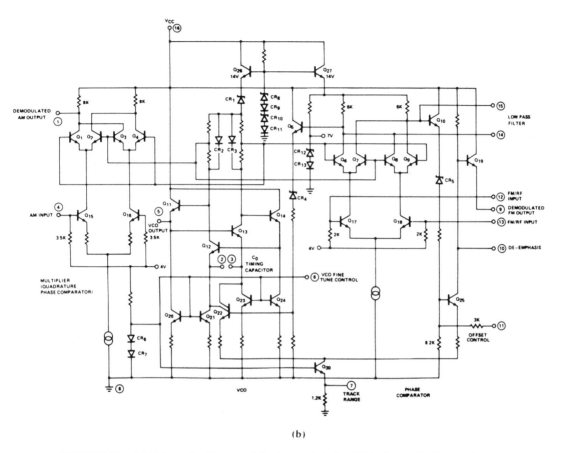

(b)

FIGURE 8.18. (b) Schematic diagram of the integrated circuit block used in the receiver. "Permission to reprint granted by Signetics Corporation, a subsidiary of U.S. Philips Corp., 811 Arques Avenue, Sunnyvale, CA 94086."

The $\cos 2\omega_c t$ term is filtered out and the modulating signal recovered from the term $(\frac{1}{2})m(t)$. This type of detection is known as *synchronous detection*.

The circuit for a receiver using a commercially available PLL is shown in Fig. 8.18(a), and the circuit for the PLL unit is shown in Fig. 8.18(b). For a complete description of the circuit, the reader is referred to the Signetics Analog Data Manual. Briefly, Q11, 12, 13, 14 form the multivibrator for the VCO. Q10, 25, 22 amplify the feedback control signal from the PLL phase detector, acting through Q21, 23 to control the emitter currents of Q12, 13 and thus the charging current to the multivibrator timing capacitor C_o (connected to pins 2 and 3) to change the VCO frequency.

The PLL phase detector consists of Q6, 7, 8, 9 and Q17, 18. Q6, 9 and Q7, 8 are coupled through a resistor network to the VCO output Q12, 13. The

AM signal (which may be the output of a conventional AM IF amplifier) is shifted by 90° by the network R_y, C_y and fed to the phase detector input on pin 13. Typical values are $R_y = 3000\ \Omega$, $C_y = 135$ pF for standard broadcast reception. Bypass capacitor C_B from pin 12 completes the shifted signal return path. The output from the phase detector from Q7 drives the feedback amplifier Q10, 25, 22. Capacitor C_L connected between pins 14, 15 provides the low-pass filtering of the control signal.

Transistors Q1, 2, 3, 4 and Q15, 16 form the multiplier circuit for the AM detector. This circuit operates in the same manner as the phase detector for the phase-locked loop. Q1, 4 and Q2, 3 are coupled to the VCO output (providing the synchronous carrier signal), and Q15, 16 are differentially driven from the unshifted AM input signal at pin 4. The demodulated output from Q1, 3 appears at pin 1, with C_x acting as a low-pass filter to remove the RF components before the signal is passed to an audio amplifier.

It is worth noting that the action of the two multiplier circuits in this chip is identical to that of the balanced modulator circuit described in Section 9.3.2, which is also a multiplier. Also, with minor circuit connection changes, the same circuit can also be used to demodulate an FM signal, operating in the manner described in Section 10.6.5.

It should also be noted that no external tuned circuits are required with the PLL detector, since, once the VCO locks onto the incoming carrier, selectivity is automatically achieved. In practice, some RF selectivity will be provided ahead of the PLL detector to prevent the VCO locking onto large unwanted carriers.

8.5 PROBLEMS

1. Amplitude modulation is applied via the output electrode of a class C amplifier. The direct voltage to the electrode is 10 V, and the peak sinusoidal modulating voltage is 4 V. Assuming ideal operating conditions for the class C stage, calculate the peak RF voltage at the output electrode.

2. A carrier of 10-V peak and frequency 100 kHz is amplitude-modulated by a sine wave of 4 V peak and frequency 1000 Hz. Determine the modulation index for the wave, and sketch the spectrum.

3. The positive RF peaks of an AM voltage wave rise to a maximum value of 15 V and drop to a minimum value of 5 V. Determine the modulation index and the unmodulated carrier amplitude, assuming sinusoidal modulation.

4. An AM transmitter has an unmodulated power output of 1 kW. Plot the power output against modulation index, for sinusoidal modulation. Also plot power output against modulation index squared. Determine the

increase in power when the modulation index is 0.5. Average powers are to be used in all cases.

5. The rms antenna current from an AM transmitter increases by 15% over the unmodulated value when sinusoidal modulation is applied. Determine the modulation index.

6. A complex modulating waveform consisting of a sine wave of amplitude 3 V and frequency 1000 Hz plus a cosine wave of amplitude 5 V and frequency 3000 Hz amplitude-modulates a 500-kHz, 10-V peak carrier voltage. Plot the spectrum of the modulated wave, and determine the average power when the modulated wave is fed into a 50-Ω load.

7. In a trapezoidal pattern displaying modulation, the length of the long vertical side is 5 cm, and of the short vertical side, 2 cm. Determine the modulation depth.

8. The dc power to a modulated class C amplifier is 500 W. Determine the power the modulator must supply for 100% sinusoidal modulation applied at the output electrode.

9. A diode detector load consists of a 0.01-μF capacitor in parallel with a 5-kΩ resistor. Determine the maximum depth of sinusoidal modulation that the detector can handle without diagonal peak clipping when the modulating frequency is (1) 1000 Hz; (b) 10,000 Hz.

10. Distinguish between *negative peak clipping* and *diagonal peak clipping* in an envelope detector. The output of a diode envelope detector is fed through a dc blocking capacitor to an amplifying stage, which has an input resistance of 10 kΩ. If the diode load resistor is 5 kΩ, determine the maximum depth of sinusoidal modulation the detector can handle without negative peak clipping.

11. If the detector in Problem 9 is coupled through a 0.1-μF capacitor into an amplifier with an input resistance of 500 kΩ, what modulation index can the detector handle without negative peak clipping occurring?

12. Explain why the driver stage of a class C collector modulator should also be partially modulated.

13. A collector modulator circuit is tuned up, a sinusoidal modulating signal of 1 kHz is applied, and the following measurements are made:

dc supply voltage = 12 V
dc supply current = 10 A
unmodulated rms current to a 50-Ω load = 1.50 A
Trapezoidal pattern monitor offsets = 1 cm and 5 cm

Find (a) the modulation index; (b) the output carrier power; (c) the output power with modulation; (d) the dissipation in the transistor with modulation (assuming the transistor is only source of losses); and (e) the collector conversion efficiency with modulation.

0-66
112·5 w
13)- 5 w
9·2 w
93·75 %

14. The modulator circuit of Fig. 8.7(a) is found to produce some envelope distortion even though it is properly adjusted. How could this nonlinearity be removed?

15. How is synchronous demodulation of an amplitude-modulated signal accomplished? Show a circuit which will do this, and explain how it works.

16. What should the minimum voltage rating of the collector of Q3 in Fig. 8.9 be? What total power with 100% modulation could the circuit be expected to put out? What power output could be expected if the battery voltage dropped to 11.5 volts?

CHAPTER 9

Single-Sideband Modulation

9.1 INTRODUCTION

Communications in the HF bands have become increasingly crowded in recent years, requiring closer spacing of signals in the spectrum. Single-sideband systems, which require only half the bandwidth of a normal AM signal and considerably less power, are used extensively in this portion of the spectrum as a result.

It was pointed out in Section 8.2.2 that all the modulation information necessary for signal transmission and recovery is present in each of the sidebands of an amplitude-modulated signal. It was also pointed out that at 100% sinusoidal modulation only 1/6 of the total power was present in one of the sidebands, while 2/3 of it is carrier signal, which contains no information. Thus, if the carrier and one of the sidebands can be eliminated from the signal before transmission, only half of the bandwidth is required for transmission (corresponding to the maximum modulating frequency to be transmitted), and only 1/6 of the total power need be transmitted for the same signal level. A comparison of the signal spectra of full AM signals (DSBFC) with double-sideband suppressed carrier (DSBSC) and with single-sideband suppressed carrier signals (SSBSC) using either sideband is shown in Fig. 9.1. Note that in

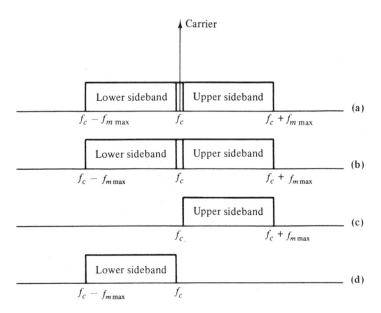

FIGURE 9.1. Amplitude-modulated signal spectra: (a) double-sideband full carrier (DSBFC); (b) double-sideband suppressed carrier (DSBSC); (c) single-sideband suppressed carrier (SSBCSC), using upper sideband (USB); (d) SSBSC using lower sideband (LSB).

(c) and (d) only one sideband is present, and that it requires only one-half the bandwidth of either (a) or (b).

9.2 SINGLE-SIDEBAND PRINCIPLES

Equation (8.9) shows that an ordinary sinusoidally modulated signal contains three components, at an upper side frequency, at a lower side frequency, and at the carrier frequency. Further, this signal is the equivalent of the carrier sinusoid being multiplied by the modulation signal with a dc component added to it:

$$e_{am} = (1 + m \sin \omega_m t)\sin \omega_c t$$

$$= \sin \omega_c t + \frac{m}{2}\cos(\omega_c - \omega_m)t - \frac{m}{2}\cos(\omega_c + \omega_m)t \qquad (8.9)$$

The two side frequencies are due entirely to the product of the carrier and the modulation signal, while the carrier term is due only to the presence of the dc offset in the modulation signal. Therefore, if the modulation circuit can be balanced so that the dc term cancels, but the product term does not cancel, the output will contain only the two sideband terms, as shown in Eq. (9.1)

$$e_{\text{bal mod}} = m \sin \omega_m t \sin \omega_c t$$

$$= \frac{m}{2} \cos(\omega_c - \omega_m)t - \frac{m}{2} \cos(\omega_c + \omega_m)t \qquad (9.1)$$

Such a circuit is called a "balanced modulator," and several balanced modulator circuits are discussed in the next section.

The first method of generating an SSB signal is derived directly from the above. It consists of a balanced modulator which cancels the carrier, followed by a bandpass filter which eliminates the unwanted sideband. Either the upper or the lower sideband may be used, provided that the receiver can be adjusted to properly demodulate either one.

The phasing method and the so-called third method both rely on multiple phase shifting and differential cancellation to eliminate both the carrier and the unwanted sideband. All three methods are discussed in detail in Section 9.4.

Demodulation of a single-sideband signal can be achieved by multiplying it with a locally generated carrier voltage at the receiver. Detectors using this principle are known as "product detectors," and most balanced modulator circuits can also be used for this purpose since they rely on the process of multiplication for their operation. It is important that the carrier be as closely synchronized in frequency and phase with the original carrier as possible, as discussed further in Section 9.5. To demonstrate that the multiplying process demodulates the incoming single sideband, consider a lower side frequency $A_L \cos(\omega_c - \omega_m)t$, where the amplitude A_L is proportional to the amplitude of the modulating signal $E_{m_{\text{max}}}$. Multiplying this by a carrier wave $\sin \omega_c t$ yields

$$A_L \cos(\omega_c - \omega_m)t \sin \omega_c t = (\tfrac{1}{2}) A_L \{ \sin[(\omega_c - \omega_m)t + \omega_c t]$$

$$- \sin[(\omega_c - \omega_m)t - \omega_c t] \}$$

$$= (\tfrac{1}{2}) A_L [\sin(2\omega_c - \omega_m)t - \sin(-\omega_m)t]$$

$$= (\tfrac{1}{2}) A_L [\sin(2\omega_c - \omega_m)t + \sin \omega_m t] \qquad (9.2)$$

The second term on the right-hand side, $(\tfrac{1}{2}) A_L \sin \omega_m t$, is the required information signal. The circuit shown in Fig. 9.3(a), with some modifications in the bias conditions, can be used as a product detector. The factor $\tfrac{1}{2}$ is not significant since the output voltage can easily be altered by the gain of the product detector as shown by Eq. (9.11). What is important is that the amplitude of the demodulated information signal is proportional to the amplitude of the received side frequency. The components around the carrier harmonics can easily be removed by low-pass filtering.

9.3 THE BALANCED MODULATOR

The balanced modulator is the heart of all methods of single-sideband modulation and demodulation. Any circuit which can produce as one of the terms in its output the *product* of two separate input signals can be used as a balanced modulator. Three such circuits are discussed here.

9.3.1 An FET Balanced Modulator Circuit

This circuit is one of a class of circuits which use the nonlinear character-istic of a semiconductor device to generate the product of two inputs. It is a form of bridge circuit in which the carrier portion of the signal is canceled in the push-pull output circuit shown in Fig. 9.2. The field-effect transistors are used because they have a transfer characteristic which is nonlinear, so that the output contains a term which is the *product* of the input voltages. The transfer curve (I_d versus V_{gs}) of a field-effect transistor is almost parabolic and may be approximated by

$$i_d = I_o + av_{gs} + bv_{gs}^2 \tag{9.3}$$

where I_o is the current for zero gate-source voltage, and $a, b, \ldots$ are constants. Since the drain currents i_{d1} and i_{d2} flow in opposite directions in the primary winding of the output transformer, the effective primary current i_p is

$$i_p = i_{d1} - i_{d2}$$
$$= a(v_{gs1} - v_{gs2}) + b(v_{gs1}^2 - v_{gs2}^2)$$
$$= a(v_{gs1} - v_{gs2}) + b(v_{gs1} + v_{gs2})(v_{gs1} - v_{gs2}) \tag{9.4}$$

Applying Kirchhoff's voltage law to the input loops of Fig. 9.2 gives

$$v_{gs1} = \tfrac{1}{2}e_m + e_c$$
$$v_{gs2} = -\tfrac{1}{2}e_m + e_c$$

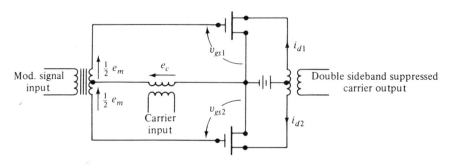

FIGURE 9.2. A balanced modulator circuit.

Substituting these in Eq. (9.4) gives

$$i_p = a(e_m) + b(2e_c)(e_m) \tag{9.5}$$

The RF output transformer will reject the low-frequency term (e_m), passing only the product term $2be_ce_m$. This term contains only the upper and lower sidebands; there is no carrier term as contained in Eq. (8.9). For example, for sinusoidal modulation, the product term contains $(\sin \omega_c t)(\sin \omega_m t)$, which can be expanded as

$$\sin \omega_c t \sin \omega_m t = \left(\tfrac{1}{2}\right)\left[\cos(\omega_c t - \omega_m t) - \cos(\omega_c t + \omega_m t)\right] \tag{9.6}$$

The appropriate side frequency may be selected from this by means of a filter. The circuit requires matched transistors; that is, I_o, a, and b for each transistor are equal.

9.3.2 An Integrated Circuit Balanced Modulator

Balanced modulators are available in integrated circuit form. Figure 9.3 shows the circuit for a very versatile commercially available unit which has many applications (see *Motorola Application Note AN531*). The circuit is one of a class of circuits which use switching to accomplish periodic signal polarity reversal, giving the effect of multiplying the signal by a square wave. An elementary analysis of the circuit, when it is used as a balanced modulator with large carrier input, follows. The carrier may be considered to be a switching voltage which alternately switches transistors Q_1, Q_4, and Q_2, Q_3 on and off, each pair being switched together. Figure 9.3(b) shows the circuit condition when the carrier has switched Q_2, Q_3 on (and Q_1, Q_4 off). Assuming that base currents are negligible in all cases, summing the currents at junctions A and B gives

$$I_2 = I + i_e \qquad \text{at junction } A$$
$$I_1 = I - i_e \qquad \text{at junction } B$$

The output voltage v_o is

$$v_o = v_2 - v_1$$
$$= R(I_2 - I_1)$$
$$= R(2i_e) \tag{9.7}$$

Applying Kirchhoff's voltage law to the loop containing e_m and R_e gives

$$e_m = V_{be5} + v_e - V_{be6}$$

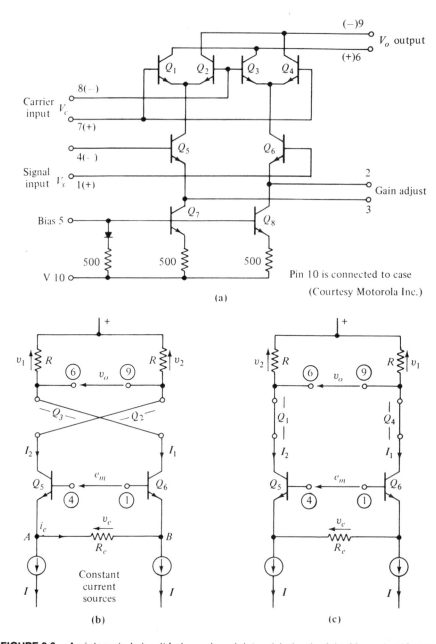

FIGURE 9.3. An integrated circuit balanced modulator: (a) circuit of the Motorola MC1596 balanced modulator; (b) equivalent circuit for the condition when negative carrier voltage has switched on transistors Q_2, Q_3; (c) equivalent circuit for the condition when positive carrier voltage has switched on transistors Q_1, Q_4. Extracted from *Motorola Application Note AN531*. (Courtesy Motorola, Inc.)

The circuit operates with $I \gg i_e$, so that $V_{be5} \approx V_{be6}$. Hence,

$$e_m \cong v_e$$

Therefore,

$$i_e = \frac{v_e}{R_e}$$

$$= \frac{e_m}{R_e} \tag{9.8}$$

where e_m is the modulating voltage. Substituting Eq. (9.8) into Eq. (9.7),

$$v_o = \frac{2\,Re_m}{R_e} \tag{9.9}$$

When the carrier input changes polarity, transistors Q_1, Q_4 are switched on, Q_2, Q_3 off. The circuit is then as shown in Fig. 9.3(c). The output voltage is now

$$v_o = v_1 - v_2$$

$$= -\frac{2\,Re_m}{R_e} \tag{9.10}$$

Thus the action of the carrier is to switch v_o, at the carrier frequency, between $\pm e_m 2R/R_e$, as shown by Eqs. (9.9) and (9.10). Representing the switching action by a square-wave function $p(t)$ at carrier frequency, Eqs. (9.9) and (9.10) can be combined to give

$$v_o = \frac{2R}{R_e} p(t) e_m \tag{9.11}$$

As shown in Eq. (2.4), a square wave can be analyzed into a sinusoidal series; thus, assuming unity amplitude, $p(t)$ can be expressed as

$$p(t) = \sin \omega_c t + \left(\tfrac{1}{3}\right)\sin 3\omega_c t + \left(\tfrac{1}{5}\right)\sin 5\omega_c t + \ldots \tag{9.12}$$

Therefore, from Eq. (9.11), on substituting Eq. (9.12), the output voltage is seen to contain a product term $e_m \sin \omega_c t$. For sinusoidal modulation this can be expanded as shown by Eq. (9.6), which shows that the output contains upper and lower side frequencies. The other components, which are side frequencies around the other harmonics of the carrier frequency (see Eq. (9.12)), are easily filtered out. The carrier itself and its harmonics do not appear in the output.

Transistors Q_1, Q_2, Q_3, and Q_4 could be switched by a low-level sinusoidal carrier such that the sidebands around the harmonics of the carrier do not appear in the output. This mode of operation has the advantage of requiring less filtering of the output, but the disadvantage is that the output level then depends on the carrier amplitude.

The advantages of the integrated circuit type of balanced modulator are that the sections can be made more nearly identical (i.e., matched for characteristics) than is the case with discrete components; also, the circuit does not require the use of transformers.

9.3.3 The Double-Balanced Ring Modulator

A circuit known as the *double-balanced ring modulator*, which is widely used in carrier telephony, is shown in Fig. 9.4(a). The name arises from the fact that the circuit is balanced for both carrier and modulating frequencies, and neither of these appear by themselves at the output; also, the diodes form a ring circuit *ABCD*. The operation is similar to that for the integrated circuit balanced modulator described in the previous section. The carrier is large and acts as a switching signal, alternately providing conducting paths as shown in Fig. 9.4(b) and (c), for the modulating signal. These are seen to be similar to the switching paths shown in Fig. 9.3(b) and (c). The analysis of the circuit is

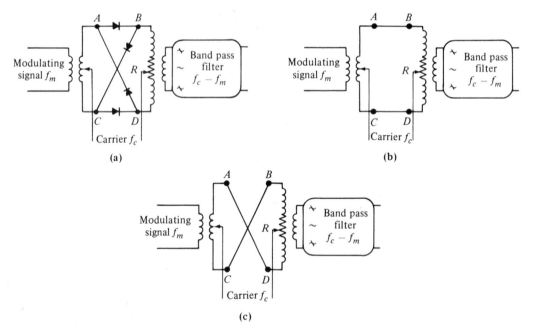

FIGURE 9.4. (a) The double-balanced ring modulator; (b) the conducting paths when diodes *AB* and *CD* are forward-biased; (c) the conducting paths when diodes *BC* and *DA* are forward-biased.

similar, resulting in an output signal proportional to $p(t) \cdot e_m$, where $p(t)$ is a square-wave switching function at carrier frequency and e_m is the modulating signal.

9.4 SSB GENERATION

9.4.1 The Balanced Modulator–Filter Method

The earliest form of SSB transmitter used a balanced modulator circuit to generate a DSBSC signal and then followed that with "sideband filters," narrow bandpass filters that only passed the desired sideband of frequencies. The system of Fig. 9.5(a) shows such a transmitter, which uses bandpass filters to eliminate the unwanted sideband. Initial modulation takes place at a low frequency, such as 100 kHz, because of the difficulty of obtaining bandpass filters with the necessary sharp characteristics at the transmitter frequency. A balanced modulator circuit performs this first modulation, which results in a signal that contains both sidebands but no carrier. Usually a single-sideband pass filter is provided, and the operator may select the sideband which gives best results by switching in a different carrier crystal to move the carrier frequency to the other edge of the passband. A balanced mixer and crystal oscillator provide up-conversion to the final transmitter frequency, and a linear RF amplifier provides the output power amplification. Linear amplifiers must be used to prevent distortion of sideband and possible regeneration of the second sideband.

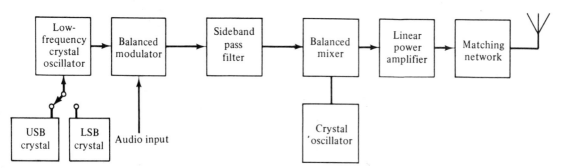

FIGURE 9.5. Single-sideband suppressed carrier transmitter using bandpass filters to eliminate the unwanted sideband.

9.4.2 The Phasing Method

Figure 9.6 shows a different means of obtaining an SSBSC signal. This circuit does not have any filters, and the primary modulation can be done at the final transmission frequency. It relies on phase shifting and cancellation to eliminate the carrier and the unwanted sideband. Expansion of the expression for a single lower side frequency, using the standard trigonometric identity, gives

$$\cos(\omega_c t - \omega_m t) = \left(\tfrac{1}{2}\right)(\cos \omega_c t \cos \omega_m t + \sin \omega_c t \sin \omega_m t) \qquad (9.13)$$

The first term on the right-hand side is the product of the carrier and the modulating signal, both phase-shifted by 90°, and the second term is the product of the unshifted carrier and the unshifted modulating signal. The circuitry required to produce the phase shifts, multiplications, and addition is relatively straightforward and is shown in block-diagram form in Fig. 9.6. (The student should be able to show that with the arrangement of Fig. 9.6, an upper side-frequency component $\sin(\omega_c + \omega_m)t$ is generated.)

The primary signal source is a crystal oscillator. Its signal drives one balanced modulator directly and the other through a circuit which shifts its phase so that it is 90° out of phase with the directly driven one. Both balanced modulators eliminate the carrier itself from the output signals. The audio signal is applied directly to the modulator with the shifted carrier, while the audio signal is shifted 90° before it is applied to the directly fed modulator. The first modulator puts out two sidebands, the upper sideband and the lower sideband, but each one is shifted in phase by +90°. The second modulator also puts out the upper and lower sidebands, but in this case the upper sideband is shifted by +90° while the lower is shifted −90°. The result is that the upper sidebands from both modulators are in phase with each other and add directly in the summing amplifier to produce the desired output

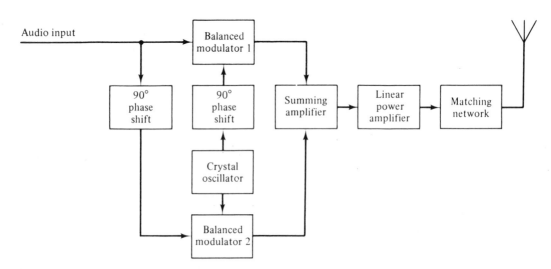

FIGURE 9.6. An SSB suppressed carrier transmitter using phase shift to obtain cancellation of sidebands.

sideband signal. The lower sidebands are shifted so that they are 180° out of phase with one another and cancel when added. The remaining sideband is amplified by linear RF power amplifiers in the final stages before transmission. Switching to the lower sideband can be accomplished by replacing the carrier phase-shift network with one with a $-90°$ phase shift.

While the system is more complex than one using filters, the individual circuits are not, and with careful adjustment of the modulators it can provide better rejection of the unwanted sideband than can the filters. The elimination of the up-converter stage also reduces the possibility of distortion of the signal after modulation. It should be noted that the shifting network in the modulation input must provide the same amount of shift at all frequencies in the modulation sideband. Such circuits are possible to build, but are complex and expensive. The "third method" eliminates the need for this wide-band shifting network.

9.4.3 The Third Method

The third method of accomplishing SSBSC modulation is called just that, the *third method*. It is attributed to D. K. Weaver and was developed during the 1950s. It is similar to the phase-shifting method already presented in that it uses shifting and cancellation for its operation, but is different in that the audio signal is first modulated onto an audio subcarrier. The system is shown in Fig. 9.7(c). The modulators 1 and 2 act to combine the audio signal with the audio subcarrier signal so that the output of modulator 1 contains the upper and lower sidebands, both shifted in phase by 90°, while the output of modulator 2 contains the unshifted upper and lower sidebands. The upper sidebands in both cases are removed by low-pass filters which cut off at the subcarrier frequency f_o. These signals are now presented to modulators 3 and 4, which are driven with the RF carrier frequency and its 90°-shifted version, respectively. The output from modulator 3 contains a sideband group $(f_c + f_o - f_m)$ shifted $+90°$ and a second sideband group $(f_c - f_o + f_m)$ shifted by $-90°$. The other modulator, modulator 4, puts out $(f_c + f_o - f_m)$ shifted by $+90°$, which is in phase with the first component of modulator 3 and adds directly to it, and also $(f_c - f_o + f_m)$ shifted by $+90°$, which is 180° out of phase with the corresponding component of modulator 3 and cancels it. The resulting output from the adder circuit is the component $(f_c + f_o - f_m)$ (the 90° phase shift can be ignored), which corresponds to the lower sideband of f_m on a carrier frequency of $(f_c + f_o)$. The other sideband and carrier have been suppressed.

If the carrier inputs to modulators 3 and 4 are interchanged, the output is the upper sideband on a carrier frequency of $(f_c - f_o)$. It should be noted that the audio subcarrier may be placed in the middle of the modulating frequency range. When this is done, the modulating frequency can be larger than f_o, and half of the lower sideband from the modulators (1 and 2) is folded over in the

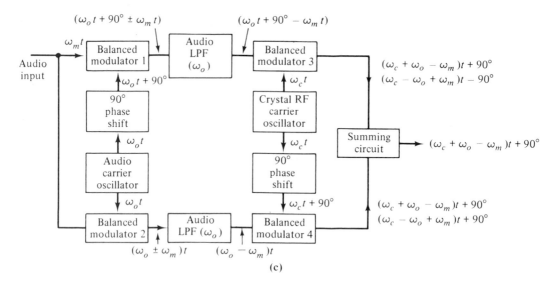

FIGURE 9.7. The "third method" of generating an SSBSC signal.

range from 0 to f_o and mixed with the other half of the sideband. This is of no consequence since the final modulators (3 and 4) act in such a manner that the sideband frequencies created by them are totally on one side of the carrier $(f_c + f_o)$.

9.5 SSB RECEPTION

It was noted in Section 9.2 that if an SSB signal is multiplied with a syncronous carrier signal, the result will contain the original modulation signal. In practice, demodulation is accomplished using either product detectors or balanced modulator circuits in conjunction with sharp cutoff IF sideband filters to select the desired sideband from the received signals.

Since both the incoming signal and the local carrier signal must remain closely synchronized in frequency to avoid severe distortion, stability in both the final demodulating oscillator and either extreme stability or frequency tracking control (AFC) on the first oscillator must be provided. Crystal oscillators or frequency synthesizers are used to obtain this stability.

Very good adjacent channel selectivity must be provided since SSB signals are usually packed closely together in the frequency spectrum. Double conversion is nearly always used in SSB receivers. The second mixer oscillator is usually a crystal oscillator that also provides the primary frequency source for the final demodulator (through multipliers or dividers).

Several variations on single sideband are used in communications. First, either the upper sideband or the lower sideband may be used for the informa-

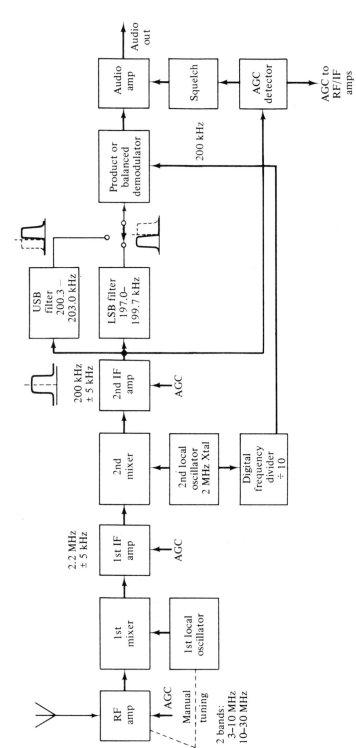

FIGURE 9.8. A single-sideband HF receiver.

tion channel, or in the case of independent sideband, both sidebands may be used, one for each of two channels. Next, a full carrier or a partially suppressed carrier may be transmitted, or the carrier may be fully suppressed.

Figure 9.8 shows the block diagram of a typical communications receiver designed for SSB reception in the HF range from 3 to 30 MHz. The circuitry is that of a standard double-conversion receiver down to the output of the second IF amplifier, except for the local oscillators. The first IF is at 2.2 MHz, with a bandwidth of 10 kHz, while the second IF is at 200 kHz, also with a 10 kHz bandwidth. The first local oscillator and RF amplifier are manually tuned, in two switched bands. The second local oscillator is a 2.0-MHz crystal oscillator. Its output is divided by 10 in a digital-counter circuit to provide the 200-kHz reference signal for the demodulator.

Two filters follow the second IF amplifier, one for each sideband with a bandpass of 2.7 kHz. The appropriate sideband is selected by a switch, and this forms the other input to the detector, which is a product detector. The output of the detector is passed through a gated audio amplifier which turns off the output signal to keep down noise when the signal level drops below a minimum value (squelch). The amplified IF signal is rectified to provide the AGC voltage for the squelch circuit and the RF and IF amplifiers.

Manually variable receivers like this one are sometimes difficult to use. The demodulator oscillator is quite stable, but the first local oscillator must also be stable and variable. Any variation in this first oscillator will cause the SSB signal to move relative to the demodulation carrier and introduce a distortion effect, which is difficult to control. The availability of digital frequency synthesizers in integrated circuit form which can be crystal controlled for stability for reasonable prices has made it possible to build very good SSB receivers for very reasonable prices. One of the largest applications of this technique is in the multichannel Citizens Band transceivers.

9.6 MODIFIED SSB SYSTEMS

9.6.1 Pilot Carrier SSB

The pilot carrier SSB system is arranged so that a low-level carrier signal is transmitted with the single sideband, in its proper place in the spectrum, but at a much lower level than would be the case for DSBFC transmission. This pilot carrier is used at the receiver to synchronize the local oscillator used for the demodulator to the original carrier, thus greatly improving the demodulator operation.

Figure 9.9(a) and (b) show an SSB pilot carrier radio transmitter and receiver, and Fig. 9.9(c) shows the frequency spectra of signals at various points within the system. The audio signal, which could be a telephone channel in the 0–4 kHz range (A), is fed to the balanced modulator to create upper and

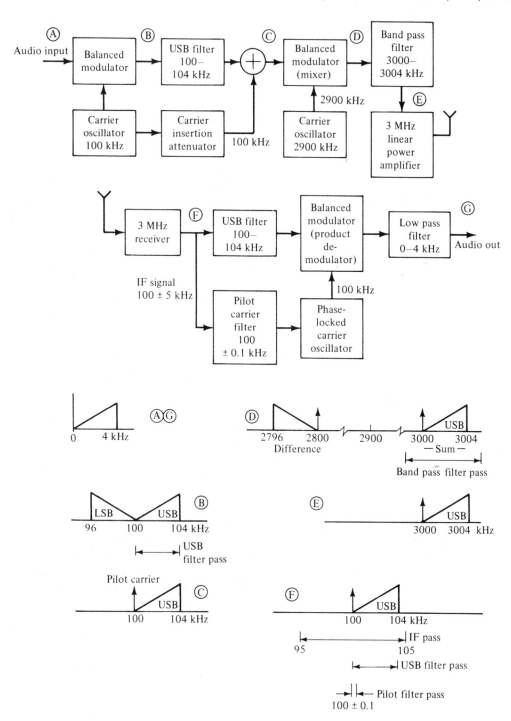

FIGURE 9.9. SSB Pilot Carrier Radio System: (a) transmitter; (b) receiver; (c) signal spectra at various points in the system.

lower sidebands around the 100-kHz carrier position (B). An upper sideband filter passes the upper sideband between 100–104 kHz, which then has a portion of the carrier added to it through an attenuator to produce the signal (C). The reinserted carrier level is adjusted through the attenuator. This signal is now modulated on a 2900-kHz carrier by a second balanced modulator which acts as a mixer to produce an upper sideband with signals between 3000 and 3004 kHz and a lower sideband between 2800 and 2796 kHz (D). A bandpass filter passes the upper sideband and rejects the lower sideband to produce the signal (E), which is then amplified and transmitted.

At the receiver, the 3 MHz signal is picked up, amplified, and down-converted to the IF at 100 ± 5 kHz, producing the pilot carrier at 100 kHz and the USB signal in the range 100–104 kHz (F). This signal is passed through an upper sideband filter to the balanced modulator, and through a very narrow bandpass filter at 100 kHz and applied to the input of the phase-locked oscillator to produce the synchronized 100-kHz carrier for the demodulator. A final low-pass filter removes the sum components of the demodulator, leaving the 0–4 kHz audio signal (G). Since the demodulator oscillator is locked to the pilot carrier, a small amount of drift in the circuits is automatically corrected for, and stable demodulation occurs. The bias signal from the phase-locked oscillator can also be used to provide automatic frequency control (AFC) to the receiver tuning oscillator.

9.6.2 Independent Sideband (ISB)

For single-sideband transmission, the carrier and one sideband are removed from the signal. It is possible to replace the removed sideband with another sideband of information created by modulating a different input signal on the same carrier, giving what is known as independent sideband (ISB) transmission. Both of the original signals have frequencies in the audio range, but in the transmitted signal each signal occupies a different group of frequencies in the spectrum. The process of distributing signals in the frequency spectrum so that they do not overlap is the process of frequency division multiplexing, or FDM. FDM is dealt with more fully in the next section.

Independent sideband modulation provides the basic mechanism for accomplishing frequency division multiplexing. Figure 9.10 shows the block diagram of an ISB transmitter which provides for four multiplexed radio-telephone channels on the same carrier. Each of the telephone channels is limited to frequencies in the 0.25–3 kHz range, somewhat less than what is usually provided for telephone channels by low-pass filters (A in Fig. 9.10(b)). Channels 2 and 3 are applied to the inputs of two balanced modulators fed from a common 6.25-kHz carrier oscillator, creating lower sideband signals from 3.25 to 6 kHz and USB signals from 6.5 to 9.25 kHz on each output. Lower sideband filters with a bandpass of 6 to 3.25 kHz remove the upper sidebands, and the lower sidebands are then added to the signals from channels 1 and 4.

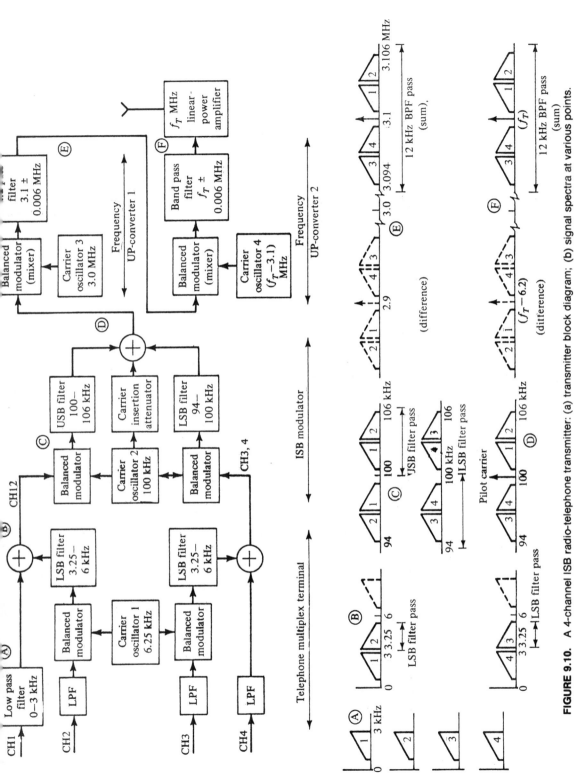

FIGURE 9.10. A 4-channel ISB radio-telephone transmitter: (a) transmitter block diagram; (b) signal spectra at various points.

The resulting two signals then have channels 1 and 2 frequency-division-multiplexed on one line and channels 3 and 4 frequency-division-multiplexed on the other line, as shown in the spectra (B). These two signals form the inputs to the ISB transmitter proper and may be applied directly or connected over telephone lines of some length if the transmitter site is remote from the multiplex terminal.

In the ISB modulator section, the two signals 12 and 34 are applied to two balanced modulators, which are also fed from a common 100-kHz carrier oscillator. The outputs are two signals, each of which contains upper and lower sidebands around the carrier frequency of 100 kHz, as shown in spectra (C). Channel 12 signal is passed through a USB filter with a passband of 100 to 106 kHz to remove its lower sideband, and channel 34 is passed through an LSB filter with a passband of 94 to 100 kHz to remove its upper sideband. The USB 12 and the LSB 34 and a small amount of 100-kHz pilot carrier (obtained through an attenuator from the carrier oscillator) are added to give the final signal (spectra (D)). This signal has the four input channel signals spread out in frequency, two on either side of the 100-kHz carrier, with none overlapping the others.

The modulated 100-kHz signal could be transmitted directly on a high-frequency cable link, or, as in this case, it can be raised to a higher radio frequency for atmospheric transmission. The remainder of the transmitter circuits provide two stages of frequency up-conversion, and power amplification to the antenna. Two stages of up-conversion are necessary because if only one is used for a frequency increase of more than about 10:1, the unwanted difference frequency components from the mixer will be too close to the wanted sum frequency components to be easily separated by filtering, especially if the final carrier frequency is more than a few MHz.

The first up-converter comprises a balanced modulator circuit used as a mixer, a 3-MHz carrier oscillator, and a bandpass sharp cutoff filter with a bandpass of 3.1 $\pm$ .006 MHz. This circuit produces the signal shown in spectra (E) which is the same as that of (D) except that it has been moved in frequency from a center of 100 kHz to a center of 3.1 MHz.

The final up-conversion works in exactly the same way, bringing the signal to its final center frequency f_T (spectra(F)). Again the sum frequency group is selected, but this time the difference frequencies are 6 MHz lower, making the final filtering operation very easy. A linear RF power amplifier tuned to the transmitting frequency f_T brings the signal up to the required antenna power level.

This circuit illustrates the use of ISB in three different areas. The first is to perform FDM at the low-frequency levels required for direct cable transmission, which will be expanded on in the next section. The second is that of ISB modulation of a radio frequency carrier, and the third is its use in frequency conversion operations, or mixing.

9.6.3 Frequency Division Multiplexing (FDM)

Frequency division multiplexing, as was noted in the previous section, is the process of combining several information channels by shifting their signals to different frequency groups within the frequency spectrum so that they can all be transmitted simultaneously on a common transmission facility. The process of frequency up-conversion and down-conversion makes this possible, and as noted in the previous section, SSBSC modulation is used extensively to accomplish these conversions.

Transmission of many signals by radiation in the radio frequency spectrum is the earliest form of FDM, and the techniques used in radio are also used in cable FDM systems. Single-sideband modulation is most often used because of its conservative use of the spectrum, although any of the radio modulation techniques may be used. Figure 9.11 shows how the various channel frequencies are assigned within the spectrum for a cable carrier system. These frequency assignments are set forth by the International Telegraph and Telephone Consultative Committee, which is based in Geneva, Switzerland (CCITT) and international agreements so that systems using them in various countries are compatible.

Modulation is performed in several cascaded stages. First, every three telephone channels (4 kHz wide) are modulated to form 3-channel pregroups. Four pregroups are then combined to form a 12-channel group. Next, five groups are combined to form a supergroup of 60 channels, and then up to 16 supergroups are combined to give a 960-channel master group for transmission on the coaxial cable. The full master group requires a bandwidth of more than 4 MHz, since spaces are left between the supergroups to make filter separation easier.

Radio microwave transmission systems also provide for two or more 6-MHz television channels, and a modified master group containing only 12 supergroups (720 channels) requiring only 3 MHz bandwidth is often used. Two such modified master groups may be transmitted stacked in each of the 6-MHz television channel slots. A microwave system with a 12-MHz modulation capacity then could carry two television channels, or 2880 telephone channels.

The pregroup modulator provides three carrier oscillators with frequencies of 12, 16, and 20 kHz. Upper sideband modulation moves the first channel from the 0–4 kHz slot into the 12–16 kHz slot; the second channel goes into the 16–20 kHz slot, and the third goes into the 20–24 kHz slot. The three signals are summed and passed on to one of the group modulators. A separate pregroup modulator set is required for every three channels, so that a 960-channel system would require 320 modulator sets. Each modulator set produces signals with the same frequency range of 12 to 24 kHz, as shown in Fig. 9.11(a). All carriers are suppressed.

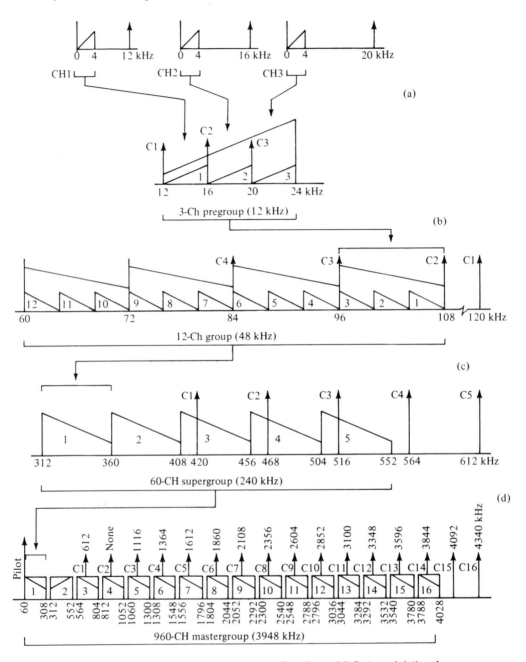

FIGURE 9.11. Cable carrier system frequency allocations: (a) first modulation forms a 3-channel pregroup; (b) 3 pregroups form a 12-channel group (12 channels may also be directly modulated to form a group); (c) 5 groups form a 60-channel supergroup; (d) 16 supergroups form a 960-channel mastergroup. (Courtesy Howard W. Sams & Co., Inc.)

The group modulator provides four carrier oscillators with frequencies of 84, 96, 108, and 120 kHz. Four pregroup signals are combined with these frequencies to move them into four consecutive frequency slots, 60–72, 72–84, 84–96, and 96–108 kHz, respectively. Lower sideband modulation is used, so that the frequencies within each channel are reversed in order. Note that the carrier associated with each slot is located 12 kHz above the upper edge of the slot, one slot away. Again, all carrier frequencies are suppressed in the modulation. The output signals from the modulators are combined and passed on to one of the supergroup modulators.

Each supergroup modulator provides carrier oscillators of 420, 468, 516, 564, and 612 kHz. These are combined again by lower sideband modulation with five group signals to move them from the 60–108 kHz slot to five consecutive slots, 312–360, 360–408, 408–456, 456–504, and 504–552 kHz. Again the order of frequencies within each group is reversed by the lower sideband modulation. The carriers are all suppressed, and the five signals are combined and passed on to a mastergroup modulator.

Each mastergroup modulator provides 16 carriers at frequencies of 612, none, 1116, 1364, 1612, 1860, 2108, 2356, 2604, 2852, 3100, 3348, 3596, 3844, 4092, and 4340 kHz. The second in the sequence is omitted because the second supergroup slot corresponds to the supergroup output frequencies, and no conversion is necessary. Lower sideband modulation moves each of the other 15 supergroup signals into 240-kHz-wide slots with 12-kHz spaces between them (to allow easier filtering), creating the final signal with frequencies from 60 to 4028 kHz (3968-kHz-wide). All carriers are suppressed, but a 60-kHz pilot signal is transmitted to provide a synchronization signal for the demodulation oscillators. This final signal contains a maximum of 960 separate 4-kHz telephone channels. The actual maximum number used in a real system may be somewhat less than this because of the necessity to keep crosstalk and noise down and to provide spare channels in case faults occur. Demodulation at the receiving end proceeds in the reverse order, going from mastergroup to supergroup to group to pregroup to channel in four separate demodulation steps.

9.7 COMPARISON OF SIGNAL-TO-NOISE RATIOS

As shown in Section 9.2, the amplitude of the output (information) signal is proportional to A_L, the received RF signal amplitude. A_L^2 is proportional to the power transmitted, P_T. The mean-square noise voltage V_n^2 for a receiver bandwidth B, is proportional to kTB (see Chapter 4), and the output signal to noise power ratio, for the same constant of proportionality relating output voltage to input power for both signal and noise will be (see Section 4.9),

$$(S/N)_{SSB} = \frac{A_L^2}{FV_n^2}$$

$$= \frac{P_T}{FkTB} \qquad (9.14)$$

where F is the noise factor of the receiver.

For normal amplitude modulation, the detector output is $mE_{c_{max}}$, so the output (information) signal power is proportional to $(mE_{c_{max}})^2$. $E_{c_{max}}^2$ is proportional to P_c, and, from Eq. (8.16),

$$P_c = \frac{P_T}{1 + \frac{m^2}{2}}$$

The noise bandwidth for double sidebands is $2B$; therefore, $V_n^2 \propto kT2B$, and the signal-to-noise power ratio is

$$(S/N)_{AM} = \frac{\left(mE_{c_{max}}\right)^2}{FV_n^2}$$

$$= \frac{m^2 P_c}{FkT2B}$$

$$= \frac{m^2 P_T}{F\left(1 + \frac{m^2}{2}\right)kT2B} \qquad (9.15)$$

Comparison of Eqs. (9.14) and (9.15) gives

$$\frac{(S/N)_{SSB}}{(S/N)_{AM}} = \left(1 + \frac{2}{m^2}\right) \qquad (9.16)$$

This assumes that the noise factor F is the same for both cases. For a modulation index $m = 1$, Eq. (9.16) gives a ratio of 3:1, or single sideband shows an improvement of 4.77 dB over 100%-modulated AM.

The comparison just made is based on the assumption of equal average powers being transmitted in each case. Sometimes the comparison is based on peak power rating of the final amplifier stage of the transmitter. For 100% sinusoidal amplitude modulation, the peak RF voltage is $4V_{CC}$, as shown in

Fig. 8.7(c), whereas for a single side frequency it will be $2V_{CC}$ (since the class C amplifier stage will be amplifying a single sinusoidal voltage and will be adjusted so that the peak-to-peak swing will be zero to $2V_{CC}$). Hence, for equal peak power ratings, the RF voltage for SSB can be doubled. This means that A_L at the receiver output can be twice the corresponding $E_{c_{max}}$, or the signal power output can be increased by a factor of four. The noise power for AM will still be twice that for SSB because of the double bandwidth, so the overall improvement in signal-to-noise power ratio for SSB over AM is 8:1, or 9 dB.

9.8 PROBLEMS

1. Explain how the principle of operation of the balanced modulator of Fig. 9.2 differs from that of Fig. 9.3.

2. Use a phasor diagram to show that the output of the transmitter in Fig. 9.7(c) contains only the upper sideband component. What change is needed in the circuit if only the lower sideband is desired?

3. Explain how the information contained in a single-sideband signal can be recovered (demodulated). If an SSB signal contains frequencies of 455.5 and 456.8 kHz, and the demodulator uses an oscillator frequency of 455 KHz, what frequencies appear in the demodulator output?

4. Show by means of trigonometric identities that the transmitter of Fig. 9.6 generates an upper sideband signal.

5. The FET's in Fig. 9.2 have a characteristic given by

$$i_d = \text{IDSS}\left(1 - \frac{V_{gs}}{V_P}\right)^2$$

where IDSS $= 10$ mA, $V_P = -0.8$ V, and the effective primary impedance of the transformer is $1\text{k}\Omega$, at high frequencies. A 0.1-V rms, 1-kHz modulating signal and a 0.5-V peak carrier at 500 kHz are applied. Find the magnitudes and frequencies of each component appearing in the transformer primary voltage.

6. The modulator of Fig. 9.3 is connected to an effective load resistance of 22 $\text{k}\Omega$ (on each side) and has a gain adjustment resistor of 1.2 $\text{k}\Omega$. A square-wave carrier of ± 1-V peak at 2 MHz and a sinusoidal modulating signal of 5-mV rms at 4 kHz are applied. Find the magnitude and frequency of each component appearing in the output voltage V_o.

7. Sketch the block diagram of an SSB generator similar to Fig. 9.6 which will generate a lower sideband signal. (Hint: see Eq. 9.13)

8. (a) Show that shifting a sine function by $+90°$ turns it into a cosine function.

   (b) Show by means of the shifted angles in Fig. 9.7 that interchanging the carrier inputs to balance modulators 3 and 4 will cause the circuit to generate an upper sideband signal.

9. The "third method" system of Fig. 9.7 uses an audio carrier frequency of 2.5 kHz and a radio carrier frequency of 250 kHz. Two audio tones of 500 Hz and 4 kHz are modulated at once. (a) Sketch the spectrum of the output signal and calculate the frequency of each component present. (b) Also calculate the band-edge frequencies and the carrier location frequency, and indicate these on the spectrum.

10. The receiver of Fig. 9.8 receives an upper sideband signal carrying the original modulating frequencies of 500 Hz and 3 kHz. As the result of mistuning, the IF signal is 1 kHz lower than it should be. What audio frequencies are present in the receiver output?

11. The receiver of Fig. 9.8 receives an upper sideband signal containing original modulating frequencies of 500 Hz and 2 kHz. However, the receiver is switched to the lower sideband filter and is tuned 3 kHz low (so that the upper sideband appears in the lower sideband window of the filter). (a) What frequencies appear in the output, and which of the original tones caused each? (b) What would this misadjustment do to a voice signal?

12. An AM transmitter with a total power output of 1000 watts is modulated by frequencies from 0 to 5 kHz at an average modulation index of 70%. This transmitter is replaced by a new SSB unit which puts out a total power of 2000 watts, with the same signal frequency range.

    (a) What is the difference in S/N created for a given receiver?

    (b) Is that change an improvement or a degradation?

    (c) What portion of the change is due only to the change of modulation technique?

13. A frequency-division multiplex system is to provide six channels, three in each direction on a cable pair. An additional pair of channels at the bottom of the spectrum, including the baseband channel, is to be provided as a spare and for supervisory use. The channels are each to have a bandwidth of 4 kHz, and a pilot carrier is to be provided 500 Hz above the upper edge of each channel, with a pad band of 1.5 kHz between each pair of channels to allow easy filtering. Alternate channels are to transmit in alternate directions. Calculate the frequencies of the upper and lower edges of each band and the frequencies of each pilot carrier. Plot the spectrum response of the system, indicating the location of each of the channels and carriers. State the upper cutoff frequency necessary for the cable.

14. Calculate the maximum number of 4-kHz channels that can be frequency-division-multiplexed onto a 6-MHz television channel if 12.5%

extra bandwidth must be allowed for each channel to prevent overlap and the system is divided into groups of 12 channels with two additional channel spaces unused, one at either end of the group, and supergroups of five groups with one group extra space between supergroups, and further into master groups of 10 supergroups, with no extra zone between. State the number of groups, supergroups, and master groups which may be used and the number of partial groups which could be included, to provide maximum usage. Compare the total to the maximum number of ideal 4-kHz channels.

CHAPTER 10

Angle Modulation

10.1 INTRODUCTION

In angle modulation, the information signal may be used to vary the carrier frequency, giving rise to *frequency modulation*, or it may be used to vary the angle of phase lead or lag, giving rise to *phase modulation*. Since both frequency and phase are parameters of the carrier angle, which is a function of time, the general term *angle modulation* covers both. Frequency and phase modulation have some very similar properties, but also some marked differences. The relationship between the two will be described in detail in this chapter.

Compared to amplitude modulation, frequency modulation has certain advantages. Mainly, the signal-to-noise ratio can be increased without increasing transmitted power (but at the expense of an increase in frequency bandwidth required); certain forms of interference at the receiver are more easily suppressed; and the modulation process can take place at a low-level power stage in the transmitter, thus avoiding the need for large amounts of modulating power.

319

10.2 FREQUENCY MODULATION

The modulating signal e_m is used to vary the carrier frequency. For example, e_m may be used to alter the capacitance of the carrier frequency oscillator circuit (see Section 10.5.1). Let the change in carrier frequency be ke_m, where k is a constant known as the *frequency deviation constant*; then the instantaneous carrier frequency is

$$f_i = f_c + ke_m \tag{10.1}$$

where f_c is the unmodulated carrier frequency.

For example, with e_m a sine wave,

$$e_m = E_{m_{max}} \sin \omega_m t \tag{10.2}$$

and the instantaneous carrier frequency becomes

$$f_i = f_c + kE_{m_{max}} \sin \omega_m t \tag{10.3}$$

A sketch of the variation of f_i with time is shown in Fig. 10.1(a). It is important to grasp that this is a frequency–time curve, not an amplitude–time curve. The amplitude–time curve for the frequency-modulated carrier is shown in Fig. 10.1(b).

The *peak frequency deviation* of the signal is defined to be

$$\Delta f = kE_{m_{max}} \tag{10.4}$$

so that Eq. (10.3) becomes

$$f_i = f_c + \Delta f \sin \omega_m t \tag{10.5}$$

Example 10.1 Sketch the instantaneous frequency–time curve for a 100-MHz carrier wave frequency-modulated by a 1-kHz square wave, the peak deviation being 90 kHz.

Solution The sketch is shown in Fig. 10.1(c).

In order to get a quantitative understanding of frequency modulation, it is first necessary to derive the equation for the modulated wave. The unmodulated carrier is a sine wave, as expressed in Eq. (8.1), for which, without loss

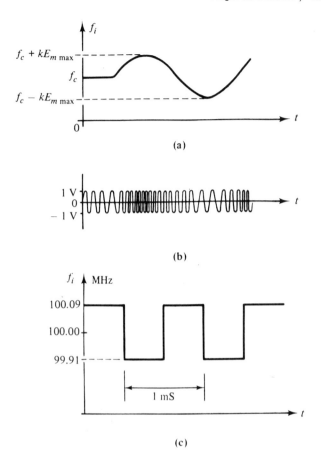

FIGURE 10.1. (a) Instantaneous frequency–time curve; (b) carrier amplitude–time curve for a sinusoidally frequency–modulated carrier; (c) instantaneous frequency-time curve for Example 10.1.

of generality, $E_{c_{max}}$ may be set equal to unity:

$$e_c = \sin(\omega_c t + \phi) \tag{10.6}$$

$\omega_c = 2\pi f_c$ = a constant angular frequency in rad/s, and ϕ is a constant phase angle in radians.

Equation (10.6) is a particular case of the more general expression,

$$e = \sin\theta(t) \tag{10.7}$$

The angular frequency of this general expression is the time rate of change of $\theta(t)$, and only when the frequency is constant is the particular form in Eq.

(10.6) valid. When the frequency is varied, as in frequency modulation, an instantaneous angular frequency may be defined as

$$\omega_i = 2\pi f_i = \frac{d\theta(t)}{dt} \tag{10.8}$$

Integrating this with respect to time gives

$$\theta(t) = \int \omega_i \, dt \tag{10.9}$$

The instantaneous frequency f_i is related to the modulation by Eq. (10.5). For example, for constant (unmodulated) angular frequency ω_c,

$$\theta(t) = \int \omega_c \, dt$$
$$= \omega_c t + \phi \tag{10.10}$$

where ϕ is the constant of integration. Equation (10.6) follows when Eq. (10.10) is substituted in Eq. (10.7).

For sinusoidal modulation, Eq. (10.5) is substituted in Eq. (10.9) to give

$$\theta(t) = \int 2\pi (f_c + \Delta f \sin \omega_m t) \, dt$$
$$= \omega_c t - \frac{\Delta f}{f_m} \cos \omega_m t + \phi \tag{10.11}$$

where $\omega_m = 2\pi f_m$. The constant ϕ may be made equal to zero by appropriate choice of reference axis, and the equation for the sinusoidally frequency-modulated wave obtained by substituting Eq. (10.11) in Eq. (10.7) is

$$e = \sin\left(\omega_c t - \frac{\Delta f}{f_m} \cos \omega_m t\right) \tag{10.12}$$

Note that Eq. (10.12) could not have been derived simply by substituting f_i for f_c in Eq. (10.6), the reason being that Eq. (10.6) is derived on the basis of *constant* frequency, and, of course, frequency modulation invalidates this.

The modulation index for frequency modulation is defined as

$$m_f = \frac{\Delta f}{f_m} \tag{10.13}$$

The equation for the sinusoidally modulated carrier then becomes

$$e = \sin(\omega_c t - m_f \cos \omega_m t) \tag{10.14}$$

Unlike amplitude modulation, the modulation index for frequency modulation can be greater than unity.

10.2.1 The Frequency Spectrum

The mathematical analysis of Eq. (10.14) leading to a frequency spectrum for a sinusoidally frequency-modulated wave is much more difficult than the corresponding amplitude-modulation analysis (see, for example, Ben Zeines, *Electronic Communications Systems*, Prentice-Hall, Inc., Englewood Cliffs, N.J., 1970, Section 9.13), and only the results will be used here. The spectrum is found to consist of a carrier component, and side frequencies at harmonics of the modulating frequency, even though no harmonics are present in the original modulating tone. The amplitudes of the various spectral components

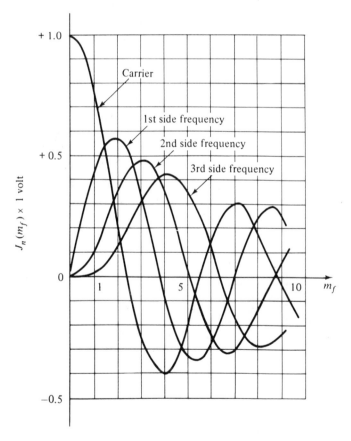

FIGURE 10.2. A plot of the magnitudes of the carrier and the first three side frequencies of a sinusoidally modulated FM wave with a carrier of one-volt magnitude versus modulation index m_f. (This is a plot of the Bessel Functions of the wave.)

are given by a mathematical function known as Bessel's Function of the First Kind, denoted here by $J_n(m_f)$; m_f is the modulation index defined in Eq. (10.13), and n is the order of the side frequency. In mathematical notation, m_f is termed the argument, and n the order, of the Bessel Function. Bessel Functions are available in both graphical and tabular form as shown in Fig. 10.2 and Table 10.1. $J_0(m_f)$ gives the amplitude of the carrier component.

As an example of the use of Table 10.1, it can be seen that, for $m_f = 0.5$, the spectral components are

Carrier (f_c)	$J_0(0.5) = 0.94$
first-order side frequencies $(f_c \pm f_m)$	$J_1(0.5) = 0.24$
second-order side frequencies $(f_c \pm 2f_m)$	$J_2(0.5) = 0.03$

The fact that the spectrum component at the carrier frequency decreases in amplitude does *not* mean that the carrier wave is amplitude-modulated. The carrier wave is the sum of all the components in the spectrum, and these add up to give a constant amplitude carrier as shown in Fig. 10.1(b). The distinction is that the modulated carrier is not a sine wave, whereas the spectrum component at carrier frequency is. (All spectrum components are either sine or cosine waves.) It will be noted from Table 10.1 that amplitudes can be negative in some instances, but it is usually not necessary to show this on a spectrum graph, only the modulus value being shown. It will also be seen that for certain values of m_f (2.4, 5.5, 8.65, and higher values not shown), the carrier amplitude

TABLE 10.1. Amplitudes of Spectrum Components for a Sinusoidally Frequency-Modulated Carrier of Unmodulated Amplitude 1.0 V (Amplitude Moduli Less than |0.01| not shown.)

Modulation Index m_f	Carrier J_0	1st J_1	2nd J_2	3rd J_3	4th J_4	5th J_5	6th J_6	7th J_7	8th J_8	9th J_9	10th J_{10}	11th J_{11}	12th J_{12}
						Side Frequencies							
0.25	0.98	0.12	0.01										
0.5	0.94	0.24	0.03										
1.0	0.77	0.44	0.11	0.02									
1.5	0.51	0.56	0.23	0.06	0.01								
2.0	0.22	0.58	0.35	0.13	0.03	0.01							
2.4	0	0.52	0.43	0.20	0.06								
3.0	-0.26	0.34	0.49	0.31	0.13	0.04	0.01						
4.0	-0.40	-0.07	0.36	0.43	0.28	0.13	0.05	0.02					
5.0	-0.18	-0.33	0.05	0.36	0.39	0.26	0.13	0.05	0.02	0.01			
5.5	0	-0.34	-0.12	0.26	0.40	0.32	0.19	0.09	0.03	0.01			
6.0	0.15	-0.28	-0.24	0.11	0.36	0.36	0.25	0.13	0.06	0.02	0.01		
7.0	0.30	0	-0.30	-0.17	0.16	0.35	0.34	0.23	0.13	0.06	0.02	0.01	
8.0	0.17	0.23	-0.11	-0.29	-0.10	0.19	0.34	0.32	0.22	0.13	0.06	0.03	0.01
8.65	0	0.27	0.06	-0.24	-0.23	0.03	0.26	0.34	0.28	0.18	0.10	0.05	0.02

goes to zero. This serves to emphasize the point that it is the sinusoidal component of the spectrum at carrier frequency, *not* the modulated carrier, which goes to zero and which varies from positive to negative peak (1 V in this case) as the frequency varies.

The spectra for various values of m_f are shown in Fig. 10.3(a), (b), and (c). In each case the spectral lines are spaced by f_m, and the bandwidth

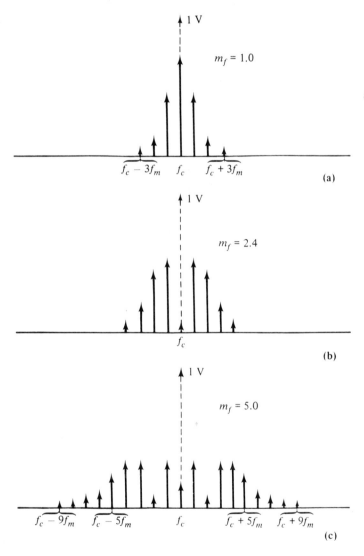

FIGURE 10.3. Spectra for sinusoidally modulated FM waves with various modulation indexes: (a) $m_f = 1.0$; (b) $m_f = 2.4$ (note missing carrier); (c) $m_f = 5.0$.

occupied by the spectrum is seen to be

$$B_{FM} = 2nf_m \qquad (10.15)$$

where n is the highest order of side frequency for which the amplitude is significant. From Table 10.1 it can be seen that where the order of side frequency is greater than $(m_f + 1)$, the amplitude is 5% or less of unmodulated carrier amplitude. Using this as a guide for bandwidth requirements, Eq. (10.15) can be written as

$$B_{FM} = 2(m_f + 1)f_m \qquad (10.16)$$

or, substituting for m_f from Eq. (10.13),

$$B_{FM} = 2(\Delta f + f_m) \qquad (10.17)$$

To illustrate the significance of this, three examples will be considered:

1. $\Delta f = 75$ kHz, $\qquad\qquad\qquad\qquad f_m = 0.1$ kHz:

$$B_{FM} = 2(75 + 0.1)$$
$$= 150 \text{ kHz}$$

2. $\Delta f = 75$ kHz, $\qquad\qquad\qquad\qquad f_m = 1.0$ kHz:

$$B_{FM} = 2(75 + 1)$$
$$= 152 \text{ kHz}$$

3. $\Delta f = 75$ kHz, $\qquad\qquad\qquad\qquad f_m = 10$ kHz:

$$B_{FM} = 2(75 + 10)$$
$$= 170 \text{ kHz}$$

Thus, although the modulation frequency changes from 0.1 kHz to 10 kHz, or by a factor of $100:1$, the bandwidth occupied by the spectrum alters very little, from 150 kHz to 170 kHz. These examples illustrate why frequency modulation is sometimes referred to as a constant-bandwidth system.

10.2.2 Average Power

The Bessel functions relate the *voltage* amplitudes of each of the sinusoidal side frequency components to the unmodulated carrier amplitude. That is,

$$E_n = J_n E_c \qquad (10.18)$$

Assuming that the amplitudes E_n and E_c are the rms values of the sinusoids, the power contained in each of the sinusoidal components (carrier residual and each side frequency) is given as

$$P_n = \frac{E_n^2}{R} \qquad (10.19)$$

Noting that there is only one carrier component, but a pair of components for each side frequency number n, the total power in the modulated signal becomes

$$P_T = P_0 + 2P_1 + 2P_2 + \ldots$$

$$= \frac{E_0^2}{R} + \frac{2E_1^2}{R} + \frac{2E_2^2}{R} + \ldots$$

$$= \frac{(J_0 E_c)^2}{R} + \frac{2(J_1 E_c)^2}{R} + \frac{2(J_2 E_c)^2}{R} + \ldots$$

$$= P_c \left(J_0^2 + 2 \left(J_1^2 + J_2^2 + \ldots \right) \right) \qquad (10.20)$$

where P_c is obtained from Eq. (8.13), and the J_n are for a constant value of modulation index m_f. The total power in the modulated waveform remains constant for all conditions of modulation. Power is extracted from the carrier component and redistributed among the side frequency components as the modulation deepens. This is indicated by the fact that the sum of the squares of the Bessel function coefficients in Eq. (10.20) for a given value of m_f is always unity.

Example 10.2 A 15-W unmodulated carrier is frequency-modulated with a sinusoidal signal such that the peak frequency deviation is 6 kHz. The frequency of the modulating signal is 1 kHz. Calculate the average power output by summing the powers for all of the side frequency components.

Solution The total average power output P is 15 W modulated. To check that this is also the value obtained from the sum of the squares of the Bessel Functions, from Eq. (10.13) we have

$$m_f = \frac{\Delta f}{f_m} = \frac{6}{1} = 6$$

The Bessel function values for $m_f = 6$ are read from Table 10.1 and substituted in Eq. (10.20) to give

$$P_T = 15[0.15^2 + 2(0.28^2 + 0.24^2 + 0.11^2 + 0.36^2 + 0.36^2 + 0.25^2$$
$$+ 0.13^2 + 0.06^2 + 0.02^2 + 0.01^2)]$$
$$= 15(1.00)$$
$$= 15 \text{ W}$$

It follows that since the average power does not change with frequency modulation, the rms voltage and current will also remain constant, at their respective unmodulated values.

10.2.3 Complex-Wave Modulation: Deviation Ratio

In the frequency-modulation process, intermodulation products are formed; that is, beat frequencies occur between the various side frequencies when the modulation signal is other than sinusoidal or cosinusoidal. It is a matter of experience, however, that the bandwidth requirements are determined by the maximum frequency deviation and maximum modulation frequency (harmonic) present in the complex modulating wave. The ratio of maximum deviation to maximum frequency component is termed the *deviation ratio* which is defined as

$$M = \frac{\Delta F}{F_m} \tag{10.21}$$

where ΔF is the maximum frequency deviation and F_m is the highest frequency component in the modulating signal. The bandwidth is then given by Eq. (10.16) on substituting M for m_f, with the same limitations on accuracy as

$$B_{\text{max}} = 2(M + 1)F_m \tag{10.22}$$
$$= 2(\Delta F + F_m)$$

This is known as *Carson's rule*.

Example 10.3 Canadian regulations state that for FM broadcast, the maximum deviation allowed is 75 kHz and the maximum modulation frequency allowed is 15 kHz. Calculate the maximum bandwidth requirements.

Solution Using Eq. (10.22),

$$B_{\text{max}} = 2(\Delta F + F_m)$$
$$= 2(75 + 15)$$
$$= 180 \text{ kHz}$$

Examination of Table 10.1 shows that side frequencies of 1% amplitude extend up to the ninth side frequency pair, so Carson's rule underestimates the bandwidth required. For M equal to 5 or greater, a better estimate is given by $B_{max} = 2(M + 4)F_m$. In this example, this would result in a maximum bandwidth requirement of 270 kHz. The economic constraints on commercial equipment limits the bandwidth capabilities of receivers to about 200 kHz.

10.2.4   Measurement of Modulation Index

The spectrum for a sinusoidally modulated wave can be measured directly on a spectrum analyzer, and the carrier frequency deviation can be measured independently on a frequency deviation meter. For a given sinusoidal modulating frequency (e.g., 1 kHz) the amplitude of the modulating signal can be adjusted until the carrier component of the spectrum, as viewed on the spectrum analyzer, disappears. The modulation amplitudes for which this happens should be in the ratio $2.4 : 5.5 : 8.65$, since the deviation should be proportional to the modulating signal amplitude, and Table 10.1 shows that these are the values of m_f for which the carrier is zero. Measurement of the deviation enables m_f to be calculated, given f_m.

10.3   PHASE MODULATION

Phase modulation results when the phase angle ϕ of the carrier is made a function of the modulating signal. The unmodulated carrier is given by

$$e_c = \sin(\omega_c t + \phi_c) \tag{10.23}$$

When phase modulated, ϕ_c is replaced by $\phi(t)$, where

$$\phi(t) = \phi_c + Ke_m \tag{10.24}$$

K is the *phase deviation constant* (analogous to k for frequency modulation) and e_m is the modulating signal, as before. Normally, ϕ_c can be dropped from the equation since it is a constant that does not affect the modulation. Also, letting $e_m = E_{m\,max} m(t)$, as before, Eq. (10.24) can be written as

$$\phi(t) = \Delta\phi m(t) \tag{10.25}$$

where

$$\Delta\phi = KE_{m\,max} \tag{10.26}$$

$\Delta\phi$ is clearly the peak phase deviation.

Substituting Eq. (10.25) in Eq. (10.23) gives the expression for the phase-modulated wave:

$$e = \sin[\omega_c t + \Delta\phi\, m(t)] \tag{10.27}$$

For sinusoidal modulation, Eq. (10.27) becomes

$$e = \sin(\omega_c t + \Delta\phi \sin \omega_m t) \qquad (10.28)$$

And, in order to emphasize the similarity with sinusoidal frequency modulation, the peak phase deviation is termed the phase modulation index and the symbol m_p is used. Equation (10.28) then becomes

$$e = \sin(\omega_c t + m_p \sin \omega_m t) \qquad (10.29)$$

Comparing Eq. (10.29) with Eq. (10.14), the similarity between phase modulation and frequency modulation for sinusoidal modulating signals will be evident, especially when it is remembered that the only difference between the sinusoidal and cosinusoidal modulating terms is a $90°$-phase difference, which would not be readily apparent when the signals are demodulated.

At this point it is instructive to compare the three methods of modulation, that is, amplitude, frequency, and phase. Figure 10.4 shows the situation

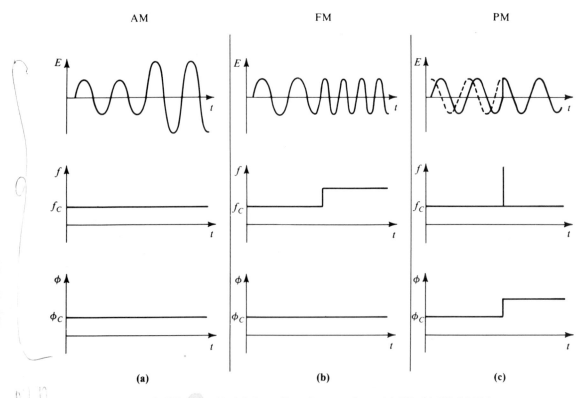

FIGURE 10.4. Modulating with a step waveform: (a) AM; (b) FM; (c) PM.

when a sinusoidal carrier is modulated by a step voltage. In the case of amplitude modulation (Fig. 10.4(a)), the amplitude follows the step change, while the frequency and phase remain constant with time. The amplitude change could be observed, for example on an oscilloscope. With frequency modulation, shown in Fig. 10.4(b), the amplitude and phase remain constant while the frequency follows the step change. Again, this change could be observed, for example on a frequency counter.

With phase modulation, the amplitude remains constant while the phase angle follows the step change with time, as shown in Fig. 10.4(c). The phase change is measured with reference to what the phase angle would have been with no modulation applied. After the step change in phase, the sinusoidal carrier appears as though it is a continuation of the dotted curve shown on the amplitude–time graph of Fig. 10.4(c). Also, from the amplitude–time graph it is seen that the frequency of the wave before the step change is the same as after the step change. However, at the step change in phase, the abrupt displacement of the waveform on the time axis makes it appear as though the frequency undergoes an abrupt change. This is shown by the spike in the frequency–time graph in Fig. 10.4(c). A phase meter could be used to measure the change in phase, but this requires the reference waveform and is not as direct as the measurement of amplitude or frequency. The spike change in frequency could be measured directly on a frequency counter. In principle, the apparent change of frequency with phase modulation will occur even where the source frequency of the carrier is held constant, for example by using a crystal oscillator. In practice, it proves to be more difficult to achieve large frequency swings using phase modulation.

The mathematical relationship between frequency and phase modulation is presented in Section 10.4.

10.3.1 Spectrum of a Sinusoidally Phase-Modulated Wave

It follows from the comparison of Eqs. (10.29) and (10.14) that for the same value of modulation index, the spectrum for the phase-modulated wave will be the same as the spectrum for the frequency-modulated wave. For example, for $m_p = 5.0$, the spectrum of Fig. 10.3(c) applies.

10.4 EQUIVALENCE BETWEEN FM AND PM

The instantaneous angular frequency as defined in Eq. (10.8) is

$$\omega_i = \frac{d\theta(t)}{dt} \tag{10.8}$$

For a phase-modulated wave, the phase modulation is expressed by Eq.

(10.24), and

$$\theta(t) = \omega_c t + \phi(t) \tag{10.30}$$

Therefore, applying Eq. (10.8), phase modulation has an equivalent angular frequency:

$$\omega_{i_{eq}} = \omega_c + \frac{d\phi(t)}{dt} \tag{10.31}$$

Let $f_{eq}(t)$ represent the equivalent frequency modulation in Eq. (10.31); then

$$\omega_c + 2\pi f_{eq}(t) = \omega_c + \frac{d\phi(t)}{dt}$$

Therefore,

$$f_{eq}(t) = \frac{1}{2\pi} \frac{d\phi(t)}{dt} \tag{10.32}$$

The importance of Eq. (10.32) lies in the fact that all angle-modulation receivers (even the "phase discriminators" described in Section 10.6) always interpret angle modulation as frequency modulation. This means that the demodulated signal output is proportional to the frequency deviation, actual or equivalent, of the received modulated signal. This is illustrated for the three separate waveforms of Fig. 10.5, for the system shown in Fig. 10.5(a). For the triangular modulation wave of Fig. 10.5(b), the phase modulation is

$$\phi(t) = +at \quad \text{for } 0 < t < T/2$$
$$= -at \quad \text{for } T/2 < t < T$$

where a is the modulus of the slope of the lines (assumed equal for positive and negative slopes). The term $\phi(t)$ has the same periodicity as $m(t)$. Applying Eq. (10.32),

$$f_{eq}(t) = \frac{1}{2\pi} \frac{d(at)}{dt}$$
$$= \frac{a}{2\pi} \quad \text{for } 0 < t < T/2$$
$$= -\frac{a}{2\pi} \quad \text{for } T/2 < t < T$$

Thus, $f_{eq}(t)$ is a square wave as shown in Fig. 10.5(b), with $\Delta f_{eq} = a/2\pi$. The output voltage has the same waveshape as $f_{eq}(t)$.

For the square-wave modulating signal of Fig. 10.5(a), the rate of change of $\phi(t)$ (the slope of the graph) is zero except at the points where the wave reverses sign when the rate of change is theoretically $\pm\infty$. Therefore, applying Eq. (10.32), $f_{eq}(t)$ will be zero except at the points of reversal, where it will be represented by impulses as shown in Fig. 10.5(c). The output voltage waveform follows the $f_{eq}(t)$ waveform.

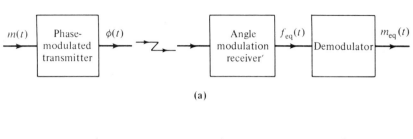

(a)

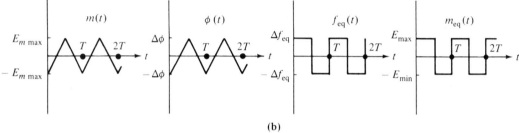

(b)

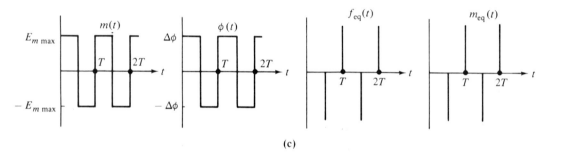

(c)

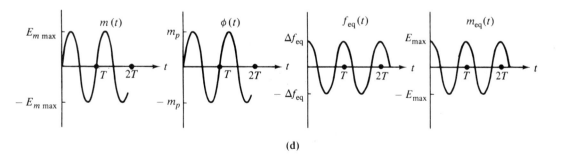

(d)

FIGURE 10.5. Phase modulation with various complex modulating signals: (a) the angle-modulation system; $m(t)$ is the modulating input and $m_{eq}(t)$ is the corresponding receiver output; (b) triangular modulating wave resulting in a square-wave output; (c) square modulating wave resulting in an impulse output; (d) a sinusoidal modulation wave resulting in a cosinusoidal output.

For the sinusoidal modulating signal of Figure 10.5(d),

$$\phi(t) = m_p \sin \omega_m t$$

Therefore,

$$f_{eq}(t) = \frac{\omega_m m_p}{2\pi} \cos \omega_m t$$

$$= f_m m_p \cos \omega_m t$$

The term $f_{eq}(t)$ is shown in Fig. 10.5(d), and it can be seen that the maximum equivalent frequency deviation is

$$\Delta f_{eq} = f_m m_p \qquad (10.33)$$

The output voltage waveform follows the $f_{eq}(t)$ curve, and therefore the amplitude of the output voltage, which is proportional to Δf_{eq}, is proportional to $f_m m_p$. For constant m_p, the output is proportional to the modulating frequency.

Example 10.4   A constant-amplitude 2-kHz sine wave is used to phase-modulate a carrier. At some later time, the frequency of the modulating signal is increased to 5 kHz. Explain what happens to the output signal from a receiver used to receive the modulated wave.

Solution   The frequency of the output voltage signal will increase from 2 to 5 kHz, and the amplitude will increase by a factor of 5/2, or 2.5 times.

The relationship expressed by Eq. (10.32) is used as the basis for obtaining frequency modulation from phase modulation. The theory is straightforward. Instead of making $\phi(t)$ proportional to $m(t)$, it is made proportional to the integral of $m(t)$, usually by means of a simple integrating network such as shown in Fig. 10.6. Thus, applying Eq. (10.24) (and dropping the constant ϕ_c),

$$\phi(t) = Ke_m$$

$$= \frac{KE_{m_{max}}}{RC} \int m(t)\,dt$$

Applying Eq. (10.32),

$$f_{eq}(t) = \frac{1}{2\pi} \frac{KE_{m_{max}}}{RC} m(t)$$

$$= \Delta f_{eq} \, m(t) \qquad (10.34)$$

where Δf_{eq} is the equivalent peak frequency variation equal to $KE_{m_{max}}/2\pi RC$.

$RC \gg T$ where T is the periodic time of $m(t)$

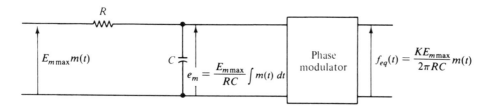

FIGURE 10.6. Conversion of phase modulation to frequency modulation.

Equation (10.34) shows that the equivalent frequency modulation from the phase modulator and integrator is proportional to the original modulating signal, which is the desired condition.

10.5 ANGLE MODULATOR CIRCUITS

In the analysis presented here, small signal conditions are assumed; it is also assumed that the modulation rate is sufficiently slow for *sinusoidal* analysis at the carrier frequency to be used.

10.5.1. The Varactor Diode

The *varactor diode* is a *pn* junction diode, the depletion capacitance of which varies with reverse bias, as shown in Fig. 10.7(a). The name *varactor* comes from *var*iable re*actor*. For tuning applications, the diode may be represented by the approximate equivalent circuit of the depletion capacitance C_d in series with a resistance R_s, which represents the resistance of the bulk *pn* regions and the lead resistance.

A fixed dc bias voltage is applied to the varactor, which adjusts the value of C_d to a center value C_{d0} which determines the unmodulated resonant frequency of a tuned circuit which includes the varactor. Modulation voltage is added to this bias to cause the capacitance to vary around the value of C_{d0} with the modulation, thus causing the resonant frequency of the LC circuit to vary about its center value of f_0.

The tuned LC circuit can be used as the frequency-determining network in one of the standard LC oscillator circuits to provide direct frequency modulation, or it may be used as a voltage-variable reactor to shift the phase of the signal from a fixed oscillator to provide direct phase modulation.

Figure 10.7(b) shows a Clapp oscillator with a varactor diode modulator allowing direct frequency modulation. This oscillator uses a common base amplifier transistor to provide the necessary gain. C_2 and C_3 provide the impedance transformer between the collector and the emitter, and tuning is accomplished by L_1 in series with the varactor diode, which replaces C_1. (See Fig. 6.4 for subscript definitions.) Since the amplifier is noninverting, impedances Z_1 and Z_3 (see Table 6.1) are interchanged. The fixed-bias voltage is provided by a resistive voltage divider, and the modulation voltage is added to it before the voltage is applied to the varactor diode through a radio frequency choke (or inductor) RFC. Adjustment of the voltage divider allows adjustment of the oscillator center frequency to the desired carrier frequency. Variation of the bias voltage about this fixed value by the modulation causes the oscillator voltage to also vary about the center frequency f_0. The frequency-modulated output is obtained from the transistor collector.

For an abrupt-type *pn* junction in silicon, semiconductor theory (see, for example, Motorola Application Note AN 178-A (1971)) gives the relationship between C_d and the reverse bias V as

$$C_d = \frac{C_0}{\left(1 + \dfrac{V}{\Phi}\right)^a} \tag{10.35}$$

where C_0 is the capacitance at zero bias and Φ is the contact potential of the junction (in volts), and the index a depends on the type of junction. For abrupt junctions, $a \cong 0.5$, and for silicon, $\Phi \cong 0.5\text{V}$.

The shape of the response of C_d vs. V is shown in Fig. 10.7(a), and, as can be seen, it is not linear. For this reason it is usually desirable to keep the modulation voltage quite small so that the variation of capacitance with voltage does not depart too far from the ideal linear slope response as discussed below. The result is that frequency deviation must also be kept small to prevent nonlinear distortion.

The frequency deviation constant k for the circuit of Fig. 10.7(b) is defined as

$$k = S_1 S_2 = \frac{df}{dV} \tag{10.36}$$

where S_1 is the slope of the varactor diode characteristic (Fig. 10.7(a)) at the operating point determined by V_0, and S_2 is the slope of the frequency/diode capacitance curve also at the operating point, determined by C_{d0}. The student should verify that the dimensions of k are Hz/V.

The slope S_1 is found by differentiating Eq. (10.35), which, for $a = 0.5$ and $\Phi = 0.5$ V, gives

$$S_1 = \frac{dC_d}{dV} = -\frac{C_d}{1 + 2V} \tag{10.37}$$

At the operating point, $C_d = C_{d0}$ and $V = V_0$, and therefore,

$$S_1 = -\frac{C_{d0}}{1 + 2V_0} \tag{10.38}$$

The frequency of the Clapp oscillator is given approximately by

$$f = \frac{1}{2\pi\sqrt{L_1}} \sqrt{\frac{1}{C_d} + \frac{1}{C_s}} \tag{10.39}$$

where

$$C_s = C_1 \text{ ser. } C_3 = \frac{C_1 C_3}{C_1 + C_3}$$

Therefore,

$$S_2 = \frac{df}{dC_d} = -\frac{f}{2\left(1 + \frac{C_d}{C_s}\right)C_d} \tag{10.40}$$

(The differentiation of these expressions is left as a mathematical exercise for the student.)

At the nominal operating frequency, $f = f_0$ and $C_d = C_{d0}$. Also, for the Clapp circuit, let $C_{d0} = nC_s$. Then

$$S_2 = -\frac{f_0}{2(1 + n)C_{d0}} \tag{10.41}$$

Combining Eqs. (10.36), (10.38), and (10.41) gives

$$k = \frac{f_0}{2(1 + n)(1 + 2V_0)} \tag{10.42}$$

Example 10.5 Calculate the frequency-deviation constant for a Clapp oscillator frequency-modulated in accordance with Eq. (10.42), for which the unmodulated frequency is 50 MHz at a fixed bias of 15 V, given that $n = 0.1$.

Solution
$$k = \frac{50 \times 10^6}{2(1 + .1)(1 + (2 \times 15))}$$
$$= 733 \text{ kHz/V}$$

The varactor diode may also be used as a phase modulator, one circuit being shown in Fig. 10.7(c). The variation of the diode capacitance alters the phase angle of the tuned-circuit admittance, and hence the phase angle of the oscillator voltage developed across it. Let the oscillator feed the circuit with a constant-current, *constant-frequency* signal; then the phase deviation constant

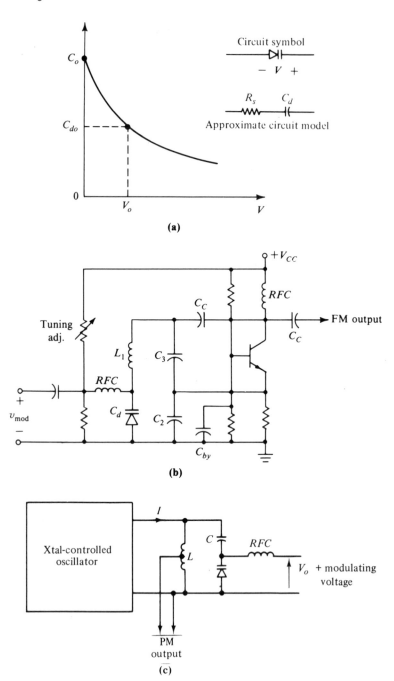

FIGURE 10.7. (a) Varactor diode characteristic; (b) varactor diode used for frequency modulation; (c) varactor diode used for phase modulation.

K [see Eq. (10.24)] for the circuit is found as follows. The total tuning capacity is

$$C_T = \frac{CC_d}{C + C_d} \tag{10.43}$$

and at the operating point let this be C_{T0}, determined by C_{d0}. The slope of the C_s/C_d curve at the operating point (as found by differentiation) is

$$\frac{dC_T}{dC_d} = S_T = \left(\frac{C_{T0}}{C_{d0}}\right)^2 \tag{10.44}$$

Next, the admittance of the tuned circuit is

$$Y = \frac{1}{R_D} + j\left(\omega C_T - \frac{1}{\omega L}\right) \tag{10.45}$$

and its phase angle is

$$\phi = \tan^{-1}\left[R_D\left(\omega C_T - \frac{1}{\omega L}\right)\right]$$

where R_D is as given by Eq. (1.85). For small angles, the angle is approximately equal to the tangent, and the slope of the ϕ/C_T curve at the operating point becomes

$$\frac{d\phi}{dC_T} = S_\phi = R_D \omega_0 \tag{10.46}$$

Substituting for R_D from Eq. (1.86) gives

$$S_\phi = \frac{Q}{C_{T0}} \tag{10.47}$$

The phase deviation constant K is then determined by

$$K = \frac{d\phi}{dV} = \frac{d\phi}{dC_T}\frac{dC_T}{dC_d}\frac{dC_d}{dV} = S_1 S_T S_\phi \tag{10.48}$$

The student should verify that K has dimensions of rad/V. Substituting from Eqs. (10.38), (10.44), and (10.46), and simplifying, gives

$$K = \frac{Q}{(1 + n)(1 + 2V_0)} \tag{10.49}$$

where $n = \dfrac{C_{d0}}{C}$

Example 10.6   Find the phase deviation constant for a phase modulator circuit which uses a varactor diode at a voltage of $V_0 = 15$ V, in series with $C = 10$ C_{d0} in a tuned circuit with $Q = 70$.

Solution   By Eq. (10.49),

$$K = -\frac{70}{(1 + 10)(1 + 2 \times 15)} = -.2 \text{ rad/V}$$

10.5.2   Transistor Angle Modulators

A transistor in which the transconductance g_m is a function of modulating voltage may be used for both frequency and phase modulation. The principle of operation is that g_m varies in accordance with the slowly varying modulating signal, and the circuit, in turn, presents a reactance at *oscillator frequency* that varies with g_m and that influences the frequency or phase of the oscillator output. Bipolar transistors or FETs may be used, but for definiteness, FETs are shown in the circuits of Fig. 10.8. For both circuits, circuit values are such that the oscillator output current I_0 is approximately equal to the drain current I_d; that is, currents I_1 and I_2 are negligible at oscillator frequency. Hence,

$$I_0 \cong I_d = g_m V_{in} \tag{10.50}$$

For the frequency modulation circuit of Figure 10.8(a), the modulator circuit has an output impedance $Z_0 = V_0/I_0$ which is capacitive, and this capacitive reactance replaces the tuning capacitor C_1 in the tuned circuit of the oscillator. Since the effective output impedance can be varied by varying the modulating voltage applied to the transistor, the oscillator output frequency can also be varied. Examining the input circuit of the transistor,

$$V_{in} = V_0 \frac{Z_2}{Z_1 + Z_2} \tag{10.51}$$

The small-signal admittance presented to the oscillator can be found by combining Eqs. (10.50) and (10.51):

$$Y = \frac{I_0}{V_0}$$

$$= g_m \frac{Z_2}{Z_1 + Z_2} \tag{10.52}$$

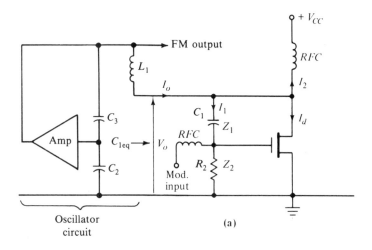

(a)

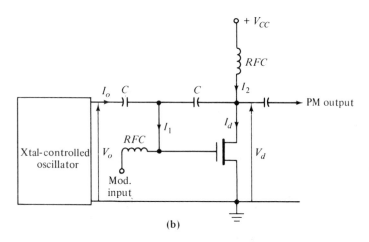

(b)

FIGURE 10.8. FET varactor circuits: (a) frequency modulation; (b) phase modulation.

By making $Z_1 = 1/j\omega_0 C_1$, $Z_2 = R_2$, and $|Z_1| \gg Z_2$, Eq. (10.52) reduces to

$$Y \cong jg_m\omega_0 C_1 R_2$$

This is equivalent to a capacitance C_{eq} replacing C_1 in the oscillator tuned circuit, where

$$C_{eq} = g_m C_1 R_2 = C_{eq0} + kV_{mod} \tag{10.53}$$

since $g_m = g_{m0}(1 + kV_{mod})$ for the field effect transistor. Thus the modulating voltage directly varies the capacitance about its center value, and this varies the oscillator frequency about its center value in the same manner that the varactor diode did in the previous section.

In the phase modulator circuit of Fig. 10.8(b), the output from a crystal-controlled oscillator is applied to the carrier input terminals of the modulator circuit as V_0. The transfer function of the modulator circuit causes a phase shift in the carrier voltage without affecting its voltage amplitude, the phase shift being directly proportional to the modulating voltage. The transfer function is defined as the ratio V_d/V_0, and it is necessary to find the phase relationship between V_d and V_0. From the circuit,

$$I_0 \cong (V_0 - V_d)\,j\frac{\omega C}{2} \tag{10.54}$$

and

$$V_{in} \cong \tfrac{1}{2}(V_0 + V_d) \tag{10.55}$$

(The approximations are valid to the extent that I_1 and I_2 can be neglected). Substituting Eq. (10.55) in Eq. (10.50) and equating to Eq. (10.54),

$$g_m\frac{(V_0 + V_d)}{2} = (V_0 - V_d)\,j\frac{\omega_0 C}{2}$$

from which

$$V_d = -V_0\frac{(g_m - j\omega_0 C)}{(g_m + j\omega_0 C)} \tag{10.56}$$

Since the modulus of the complex term in the numerator is equal to that in the denominator, it follows that the modulus of V_d is equal to that of V_0; that is,

$$|V_d| = |V_0| \tag{10.57}$$

This means that the circuit does not introduce amplitude modulation, which is an advantage.

Eq. (10.56) can be written as

$$V_d = V_0 \frac{\underline{/-\theta}}{\underline{/\theta}} = V_0 \underline{/-2\theta} \tag{10.58}$$

where $\theta = \tan^{-1}\left(\dfrac{\omega_0 C}{g_m}\right)$ $\hspace{3cm}$ (10.59)

Hence the phase angle of V_d with respect to V_0 is

$$\phi = -2\theta$$

$$= -2\tan^{-1}\left(\frac{\omega_0 C}{g_m}\right) \tag{10.60}$$

This shows that the phase shift is a function of g_m, which in turn is a function of the modulating voltage. Although the relationship between ϕ and g_m, expressed by Eq. (10.60), is nonlinear, it is found in practice that the phase modulation characteristic is reasonably linear up to about $\pm 45°$ phase shift.

10.5.3 Directly Modulated FM Transmitters

Direct frequency modulation (Fig. 10.9(a)) can be achieved by use of the circuits of Fig. 10.7(b) or Fig. 10.8(a). The peak frequency deviation is kept small, and the modulated signal from the oscillator is then passed to a frequency multiplier circuit which raises the output frequency to the desired carrier frequency. A power amplifier drives the antenna.

Direct frequency modulation at the final carrier frequency could be used, with the multiplier stages omitted, but a conflict immediately arises between obtaining adequate frequency deviation and maintaining high frequency stability. Crystal oscillators can be directly frequency-modulated since the crystal frequency can be "pulled" by a small amount, but to achieve usable final deviations, a high frequency multiplication factor is necessary, as described below. Even so, direct frequency modulation is only used for narrow-band FM, which employs relatively small deviations.

When an FM signal is passed through a frequency multiplier circuit such as a class C amplifier with its output tank tuned to the second or third harmonic, not only will the carrier frequency be multiplied, but the frequency deviation will also be multiplied. The multiplication ratio will be equal to the harmonic number to which the output is tuned.

The distinction between frequency multiplication and up-conversion as obtained by mixing is important in the operation of FM systems. The general expression for a frequency-modulated carrier is given by Eq. (10.12). As shown in Section 5.7.1, a class C amplifier produces harmonics of the input signal and, provided the harmonic-tuned output circuit is sufficiently broadband to accept the sidebands, the nth harmonic of the input (Eq. (10.7)) will be selected, giving as the output voltage

$$v_{\text{out}} = \sin(n\theta(t)) \tag{10.61}$$

$\theta(t)$ is given by Eq. (10.9), and therefore the output frequency is

$$\begin{aligned} f_{\text{out}} &= nf_i \\ &= nf_c + n\,\Delta fm(t) \end{aligned} \tag{10.62}$$

This shows that the deviation is multiplied by the harmonic number n.

When the signal is up-converted by mixing with an oscillator frequency f_0, the final frequency is given by

$$\begin{aligned} f_{\text{out}} &= f_0 + f_i \\ &= (f_0 + f_c) + \Delta fm(t) \end{aligned} \tag{10.63}$$

The carrier frequency is raised to $f_0 + f_c$, and the modulation is unaltered.

The foregoing theory is used in the scheme shown in Fig. 10.9(b) to achieve an increase in the deviation without increasing the carrier frequency. The input frequency f_i, given by Eq. (10.5), is multiplied up to nf_i. When this is applied to a down-converter along with an oscillator frequency $f_0 = (n - 1)f_c$, the resulting output frequency is

$$\begin{aligned} f_{\text{out}} &= nf_i - f_0 \\ &= f_c + n\,\Delta fm(t) \end{aligned} \tag{10.64}$$

This shows that the carrier is converted back to its original frequency with the deviation multiplied by n times.

Where the main oscillator is only an LC oscillator, the direct-modulation scheme is not capable of meeting frequency stability regulations. Stability is improved in Fig. 10.9(c) by an AFC circuit. A sample of the final output signal is mixed with the signal from a stable crystal oscillator. The IF produced contains the difference frequency between the carrier and the fixed oscillator. A discriminator circuit generates a voltage which is proportional to this difference frequency. It also contains the modulation signal, and a low-pass filter is used to remove this, leaving a varying dc level which is proportional to the difference between the carrier frequency and the oscillator. This voltage is added to the modulating audio signal and applied to the reactance modulator in a manner so as to correct any drift in the main oscillator frequency. The gain of the frequency feedback loop is determined by the frequency multiplication constant and by the modulator and discriminator gains. Care must be taken to be sure that the feedback loop is stable; otherwise, oscillations may occur at modulating frequencies.

Class C power amplifiers can be used for FM transmitters because any small variations in the amplitude of the FM signal are usually removed in the receiver circuits by limiting amplifiers. Further, they do not have any signifi-

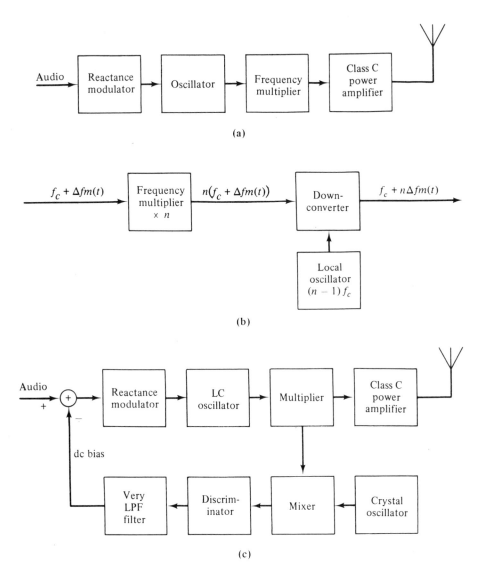

FIGURE 10.9. (a) A directly modulated FM transmitter; (b) system for increasing the deviation of an FM signal; (c) frequency-stabilized reactance modulator.

cant effect on the modulation itself, and interference from noise is greatly reduced. The result is that the FM transmitter is much more efficient than an equivalent AM transmitter.

10.5.4 Indirectly Modulated FM Transmitters

Phase modulation may be used to achieve frequency modulation by the indirect method, the theory being given in Section 10.4. It is necessary only to integrate the modulating signal prior to applying it to the phase modulator, the phase modulation being accomplished by either of the circuits of Fig. 10.7(c) or Fig. 10.8(b). The block diagram for such a transmitter is shown in Fig. 10.10(a). This transmitter is widely used in VHF and UHF radio telephone equipment.

Another indirect method that is widely used is the Armstrong method, in which FM is achieved from a combination of DSB suppressed carrier and phase modulation. Fig. 10.10(b) shows the block diagram of an Armstrong-modulated FM transmitter. The carrier source is a crystal oscillator with a frequency that is easy to handle in the modulators. Generally this will be quite low, say 100 kHz, because of the need to do several stages of frequency multiplication. A sample of the carrier is separated and shifted by 90° before application to a balanced modulator. The audio is passed through an integrator circuit before being applied to the modulator.

The output of the balanced modulator is reduced in amplitude so that it is very small compared to the oscillator output. This output is the phasor sum of the two sidebands, without any carrier, and is displaced in phase by 90° from the oscillator output. When the sidebands are added to the oscillator output, phase modulation occurs. This can be seen from the phasor diagram of Fig. 10.10(c). The sideband phasor sum leads the carrier by 90° and varies in amplitude. It is phasor-added to the carrier to produce the voltage E_{pm}. As the magnitude of the sidebands changes, so does the angle ϕ between E_{pm} and the carrier. This angle is given by

$$\phi = \tan^{-1}\left(\frac{E_{dsb}}{E_c}\right) \tag{10.65}$$

However, if the angle ϕ can be kept less than about 10°, the tangent is approximately equal to the angle itself:

$$\tan\phi \cong \phi < 0.17 \text{ rad} \tag{10.66}$$

Therefore, for small angular deviations, say, less than 10°, an error of less than

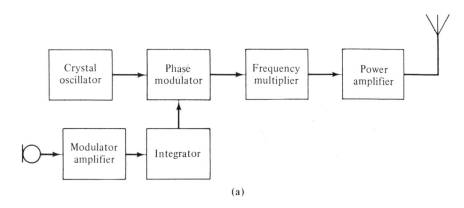

(a)

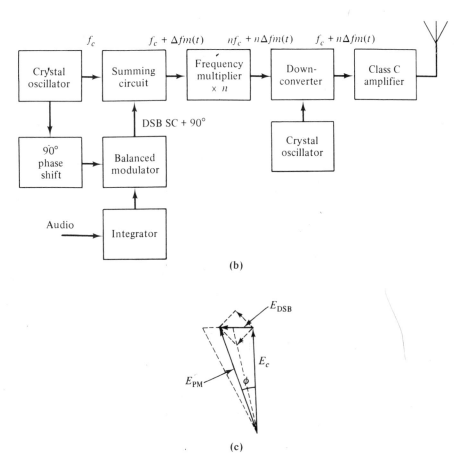

(b)

(c)

FIGURE 10.10. Indirect frequency modulation: (a) phase to frequency modulation; (b) the Armstrong method; (c) the phasor diagram illustrating the principle of the Armstrong method.

1% in the modulation will occur, and

$$\phi \cong \frac{E_{dsb}}{E_c} \tag{10.67}$$

This shows that the phase modulation ϕ is proportional to the double-sideband signal which, in turn, is proportional to the integrated modulation signal, and, as shown previously, this produces the equivalent frequency modulation.

The wave at this point has a very small frequency deviation, and this must be increased before transmission. This is accomplished by passing the FM signal through a series of frequency multiplier amplifiers until the deviation is sufficiently large. The signal is then up- or down-converted to the final transmitting frequency and applied to drive a class C power amplifier.

Example 10.7 An Armstrong transmitter is to be used for transmission at 152 MHz in the VHF band, with a maximum deviation of 15 kHz at a minimum audio frequency of 100 Hz. The primary oscillator is to be a 100-kHz crystal oscillator, and the initial phase-modulation deviation is to be kept to less than 12°, to avoid audio distortion. Find the amount by which the frequency must be multiplied to give proper deviation, and specify a combination of doublers and triplers which will give this. Also, specify the mixer crystal and any multiplier stages needed.

Solution The maximum phase deviation of the modulator is

$$\Delta\Phi_{max} = 12° = 0.2096 \text{ rad} \equiv m_p$$

From Eq. (10.33),

$$\Delta f_{max} = \Delta\Phi_{max} f_{min} = 0.2094 \times 0.1 = 0.02094 \text{ kHz}$$

The frequency deviation increase required is

$$N = \frac{\Delta f_{max \text{ allow}}}{\Delta f_{max}} = \frac{15 \text{ kHz}}{0.0209 \text{ kHz}} = 716.33 \left(\text{use } 729 = 3^6\right)$$

The modulated waveform will be passed through a chain of six tripler stages, giving a final deviation of

$$0.02094 \times 729 = 15.265 \text{ kHz}$$

at a frequency of $0.1 \times 729 = 72.9$ MHz. The deviation is slightly high, but a slight attenuation of the audio signal will compensate for this. The mixer oscillator signal is

$$f_0 = 152 - 72.9 = 79.1 \text{ MHz}$$

f_0 is best obtained by using two tripler stages from an 8.7889-MHz crystal oscillator.

10.5.5 Preemphasis

Preemphasis of the modulating signal is necessary to compensate for the deemphasis introduced in the receiver to improve the signal-to-noise ratio (see Sections 10.7 and 10.6.9). A simple preemphasis network is shown in Fig. 10.11, along with the modulus of the transfer function. Over the preemphasis range, $\omega_1 < \omega < \omega_2$,

$$\left| \frac{V_0}{V_i} \right| = 20 \log \left\{ \frac{R_2}{R_1 + R_2} \sqrt{1 + (f//f_1)^2} \right\} \quad (\text{in dB}) \qquad (10.68)$$

where $f_1 = 1/2\pi C R_1$ and f_1 is standardized at 2.1 kHz. The frequency (f_2) at which preemphasis levels off is chosen to be above the bandpass. It should be noted that phase modulation by itself provides the required preemphasis characteristic if the integrator is modified to be ineffective above the 2.1-kHz preemphasis cutoff frequency.

10.5.6 FM Broadcast

The prime requisite of FM broadcasting is excellent fidelity, since music provides the chief program material. Frequency modulation acts in several ways to improve this fidelity. First, since FM broadcasting takes place in the

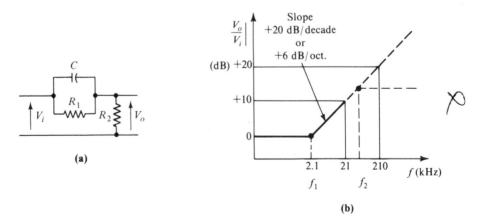

FIGURE 10.11. A preemphasis network and its transfer function.

VHF band from 88 to 108 MHz, a much wider baseband can be used. The baseband width presently in use is 50 Hz to 15 kHz, with a maximum allowable deviation of ± 75 kHz. Channel spacing is 200 kHz, and power outputs of as much as 100 kW are used.

Until recently single-channel monophonic FM broadcasts were common, but nearly all FM stations are now transmitting two-channel stereo programs. In the near future this may be further changed to provide four-channel stereo. Further, some FM stations are frequency-division-multiplexing an additional channel on their carrier for the purpose of providing background music for public buildings, a system licensed as a "subsidiary communications authorization," or SCA.

The transmitters used are typically of the Armstrong type discussed in Section 10.5.4. Initial deviations are kept small to limit modulation distortion, and many stages of frequency multiplication are used to bring the deviation up to that required at the output. Again, to provide the necessary power output, the output stage is a push-pull parallel class C amplifier. Water-cooled vacuum tubes are used in this stage.

The main differences between the FM systems already discussed and the broadcast system lies in the composition of the audio signal presented to the modulator. It is a composite signal carrying several signals, and these will be discussed.

First, if monophonic FM transmission is to be used, only one channel is needed, and the single audio channel is applied directly to the modulator input. It must be fully compensated to provide good fidelity in the band from 50 to 15,000 Hz.

Two-channel stereo is accomplished as follows. The two audio channels are not simply frequency-division-multiplexed before modulation. They are first mixed to provide two new signals, one of which is a *balanced* monophonic signal. The first is the sum of the two input channels, and the second is the difference of the two. The sum channel is modulated directly in the baseband assignment between 50 and 15,000 Hz. The difference signal is DSBSC modulated in the 23–53 kHz slot about a carrier at 38 kHz. A pilot carrier at 19 kHz is also transmitted. The sum signal in the audio portion of the band can be demodulated by a monophonic FM receiver to provide normal monophonic reception. A receiver with a stereo demodulator can also retrieve the difference signal and combine the two to produce the original L and R channel signals. Fig. 10.12(a) shows the block diagram of the premodulation mixing circuits.

The L and R channels are summed and passed through a low-pass 15-kHz filter to form the monophonic portion of the baseband signal. The R channel is inverted and then added to the L signal to yield the difference signal L − R. This signal is DSBSC-modulated on the 38-kHz carrier by a balanced modulator and passed through a 23–53 kHz bandpass filter to remove any unwanted signal components. The two channel bands and the 19-kHz pilot carrier are added together to produce the final baseband signal (Fig. 10.22(b)).

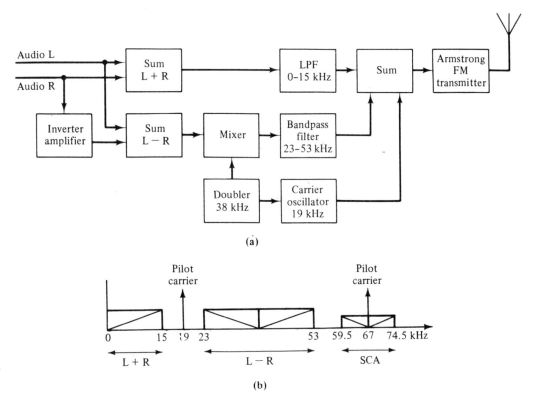

FIGURE 10.12. An FM stereo broadcast transmitter: (a) block schematic; (b) baseband spectrum showing position of the auxiliary 7.5-kHz channel.

This final composite signal is presented to the modulator input of the FM transmitter.

An additional channel is often modulated on the same carrier for service to commercial operations, such as music to stores and public buildings. This channel is limited to a signal bandwidth of 7.5 kHz and is multiplexed to lie in the range from 53 to 75 kHz, with a pilot carrier at 67 kHz. The sideband frequencies are shown in Fig. 10.12(b). This auxiliary channel does not interfere in any way with the ordinary broadcast.

10.6 ANGLE MODULATION DETECTORS

10.6.1 Basic Detection of FM Signals

In order to detect an FM signal, it is necessary to have a circuit whose output voltage varies *linearly* with the frequency of the input signal. The slope detector is a very basic form of such a circuit, although its linearity of response

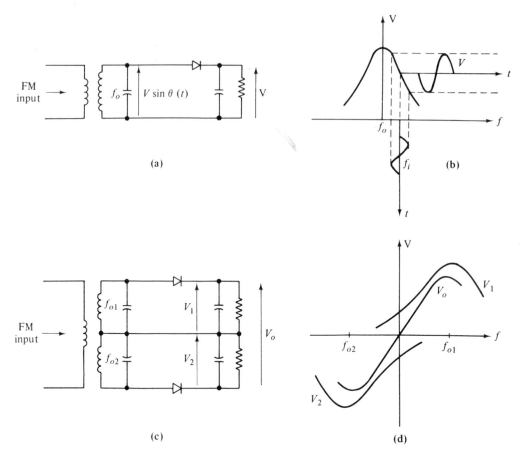

FIGURE 10.13. (a) FM slope detector; (b) magnitude of output voltage plotted against frequency; (c) balanced double-tuned slope (or Round-Travis) detector; (d) output voltage plotted against frequency, showing the S-shaped transfer function curve.

is not good. Fig. 10.13(a) shows the basic arrangement. By tuning the circuit to receive the signal on the slope of the response curve Fig. 10.13(b), the carrier amplitude V is caused to vary with frequency. In this case the circuit is tuned so that its resonant frequency f_0 is lower than the carrier frequency f_i. When the signal frequency increases above f_i, with modulation the amplitude of the carrier voltage drops. When the signal frequency decreases below f_i, the carrier voltage rises. The change of voltage results because of the change in the magnitude of the impedance in the tuned circuit as a function of frequency and results in an effective conversion of frequency modulation into amplitude modulation. The modulation is recovered from the amplitude modulation by means of a normal envelope detector. However, the linear range on the voltage/frequency-transfer characteristic is limited.

Linearity can be improved by utilizing the arrangement of Fig. 10.13(c), a circuit known as the *Round-Travis detector,* or *balanced slope detector.* This circuit combines two circuits of the type shown in Fig. 10.13(a) in a balanced configuration. One slope detector is tuned to f_{01} above the incoming carrier frequency, and the other to f_{02} below the carrier frequency, and the envelope detectors combine to give a differential output. This means that by showing the V_1 response as positive, the V_2 response can be shown negative on the same axes, and the output $V_0 = |V_1| + |V_2|$ will have an S shape when plotted against frequency, as shown in Fig. 10.13(d). This S-curve is characteristic of FM detectors. When the incoming signal is unmodulated, the output is balanced to zero; when the carrier deviates toward f_{01}, $|V_1|$ increases while $|V_2|$ decreases, and the output goes positive; when it deviates toward f_{02}, $|V_1|$ decreases while $|V_2|$ increases, and the output goes negative.

10.6.2 Foster–Seeley Discriminator

The Foster–Seeley discriminator recovers the modulating voltage from the frequency modulation by utilizing the phase-angle shift between primary and secondary voltages of a tuned transformer. The phase angle is a function of frequency, and by arranging for phasor-sum and phasor-difference components of primary and secondary voltages to be applied to two envelope detectors, the outputs of which are then combined, demodulation is achieved. Because the circuit relies on a phase-angle variation, it is also known as a *phase discriminator,* but it must not be thought that it detects phase modulation directly; it converts a frequency variation, real or equivalent, to a circuit phase-angle variation, which in turn is converted to an amplitude variation.

The basic circuit is shown in Fig. 10.14(a). Capacitor C_c has negligible reactance at carrier frequency; therefore, approximately the full primary voltage appears across the radio-frequency choke (RFC), which has a high reactance at carrier frequency. The primary voltage is connected from the RFC to the center tap on the transformer secondary, with the result that the radio-frequency voltage applied to diode D_1 is $V_1 + \frac{1}{2}V_2$, and to diode D_2, $V_1 - \frac{1}{2}V_2$. The phase relationship between V_1 and V_2 can be found as follows. The EMF induced in the secondary of the transformer E_2 is related to the primary voltage by the expression

$$E_2 \cong V_1 k \sqrt{\frac{L_2}{L_1}} \qquad (1.125)$$

and for frequencies very near resonance, it experiences practically no phase shift. This secondary EMF is applied to the series resonant circuit formed by R_2, L_2, and C_2, whose impedance is

$$Z_2 = R_2(1 + jyQ)$$

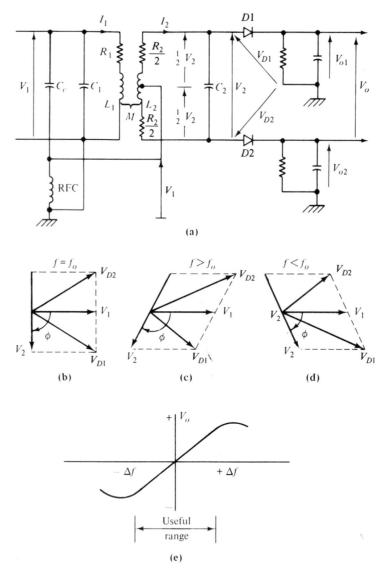

FIGURE 10.14. (a) Foster-Seeley discriminator; (b) phasor sums of $V_1 \pm V_2/2$ at resonance $(f = f_0)$; (c) phasor sums of $V_1 \pm V_2/2$ at $f < f_0$; (e) output voltage versus frequency deviation.

where

$$y = \frac{\omega}{\omega_0} - \frac{\omega_0}{\omega} \qquad (1.65)$$

Let the frequency deviation of the carrier be $\delta\omega$, so that at any time,

$$\omega = \omega_0 + \delta\omega$$

Then

$$y = \frac{\omega_0 + \delta\omega}{\omega_0} - \frac{\omega_0}{\omega_0 + \delta\omega} \approx 2\frac{\delta\omega}{\omega_0} \qquad (10.69)$$

when $\delta\omega \ll \omega_0$.

Now the secondary current I_2 becomes

$$I_2 = \frac{E_2}{Z_2} = \frac{V_1 k}{R_2}\sqrt{\frac{L_2}{L_1}}\frac{1}{\sqrt{1 + y^2 Q^2}}\underline{/\tan^{-1}(-yQ)}$$

$$\approx \frac{V_1 k}{R_2}\sqrt{\frac{L_2}{L_1}}\underline{/\tan^{-1}\left(-2\frac{Q\delta\omega}{\omega_0}\right)} \qquad (10.70)$$

after substitutions from Eqs. (1.125), (1.65), and (10.69) and after noting that near resonance, $y^2 Q^2 \ll 1$ so that the current magnitude is determined mainly by the secondary resistance. The current *at resonance* experiences no phase shift from V_1, but the phase shift of I_2 varies about zero directly with the frequency deviation of the carrier. The voltage V_2 is the drop across the capacitor C_2 due to current I_2 and becomes, upon substitution,

$$V_2 = V_1 Q k \sqrt{\frac{L_2}{L_1}}\underline{/\phi} \qquad (10.71)$$

where

$$\phi = \tan^{-1}\left(\frac{-2Q\delta\omega}{\omega_0}\right) - \frac{\pi}{2} \approx -\left(\frac{\pi}{2} + 2\frac{Q\delta\omega}{\omega_0}\right) \qquad (10.72)$$

and where the deviation angle is less than about 0.5 radian and Q is obtained from Eq. (1.62). This shows that the secondary voltage V_2 is shifted from the primary voltage V_1 by $-90°$ and has a further phase shift which is directly proportional to the frequency deviation.

The phasor sum of $V_1 \pm \frac{1}{2}V_2$ is then as shown in Fig. 10.14(b), (c), and (d) for three different conditions of carrier frequency. The envelope detector D_1 will produce an output voltage proportional to $|V_{D1}|$, and that of D_2 an output voltage proportional to $|V_{D2}|$. (For an explanation of the action of the envelope detector, see Section 8.2.7.) The output voltage is seen to be

$$V_0 = V_{02} - V_{01}$$
$$= K(|V_{D2}| - |V_{D1}|) \qquad (10.73)$$

where K is a constant of the detector circuits.

To summarize, as the frequency increases, the phase shift becomes more negative and $|V_{D2}|$ increases while $|V_{D1}|$ decreases; hence, V_0 becomes more positive. When the frequency decreases, the phase shift becomes more positive and $|V_{D2}|$ decreases while $|V_{D1}|$ increases; therefore, V_0 becomes more negative.

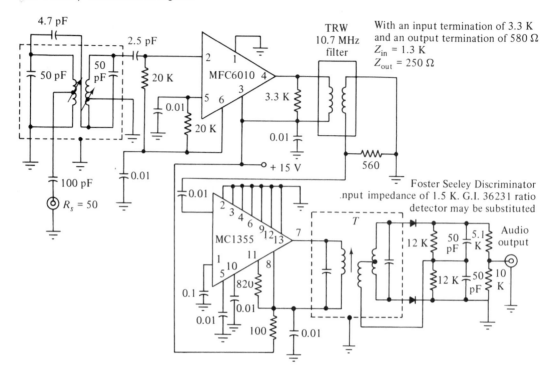

FIGURE 10.15. FM receiver unit utilizing integrated circuit amplifiers and a Foster – Seeley discriminator. (Courtesy Motorola, Inc.)

The curve of output voltage V_0 against frequency deviation Δf is sketched in Fig. 10.14(e). This S-curve is characteristic of discriminators, the curving at the extremes resulting from the secondary voltage becoming very small at frequencies well removed from resonance. Amplitude variations are reduced to negligible proportions by amplitude-limiting the signal before applying it to the discriminator. There are many practical variations of the circuit, and Fig. 10.15 shows it being used with integrated circuits. In this application, amplifier block MFC6010 provides gain and amplifier block MC1355 provides gain limiting. Note that the primary voltage is coupled into the secondary by means of a tertiary winding which is very closely coupled to the primary. Also, the ground point is shifted so that the output may be taken with respect to ground, a much more practical arrangement.

10.6.3 Ratio Detector

A very simple change can be made to the Foster–Seeley discriminator which improves the limiting action, but at the expense of reducing the output. One of the diodes is reversed so that the two diodes conduct in series (Fig. 10.16(a)); the detector circuits then provide a damping action, which tends to maintain a constant secondary voltage, as will be shown.

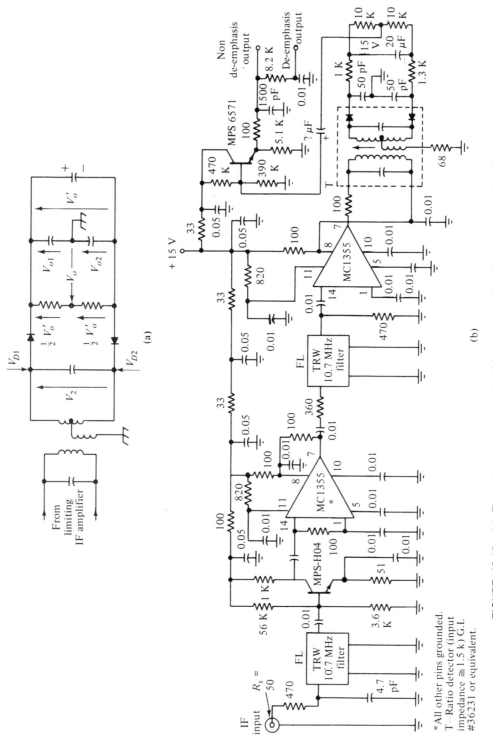

FIGURE 10.16. (a) The ratio-detector circuit; (b) a high-quality FM receiver unit utilizing integrated circuit amplifiers and a ratio detector (Courtesy Motorola, Inc.).

357

Diodes D_1 and D_2 and associated loads RC form envelope detectors as before, and the frequency-to-phase-to-amplitude chain of events occurs as in the Foster–Seeley discriminator. Now, however, the polarity of voltage in the lower capacitor is reversed, so the sum voltage appears across the combined loads (rather than the difference voltage as in the Foster–Seeley). Hence, as V_{01} increases, V_{02} decreases and V_0' remains constant (and, of course, V_0' remains constant as V_{01} decreases and V_{02} increases). Therefore, a large capacitor (in practice usually an electrolytic type) can be connected across the V_0' points without affecting the voltage, except to improve the "constancy."

From the circuit of Fig. 10.16(a), two equations can be written for the output voltage V_0, as follows:

$$V_0 = \tfrac{1}{2}V_0' - V_{02}$$

and

$$V_0 = -\tfrac{1}{2}V_0' + V_{01}$$

Adding,

$$2V_0 = V_{01} - V_{02}$$

Therefore,

$$V_0 = \tfrac{1}{2}(V_{01} - V_{02}) \tag{10.74}$$

or

$$V_0 = \tfrac{1}{2}K(|V_{D1}| - |V_{D2}|)$$

The output voltage is one-half that of the Foster–Seeley circuit.

Limiting action occurs as a result of variable damping on the secondary of the transformer. For example, if the input voltage amplitude $(V_{1\max})$ were to suddenly increase, as would occur with a noise spike of voltage, the voltage V_0' could not follow immediately, since it is held constant by means of the large capacitor. The voltage across the diodes in series is $V_2 - V_0'$, and since V_2 will increase with V_1, the diodes conduct more heavily (i.e., more current flows). This results in heavier damping of the secondary (which is also reflected into the primary), which reduces the Q factor. This, in turn, tends to offset the increase in V_2 by reducing the gain of the limiting amplifier feeding the circuit.

If the input voltage should decrease suddenly, diode conduction is reduced and so is the damping. The gain of the limiting amplifier therefore increases, tending to offset the original decrease in voltage. The fact, too, that the diode load, as seen by the two diodes in series, has a long time constant and therefore cannot respond to fast changes in voltage helps in the limiting process.

Of course, the limiting action will not be effective for slow changes in amplitude, as V_0' will gradually adjust to the new level determined by input voltage amplitude.

Figure 10.16(b) shows the circuit diagram of a high-quality FM stage utilizing a ratio detector in conjunction with integrated circuit amplifiers. Both amplifying blocks MC1355 provide limiting gain.

10.6.4 Quadrature Detector

This type of FM detector also depends on the frequency/phase angle relationship of a tuned circuit, but has the advantage that only one tuned circuit is required. As a result, it is becoming increasingly popular in integrated FM strips.

Consider first the general principle. If a phase-modulated signal is multiplied by a signal which is the same FM signal shifted by 90°, one of the components in the multiplier output signal will be the modulation itself. When the carrier current flows through the inductor in Fig. 10.17(a), the voltage V_L will be the modulated carrier shifted by 90°. The same current flows through Z, which is a parallel resonant circuit tuned to the carrier center frequency. The voltage across this circuit experiences a phase shift which is directly related to the frequency deviation of the carrier, as shown below, resulting in a frequency modulation to phase modulation conversion. Now when the voltage across the inductor is multiplied by the voltage across the tuned circuit, the modulation appears in the output of the multiplier.

Now assuming that the carrier frequency varies slowly enough, the carrier voltage (or current) may be assumed to be sinusoidal. Thus, the current

$$i = I_m \sin \omega t \tag{10.75}$$

flows in the circuit of Fig. 10.17(a). The voltage across the inductor leads the current by 90°:

$$v_L = V_{L_{max}} \cos \omega t \tag{10.76}$$

The voltage across the parallel tuned circuit will be in phase with the current at resonance. Assuming that $\omega_0/\omega \cong 1$, at frequencies within a range of about $\pm 1\%$ of resonance, the admittance phase angle of the circuit is, by Eq. (1.81),

$$\Phi \cong \tan^{-1} yQ \tag{10.77}$$

The impedance phase angle is $-\Phi$, and the voltage across the tuned circuit is

$$v_Z = V_{Z_{max}} \sin(\omega t - \Phi) \tag{10.78}$$

The tuned circuit is externally damped (see Fig. 10.17(c)) such that the damped

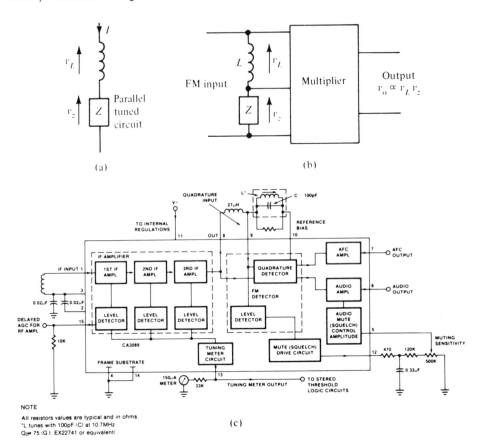

FIGURE 10.17. (a) Circuit used to obtain phase shift in a quadrature detector; (b) block schematic of the basic quadrature detector; (c) block diagram of Signetics CA3089 FM IF system incorporating the quadrature detector. The system is available in a single 16-lead dual-in-line package. Externally required components are shown outside the main block. "Permission to reprint granted by Signetics Corporation, a subsidiary of U.S. Philips Corp., 811 E. Arques Avenue, Sunnyvale, CA 94086."

Q factor is on the order of 15–20, and the variable y, as given by Eq. (1.65), can be approximated (see Eq. (10.69)) as

$$y = \frac{f}{f_0} - \frac{f_0}{f}$$

$$\cong 2\frac{\delta f}{f_0} \qquad (10.79)$$

where $\delta f = f - f_0$, f_0 is the resonant frequency of the circuit, and $\delta f \ll f_0$. Now consider the two voltages v_L and v_z applied as inputs to a multiplier (Fig.

10.17(b)). The output voltage, expressed as a proportionality, is given by

$$v_0 \propto v_L v_Z$$

$$\propto \cos \omega t \sin(\omega t - \Phi) \tag{10.80}$$

By means of trigonometric identities, this can be expressed as

$$v_0 \propto \sin(2\omega t + \Phi) + \sin \Phi \tag{10.81}$$

A low-pass filter is used to select the $\sin \Phi$ component and reject the $\sin(2\omega t + \Phi)$ component.

For small angles, $\sin \Phi \approx \tan \Phi$, and as shown by Eq. (10.77), $\tan \Phi \cong yQ$. Therefore, the output is proportional to y:

$$v_0 \propto y$$

$$\propto \frac{\delta f}{f_0} \tag{10.82}$$

From Eq. (10.1), $\delta f = f_i - f_c = k e_m$, where e_m is the modulating signal. Hence, $v_0 \propto e_m$, i.e., the output voltage is proportional to the original modulating signal. Observe here, as in previous FM analyses, that the frequency modulation is assumed to be at a slow enough rate to allow *sinusoidal* analysis to be used.

Figure 10.17(c) shows in block diagram form a commercially available integrated circuit, Signetics CA3089, which incorporates the quadrature detector. The single chip also incorporates a three-stage IF amplifier/limiter circuit with level detectors, an audio amplifier, and an automatic frequency control output. Automatic frequency control and limiting are discussed in Sections 10.6.6 and 10.6.7.

10.6.5   Phase-Locked Loop Discriminator

A phase-locked oscillator can be used to demodulate a frequency-modulated signal. The voltage-controlled oscillator (VCO) in Fig. 10.18 is basically a frequency modulator and oscillates at the center IF frequency when no signal is being received or when the modulation on the received carrier is zero. Under the latter condition, the oscillator output is exactly the same as the received IF in frequency, and the phase comparator circuit puts out a zero signal. When the incoming frequency rises because of modulation to $f_{IF} + \delta f$, the phase comparator output creates an output signal which drives the VCO frequency up until it again matches the IF, this time at $f_{IF} + \delta f$. The signal appearing at the input to the VCO is the sum of a fixed dc bias plus the comparator output signal. Since the oscillator shifts to a higher-frequency, and for this to be true the input to it must be larger than the bias value, there must be an output from

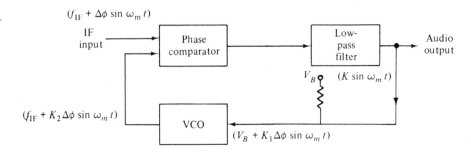

FIGURE 10.18. The phase-locked loop frequency discriminator.

the comparator when the oscillator is tracking. This means that the oscillator must be out of phase with the IF by an amount proportional to the deviation.

Now if the modulation drives the received-signal frequency low, the oscillator will also be forced to move low in frequency, and the comparator output will adjust itself to the value necessary to produce this frequency. If the received-signal frequency is varying in accordance with a modulation signal, the value of the voltage at the input to the VCO will vary about the bias value in accordance with the modulating signal. A coupling capacitor and a low-pass filter in the output lead remove the dc component and the carrier components from the bias voltage, leaving only the modulation signal.

10.6.6 Automatic Frequency Control

Automatic frequency control is used in FM receivers to "lock onto" the received signal and stabilize reception. The circuit is an application of negative feedback. A signal is derived that is proportional in amplitude to the average deviation of the received center frequency from the center of the receiver IF bandpass. This signal is used to vary the reactance of a tuning diode (voltage-variable capacitor) in the oscillator tank circuit to shift its frequency enough to compensate for the deviation and bring the signal back into the center of the IF bandpass.

Voltage-controlled oscillators (VCOs) are basically FM-modulator circuits. Earlier FM-modulator circuits such as the variable reactance tube modulator were quite complex, and their application in receivers for AFC was limited. The varactor diode became available in quantity and quality in the 1960's, and this made feasible the application of AFC to even the cheapest of FM receivers. The varactor diode is discussed in Section 10.5.1. Typical tuning varactors have ranges of tuning (f_{max}/f_{min}) of 2:1 up to 10:1 and maximum capacities of 10 pF (suitable for AFC circuits) to 500 pF (suitable for directly tuning AM broadcast receivers in the MF band).

The capacitance is maximum at zero bias and decreases as voltage increases according to Eq. (10.35).

The circuit shown in Fig. 10.19(a) uses a Foster–Seeley discriminator as the FM detector, although the circuit will work equally well with a slope detector or a ratio detector. The voltage divider R_1, R_2 provides a bias voltage V_B which is added to the discriminator output voltage V_o to give the reverse bias voltage on the varactor diode V_d. The oscillator is adjusted so that an RF signal at the dial frequency will produce an IF equal to the center frequency IF_o and an output V_o of zero. Under these conditions the varactor diode bias V_d equals V_B and the oscillator is set to IF above the dial frequency.

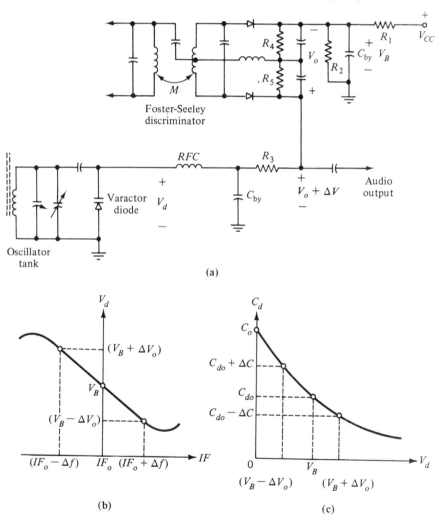

FIGURE 10.19. Automatic frequency control using a varactor diode: (a) circuit; (b) plot of bias output versus input frequency; (c) plot of varactor diode capacitance versus applied voltage.

Now, referring to Figs. 10.19(b) and (c), if the signal frequency is higher than the dial frequency, the IF will be below IF_o at $IF_o - \Delta f$, V_o will be $+\Delta V_o$, V_d will be $V_B + \Delta V_o$, C_d will be $C_{do} - \Delta C$ and f_o will be $f_o + \Delta f_o$. The IF will be raised toward IF_o. If the signal frequency is lower than the dial frequency, the IF will be above IF_o at $IF_o + \Delta f$, V_d will be lowered to $V_B - \Delta V_o$ and the oscillator frequency will be lowered to $f_o - \Delta f$. This will lower IF toward IF_o.

The oscillator will "lock onto" a signal, which is near the dial frequency, and follow it, adjusting automatically to keep the IF near IF_o and to the center of the discriminator characteristic. It will let go if Δf drives V_o beyond either end of the discriminator curve. If no signal is present, the oscillator will be kept at IF above the dial frequency by the bias V_B.

10.6.7 Amplitude Limiters

Amplitude limiters are amplifier circuits that are used to eliminate amplitude modulation and amplitude-modulated noise from received FM signals before detection. This step is necessary because most of the discriminator circuits respond to a greater or lesser degree to amplitude and frequency variations in the FM signal, introducing an unwanted source of noise.

The limiter works in such a manner that limiting begins when the input signal becomes larger than that required to drive the amplifier from cutoff to saturation (i.e., across its active range). Figure 10.20(a) shows such an amplifier. It uses a bipolar transistor and double-tuned IF transformers, with the output transformer supplying the detector circuit. Resistors R_1, R_2, and R_E provide dc bias under zero signal conditions to maintain the transistor in the active region. R_{dc} acts to reduce the effective supply voltage and limit the saturation value of the collector current to a low value, so that saturation will occur at a low-input-signal level.

For very low signal levels, the circuit acts as a normal class A amplifier. For larger signals, which drive the transistor into cutoff, the baseleak circuit formed by C_2 and R_2 charges up, driving the operating point toward cutoff. At the same time, the positive peaks drive the base current beyond the saturation point, and the collector-current waveform has both the positive and negative peaks clipped off to produce an approximately rectangular waveform. The fundamental of this rectangular current waveform drives the tuned circuit of the output transformer in the "flywheel" manner typical of the class C amplifier. The point at which the signal-input range exceeds the active range of the transistor forms the threshold of the limiter, beyond which limiting action takes place. Further increase of input signal does not significantly increase the collector-current magnitude, and its fundamental component remains almost constant. If a very large increase in input occurs, such as by a large noise pulse, the bias circuit will drive even further into cutoff, and the conduction angle of the amplifier will decrease. The limiter is then said to have been "captured" by the noise and will not be released until the bias capacitor discharges. This effect puts an upper limit on the limiting range, and it means that if large noise pulses are received, they could cause the limiter to respond and reduce the desired input signal. The shape of the limiter response is indicated in Fig. 10.20(b).

If the receiver is to respond to a very large range of input signals and noise signals, it may be necessary to increase the range of the limiter. This may be accomplished in two ways. First, for large-signal conditions, AGC may be provided so that the limiter is not driven beyond its limiting range. Second, a

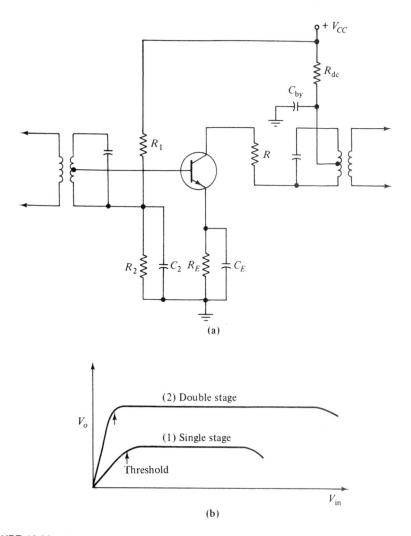

FIGURE 10.20. Amplitude limiter circuit: (a) schematic; (b) limiter response showing the expanded range of a double-stage limiter.

second stage of limiting may be provided. Two-stage limiting is much more effective and provides very good protection against large-amplitude noise pulses. Since there is more gain in the limiter circuits, the threshold at which limiting occurs is lower, and less preamplifier gain is required.

It was stated that most types of FM detectors require amplitude limiting in order to function properly. The ratio detector (see Section 10.6.3) is an exception to this, because it has a degree of inherent limiting built into it. In critical applications it is still necessary to provide additional limiting, but the ratio detector performs well enough otherwise.

10.6.8 FM Broadcast Receivers

FM commercial broadcasting in North America takes place in the VHF band between 88 and 108 MHz. Within this band, allotted frequencies are spaced 200 kHz apart and are allowed a maximum frequency deviation of ± 75 kHz around the carrier frequency. Propagation at VHF is restricted to line of sight, and coverage is usually only for a radius of about 50 miles around the transmitter location. The programs broadcast on these channels in the past have been mostly music, and the basic modulating frequency bandwidth is 15 kHz, as opposed to the 5 kHz used on AM stations.

Figure 10.21(a) shows the block schematic of a typical FM broadcast receiver of the monaural or single-channel variety. It is a superheterodyne circuit, with a tuned RF amplifier so that maximum signal sensitivity is typically between 1 to 10 μV. The RF-stage-tuned circuits and the local oscillator are tuned by a three-ganged variable capacitor controlled from a panel knob. The oscillator frequency can be varied from 98.7 to 118.7 MHz, yielding an intermediate frequency (IF) of 10.7 MHz.

The IF amplifier section is comprised of several high-gain stages, of which one or more are amplitude limiters. The schematic shown here has one high-gain nonlimiting input stage, followed by one amplitude-limiting stage. All stages are tuned to give the desired bandpass characteristic, which is shown in Fig. 10.21(b). This is centered on 10.7 MHz and has a 180-kHz bandwidth to pass the desired signal. Amplitude limiting is usually arranged to have an onset threshold of about 1 mV at the limiting-stage input, corresponding to the quieting level of input signal, which may be set at 10 μV or lower.

The FM detector may be any one of several types of FM detectors described in Section 10.5, perhaps incorporating automatic frequency control as described in Section 10.6.6.

10.6.9 Deemphasis

It will be shown in Section 10.7 that the noise voltage output of an FM detector resulting from noise phase modulation increases directly in proportion to the frequency, or at 6 dB per octave. By introducing a filter termed a *deemphasis network* just after the detector which attenuates at 6 dB per octave,

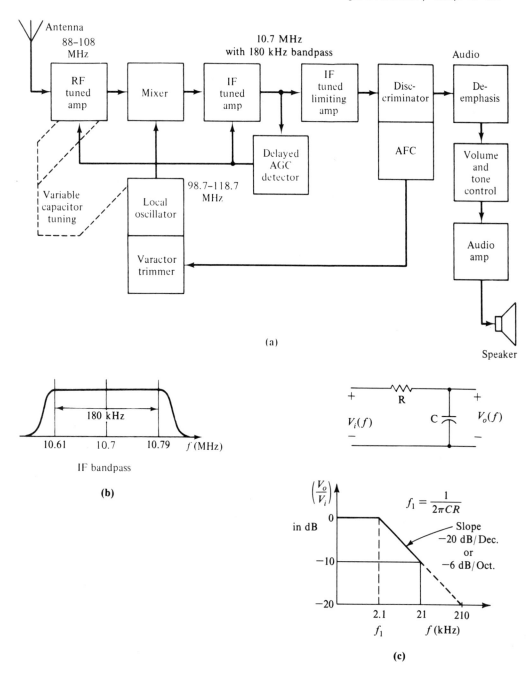

(a)

(b)

(c)

FIGURE 10.21. (a) Block diagram of a typical FM monaural broadcast receiver; (b) IF amplifier band-pass characteristics; (c) deemphasis network and transfer function.

the noise spectrum can be leveled off, thus improving the signal-to-noise ratio. A simple, but widely used deemphasis network, along with its response curve, is shown in Fig. 10.21(c). The transfer function is given by

$$\left| \frac{V_0}{V_i} \right| = 20 \log\left(\frac{1}{\sqrt{1 + (f//f_1)^2}} \right) \quad \text{(in dB)} \qquad (10.83)$$

where $f_1 = 1/2\pi CR$. The time constant CR is standardized at 75 μs, and therefore, $f_1 = 2.1$ kHz. This should be compared with the preemphasis discussed in Section 10.5.5.

The reduction in average noise power is determined by the ratio of the area of the power output spectrum with and without deemphasis, and is found to be

$$\text{deemphasis improvement} = \frac{1}{3}\left(\frac{B}{f_1} \right)^2 \qquad (10.84)$$

where B is the audio bandwidth. For $f_1 = 2.1$ kHz and $B = 15$ kHz, the improvement is calculated to be 12.3 dB.

To compensate for deemphasis of the modulating signal, a matching preemphasis network must be used at the transmitter (see Section 10.5.5).

Deemphasis can also be used with other systems—for example, AM—but the improvement offered is not as great as with FM, and there are other difficulties which make its use in other systems not worthwhile.

10.6.10 FM Stereo Receivers

All new FM broadcast receivers are being built with provision for receiving stereo, or two-channel broadcasts. The left (L) and right (R) channel signals from the program material are combined to form two different signals, one of which is the left-plus-right signal and one of which is the left-minus-right signal. The (L − R) signal is double-sideband suppressed carrier (DSBSC) modulated about a carrier frequency of 38 kHz, with the LSB in the 23–38 kHz slot and the USB in the 38–53 kHz slot. The (L + R) signal is placed directly in the 0–15 kHz slot, and a pilot carrier at 19 kHz is added to synchronize the demodulator at the receiver. The composite signal spectrum is shown in Fig. 10.22(b).

Figure 10.22(a) shows the block schematic of a stereo channel decoder circuit. The output from the FM detector is a composite audio signal containing the frequency-multiplexed (L + R) and (L − R) signals and the 19-kHz pilot tone. This composite signal is applied directly to the input of the decode matrix.

The composite audio signal is also applied to one input of a phase-error detector circuit, which is part of a phase-locked loop 38-kHz oscillator. The output drives the 38-kHz voltage-controlled oscillator, whose output provides the synchronous carrier for the demodulator. The oscillator output is also frequency-divided by 2 (in a counter circuit) and applied to the other input of the phase comparator to close the phase-locked loop. The phase-error signal is also passed to a Schmitt Trigger circuit which drives an indicator lamp on the panel which lights when the error signal goes to zero, indicating the presence of a synchronizing input signal (the 19-kHz pilot tone).

The outputs from the 38-kHz oscillator and the filtered composite audio signals are applied to the balanced demodulator, whose output is the (L − R) channel. The (L + R) and (L − R) signals are passed through a matrix circuit which separates the L and R signals from each other. These are passed through deemphasis networks and low-pass filters to remove unwanted high frequency components and are then passed to the two channel audio amplifiers and speakers. On reception of a monaural signal, the pilot-tone indicator circuit goes off, indicating the absence of pilot tone, and closes the switch to disable the (L − R) input to the matrix. The (L + R) signal is passed through the matrix to both outputs. An ordinary monaural receiver tuned to a stereo signal would produce only the (L + R) signal, since all frequencies above 15 kHz are removed by filtering, and no demodulator circuitry is present. Thus, the stereo signal is compatible with the monaural receivers.

Much of the circuitry for FM receivers is now available in integrated-circuit form. One of the first pieces of circuitry to be produced in ICs was the stereo decoder. Figure 10.23 shows a typical IC decoder chip, the RCA CA3090AQ. This chip contains all the circuits necessary to accomplish the functions of Fig. 10.22(a) and comes with the chip mounted in a 14-pin DIP. Only a few external components are needed to make the circuit work. These include bias resistors and deemphasis capacitors on the output, reactive components (L and C) for the VCO, and some bypass capacitors and bias resistors. The rest is built into the chip.

10.7 NOISE IN FREQUENCY MODULATION

Noise voltages and currents at the input to the receiver cannot directly frequency-modulate the incoming carrier since its frequency is fixed at a distant transmitter, which may, in fact, be crystal-controlled. The noise, however, does produce both amplitude and phase modulation. In a frequency-modulation receiver, the amplitude modulation is removed by limiters, while the phase modulation is detected as output noise.

The noise voltage at the limiter input may be represented by a phasor at the carrier frequency which has a randomly varying amplitude E_n and a randomly varying phase Φ_n with respect to the carrier phasor E_c, as shown in

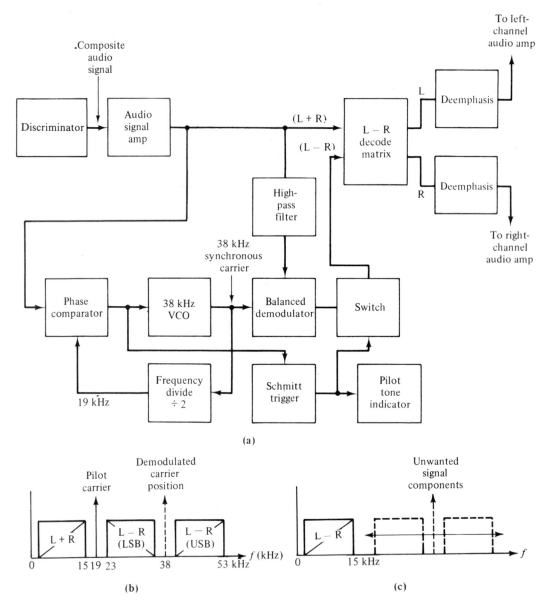

FIGURE 10.22. (a) Block diagram of an FM signal stereo channel decoder system; (b) spectrum of the composite audio signal from the FM detector; (c) spectrum of the demodulated left minus right (L − R) signal after removal of higher frequencies.

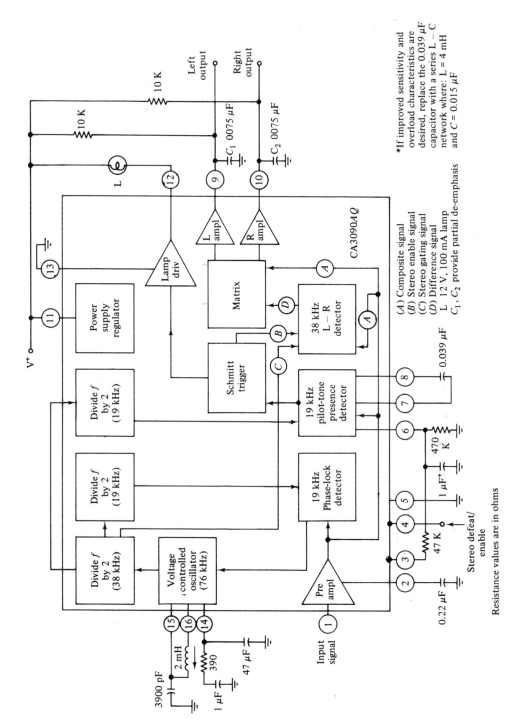

FIGURE 10.23. Functional block diagram of a typical stereo decoder IC chip — the RCA CA3090AQ. The circuit is mounted in a 14-pin dual in-line package (DIP). (Courtesy of RCA — Data File No. 684)

*If improved sensitivity and overload characteristics are desired, replace the 0.039 μF capacitor with a series L — C network where: L = 4 mH and C = 0.015 μF

(A) Composite signal
(B) Stereo enable signal
(C) Stereo gating signal
(D) Difference signal
L 12 V, 100 mA lamp
C_1, C_2 provide partial de-emphasis

Resistance values are in ohms

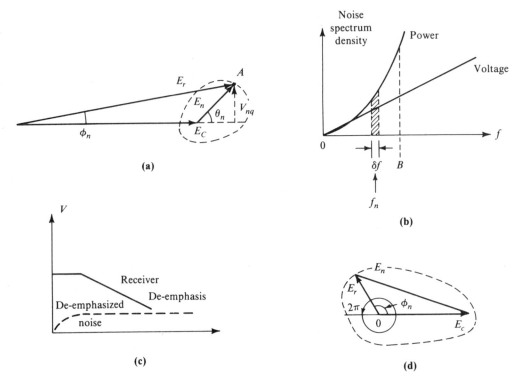

FIGURE 10.24. Noise in angle modulation circuits: (a) phasor diagram for carrier voltage very much greater than noise; (b) power and voltage spectrum density curves; (c) the effect of deemphasis on noise; (d) phasor diagram for carrier voltage comparable to noise voltage.

Fig. 10.24(a). The resultant phasor is E_r, and the tip of this, point A, will trace out a random path shown dotted. The random phase Φ_n results.

For the situation where $E_c \gg |E_n|$ most of the time (remembering that E_n varies randomly in time), the phase modulation is given approximately by

$$\Phi_n = \tan^{-1}\left(\frac{V_{nq}}{E_c}\right)$$

$$\cong \frac{V_{nq}}{E_c} \tag{10.85}$$

where V_{nq} is the quadrature component of the noise phasor (Figure 10.24(a)).

As shown in Section 10.4, for phase modulation the demodulated output voltage is proportional to the rate of change of the phase modulation or equivalent frequency modulation (Eq. (10.32)), and this makes the spectrum density for the noise voltage proportional to frequency (rate of change is proportional to frequency); the power-spectrum density for the noise output is, therefore, proportional to frequency squared. The spectrum densities are sketched in Fig. 10.24(b). The noise voltage or power over a small bandwidth δf at frequency f_n is obtained by multiplying the respective spectrum density by δf; the average value over the total bandwidth B requires a summation (or integration).

A consequence of noise spectrum density increasing with frequency is that the signal-to-noise ratio is degraded at high audio frequencies, which is unfortunate, as it is shown in Section 3.2 that good articulation efficiency of speech needs the high-frequency content. In practice, the situation is easily remedied by preemphasizing the high frequencies at the transmitter and using a deemphasis filter in the receiver audio section which corrects for the preemphasis while leveling off the noise, as shown in Fig. 10.24(c). Preemphasis and deemphasis are discussed further in Sections 10.6.9 and 10.5.5.

From Eq. (10.85), the noise phase modulation is seen to be inversely proportional to E_c, and, therefore, the noise power will be inversely proportional to E_c^2, that is, to P_c, the unmodulated carrier power. Because of the spectrum density dependence on f_n^2, the average noise power obtained by integrating the spectrum-density curve will be proportional to f_n^3 and, therefore, to B^3 for a base-bandwidth B. Thus, the noise power is proportional to B^3/P_c. The signal power for sinusoidal modulation is proportional to $(\Delta f)^2$ where Δf is the peak signal deviation. (It is assumed that the signal modulation does not interact with the noise modulation, which is nearly always the case.) Hence, the power signal-to-noise ratio is proportional to $P_c(\Delta f)^2/B^3$, the detailed mathematical analysis giving

$$(S/N)_{\text{FM}} = \left(\frac{3}{2kT}\right)\left(\frac{P_c(\Delta f)^2}{FB^3}\right) \qquad (10.86)$$

where F is the receiver noise factor.

Comparing this to the signal-to-noise ratio for a single-sideband transmission, Eq. (9.14) (where the carrier power P_c is equal to the transmitted power P_T for the single sideband), it is seen that Eq. (10.86) can be rearranged as

$$(S/N)_{\text{FM}} = 1.5\left(\frac{\Delta f}{B}\right)^2 (S/N)_{\text{SSB}} \qquad (10.87a)$$

Compared to a 100% amplitude-modulation transmission, the result, from Eq. (9.15), is

$$(S/N)_{\text{FM}} = 4.5\left(\frac{\Delta f}{B}\right)^2 (S/N)_{\text{AM}} \qquad (10.87b)$$

From this, it can be seen that the FM offers an improvement when $\Delta f/B$ exceeds about 0.5, but, of course, only at the expense of an increase in bandwidth. The ratio $\Delta f/B$ corresponds to M of Eq. (10.22); thus, the increase in bandwidth required is given by Eq. (10.22) for a given increase in M. The value $M = 0.5$ is a convenient dividing line between what is termed "narrow-band" and "wideband" FM. (It must be kept in mind that Eq. (10.16) gives the RF bandwidth, which is that required by the modulated carrier, while here B is the base bandwidth, fixed by the highest audio frequency.)

The improvement factor contained in Eq. (10.87) only applies when the carrier is much greater than the noise, as shown by the phasor diagram of Fig. 10.13(a). This would be the situation, for example, well within the reception area of an FM broadcast transmission; the specified values for broadcast are $\Delta f_{max} = 75$ kHz, $B = 15$ kHz, and the signal-to-noise improvement, therefore, for a wideband FM broadcast system compared to an AM is about 20 dB, neglecting deemphasis. There are other situations, however, for example, in mobile communications systems, where the received signal power may be very small, and the phasor diagram is as shown in Fig. 10.23(d). Under these conditions, the noise phase modulation can occasionally change abruptly by 2π radians, when the tip of the resultant phasor moves in a path to enclose the origin as shown. This gives rise to a "spike" of noise at the output (resulting from the high rate of change of noise phase) which has a high energy content and which rapidly degrades the signal-to-noise ratio. The level of input at which the spikes cause the signal-to-noise ratio to drop 1 dB below the level given by Eq. (10.87) is termed the FM *threshold*. A number of special threshold detectors are available, one of which utilizes the phase-locked loop described in Section 10.6.5.

10.8 PROBLEMS

1. In a frequency-modulating system, the frequency deviation constant is $k = 1$ kHz/V. A sinusoidal modulating signal of amplitude 15 V and frequency 3 kHz is applied. Calculate (a) the peak frequency deviation; and (b) the modulating index.

2. Sketch the instantaneous frequency–time curve and the amplitude–time curve corresponding to Fig. 10.1 but with a sawtooth modulating wave.

3. A carrier wave of amplitude 5 V and frequency 90 MHz is frequency-modulated by a sinusoidal voltage of amplitude 5 V and frequency 15 kHz. The frequency-deviation constant is 1 kHz/V. Sketch the spectrum of the modulated wave.

4. Explain why it is that in some circumstances the bandwidth required for a sinusoidally frequency-modulated carrier is greater than twice the modulat-

ing frequency. Determine the bandwidth occupied by a sinusoidally frequency-modulated carrier for which the modulating index is 2.4.

5. The phase deviation constant in a phase-modulation system is $K = 0.01$ rad/V. Calculate the maximum phase deviation when a modulating signal of amplitude 10 V is applied.

6. Explain why, in an angle-modulation system, sinusoidal phase modulation gives rise to a receiver output which increases as the modulating frequency increases, even though the modulating signal amplitude remains constant.

7. Explain in detail the operation of the phase-modulator circuit of Fig. 10.8(b). How could this circuit be modified to obtain frequency modulation?

8. Explain in detail the operation of the phase discriminator of Fig. 10.16(b). Can this circuit detect direct phase modulation?

9. Sketch the graphs of (a) equivalent frequency deviation: and (b) average noise power output, for noise in a frequency-modulation receiver.

10. Explain how, in a frequency-modulation system, the signal-to-noise ratio can be improved without having to increase the transmitted power. How are other factors of the system affected?

11. It is required to transmit a complex modulating signal in which the highest-frequency component of the spectrum is 3 kHz. Compare the bandwidths required for AM, suppressed-carrier single-sideband, and FM systems, given that the maximum frequency deviation in the latter system is 15 kHz. Also, compare the signal-to-noise ratios when the modulating signal is a 1-kHz tone, the deviation in the FM system remaining at 15 kHz, the modulation index for the AM system being unity, and the signal-to-noise ratio for the SSB system being 30 dB.

12. Prove that frequency multiplication of an FM signal increases the deviation while frequency conversion does not increase the deviation.

13. An Armstrong FM modulation system uses a primary oscillator of 1 MHz, and maximum phase deviation for good linearity is limited to 10°. (a) Find the corresponding frequency deviation at the modulator output. (b) If the transmitted signal is to have a maximum frequency deviation of 30 kHz, at a carrier frequency of 120 MHz, specify the frequency-multiplication factor needed and suggest a chain of doublers and triplers to accomplish this. (c) For the chain you suggest, specify the local oscillator frequency needed for the final conversion operation. The minimum modulating frequency is 100 kHz.

14. Explain the method by which two-channel stereo signals are transmitted over an FM broadcast transmitter. Suggest a way in which four-channel stereo might be broadcast.

CHAPTER 11

Pulse Modulation

11.1 INTRODUCTION

Previous chapters have discussed the modulation of the amplitude, the frequency, or the phase of a sinusoidal carrier signal to transmit a low-frequency analog information signal, such as voice. This chapter will discuss several systems which modulate some characteristics of a continuous train of discrete pulses as a carrier instead of a sine wave.

Pulse modulation is an analog system, even though *discrete*-valued samples of continuous analog signal are used in the modulation process. These discrete samples can take on any value within the range of the continuous signal. This is one of the ways in which pulse transmission differs from digital transmission, in which a discrete value can only be one of a finite set of numbers. Digital systems are described in Chapter 17, but it should be borne in mind that digital systems also employ pulses, so that many of the characteristics of the transmission channel described in the present chapter apply also to digital systems.

In the absence of noise and distortion it is possible to completely recover a continuous analog signal from its discrete samples, providing the samples are taken at greater than a certain minimum rate. This minimum rate (for samples equispaced in time) is equal to twice the highest frequency contained in the continuous signal. For example, it is shown in Chapter 3 that telephone speech may contain frequencies up to 3400 Hz. Assuming that the highest frequency is 3400 Hz, then the telephone signal would have to be sampled at greater than 6800 samples per second. In practice, a sampling rate of 8000 Hz would be used. The minimum sampling rate is known as the *Nyquist Sampling Rate*.

Before describing specific methods of pulse modulation, some general properties of pulse transmission will be described.

11.2 PULSE TRANSMISSION

A perfect rectangular pulse is shown in Fig. 11.1(a). It has a defined time duration or width τ and a constant amplitude during that time period. A Fourier analysis of this pulse shows that it has a continuous frequency spectrum (see Fig. 2.3(d)). If such a pulse is passed through a low-pass filter, then all the frequencies above the filter cutoff frequency will be eliminated, and the pulse at the output will be rounded off, as in Fig. 11.1(b). This is discussed in more detail in Chapter 2. Since pulse-generating equipment is far from perfect, this is the type of pulse that will be produced in practice. It will still have a constant amplitude during the pulse, except for the lead and tail edges. If the rise and fall times of the edges are very much shorter than the pulse width τ, we can treat the pulse as a perfect one, and ignore the rise and fall times.

If the transmission channel has a bandwidth that is too narrow or that introduces excessive phase distortion, the pulse will be smeared and reduced in amplitude, as shown in Fig. 11.1(c). If the distortion is too severe, it will be impossible to detect the pulse at the receiving end. Noise introduced during transmission may produce an unwanted pulse or cancel a wanted pulse, resulting in transmission error, as well as increasing the pulse distortion.

An estimate of the bandwidth required can be obtained from the results of the frequency analysis presented in Chapter 2. Consider first a periodic pulse waveform. As shown in Fig. 2.3(b), the higher amplitude components of the spectrum are contained between zero and $10f$, for the example given. In general, the first zero after the origin can be obtained by equating Eq. (2.7) to zero:

$$V_n = \frac{2V_\tau}{T}\frac{\sin(nx)}{nx} = 0 \qquad (11.1)$$

This requires that $\sin(nx) = 0$ or $nx = \pi$. But $x = \pi\tau/T$ (see Eq. (2.6)), and

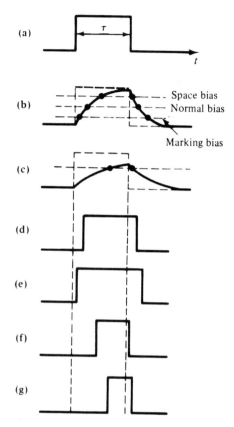

FIGURE 11.1. Pulse transmission distortion: (a) a rectangular pulse; (b) a rectangular pulse which has not been severely distorted by bandwidth limitations; the intercepts for normal, marking, and spacing bias are shown; (c) a rectangular pulse distorted by transmission over a facility with too low a bandwidth; the threshold intercepts for normal bias are shown; (d) a reconstructed pulse with normal bias delayed but unchanged in length from (b); (e) a reconstructed pulse lengthened by marking bias; (f) a reconstructed pulse shortened by spacing bias; (g) a reconstructed pulse resulting because of reduced pulse amplitude with normal bias in (c).

therefore the first zero occurs when

$$\frac{n\pi\tau}{T} = \pi$$

Therefore,

$$\frac{n}{T} = \frac{1}{\tau} \qquad\qquad (11.2)$$

Let $f_0 = 1/T$ be the frequency of the pulsed wave; then nf_0 is the frequency of the nth harmonic at which the spectrum goes to zero, and this is equal to $1/\tau$.

For a single pulse, the spectrum is a continuous function of frequency, as shown by Eq. (2.8) and Fig. 2.3(d). From Eq. (2.8), the first zero after the origin is obtained when $\sin(\pi f \tau) = 0$, or $(\pi f \tau) = \pi$, and this gives

$$f = \frac{1}{\tau} \tag{11.3}$$

Again, it is seen that most of the spectrum is contained in the frequency interval from zero to $1/\tau$, where τ is the pulse width. For satisfactory pulse transmission in a communications system, the bandwidth of the channel should be at least $1/\tau$. If, however, it is desired to accurately display the pulse waveform on an oscilloscope, the oscilloscope bandwidth should be at least four times $1/\tau$ in order to pass more of the high-frequency detail of the spectrum.

Referring again to Eq. (11.2), for the bandwidth estimate to be useful for the periodic pulse waveform, sufficient harmonic content must be included in the first spectrum peak, which requires that $(T/\tau) \gg 1$. From Eq. (2.7), it is found that the second spectrum peak occurs at $n = 1.5(T/\tau)$. For example, with $T/\tau = 10$, the first zero is at $n = 10$, and the next peak is at $n = 15$, as shown in Fig. 2.3(b). From Eq. (2.7), the magnitude of the second spectrum peak, which occurs at the 15th harmonic, is

$$V_{15} = \frac{\sin(1.5\pi)}{1.5\pi} = 0.212$$

This is relative to the fundamental amplitude and gives the amplitude of the largest component which would be lost by low-pass filtering with the cutoff at $n = 10$.

Now consider a square wave, which is a special case of a pulse waveform with $T/\tau = 2$. From Eq. (11.1), zeroes occur at $n = 2, 4, 6, \ldots$; that is, all even harmonics are zero, which is in agreement with the results of Section 2.2. The second spectrum peak occurs at $n = 1.5(T/\tau) = 1.5 \times 2 = 3$, or at the third harmonic, and in fact, all the spectrum peaks coincide with the odd harmonics, as shown in Fig. 2.1. Thus, in this case, passing the square wave through a low-pass filter with cutoff at $1/\tau$ would only allow the first harmonic to pass, and the square wave would be converted to a sine wave. Since the harmonic amplitudes for the square wave decrease according to $1/n$, any low-pass filtering should pass up to at least the ninth harmonic.

To reduce the effects of distortion, it is necessary to regenerate pulse signals at repeater stations or at the receiver. If the repeater or receiver is improperly adjusted, a further form of pulse distortion occurs. This is illustrated in Fig. 11.1(b), (d), (e), and (f), and is termed *bias distortion*. The pulse of Fig. 11.1(a) is transmitted and becomes distorted because of the bandwidth and phase characteristics of the cable. Receiver *bias* is the minimum level of pulse amplitude to which the receiver will respond. If the receiver is adjusted for normal bias (Fig. 11.1(b)), then for equally spaced (square-wave) pulses, the pulse area above the bias level will be equal to the area below for symmetrical pulses, and the regenerated pulse shown in Fig. 11.1(d) will be delayed, but not lengthened or shortened. If the bias is set too low, then the pulse area above bias will be larger than that below, and the period of the regenerated pulse will be both lengthened and delayed, as shown in Fig. 11.1(e). This is referred to as *marking bias distortion*. If the bias is set too high, then the pulse is both delayed and shortened, as shown in Fig. 11.1(f). This is called *spacing bias distortion*. A similar type of distortion is produced when the received pulse is severely reduced in amplitude, but the bias is kept at the normal level. The result is that the pulse area above bias is smaller than that below, causing the same type of delay and shortening as spacing bias distortion, shown in Fig. 11.1(g).

The effects of distortion can be minimized by proper receiver adjustment. This is done by sending over the transmission line a standard squarewave signal at the proper transmission speed. The sampling speed and bias of the receiver are then adjusted until the receiver produces an undistorted signal from this transmission.

11.2.1 Intersymbol Crosstalk (Intersymbol Interference)

Unwanted signals such as noise and crosstalk can be added to the signal during transmission, which may mask the pulse so that it cannot be retrieved. Figure 11.1(b) shows the typical shape of a pulse which has been passed through a bandwidth-limited channel. When a pulse is passed through a low-pass filter or limited channel, its corners are rounded off, its amplitude is reduced, and it is smeared by unequal time delays into the adjacent time slot. Figure 11.2(b) shows the result of increasingly lower filter cutoff frequency on the pulse shape. Note that the pulse "tail" endures long after the nominal pulse period and can interfere with a succeeding pulse. This effect is known as *intersymbol crosstalk* or *intersymbol interference*. The succeeding pulse has energy from the first pulse added to it. In a pulse-amplitude-modulated system, this would cause serious distortion.

A similar form of intersymbol crosstalk results if the channel bandpass is low-frequency-limited as well, causing the top of the pulse to drop. When the pulse ends, the signal falls below the zero axis and may not return to zero before the next pulse appears. Energy is subtracted from the succeeding pulse, making it look lower than it is.

In a time-division-multiplexed pulse-amplitude-modulated system, this form of crosstalk can couple signal energy from one channel into another which may be of sufficient amplitude to be intelligible, hence the name "crosstalk."

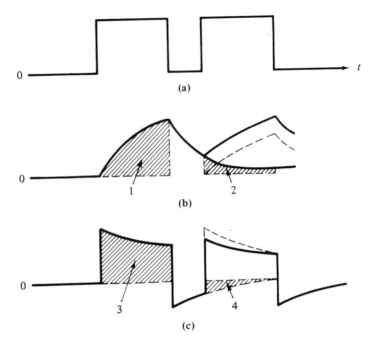

FIGURE 11.2. Intersymbol crosstalk: (a) two adjacent pulses; (b) upper frequency limitation causes energy from area 2 of the first pulse to be added to the second pulse, increasing its amplitude; (c) lower frequency limitation causes energy from area 4 of the first pulse to be subtracted from the second pulse, decreasing its amplitude.

11.3 PULSE AMPLITUDE MODULATION

Pulse amplitude modulation (PAM) results when a train of constant-amplitude very short pulses occurring at a fast pulse repetition rate is caused to vary in amplitude with a slower modulating wave $m(t)$. The effect is that of multiplying the pulse train by the modulating signal. Figure 11.3(a), (b), and (c) shows a pulse train, a modulating sine wave, and the resulting PAM waveform. The envelope of the pulse heights corresponds to the modulating wave. If the original waveform is to be recovered, it is necessary only to pass the PAM waveform through a low-pass filter.

11.3.1 Natural PAM Sampling

Natural PAM sampling occurs when finite-width pulses are used in the modulator, and the tops of the pulses are forced to follow the modulating waveform, as shown in Fig. 11.3(c). The result is a spectrum containing a pair of sidebands around each harmonic of the pulse-sampling frequency, and the modulating base-band itself.

Letting $m(t)$ represent the modulating waveform and $p(t)$ the pulse train, the sampled wave is represented by the product $m(t)p(t)$. Now, Eqs.

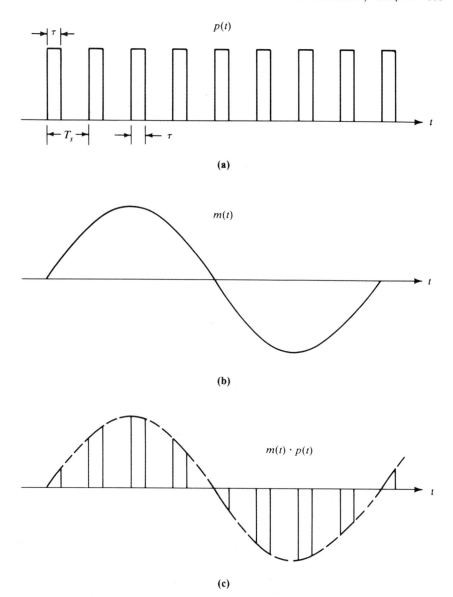

FIGURE 11.3. (a) A uniform pulse train $p(t)$; (b) the modulating waveform $m(t)$; and (c) the PAM sampled waveform. This is referred to as Natural PAM Sampling because the pulse heights follow the modulating envelope.

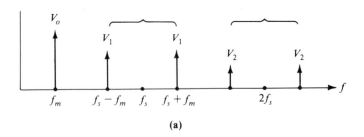

(a)

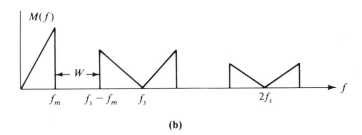

(b)

FIGURE 11.4. (a) Part of the spectrum for the PAM waveform of Figure 11.3(c), with $m(t) = \sin \omega_m t$; (b) the more general form of the spectrum.

(2.6) and (2.7) show that $p(t)$ can be analysed into a dc component $V_0 = V\tau/T_s$ and harmonic components of the form $V_n \cos(n\omega_s t)$, where $\omega_s = 2\pi/T_s$ is the radian sampling frequency. Therefore, the product $m(t)p(t)$ contains the original signal waveform in the low-frequency component $V_0 m(t)$, and this can be recovered from the sampled PAM waveform by means of low-pass filtering. The harmonic terms are of the form $V_n m(t)\cos(n\omega_s t)$, and this is similar to the double-sideband suppressed carrier (DSBSC) described in Chapter 9. For $m(t) = \sin \omega_m t$, the spectrum is as shown in Fig. 11.4(a). The spectrum for the more general case is shown in Fig. 11.4(b), where $M(f)$ is the spectrum for the original baseband signal, and f_m in this case is the highest baseband frequency.

Figure 11.4 illustrates why the sampling frequency f_s must be greater than twice the highest frequency f_m in the baseband signal. If $M(f)$ is to be recovered by low-pass filtering, the separation width W to the next lower sideband must be greater than zero. But W is zero when $f_m = f_s - f_m$, as shown in Fig. 11.4(b), and hence $f_s = 2f_m$. Therefore, to have W greater than zero requires that $f_s > 2f_m$.

11.3.2 Flat-topped PAM Sampling

Flat-topped PAM sampling is a system quite often used because of the ease of generating the modulated wave. The waveform and spectrum for this case is shown in Figure 11.5. Finite-width pulses are used, but they have flat tops *after* modulation. When the modulating waveform is recovered from the

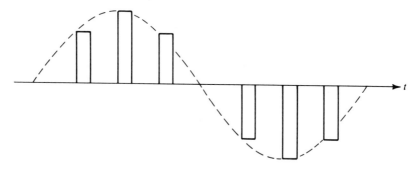

FIGURE 11.5. Flat-top sampling.

flat-top sampled waveform by means of low-pass filtering, it will show the distortion introduced by the flat-top segments of the waveform. This distortion is negligible when the pulse width is much less than the sampling period T_s, which is usually the case. Even when the pulse width is comparatively large, the form of the distortion is known and can be corrected by passing the recovered waveform through a compensating network.

PAM signals are very rarely used for transmission purposes directly. The reason for this lies in the fact that the modulating information is contained in the amplitude factor of the pulses, which can easily be distorted during transmission by noise, crosstalk, or other forms of distortion. PAM is used frequently as an intermediate step in other pulse-modulating methods, especially where time-division multiplexing is used.

11.4 TIME-DIVISION MULTIPLEXING

PAM waveforms can be generated which use a very small duty cycle so that if only one PAM wave were sent over a channel, much of the transmission time available would go unused. This time can be utilized by sandwiching the pulse trains of other PAM signals into the unused time intervals between pulses. Figure 11.6(a) shows how this would work for the *time-division multiplexing* (TDM) of five PAM signals, each with the same sampling rate. Only two of the five waves are shown for simplicity.

Figure 11.6(b) shows the block diagram of a TDM system which could be used to multiplex the five signals. A rotating switch called a commutator (realized with electronic switches) connects the output of each PAM channel modulator to the communication channel input in turn, dwelling on each contact only for the duration time of one pulse. The channel must have sufficient bandwidth to handle these pulses, which now occur at five times the sampling rate of one channel. The bandwidth required should be much greater than the system pulse repetition rate to make intersymbol crosstalk negligible. This includes a guard time between pulses as discussed below.

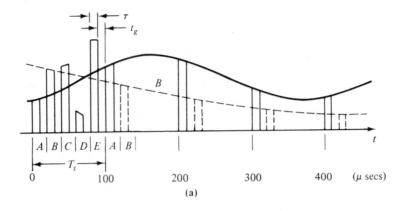

(a)

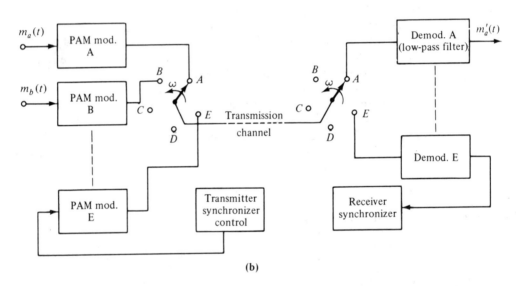

(b)

FIGURE 11.6. (a) A five-channel time-division-multiplexed (TDM) PAM signal; (b) a five-channel time-division-multiplexed system.

At the receiving end, another switch, rotating in synchronism with the sending commutator, decommutates the pulse trains and connects them to the appropriate demodulator circuits. Since it is possible for the receiver to get out of step with the transmitter, it is necessary to send synchronizing signals along with the information signals. One channel of the group is usually reserved for this purpose.

Because of the problem of intersymbol crosstalk, the pulses from the PAM modulators are made shorter than the allotted pulse duration time, so that a "guard time" exists between each pair of pulses. This extra guard time means that the effective pulse width is about half of the pulse period.

The subject of multiplexing is also dealt with in Section 9.6.3.

11.5 PULSE-TIME MODULATION

Pulse-time modulation (PTM) is accomplished by four distinct methods. The first three of these comprise a group known as *pulse-duration* or *pulse-width modulation* (PDM or PWM); the fourth is pulse-position modulation (PPM). Three variations of pulse-width modulation are possible, depending on whether the lead edge, the tail edge, or the center of the pulse is kept fixed in phase as

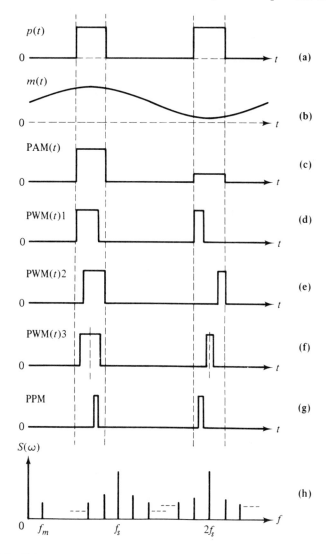

FIGURE 11.7. Pulse-time modulation (PTM) signals: (a) carrier sampling pulses; (b) modulating signal; (c) PAM signal; (d) PWM tail-edge modulated signal; (e) PWM lead-edge modulated signal; (f) PWM symmetrical signal; (g) PPM signal; (h) spectrum of a PWM signal.

the pulse width is varied with the modulation. This form of modulation is also known as class D modulation. PPM, the fourth mode, is generated by changing the position of a fixed-length pulse within a fixed time slot. Figure 11.7(a)–(g) shows the results of each of these modes for two sample periods.

It should be noted that the same restrictions as to sampling rate, guard-time allowances, and time-division multiplexing apply to all forms of pulse-time modulation as apply to PAM signals. Added noise is not as much trouble, since PTM corresponds to FM and PM in form, and thus can run at fixed amplitude and be regenerated relatively easily to reduce noise. Figure 11.7(h) shows the form of the spectrum of a pulse-width-modulated signal for a sinusoidal modulating signal with a frequency f_m near the upper edge of the baseband. The spectrum is that of a frequency-modulated carrier, with the baseband and many harmonics present. Since the sampling frequency is near the maximum baseband frequency ((2 to 10) $\times f_m$), then if large deviations are allowed, the lower sidebands will appear in the baseband and cause distortion. This can only be overcome by increasing the sampling frequency or by keeping the deviation low (i.e., by keeping the pulse-width deviation to a small percentage of the pulse slot width). The same is true of the PPM signal.

11.5.1 Pulse-Width Modulation

Figure 11.8 shows a *pulse-width-modulation* (PWM) system. A PAM signal is generated first and may be multiplexed with other PAM signals. The PWM signal is generated from the PAM signal and transmitted. At the receiver, the PAM signal is regenerated and then demodulated by low-pass filtering.

The modulating signal $m(t)$ (A in the diagram) is applied to the input of a PAM modulating circuit, to generate the PAM signal (B). The same pulse train (C) which supplies the PAM modulator is used to gate on a ramp generator, to generate a train of ramp pulses (D) which all have equal slopes, amplitudes, and durations. These ramp pulses are added directly to the PAM pulses to produce varying-height ramps (E). The varying-height ramps gate a Schmitt trigger circuit to generate the varying-width rectangular pulses of the PWM wave (F). These PWM pulses can be transmitted directly or used as the input to a pulse-position modulator.

At the receiver, the received pulses (G) are put through a regenerating circuit to remove some of the noise and square up the pulses. These regenerated pulses (H) drive a reference pulse generator to produce a train of constant width–constant height pulses synchronized to the leading edge of the received pulses, but delayed by a fixed interval (I). The regenerated pulses also gate a ramp generator, which produces a constant-slope ramp for the duration of the pulse. At the end of the pulse, a sample-and-hold amplifier retains the final ramp voltage until it is reset at the end of the period, resulting in the ramp-and-pedestal waveform (J). The constant-amplitude pulses (I) are ad-

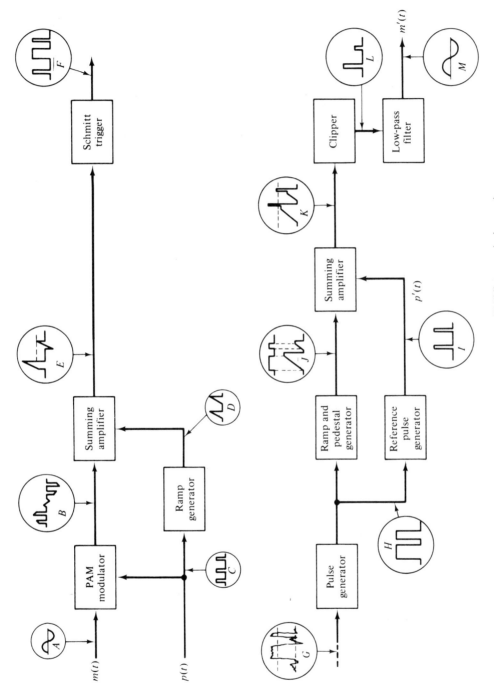

FIGURE 11.8. Pulse-width-modulation (PWM) transmission system.

ded to the pedestals to form K, which is then clipped off at a threshold level to form the PAM signal at L. The signal waveform $m'(t)$, shown at M, is recovered by low-pass filtering.

11.5.2 Pulse-Position Modulation

Pulse-position modulation (PPM) signals can be readily generated from the PWM signals by using the modulated edge of the PWM pulse to trigger a monostable multivibrator circuit which generates fixed-width, fixed-amplitude pulses when triggered. At the receiver, a fixed-period reference pulse is generated from the incoming PPM waveform, and a flip-flop (bistable multivibrator) is set by the reference pulse and reset by the PPM pulse. The result is a rectangular width-modulated pulse at the output of the flip-flop.

11.6 PROBLEMS

1. What sampling rate would be appropriate for each of the following?
   (a) A telephone channel limited to 4-kHz bandwidth.
   (b) A music channel with a maximum signal frequency of 20 kHz.
   (c) A television video channel with a maximum bandwidth of 4.5 MHz.

2. Describe how spacing bias on a telegraph circuit can cause pulse shortening.

3. Describe how transmission distortion of a time-division-multiplexed signal can cause crosstalk between two adjacent channels.

4. Carefully plot the spectrum of a flat-top sampled PAM signal which has a 1-kHz sine modulating signal, a sampling frequency of 8 kHz and a pulse width of 31.25 μS, up to the 6th harmonic of the sampling frequency. Draw in with a dotted line the amplitude envelope curve.

5. (a) Sketch a channel-interleaving scheme for time-division-multiplexing the following PAM channels: five 4-kHz telephone channels and one 20-kHz music channel. (b) Find the pulse repetition rate of the multiplexed signal. (c) Estimate the minimum system bandwidth required.

PART 4

TRANSMISSION AND RADIATION OF SIGNALS

CHAPTER 12

Transmission Lines and Cables

12.1 INTRODUCTION

Transmission of information as an electromagnetic signal always occurs as a *transverse electromagnetic* (TEM) wave or as the combination of such waves as in a waveguide (see Chapter 13). The basic properties of the TEM wave are outlined in Appendix B, and propagation of these as radio waves is described in Chapter 14.

With transmission lines, the metallic conductors confine the TEM wave to the vicinity of the dielectric surrounding the conductors; as a result, some aspects of the transmission are best treated in terms of the *distributed circuit* parameters of the line, while others require the wave properties of the line to be taken into account. It must be clearly understood, however, that these are complementary views of the transmission; the voltages and currents on the line always accompany the TEM wave, and the particular viewpoint adopted usually depends on which properties are most easily measured.

Transmission lines may be *balanced* or *unbalanced* with respect to ground. The two basic types of transmission lines are the two-wire line, which is usually operated in the balanced mode (Fig. 12.1(a)), and the coaxial line, which is always operated in the unbalanced mode (Fig. 12.1(b)). The electromagnetic

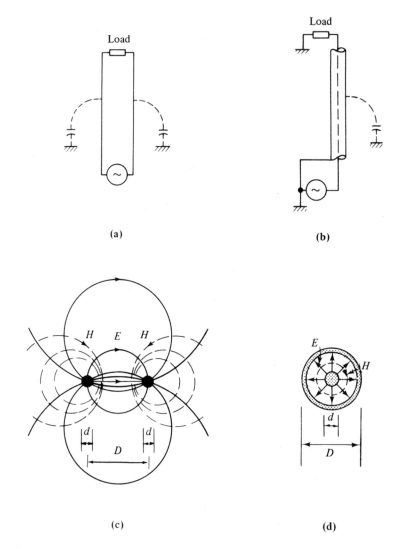

FIGURE 12.1. (a) A two-wire line; (b) a coaxial line; (c) electromagnetic fields around a two-wire line; (d) electromagnetic fields around a coaxial line.

field configurations for each type are shown in Fig. 12.1(c) and (d). In each case the direction of propagation of the TEM wave is into the page, and inspection shows that the E (electric) field is at right angles to the H (magnetic) field, and that both are at right angles to the direction of propagation, as required of a TEM wave.

With the coaxial line, the outer conductor forms a shield which confines the wave to the space between the conductors so that radiation from the line is negligible. However, the line is basically unbalanced since the external capacitance is between the outer conductor only and ground.

The two-wire line is normally operated in the balanced mode, the conductors being arranged so that they present equal capacitances to ground. (Of course, this may be a difficult condition to maintain in practice.) Radiation can occur from a two-wire line since the TEM wave can radiate out from the line as well as along it. The two-wire line is less expensive than the coaxial line and is used for the majority of low-frequency telephone circuits. Care must be taken to maintain balance conditions, as described in Section 12.14. For high-frequency circuits (including multichannel telephony and radio feeders), the coaxial line is used to minimize radiation; and where balanced radio antennas have to be connected to coaxial lines, special transmission line transformers known as *baluns* (*bal*anced to *un*balanced) are employed.

12.2 PRIMARY LINE CONSTANTS

From the circuit point of view, a transmission line will have series resistance and inductance, which together go to make up the series impedance of the conducting wires, and shunt conductance and capacitance of the dielectric between the conductors, which go to make up the shunt admittance of the line. A small length δx of the line may therefore be represented approximately by the filter section (Fig. 12.2), where, as part of the approximation, the values for both the *go* and *return* wires are included in the lumped components. The parameters R, L, G, and C shown in Fig. 12.2 are known as the *primary line constants*; these are the series resistance R, in ohms/meter; the series inductance L, in henries/meter; the shunt conductance G, in siemens/meter; and the shunt capacitance C, in farads/meter. The primary constants take into account both the *go* and *return* lines. They are constant in that they do not vary with voltage and current; however, they are frequency-dependent to some extent. The series resistance R increases with frequency as a result of skin effect (see Section 1.10). The inductance L is almost independent

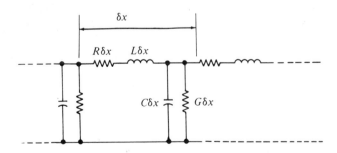

FIGURE 12.2. Circuit approximations for a short length δx of transmission line.

of frequency for open lines, but tends to decrease with increasing frequency for screened cables. The capacitance C is almost independent of frequency, whereas the conductance G tends to increase with the frequency (i.e., the shunt resistance decreases) because of increasing dielectric loss with increase in frequency.

No simple formulas can be given for the primary constants that cover all frequency ranges, but simplifications are possible for particular, well-defined operating ranges, e.g., low- or audio-frequencies and high- or radio-frequencies. As will be shown in Section 12.4, the most practically useful characteristic of a line is the *characteristic impedance*, which at high frequencies is determined by the series inductance and shunt capacitance. These, as shown below, depend on line geometry, which therefore sets a limit on the range of characteristic impedances possible in practice.

For the two-wire line of Fig. 12.1(c), with the conductors embedded in a medium of permittivity ε (F/m) and permeability μ (H/m), and with the line dimensions in meters, the primary inductance and capacitance per unit length are given approximately by

$$\text{Two-wire line:} \qquad L \cong \frac{\mu}{\pi} \ln\left(\frac{2D}{d}\right) \qquad \text{H/m} \qquad (12.1)$$

$$C \cong \frac{\pi\varepsilon}{\ln\left(\frac{2D}{d}\right)} \qquad \text{F/m} \qquad (12.2)$$

For the coaxial line of Fig. 12.1(d), with dielectric of permittivity ε (F/m) and permeability μ (H/m), and again, with the line dimensions in meters, the approximate forms of the equations are

$$\text{Coaxial line:} \qquad L \cong \frac{\mu}{2\pi} \ln(D/d) \qquad \text{H/m} \qquad (12.3)$$

$$C \cong \frac{2\pi\varepsilon}{\ln(D/d)} \qquad \text{F/m} \qquad (12.4)$$

12.3 PHASE VELOCITY AND LINE WAVELENGTH

In Appendix B the phase velocity of a TEM wave is given by Eq. (B.11) as

$$v_p = \frac{1}{\sqrt{\mu\varepsilon}} \qquad (12.5)$$

For free space the values are $\mu = \mu_0 = 4\pi \times 10^{-7}$ H/m and $\varepsilon = \varepsilon_0 = 8.85 \times 10^{-12}$ F/m, giving $v_p = 3 \times 10^8$ m/s. This is the velocity of light, normally denoted by the letter c. For a transmission line the permeability may be assumed equal to the free-space value, but the permittivity may differ from the free-space value, depending on the dielectric used. The expression for permittivity is $\varepsilon = \epsilon_r \epsilon_0$ where ε_r is the relative permittivity (or dielectric constant). Substituting this in Eq. (12.5) along with the free-space values gives

$$v_p = \frac{c}{\sqrt{\varepsilon_r}} \tag{12.6}$$

Typically, ε_r may range from 1 to 5, and thus the phase velocity of the TEM wave on the line may be less than the free-space value.

The wavelength of the wave is given by Eq. (B.4) as

$$\lambda = \frac{v_p}{f} \tag{12.7}$$

Substituting for v_p from Eq. (12.6) gives

$$\lambda = \frac{c}{f\sqrt{\varepsilon_r}}$$

$$= \frac{\lambda_0}{\sqrt{\varepsilon_r}} \tag{12.8}$$

where λ_0 is the free-space wavelength. The wavelength, as given by Eq. (12.8), is the value which must be used in transmission-line calculations; for example, the phase shift coefficient, which is the phase shift per unit length, is given by

$$\beta = \frac{2\pi}{\lambda} \tag{12.9}$$

where λ is as given by Eq. (12.8).

Another useful expression for the phase velocity can be obtained in terms of the inductance per unit length L and the capacitance per unit length C. From Eqs. (12.1) and (12.2) for the two-wire line, and Eqs. (12.3) and (12.4) for the coaxial line, it can be seen that

$$LC = \mu\varepsilon \tag{12.10}$$

Hence, on substituting Eq. (12.10) in Eq. (12.5), the result is

$$v_p = \frac{1}{\sqrt{LC}} \tag{12.11}$$

12.4 CHARACTERISTIC IMPEDANCE

Energy travels along a transmission line in the form of an electromagnetic wave, the wave set up by the signal source being known as the incident (or forward) wave. Only when the load impedance at the receiving end is a reflectionless match for the line, as discussed in Section 1.14, will all the energy be transferred to the load. If reflectionless matching is not achieved, energy will be reflected back along the line in the form of a reflected wave (hence the name reflectionless matching). Because of the distributed nature of a transmission line, the question may be asked: Exactly to what impedance of the line must the load be matched? This can be answered by considering a hypothetical line, infinite in length and for which no reflection can occur, since the incident wave never reaches the end. The ratio of maximum voltage to maximum current at any point on such a line is found to be constant, i.e., independent of position. This ratio is known as the characteristic impedance Z_0. Now, if a finite length of line is terminated in a load impedance $Z_L = Z_0$, this will appear as an infinite line to the incident wave since at all points, including the load termination, the ratio of voltage to current will equal Z_0. Thus, the characteristic impedance of a transmission line is the ratio of voltage to current at any point along the line on which no reflected wave exists.

With a sinusoidal signal of angular frequency ω rad/sec, the characteristic impedance in terms of the primary constants is found to be

$$Z_0 = \sqrt{\frac{R + j\omega L}{G + j\omega C}} \quad \Omega \quad (12.12)$$

At low frequencies such that $R \gg \omega L$ and $G \gg \omega C$, the expression for Z_0 reduces to

$$Z_0 \cong \sqrt{\frac{R}{G}} \quad \Omega \quad (12.13)$$

and at high frequencies such that $R \ll \omega L$ and $G \ll \omega C$, it becomes

$$Z_0 \cong \sqrt{\frac{L}{C}} \quad \Omega \quad (12.14)$$

It will be observed that each limiting value is purely resistive (there is no j coefficient) and independent of frequency. Between these limits Z_0 is complex and frequency-dependent, and it is found that for most practical lines it is capacitive. However, above a few tens of kilohertz for two-wire lines and a few hundred kilohertz for coaxial lines, the high-frequency approximation for Z_0 is sufficiently accurate for most practical purposes, and this will be the expression used throughout this chapter.

Substituting Eqs. (12.1) and (12.2) into Eq. (12.14) gives Z_0 in terms of line dimensions, permittivity, and permeability for the two-wire line:

$$Z_0 = \frac{1}{\pi} \sqrt{\frac{\mu}{\varepsilon}} \ln\left(\frac{2D}{d}\right) \quad \Omega \tag{12.15}$$

And for the coaxial line, from Eqs. (12.3) and (12.4), Eq. (12.14) gives

$$Z_0 = \frac{1}{2\pi} \sqrt{\frac{\mu}{\varepsilon}} \ln\left(\frac{D}{d}\right) \quad \Omega \tag{12.16}$$

For the dielectrics encountered in practice, the permeability will be equal to that of free space, $\mu = \mu_0 = 4\pi \times 10^{-7}$ H/m; and the permittivity will be given by $\varepsilon = \varepsilon_r \varepsilon_0$, where $\varepsilon_0 = 8.854 \times 10^{-12}$ F/m is the permittivity of free space and ε_r is the relative permittivity or dielectric constant. Substituting these into the impedance equations, Eqs. (12.15) and (12.16), gives

Two-wire line: $\qquad\qquad Z_0 = \dfrac{120}{\sqrt{\varepsilon_r}} \ln\left(\dfrac{2D}{d}\right) \quad \Omega \tag{12.17}$

Coaxial line: $\qquad\qquad Z_0 = \dfrac{60}{\sqrt{\varepsilon_r}} \ln\left(\dfrac{D}{d}\right) \quad \Omega \tag{12.18}$

In each case, it will be seen that for a given dielectric constant, the characteristic impedance is determined by the ratio D/d (Fig. 12.1(c) and (d)). For dielectrics in common use, the dielectric constant will be within the range 1–5, and practical limitations on the ratio D/d for each type of line limit Z_0 to the range of about 40 to 150 ohms for the coaxial line, and 150 to 600 ohms for the two-wire line.

12.5 THE PROPAGATION COEFFICIENT

The propagation coefficient γ determines the variation of current or voltage with distance x along a transmission line. The current (and voltage) distribution along a matched line is found to vary exponentially with distance, the equations being

$$I = I_s e^{-\gamma x} \tag{12.19}$$
$$V = V_s e^{-\gamma x} \tag{12.20}$$

where I_s is the magnitude of the current and V_s is the magnitude of the voltage of the input or sending end of the line.

Like the characteristic impedance, the propagation coefficient also depends on the primary constants and the angular velocity of the signal. It is given by

$$\gamma = \sqrt{(R + j\omega L)(G + j\omega C)} \qquad (12.21)$$

This is also a complex quantity, and can be written as

$$\gamma = \alpha + j\beta \qquad (12.22)$$

where α is known as the *attentuation coefficient* and determines how the voltage or current decreases with distance along the line, and β is known as the *phase-shift coefficient* and determines the phase angle of the voltage (or current) variation with distance. The phase-shift coefficient is the phase shift per unit length, and since a phase shift of 2π radians occurs over a distance of one wavelength λ, then

$$\beta = \frac{2\pi}{\lambda} \qquad (12.23)$$

To see how the propagation coefficient affects the current, consider the current equation

$$I = I_s e^{-(\alpha + j\beta)x} \qquad (12.24)$$

$$= I_s e^{-\alpha x}\left(e^{-j\beta x}\right) \qquad (12.25)$$

The last equation can be represented graphically as shown in Fig. 12.3(a). The length of the phasor line represents $I_s e^{-\alpha x}$, and the angle of rotation from the reference line represents βx.

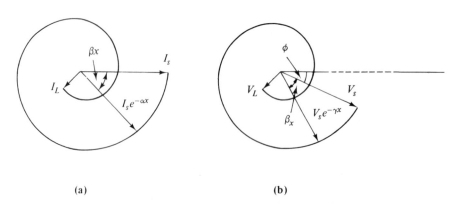

(a)   (b)

FIGURE 12.3. (a) Current and (b) voltage phasor diagrams.

The attenuation is often expressed in terms of a unit known as the *neper*. The magnitude of the current (on the matched line) is

$$|I| = I_s e^{-\alpha x} \tag{12.26}$$

and the attenuation of the current, in nepers, is defined as

$$N = -\ln\left(\frac{|I|}{I_s}\right) \tag{12.27}$$

$$= -\ln(e^{-\alpha x})$$

$$= \alpha x \text{ nepers} \tag{12.28}$$

The attentuation coefficient α can therefore be alternatively expressed as the attenuation in nepers per unit length of line:

$$\alpha = \frac{N}{x} \tag{12.29}$$

Example 12.1   The input current in a matched line is 50 mA and the load current is 1 mA. The line is 1 km long. Calculate (a) the total attenuation in nepers, and (b) the attenuation coefficient.

Solution   (a)
$$N = -\ln(1/50)$$
$$= 3.9 \text{ nepers}$$

(Note that by including a minus sign in the defining equation for the neper, the attenuation comes out a positive number.)

(b)
$$\alpha = \frac{3.9}{1000}$$
$$= 3.9 \text{ mN/m}$$

The neper is a useful unit in theoretical work, but in practical work the decibel is more commonly used. In Appendix A, the relationship between the decibel value D and the neper value N of a given current ratio is shown to be

$$D = 8.686N \tag{12.30}$$

Hence, for the transmission line,

$$D = 8.686\alpha x \quad \text{decibels} \tag{12.31}$$

where $N = \alpha x$. Therefore, if the attenuation coefficient is *defined* in decibels per unit length, say $[\alpha]$, then

$$[\alpha] = 8.686\alpha \qquad dB/m \qquad (12.32)$$

Example 12.2   The attenuation coefficient of a line is 0.0006 N/m. Determine the attenuation coefficient in (a) dB/m, and (b) dB/mile.

Solution   (a) $[\alpha] = 8.686\alpha$

and since α numerically equals the attenuation in N/m,

$$[\alpha] = 8.686 \times 0.0006$$
$$= 0.00521 \qquad dB/m$$

(b) Since there are 1609 m in 1 mile, the attenuation coefficient in dB/mi is

$$0.00521 \times 1609 = 8.4 \, dB/mi$$

For a matched line, $V = IZ_0$ at any point along the line, and for $Z_0 = Z_0/\!-\phi$ (assuming a leading phase angle), the voltage–distance phasor diagram will be similar to that for current, as shown in Fig. 12.3(b). The length of the phasor is modified by Z_0, and the reference line for the voltage is displaced by $-\phi$.

Both α and β are determined by the primary constants and the frequency, since

$$\gamma = \alpha + j\beta = \sqrt{(R + j\omega L)(G + j\omega C)} \qquad (12.33)$$

The square-root expression can be expanded by means of the binomial expansion, and an approximation sufficiently accurate for most practical purposes is to take the expansion to the third term. This results in the expression

$$\alpha \approx \frac{R}{2Z_0} + \frac{GZ_0}{2} \qquad (12.34)$$

where Z_0 in this particular case is $\sqrt{L/C}$ (i.e., the high-frequency limit to the characteristic impedance). In most practical lines G is very small (almost zero), so the further approximation

$$\alpha \approx \frac{R}{2Z_0} \qquad (12.35)$$

is usually valid.

As an example, the primary constants for a coaxial cable at a frequency of 10 MHz were determined approximately as

$$L = 234 \text{ nH/m}$$
$$C = 93.5 \text{ pF/m}$$
$$R = 0.568 \text{ }\Omega/\text{m}$$
$$G = 0$$

Therefore, at a frequency of 10 MHz,

$$Z_0 \cong \sqrt{\frac{234 \times 10^{-9}}{93.5 \times 10^{-12}}} = 50 \text{ }\Omega \qquad \text{(from Eq. (12.14))}$$

and

$$\alpha = \frac{0.568}{2 \times 50} = 0.00568 \text{ N/m} \qquad \text{(from Eq. (12.35))}$$

The attenuation, in dB/m [α], is $0.00568 \times 8.686 = 0.0493$ dB/m. The major variation in R with frequency is due to skin effect, for which R is proportional to the square root of frequency, and this will cause the attenuation to vary likewise.

The binomial expansion for γ results in the following expression for β:

$$\beta \cong \omega\sqrt{LC}\left[1 + \frac{1}{8}\left(\frac{R}{\omega L} - \frac{G}{\omega C}\right)^2\right] \tag{12.36}$$

An interesting situation arises when

$$\frac{R}{\omega L} = \frac{G}{\omega C} \tag{12.37}$$

The phase-shift coefficient is then seen to be equal to

$$\beta = \omega\sqrt{LC} \tag{12.38}$$

This meets the condition, discussed in the following section, required for distortionless transmission (i.e., β is proportional to ω). The condition for distortionless transmission is seen to be

$$\frac{R}{\omega L} = \frac{G}{\omega C}$$

and this can be rearranged as

$$\frac{R}{G} = \frac{L}{C} \tag{12.39}$$

In any practical line R and G both will be as small as possible in order to keep losses at a minimum. The ratio L/C will not, in general, be equal to R/G, and therefore, if distortionless transmission is desired, the ratio L/C will have to be altered to equal R/G. In all practical cases this is achieved by increasing L, a technique known as *loading*. One common form of loading is to add inductance coils in series with the line at regularly spaced intervals. However, it is not practical to load to achieve completely distortionless conditions, as it is found that this would require an inordinately large value of inductance. Furthermore, large inductance decreases phase velocity (see Eq. (12.11)), which may introduce unacceptable delays on long distance telephone circuits. It is interesting to note that the introduction of pulse code modulation (see Section 17.6) on normal telephone lines requires the removal of the loading coils in order to increase the bandwidth of the lines.

12.6 PHASE AND GROUP VELOCITIES

The phase velocity of an electromagnetic wave has already been discussed briefly in Section 12.3. For wave motion generally, the following simple relationship exists for frequency f, wavelength λ, and phase velocity v_p (see Appendix B):

$$\lambda f = v_p \tag{12.40}$$

Since $\beta = 2\pi/\lambda$ and $\omega = 2\pi f$, multiplying the left-hand side of Eq. (12.40) by $2\pi/2\pi$ results in

$$\frac{\omega}{\beta} = v_p \tag{12.41}$$

It can be seen therefore that while β is proportional to ω, the phase velocity will be constant (i.e., independent of frequency), and therefore all component waves making up a signal will be transmitted at the same velocity v_p. This is the distortionless transmission condition referred to in the previous section.

The situation can occur where β is not proportional to ω, as, for example, in the general case given by Eq. (12.36). Component sine waves of a signal will be transmitted with different velocities, and the question then arises, at what velocity does the signal wave travel? An answer to this question can be obtained by considering two sinusoidal waves differing in frequency by a small amount $\delta\omega$. The combined wave will be shown in Fig. 12.4, where, for clarity, the individual sine waves are assumed to have equal amplitudes, and line attenuation is ignored. The composite signal is seen to consist of high-frequency waves (many zero crossings on the time axis) modulated by a low-frequency envelope. Detailed analysis shows that the envelope travels along the line with

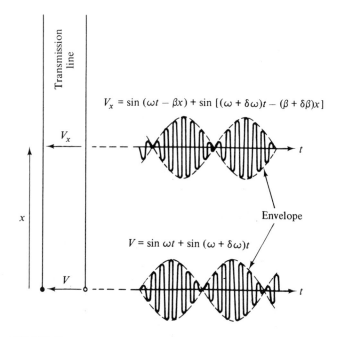

$$V_x = \sin (\omega t - \beta x) + \sin [(\omega + \delta\omega)t - (\beta + \delta\beta)x]$$

$$V = \sin \omega t + \sin (\omega + \delta\omega)t$$

FIGURE 12.4. Wave group, used in determination of group velocity.

a velocity given by

$$v_g = \frac{\delta\omega}{\delta\beta} \tag{12.42}$$

where the subscript g signifies *group velocity* (i.e., the velocity at which the group (of two sine waves) travels). It is also the velocity at which the energy is propagated along the line.

In the limit the group velocity is given by the differential coefficient of ω with respect to β, and if β varies rapidly with ω, serious distortion will result.

The condition stated previously for distortionless transmission was that v_p be constant. A more general condition will now be stated, namely, that for distortionless transmission, the β/ω graph should be a straight line, which may be expressed as

$$\beta = t_d\omega + \beta_0 \tag{12.43}$$

where t_d, known as the *group delay time*, is the slope of the line, and β_0 is the intercept, which must be equal to $\pm n\pi$ for distortionless transmission. This graph is sketched in Fig. 12.5, along with the line representing the simpler condition, $\beta = \omega/v_p$.

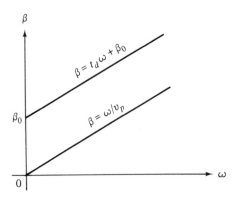

FIGURE 12.5. β/ω graph for distortionless transmission.

It will be seen that t_d is the reciprocal of the group velocity, and therefore, in addition to the constraint that $\beta_0 = \pm n\pi$, v_g must be constant (for constant t_d) for distortionless transmission. This more general set of conditions cover situations where distortionless transmission is achieved even though the simpler condition (β/ω = constant) is violated.

12.7 STANDING WAVES

When the load impedance does not match the line impedance, part of the energy in the incident wave is reflected at the load. This gives rise to a reflected wave traveling back along the line toward the source. If the source impedance does not match the line, a further reflection will take place, and in this way, multiple reflections can be set up at both load and source. The overall effect can be treated as the resultant of a single incident and a single reflected wave. These can be represented by rotating phasors in the manner of Fig. 12.3, but with the reflected wave phasor rotating in the opposite direction to the incident wave phasor (Figure 12.6(a)). This is because from a given point, an increase in x away from the source is a decrease in distance x toward the point of reflection.

Clearly, at certain values of x, the phasors will be in opposition, giving rise to voltage minima (Figure 12.6(b)), while at other values of x they will coincide (Figure 12.6(c)), giving rise to voltage maxima. The voltage as a function of distance x is sketched in Fig. 12.6(d), where it can be seen to go through a series of maxima and minima. This voltage pattern is stationary as regards position and is therefore referred to as the *voltage standing wave* (VSW). It is essential to grasp, however, that at any given point, the voltage is time-varying at the frequency of the input signal. Thus, at point x_1, corresponding to the first minimum, the voltage alternates sinusoidally between the peak values $\pm V_{1\,min}$, as shown in Fig. 12.6(e).

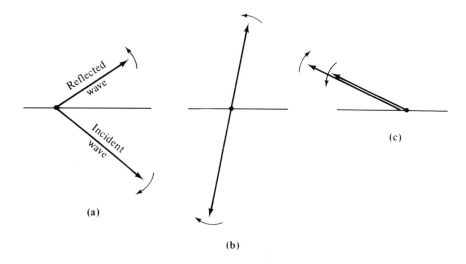

(a)

(b)

(c)

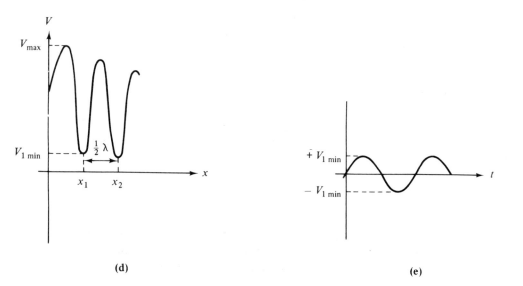

(d)

(e)

FIGURE 12.6. (a), (b) and (c) Phasor representation of forward and reflected waves; (d) voltage standing wave; (e) voltage time variation.

Another important feature of the voltage standing-wave pattern is that the distance between successive minima (and between successive maxima) is $\frac{1}{2}\lambda$. This is easily shown from the fact that when they are at a minimum, the phasors form a straight line (Fig. 12.6(b)). On moving from position x_1 to x_2, the phasors must rotate to interchange positions, (i.e., each phasor rotates by a phase angle of π radians). But each phasor also rotates by an angle given by

$\beta(x_2 - x_1)$, where $\beta = 2\pi/\lambda$, the phase-shift coefficient. Therefore,

$$\frac{2\pi}{\lambda}(x_2 - x_1) = \pi \tag{12.44}$$

$$\therefore (x_2 - x_1) = \tfrac{1}{2}\lambda \tag{12.45}$$

In a similar way, it is easy to show that the distance between a minimum and the following maximum is $\tfrac{1}{4}\lambda$.

A current standing wave will also occur. For a purely resistive Z_0, the incident current wave will always be in phase with the incident voltage wave, while the reflected current wave will always be anti-phase to the reflected voltage wave. This happens because either the electric or the magnetic field of the reflected wave must reverse direction, as shown in Fig. B.1(b) of Appendix B. A consequence of the anti-phase relationship between reflected voltage and current is that the current maxima occur along with the voltage minima, and the current minima along with the voltage maxima, for the standing wave patterns (bearing in mind that these are the phasor sums of incident and reflected waves). As already mentioned, this condition only holds when Z_0 is purely resistive. It enters into the theory of slotted-line measurements discussed in Section 12.10.

12.8 LOSSLESS LINES AT RADIO FREQUENCIES

For many applications at radio frequencies, the losses in a transmission line are small enough to be ignored (for example, in short lengths of good-quality cable), and therefore the attenuation coefficient α can be set equal to zero. The propagation coefficient is then given by

$$\gamma = j\beta \tag{12.46}$$

It is often more convenient to measure distances from the load end rather than from the sending end. Denoting this by ℓ, an equation for the incident voltage, similar in form to Eq. (12.20), may be written:

$$V_i = V_I e^{j\beta\ell} \tag{12.47}$$

Here, V_i is the incident wave voltage at any distance ℓ from the load; V_I, which may be complex, is the value at the load ($\ell = 0$); and $j\beta$ replaces γ of Eq. (12.20). A positive exponential is used in Eq. (12.47) compared to the negative exponential in Eq. (12.20) since ℓ is measured in the opposite direction to x.

A similar equation can be written for the reflected voltage wave:

$$V_r = V_R e^{-j\beta\ell} \tag{12.48}$$

Here, V_R (which may be complex) is the value of the reflected voltage at the load, and, of course, a negative exponential is used since the reflected wave's phase changes in the opposite sense to that of the incident wave. At any point on the line,

$$V = V_i + V_r \qquad (12.49)$$

In particular, at the load ($\ell = 0$),

$$V_L = V_I + V_R \qquad (12.50)$$

The equations for current are

Incident wave:
$$I_i = \frac{V_i}{Z_0} \qquad (12.51)$$

Reflected wave:
$$I_r = -\frac{V_r}{Z_0} \qquad (12.52)$$

Note the minus sign for the reflected current wave, signifying the 180° phase change discussed in the previous section.

The resultant current at any point ℓ from the load is

$$I = I_i + I_r \qquad (12.53)$$

$$= \frac{V_i - V_r}{Z_0} \qquad (12.54)$$

In particular, the load current is

$$I_L = \frac{V_I - V_R}{Z_0} \qquad (12.55)$$

Normally it is not necessary to know V_I and V_R separately, the ratio V_R/V_I entering more into calculations. This ratio is termed the *voltage reflection coefficient* Γ_L:

$$\Gamma_L = \frac{V_R}{V_I} \qquad (12.56)$$

Note that Γ_L is defined in terms of the incident and reflected waves *at the load*.

From Eqs. (12.50) and (12.55), the load impedance may be expressed as

$$Z_L = \frac{V_L}{I_L}$$

$$= \frac{V_I + V_R}{V_I - V_R} Z_0 \qquad (12.57)$$

from which

$$\Gamma_L = \frac{Z_L - Z_0}{Z_L + Z_0} \tag{12.58}$$

The fact that the voltage reflection coefficient is determined solely by the load impedance and the characteristic impedance makes it an important practical parameter.

Three specific load conditions which occur frequently are (i) matched load; (ii) short-circuit load, and (iii) open-circuit load. For each of these, the following obtain:

(i) Matched load, $Z_L = Z_0$: $\Gamma_L = 0$
(ii) Short-circuit, $Z_L = 0$: $\Gamma_L = -1$
(iii) Open-circuit, $Z_L = \infty$: Here it is first necessary to rearrange the equation as

$$\Gamma_L = \frac{1 - Z_0/Z_L}{1 + Z_0/Z_L}$$

$$= 1 \qquad (\text{as } Z_L \text{ goes to infinity}) \tag{12.59}$$

It is left as an exercise for the student to show that the corresponding load conditions for the three cases are

(i) $\quad V_L = V_I$ and $I_L = V_I/Z_0$
(ii) $\quad V_L = 0$ and $I_L = 2V_I/Z_0$
(iii) $\quad V_L = 2V_I$ and $I_L = 0$

12.9 VOLTAGE STANDING-WAVE RATIO

The *voltage standing-wave ratio* (VSWR) is defined as

$$\text{VSWR} = \frac{V_{\max}}{V_{\min}} \tag{12.60}$$

where $V_{\max}$ and $V_{\min}$ are as shown in Fig. 12.6. The line is assumed lossless so that the maxima all have the same value $V_{\max}$, and the minima all have the value $V_{\min}$.

As previously shown (Figure 12.6(c)), a maximum occurs when the phasors are in phase:

$$V_{\max} = |V_I| + |V_R| \tag{12.61}$$

$$= |V_I|(1 + |\Gamma_L|) \tag{12.62}$$

(Note: $|V_I| + |V_R|$ is not the same as $|V_I + V_R|$, which is the modulus of the load voltage.)

A minimum occurs when the phasors are antiphase:

$$V_{min} = |V_I|(1 - |\Gamma_L|) \tag{12.63}$$

Therefore,

$$\text{VSWR} = \frac{|V_I|(1 + |\Gamma_L|)}{|V_I|(1 - |\Gamma_L|)}$$

or

$$\text{VSWR} = \frac{(1 + |\Gamma_L|)}{(1 - |\Gamma_L|)} \tag{12.64}$$

The VSWR can range in value from unity to infinity; that is,

$$1 \le \text{VSWR} \le \infty \tag{12.65}$$

Ideally, the VSWR should equal 1, as this represents a matched condition, and practical adjustments on RF transmission lines are often aimed at minimizing the VSWR. Note that the VSWR is always a real number (i.e., it has no imaginary part).

The equation for VSWR can be rearranged to give

$$|\Gamma_L| = \frac{\text{VSWR} - 1}{\text{VSWR} + 1} \tag{12.66}$$

12.10 SLOTTED-LINE MEASUREMENTS AT RADIO FREQUENCIES

Use is made of the voltage standing wave at high frequencies to determine unknown impedances which would otherwise be difficult to measure. In principle, the method requires only a determination of the VSWR and the distance of the first minimum from the load (the minimum rather than the maximum is chosen in practice because it is more sharply defined). The apparatus consists of a slotted section of coaxial line through which a probe can sample the electric field and hence the voltage standing wave. The output signal from the probe is often converted to a signal of lower frequency and then amplified and rectified to produce dc output proportional to the voltage standing-wave amplitude; alternatively, for simpler measurements, the probe output may be directly rectified and the dc read by a microammeter. The distance of the probe from the load can be read directly off a calibrated distance scale. There will always be specific corrections to make for a given apparatus to allow for end effects, details of which should be included in the manufacturer's handbook. The theory of the measurement technique is as follows.

At distance ℓ from the load, the voltage reflection coefficient is found, using Eqs. (12.47) and (12.48), to be

$$\Gamma = \frac{V_r}{V_i}$$

$$= \Gamma_L e^{-j2\beta\ell} \tag{12.67}$$

Since Γ_L is complex, it may be written in terms of its modulus and phase angle as

$$\Gamma_L = |\Gamma_L|e^{j\phi_L} \tag{12.68}$$

Combining Eqs. (12.67) and (12.68) gives

$$\Gamma = |\Gamma_L|e^{j(\phi_L - 2\beta\ell)}$$

$$= |\Gamma_L|\underline{/\phi_L - 2\beta\ell} \tag{12.69}$$

In particular, let ℓ_{min} represent the distance from the load to the first voltage minimum; then the voltage reflection coefficient at the minimum is

$$\Gamma_{min} = |\Gamma_L|\underline{/\phi_L - 2\beta\ell_{min}} \tag{12.70}$$

The angle $(\phi_L - 2\beta\ell_{min})$ is the phase angle of the reflected voltage with respect to the incident voltage at the first voltage minimum. At a voltage minimum, the two voltages are in anti-phase; therefore, the reflected voltage lags the incident voltage by π radians. This is because the incident voltage advances in phase as ℓ increases (shown by the $+\beta\ell$ term), while the reflected voltage lags (as shown by the $-\beta\ell$ term). Therefore,

$$\phi_L - 2\beta\ell_{min} = -\pi$$

from which it follows that

$$\phi_L = \left(\frac{4\ell_{min}}{\lambda} - 1\right) \tag{12.71}$$

This shows that the angle ϕ_L can also be determined by slotted line measurements, and since the modulus $|\Gamma_L|$ is determined from the VSWR measurement, as shown by Eq. (12.66), the reflection coefficient at the load is known. This in turn allows the load impedance to be found, using Eq. (12.58), to be

$$Z_L = Z_0 \frac{1 + \Gamma_L}{1 - \Gamma_L} \tag{12.72}$$

Although $\ell_{\min}$ was specified as the distance to the first voltage minimum, in fact any voltage minimum can be used (assuming losses can be ignored) since minima are separated by $n\lambda/2$, and this will simply add $n2\pi$ radians to ϕ_L, which has no effect on the principal value.

Example 12.3 Measurements on a 50-Ω slotted line gave a VSWR of 2.0 and a distance from load to first minimum of 0.2λ. Determine both the equivalent series and equivalent parallel circuits for Z_L.

Solution From Eq. (12.66),

$$|\Gamma_L| = \frac{\text{VSWR} - 1}{\text{VSWR} + 1} = \frac{2 - 1}{2 + 1} = \frac{1}{3}$$

Also,

$$\phi_L = \pi(4 \times 2 - 1)$$

$$= -36°$$

But

$$e^{j\phi_L} = \cos \phi_L + j \sin \phi_L \qquad (\text{Euler's identity})$$

$$= -0.809 + j0.588$$

and

$$|\Gamma_L|e^{j\phi_L} = \frac{-0.809 + j0.588}{3}$$

$$= -0.27 + j0.196$$

From Eq. (12.72),

$$Z_L = 50\frac{1 - (-0.27 + j0.196)}{1 + (-0.27 + j0.196)}$$

$$= 50\frac{1.27 - j0.196}{0.73 + j0.196}$$

The series equivalent is $Z_L = R_s + jX_s$, and, therefore,

$$R_s + jX_s = 50\frac{(1.27 - j0.196)(0.73 - j0.196)}{0.73^2 + 0.196^2}$$

$$= 77.8 - j34.3$$

from which

$$R_s = 77.8 \ \Omega$$

$$|X_s| = 34.3 \ \Omega \ (\text{capacitive})$$

The parallel equivalent is $Y_L = G + jB$, where $Y_L = 1/Z_L$, $G = 1/R_p$, and $B = 1/X_p$. Therefore, from the above expression for Z_L:

$$G + jB = \frac{0.73 + j0.196}{50(1.27 - j0.196)}$$

$$= \frac{(0.73 + j0.196)(1.27 + j0.196)}{50(1.27^2 + 0.196^2)}$$

$$= 10.76 + j4.75 \text{ mS}$$

from which

$$R_p = \frac{1}{G}$$

$$= \frac{10^3}{10.76}$$

$$= 92.9 \ \Omega$$

and

$$|X_p| = \left|\frac{1}{B}\right|$$

$$= \frac{10^3}{4.75}$$

$$= 211 \ \Omega \text{ (capacitive)}$$

Note that R_p and X_p can be obtained by direct application of Eqs. (1.50) and (1.51).

12.11 TRANSMISSION LINES AS CIRCUIT ELEMENTS

The voltage reflection coefficient at any point on the line is defined as:

$$\Gamma = \frac{V_r}{V_i} \tag{12.73}$$

It will be seen that Eq. (12.57) for the voltage reflection coefficient at the load is a particular case of this.

Following the same line of reasoning used to derive Eq. (12.72) for the load impedance, the impedance at any point on the line any distance from the load can be written in terms of the voltage reflection coefficient at that point as

$$Z = Z_0 \frac{1 + \Gamma}{1 - \Gamma} \tag{12.74}$$

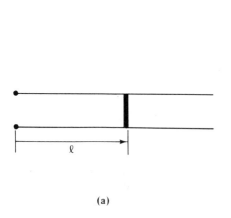

(a)

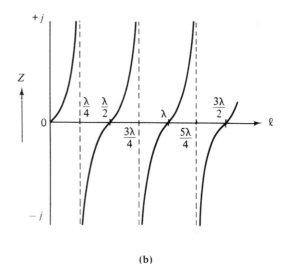

(b)

FIGURE 12.7. (a) A short-circuited line; (b) reactance variation with line length.

Certain cases are of particular importance:

Case (*i*): $Z_L = 0$ (*a short circuit*), Fig. 12.7(*a*). From Eq. (12.58), $\Gamma_L = -1$, and hence, from Eq. (12.67), $\Gamma = -e^{-j2\beta\ell}$. Substituting this in Eq. (12.74) gives

$$Z = Z_0 \frac{1 - e^{-j2\beta\ell}}{1 + e^{-j2\beta\ell}}$$

$$= jZ_0 \tan \beta\ell \qquad (12.75)$$

The latter step makes use of the trigonometric identity $j \tan \theta = 1 - e^{-j2\theta}/1 + e^{-j2\theta}$. The graph of Z/ℓ is shown in Fig. 12.7(b), where it is seen that for $0 \leq \ell \leq \lambda/4$ and $\lambda/2 \leq \ell \leq 3\lambda/4$, etc., Z is inductive ($+j$). In practice, to obtain a variable inductive reactance, the short circuit on the line would be variable in such manner that ℓ could be varied between 0 and $\lambda/4$. (The other possible ranges are usually avoided since the longer line lengths are undesirable practically.)

The equivalent inductance L_{eq} is obtained by equating

$$j\omega L_{eq} = jZ_0 \tan \beta\ell \qquad \left(0 \leq \ell \leq \frac{\lambda}{4}\right) \qquad (12.76)$$

or

$$L_{eq} = \frac{Z_0}{\omega} \tan \beta\ell \qquad (12.77)$$

Example 12.4 A 50-Ω short-circuited line is 0.1λ in length, at a frequency of 500 MHz. Calculate (a) the equivalent inductive reactance, and (b) the equivalent inductance.

Solution

$$\beta\ell = 2\pi \times 0.1$$
$$= 36°$$

(a)
$$Z = j50 \times \tan 36°$$
$$= j50 \times 0.7265$$
$$= j36.33 \ \Omega$$

(b)
$$L_{eq} = \frac{36.33}{2\pi \times 500 \times 10^6} \times 10^9 \ nH$$
$$= 11.6 \ nH$$

Because the inductive reactance is not directly proportional to frequency, but to $\tan \beta\ell$, L_{eq} shows a frequency dependence. In the previous example, let the frequency be doubled; then

$$\ell = 0.2\lambda$$
$$= 72°$$
$$Z = j50 \times \tan 72°$$
$$= j153.9 \ \Omega \ (\text{i.e., well over four times the previous value})$$
$$L_{eq} = \frac{153.9}{2\pi \times 1000 \times 10^6} \times 10^9 \ nH$$
$$= 24.5 \ nH$$

From Fig. 12.7(b) it is also seen that $\frac{1}{4}\lambda \leq \ell \leq \frac{1}{2}\lambda$ (or $\frac{3}{4}\lambda \leq \ell \leq \lambda$, etc.), i.e., the impedance appears capacitive $(-j)$. The equivalent capacitive impedance is given by

$$-j\frac{1}{\omega C_{eq}} = -jZ_0 \tan \beta\ell \qquad \left(\tfrac{1}{4}\lambda \leq \ell \leq \tfrac{1}{2}\lambda\right) \qquad (12.78)$$

or

$$C_{eq} = \frac{1}{\omega Z_0 \tan \beta\ell} \qquad (12.79)$$

For example, let $\ell = 3\lambda/8$ at a frequency of 500 MHz; then

$$\beta\ell = 135°$$
$$Z = j50 \times \tan 135°$$
$$= -j50 \times \tan 45°$$
$$= -j50 \ \Omega$$
$$C_{eq} = \frac{10^{12}}{2\pi \times 500 \times 10^6 \times 50} \ pF$$
$$= 6.4 \ pF$$

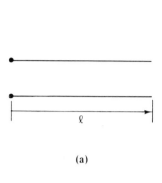

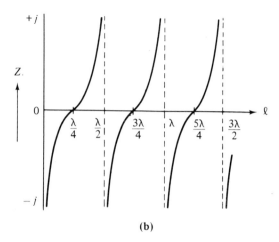

FIGURE 12.8.   (a) An open-circuited line; (b) reactance variation with line length.

Case (*ii*): $Z_L = \infty$ (*an open circuit*), Fig. 12.8(*a*). In this case, $\Gamma_L = +1$, so Z becomes

$$Z = Z_0 \frac{1 + e^{-j2\beta\ell}}{1 - e^{-j2\beta\ell}}$$

$$= -jZ_0 \cot \beta\ell \qquad (12.80)$$

The graph of Z/ℓ is shown in Fig. 12.8(b), where it will be seen that when $0 \le \ell \le \lambda/4$ the open-circuited line appears capacitive, and when $\lambda/4 \le \ell \le \lambda/2$, it appears inductive (with the pattern repeating at $\frac{1}{2}\lambda$ intervals). Although the open-circuited line finds some use, it is not as practically convenient as the short-circuited line. The short circuit provides a mechanical support for the end of the line, it is easily adjustable, and the length ℓ is well defined. However, open-circuited lines are readily fabricated in microstrip and stripline construction.

Case (*iii*): $\ell = \lambda/4$.

$$\beta\ell = \frac{2\pi}{\lambda} \frac{\lambda}{4} = \frac{\pi}{2}$$

$$Z = Z_0 \frac{1 + \Gamma_L e^{-j\pi}}{1 - \Gamma_L e^{-j\pi}}$$

$$= Z_0 \frac{1 - \Gamma_L}{1 + \Gamma_L}$$

$$= Z_0 \frac{Z_0}{Z_L}$$

$$= \frac{Z_0^2}{Z_L} \qquad (12.81)$$

This important relationship shows that the impedance as seen at the input to the line is Z_0^2/Z_L, and therefore a $\lambda/4$ section of line may be used to transform an impedance value from Z_L to Z_0^2/Z_L. For this reason, the $\lambda/4$ section is often known as a *quarter-wave transformer*.

Suppose, for example, it is required to match a 73-Ω antenna to a 600-Ω feeder line at a frequency of 150 MHz, shown in Fig. 12.9. In order to match the main feeder, the impedance Z_L' must equal 600 Ω. Therefore,

$$Z_L' = \frac{\left(Z_0'\right)^2}{Z_L}$$

and

$$Z_0' = \sqrt{Z_L Z_L'}$$

$$= \sqrt{73 \times 600}$$

$$= 209 \ \Omega$$

With this arrangement, there will be no standing waves on the main feeder line (600 Ω); that is, all of the power transmitted along the feeder will pass smoothly into the $\lambda/4$ section. There will, of course, be a mismatch between the $\lambda/4$ section and the 73-Ω load, so a standing wave will occur on the $\lambda/4$ section; but, in practice, this can be tolerated because the matching section is considerably shorter than the main feeder. As can be easily verified, the reflection coefficient is -0.52 with, and -0.78 without, the $\lambda/4$ matching section.

Since the $\lambda/4$ section is a resonant element (i.e., it is only fully effective at the frequency which makes the length exactly $\lambda/4$), it is only useful over a narrow range of frequencies (e.g., voice-modulated radio waves). However, special broadbanding techniques are available.

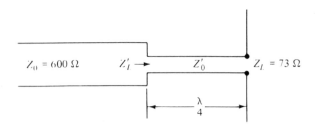

FIGURE 12.9. $\frac{1}{4}\lambda$ transformer matching.

Two cases of special interest arise when $Z_L = 0$ (short circuit) and $Z_L = \infty$ (open circuit). For the short-circuit load,

$$Z = \frac{Z_0^2}{0}$$

$$= \infty \qquad (12.82)$$

That is, the $\frac{1}{4}\lambda$ short-circuited section appears as a parallel resonant circuit (high impedance). In practice, line resistance will restrict Z to a finite, but high, value; the equivalent Q factor of the circuit will be high—on the order of 3000.
For the open-circuit line,

$$Z = \frac{Z_0^2}{\infty}$$

$$= 0 \qquad (12.83)$$

That is, the open-circuit $\frac{1}{4}\lambda$ section appears as a series resonant circuit (low impedance). Again, in practice, line resistance will result in Z being a small positive resistance, not zero.

12.12 SMITH CHART

The work involved in transmission-line calculations may be considerably reduced by using a special chart known as a *Smith chart* (devised by P. H. Smith in 1944). The theory behind the chart is too involved to be given here, but its use will be explained.
The chart is based on the relationship given by Eq. (12.74). Normalized values of impedance are frequently used. Normalized impedance is simply the ratio of Z/Z_0 and is denoted by the lower case letter z:

$$z = \frac{Z}{Z_0} \qquad (12.84)$$

It is important to note that normalized impedance is dimensionless. In terms of normalized impedance, Eq. (12.74) becomes

$$z = \frac{1 + \Gamma}{1 - \Gamma} \qquad (12.85)$$

In terms of normalized resistance r and normalized reactance x, the normalized impedance is $z = r + jx$. Any point on the Smith chart shows the four quantities r, x, $|\Gamma|$, and ϕ. Point z_1 on Fig. 12.10(a) shows $r_1 = .55$, $x_1 = 1.6$,

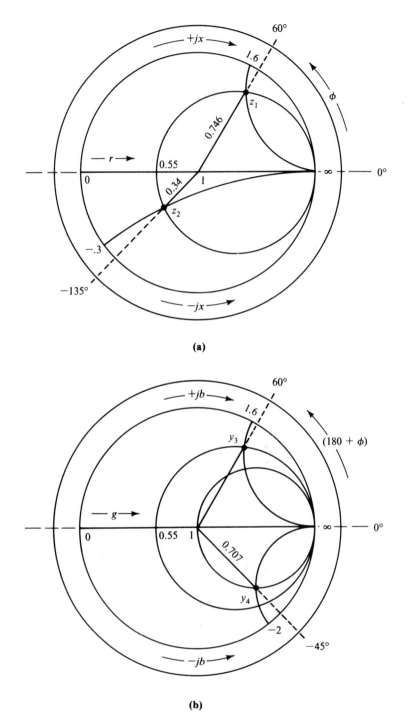

FIGURE 12.10. (a) Impedance points z_1 and z_2 on the Smith Chart; (b) admittance points y_3 and y_4 on the Smith Chart.

$|\Gamma_1| = .746$, and $\phi_1 = 60°$, i.e.,

$$z_1 = .55 + j1.6$$
$$\Gamma_1 = .746\underline{/60°}$$

Similarly, point z_2 represents $z_2 = .55 - j.3$, and $\Gamma_2 = .34\underline{/- 135°}$. In order to find the actual impedance values corresponding to z_1 and z_2, the characteristic impedance of the line must be known. Suppose this is 50 ohms; then $Z_1 = 50(.55 + j1.6) = 27.5 + j80$ ohms, and $Z_2 = 50(.55 - j.3) = 27.5 - j15$ ohms.

Admittance values may also be shown in the Smith chart. Defining normalized admittance by

$$y = \frac{1}{z} \qquad 2 = 2 \cdot \frac{1+\rho}{1-\rho} \qquad (12.86)$$

and then, using Eq. (12.85), we have

$$3 = \frac{1+\rho}{1-\rho}$$

$$y = \frac{1 - \Gamma}{1 + \Gamma} \qquad (12.87)$$

Noting that $-\Gamma = |\Gamma|\underline{/180° + \phi}$, Eq. (12.87) can be written as

$$y = \frac{1 + |\Gamma|\underline{/180° + \phi}}{1 - |\Gamma|\underline{/180° + \phi}} \qquad (12.88)$$

This is similar to the relationship (12.85) on which the chart is based, except that the angle coordinate must be read as $(180° + \phi)$, where ϕ is the phase angle for the voltage reflection coefficient. Normalized admittance may be written in terms of normalized conductance g and normalized susceptance b as $y = g + jb$, so that the same scales as used for r, x, $|\Gamma|$, and ϕ may also be used for g, b, $|\Gamma|$, and $(180° + \phi)$.

In Fig. 12.10(b) the point $y_3 = .55 + j1.6$ is shown. The corresponding voltage reflection coefficient values are $|\Gamma| = .746$ and $\phi_3 = -120°$. The angle relationship shown on the chart is $180° + \phi_3 = 60°$. Another example shown is $y_4 = 1 - j2$. The corresponding voltage reflection coefficient is $.707\underline{/- 225°}$. This can also be expressed as $.707\underline{/135°}$. These are normalized values where $y = Y/Y_0$ and $Y_0 = 1/Z_0$. Again, assuming a value of $Z_0 = 50$ ohms, the actual admittance for y_3 is $Y_3 = Y_0 Y = (.55 + j1.6)/50 = 1.1 + j32$ mS. The equivalent *parallel* components of resistance and reactance are

$$R_p = \frac{1}{1.1 \times 10^{-3}} = 909 \text{ ohms}$$

$$X_p = -\frac{1}{32 \times 10^{-3}} = -31.25 \text{ ohms}$$

The negative sign shows that the reactance is capacitive.

Two important points on the chart are the impedance at a voltage maximum and the impedance at a voltage minimum. At a voltage maximum the phase angle of the voltage reflection coefficient is zero; hence,

$$\Gamma_{max} = |\Gamma_L|\underline{/0^\circ} = |\Gamma_L|$$

Therefore,

$$z_{max} = \frac{1 + |\Gamma_L|}{1 - |\Gamma_L|}$$

$$= \text{VSWR} \qquad (12.89)$$

This last relationship is obtained from Eq. (12.64). Thus, the normalized impedance at a voltage maximum is seen to be purely resistive and equal in magnitude to VSWR, and so must lie on the r-axis as shown in Fig. 12.11(a). At a voltage minimum, the phase angle of the voltage reflection coefficient is 180°; and therefore,

$$z_{min} = \frac{1 - |\Gamma_L|}{1 + |\Gamma_L|}$$

$$= \frac{1}{\text{VSWR}} \qquad (12.90)$$

The normalized impedance at a voltage minimum, therefore, is also purely resistive, but is equal in magnitude to $1/\text{VSWR}$. This point must also lie on the r-axis, as shown in Fig. 12.11(a). The Smith chart axes are arranged so that a circle centered on $r = 1$ passes through both of these points. The VSWR value for such a circle *applies for every point cut by the circle*. For example, the normalized impedance of $.4 + j.75$ results in a VSWR of 4, as shown in Fig. 12.11(b). Care must always be taken not to confuse the "r-circles" with the "VSWR-circles." The VSWR circle in Fig. 12.11(b) is shown dotted.

Another use to which the VSWR circle is put is in finding the admittance corresponding to a given impedance value, and vice versa. The normalized admittance y is found diametrically opposite z on the VSWR circle, as shown in Fig. 12.11(b). For example, for $z = 1.4 + j1.7$ in the figure, y would be found as $y = .29 - j.35$. The impedance–admittance transformation provided on the Smith chart need not be associated with transmission lines, and in fact the chart provides a graphical means of solving the series–parallel equivalence equations given in Section 1.7. Some specific examples will be shown shortly.

The Smith chart is used with slotted line measurements to determine impedance and admittance. A distance scale is added to the chart, the zero reference for this scale being the 180° phase angle point. As shown by Eq. (12.69), the voltage reflection coefficient phase angle is given by $\phi_L - 4\pi\ell/\lambda$. The normalized distance is ℓ/λ, and as ℓ increases, the phase angle decreases,

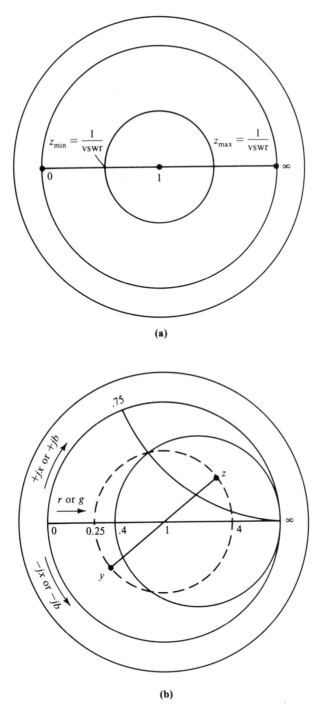

(a)

(b)

FIGURE 12.11. (a) The impedance points corresponding to voltage maxima and minima; (b) a VSWR = 4 circle shown dotted, and the impedance-admittance z-y transformation on this.

so that from the 180° reference point, the decreasing phase angle scale can also be calibrated in *wavelengths towards generator*. Likewise, for ℓ decreasing, i.e., going towards load from a voltage minimum, the corresponding increasing scale for phase angle can be calibrated in *wavelengths towards load*. These scales are shown in Fig. 12.12(b).

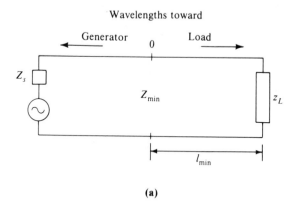

(a)

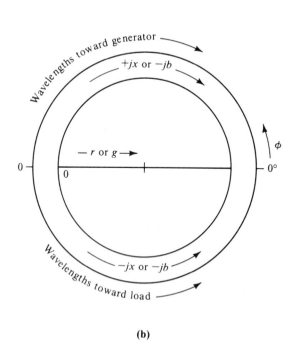

(b)

FIGURE 12.12. (a) Line lengths from a voltage minimum; (b) the normalized distance scale on the Smith Chart.

The total circumferential length of the chart is limited to 0.5 since the standing-wave pattern repeats itself at this interval.

Example 12.5   Rework Example 12.3 using the Smith chart. Data from Example 12.3 are VSWR $= 2:1$, $\ell_{min} = .2$, and $Z_0 = 50 \; \Omega$.

Solution   Draw the VSWR $= 2$ circle to cut the r-axis at 2 and .5 as shown in Fig. 12.13. Move along the *wavelengths-towards-load* scale a distance of .2, and draw a straight line from this point to the chart center. Where this line cuts the VSWR circle, point z_L in Fig. 12.13 gives the load impedance.
From the chart,

$$z_L = 1.55 - j.7$$

Thus,

$$Z_L = 50(1.55 - j.7)$$
$$= 77.5 - j35 \; \Omega$$

This compares with $77.8 - j34.3 \; \Omega$ obtained by calculation in Example 12.3.
To find the equivalent admittance, move diametrically opposite from z_L to y_L on the VSWR circle, as shown in Fig. 12.13. From the chart,

$$y_L = .54 + j.24$$

Therefore,

$$Y_L = \frac{.54 + j.24}{50}$$
$$= 10.8 + j4.8 \text{ mS.}$$

This compares with $10.76 + j4.75$ mS obtained by calculation in Example 12.3.

The above example shows that in order to find the load impedance point, given the position of a voltage minimum, it is necessary only to move the distance ℓ_{min}/λ *towards load* on the chart, and the z_L point is located where the radius line cuts the VSWR circle. It also shows how much less work is involved in using the Smith chart to solve the problem, compared to calculation.

Example 12.6   Rework Example 12.4 using the Smith chart. Data from Example 12.4 are $Z_0 = 50 \; \Omega$, $\ell/\lambda = .1$, load is a short circuit.

Solution   With a short-circuited load, the VSWR $= \infty$, and therefore the VSWR circle coincides with the normalized reactance (or susceptance) scale. Part of the scale is shown in heavy outline in Fig. 12.13. The problem is to find the input reactance at $.1\lambda$ towards generator. Therefore, moving along the VSWR circle

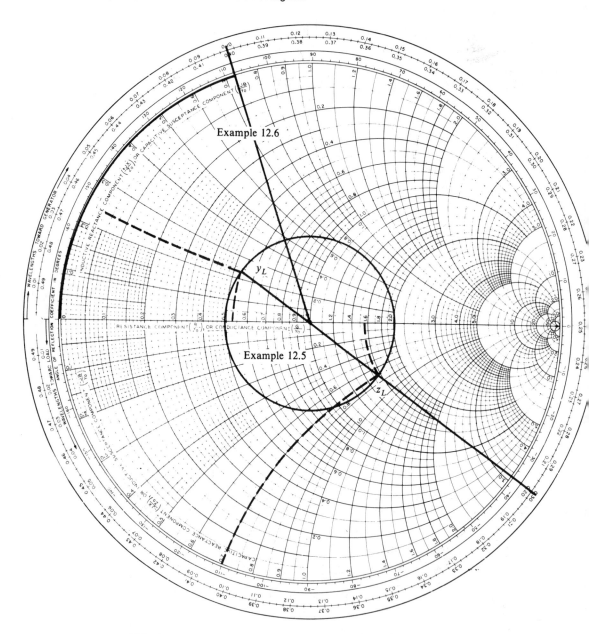

FIGURE 12.13. Smith Chart solutions for worked examples 12.5 and 12.6.

a distance 0.1 *towards generator* gives a normalized input impedance of

$$z_{in} = j.725$$

Thus,

$$Z_{in} = 50 \times j.725$$
$$= j36.25 \ \Omega$$

This compares with $j36.33 \ \Omega$ obtained by calculation in Example 12.4.

It is important to observe that *clockwise movement from any position on the chart gives a shift in position towards generator*, while *counterclockwise movement gives a shift in position towards load*. It is not necessary that such shifts take place from the zero origin on the wavelengths scale.

Reflectionless matching of transmission lines can be achieved by using a reactive stub to tune out the reactance or susceptance at the correct position on the line. The position of the stub, its length, its characteristic impedance, and whether it should be open or short-circuited, are all parameters which must be taken into consideration. In the method to be described here, the characteristic impedance of the stub is taken to be the same as that of the main feeder to be matched. Where the stub is connected in series with the main line, it is best to work with impedances and reactances. Where the stub is connected in parallel, as shown in Fig. 12.14(a), which is the most convenient arrangement for coaxial lines, it is best to work with admittances and susceptances. The problem to be solved is that of finding the position ℓ_1, the length ℓ_2, and whether the stub should be open or short-circuited. This problem is easily solved using the Smith chart. In terms of normalized admittances, the length ℓ_1 must transform the load admittance y_L into an admittance $y = 1 + jb$. The stub must add a susceptance $-jb$ so that the effective load admittance as seen by the main feeder is $(1 + jb) - jb = 1$. The actual admittance at this point is therefore $Y_0 \times 1 = Y_0$, i.e., the line is matched.

Referring to Fig. 12.14(b), the first step on the Smith chart is to enter the point z_L, draw the VSWR circle, and locate the y_L point diametrically opposite. Of course, if y_L is known, it may be entered directly on the chart, and the VSWR circle may be drawn. The second step is to move from y_L in the direction *wavelengths towards generator*, to the point where the VSWR circle cuts the $r = 1$ circle. This gives the length ℓ_1, as shown on Fig. 12.14(b), and the stub susceptance, jb. The third step is to go to the $-jb$ point on the chart and move in the direction *wavelengths towards load*, keeping in mind that it is the stub load which is referred to. The load point is reached when either zero susceptance or infinite susceptance is reached. Zero susceptance means that the stub load must be an open circuit, and infinite susceptance means that it must be a short circuit. For the conditions shown in Fig. 12.14(b), the infinite susceptance is reached first. The length ℓ_2 is determined as shown, and the stub load required in this case is a short circuit.

Single stub matching suffers from the disadvantage that the position ℓ_1 cannot easily be changed, and therefore the method is really suitable only for a fixed load. Where various loads may be encountered, double stub matching is often used, one such arrangement being shown in Fig. 12.15(a). As with single stub matching, the objective is to transfer the load point onto the $g = 1$ circle and then tune out the susceptance using stub ℓ_2. In the arrangement shown, an arbitrary length ℓ of line transforms the load admittance from y_L to y_L', this

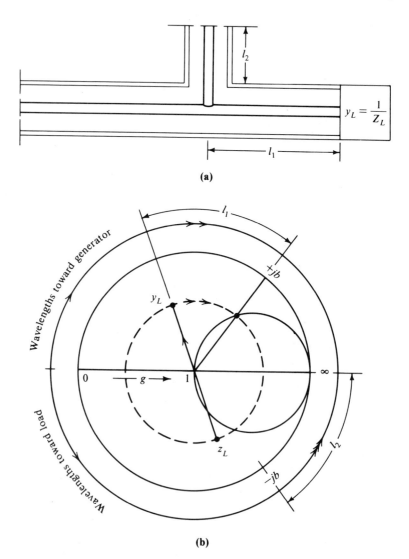

(a)

(b)

FIGURE 12.14. (a) Single-stub matching; (b) indicating the steps required on a Smith Chart to determine the lengths of l_1 and l_2.

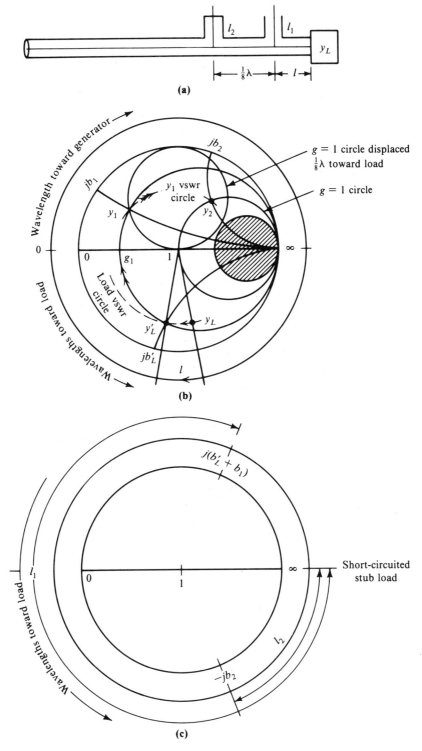

FIGURE 12.15. (a) Double-stub matching; (b) admittance transformations; (c) short-circuited stub lengths.

being a fixed transformation as shown in Fig. 12.15(b). Stub length ℓ_1 is now adjusted to bring y'_L onto the $\frac{1}{8}\lambda$-displaced $g = 1$ circle at point y_1. To achieve this, the length ℓ_1 is adjusted so that the stub adds susceptance $j(b'_L + b_1)$ in parallel with the existing $-jb'_L$. The conductance remains constant at g_1. At this stage, there is no movement along the main line, the stub ℓ_1 being adjusted to move the total susceptance from $-jb'_L$ to $+jb_1$. Any point on the $\frac{1}{8}\lambda$-displaced $g = 1$ circle is automatically transferred to the $g = 1$ circle on moving the $\frac{1}{8}\lambda$ distance from stub 1 to stub 2 in the *towards generator* direction. Thus, point y_1 is transformed into point y_2. Length ℓ_2 is now adjusted to tune out the susceptance $+jb_2$. The required stub lengths are shown on Fig. 12.15(c). Since both stubs are shown short-circuited, the stub load admittance in each case is infinite. Stub 1 has to provide a positive susceptance $+j(b'_L + b_1)$ for the example shown, and the displacement must be *towards (stub) load*, which is infinite admittance. In this example, ℓ_1 is seen to be greater than .25λ, since half the circumference of the chart is 0.25λ. Stub 2 has to provide a susceptance $-jb_2$ from the short circuit, and so ℓ_2 is found as shown on Fig. 12.15(c).

Since stub lengths ℓ_1 and ℓ_2 are both adjustable, the system can be readjusted for different loads. One disadvantage is that if the load point y'_L should fall within the circle shown shaded, it could not be brought onto the $\frac{1}{8}\lambda$-displaced $g = 1$ circle, and so points within the shaded circle cannot be matched. For the $\frac{1}{8}\lambda$ spacing, examination of the Smith Chart shows that the forbidden region is enclosed by the $g = 2$ circle.

12.13   TIME-DOMAIN REFLECTOMETRY

The basic arrangement for a *time-domain reflectometry* (TDR) system is shown in Figure 12.16(a). The method utilizes direct measurement of the incident wave and the reflected wave as functions of time, hence the name "time-domain reflectometry." The incident wave is supplied by a step generator, the step voltage having a very fast rise time. The oscilloscope records the outgoing step V_i, and, at a somewhat later time, the reflected voltage V_r. The time between V_i and V_r is a measure of the length of the cable.

The nature of the load shows up in the shape of the reflected wave. As shown in Section 12.8, for an open circuit $\Gamma_L = 1$, and the incident wave is totally reflected at the load. Therefore, assuming a lossless line, the oscilloscope trace will be as shown in Fig. 12.16(b). A short circuit results in a $\Gamma_L = -1$, so the incident step is totally reflected with a phase reversal, and therefore the oscilloscope trace is shown in Fig. 12.16(c).

A purely resistive load of value $2Z_0$ results in a reflection coefficient of

$$\Gamma_L = \frac{2Z_0 - Z_0}{2Z_0 + Z_0}$$

$$= \frac{1}{3} \tag{12.91}$$

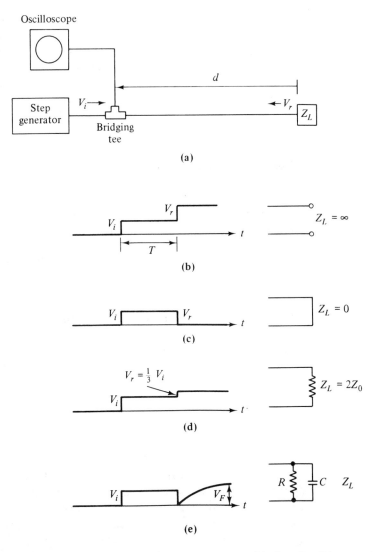

FIGURE 12.16. (a) A time-domain reflectometry system. Displays for: (b) open-circuited termination; (c) short circuit; (d) resistive load = $2Z_0$; (e) parallel RC load.

That is,

$$V_R = \frac{V_I}{3} \tag{12.92}$$

The trace will therefore be as shown in Fig. 12.16(d).

An example of a complex load is the parallel RC circuit shown in Fig. 12.16(e). Initially, the capacitor behaves as a short circuit (assuming it is initially uncharged). However, C will immediately start to charge up, the voltage following an exponential law. When C is fully charged, it can be considered an open circuit, so the line is effectively terminated in R. The reflection coefficient for the final condition is, therefore,

$$\Gamma_L = \frac{R - Z_0}{R + Z_0} \tag{12.93}$$

and the final voltage V_F is

$$
\begin{aligned}
V_F &= V_I + V_R \\
&= V_I(1 + \Gamma_L) \\
&= V_I\left(1 + \frac{R - Z_0}{R + Z_0}\right)
\end{aligned}
\tag{12.94}
$$

The oscilloscope trace is shown in Fig. 12.16(e).

Knowing the elapsed time T between the incident and reflected voltages, and the phase velocity on the cable, the cable length between bridging tee and point of reflection can be found. Thus, d (Fig. 12.16(a)) is given by

$$d = v_p \frac{T}{2} \tag{12.95}$$

T is the "round-trip" time; therefore, it is necessary to divide by 2 as shown.

This feature is particularly important in locating faults and sources of mismatch on a cable, which show up as reflections. The cable shown in Fig. 12.17(a), for example, may result in a trace as shown in Fig. 12.17(b). The phase velocity can be found by measuring T for a known length of cable. The

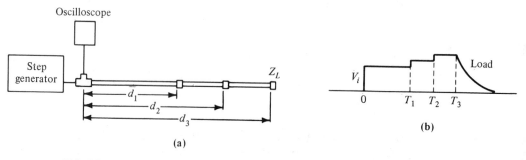

FIGURE 12.17. (a) TDR system used to determine discontinuities; (b) display.

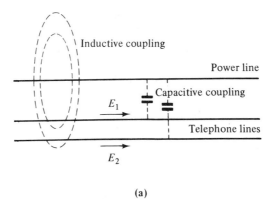

(a)

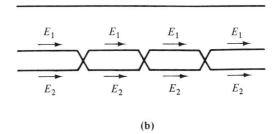

(b)

FIGURE 12.18. (a) Induced interference; (b) line transposition to eliminate interference.

various distances (Fig. 12.17(a)) are, then,

$$d_1 = v_p \frac{T_1}{2} \tag{12.96}$$

$$d_2 = v_p \frac{T_2}{2} \tag{12.97}$$

$$d_3 = v_p \frac{T_3}{2} \tag{12.98}$$

TDR is a very versatile system of measurement for determining transmission-line properties, and fuller details will be found in the *Application Notes 62* and *67* issued by the Hewlett-Packard Company.

12.14 TELEPHONE LINES AND CABLES

The simplest type of line (apart from the single overhead wire with ground return) is the two-wire overhead line. The wire is usually made of cadmium copper, which provides better mechanical strength than hard-drawn copper. Bare wires are used where possible, but where insulation or protection is required, a polyvinyl chloride (PVC) covering is used.

Power lines (60 Hz) can induce interference in open-wire lines, but this can be eliminated by *transposing* the telephone wires. Figure 12.18(a) shows how interference may be picked up through both inductive and capacitive coupling. The induced interference voltages E_1 and E_2 act in opposition, but because they are not in general equal in magnitude, the resultant interference voltage $E_1 - E_2$ will be different from zero. By transposing the wires as shown in Fig. 12.18(b), the induced voltages in each wire can be equalized so that they cancel.

Where open-wire lines are not feasible, the wires are formed into a multicore cable assembly which is usually carried underground in a ducting. For this type of cable the wires are annealed copper, which is flexible, and each wire is insulated with high-grade paper tape. A lead-alloy sheath is extruded over the wires to form the cable as shown in Fig. 12.19(a).

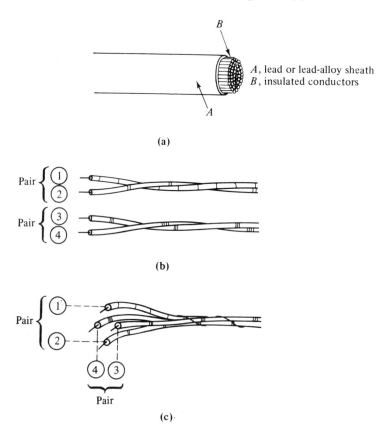

FIGURE 12.19. (a) Lead-sheathed multicore cable; (b) twin-cable assembly; (c) star-quad cable assembly. (Courtesy BICC Telephone Cables Div.)

For larger cable assemblies, for example those required between exchanges, the wires of a given pair are either twisted together in pairs to form a *twin cable* (Fig. 12.19(b)), or they may be twisted to form a *star quad* (Fig. 12.19(c)). These are then assembled in larger units either as a concentric assembly or in unit construction. The latter is favored for use in local distribution networks, as branching into smaller units is then conveniently and tidily achieved.

For higher-frequency signals (e.g., multicarrier telephony or television), coaxial cables are normally employed, the constructional features of which are shown in Fig. 12.20(a). These, in turn, may form part of a composite cable assembly as shown in Fig. 12.20(b).

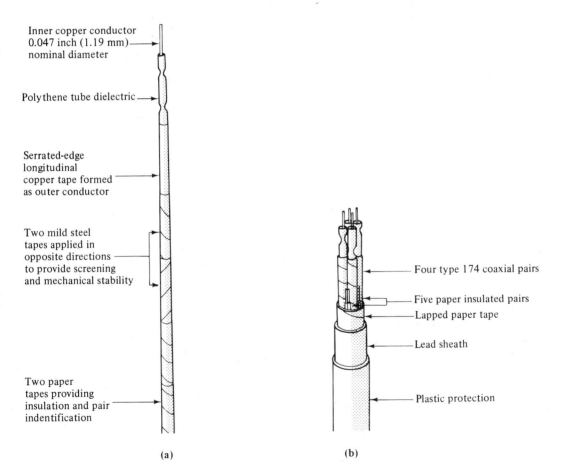

Inner copper conductor
0.047 inch (1.19 mm)
nominal diameter

Polythene tube dielectric

Serrated-edge
longitudinal
copper tape formed
as outer conductor

Two mild steel
tapes applied in
opposite directions
to provide screening
and mechanical stability

Two paper
tapes providing
insulation and pair
indentification

(a)

Four type 174 coaxial pairs

Five paper insulated pairs

Lapped paper tape

Lead sheath

Plastic protection

(b)

FIGURE 12.20. (a) A coaxial cable; (b) a composite cable assembly. (Courtesy BICC Telephone Cables Div.)

12.15 RADIO-FREQUENCY LINES

For low-power applications, lines with solid dielectrics, such as the twin feeder or the coaxial line, are normally used. With high-power lines (such as would be used to feed radio-frequency power to antennas in the kilowatt range), the solid dielectric is omitted to keep losses at a minimum, and an open-type construction is used with well-spaced support insulators, so that the main dielectric is air.

12.16 MICROSTRIP TRANSMISSION LINES

Microwave integrated circuit assemblies utilize special forms of transmission lines, the two most common types being the microstrip line and the stripline. The basic structures for both types, along with their electromagnetic field configurations, are shown in Fig. 12.21. Because the microstrip assembly is open on the top, the field distribution tends to be complex, and this affects the characteristic impedance of the lines. However, the open structure is easier to fabricate, and discrete components are readily added to the circuit. With the stripline, the fields are confined to the dielectric region, and this is more like a distorted version of the coaxial line field distribution shown in Fig. 12.1.

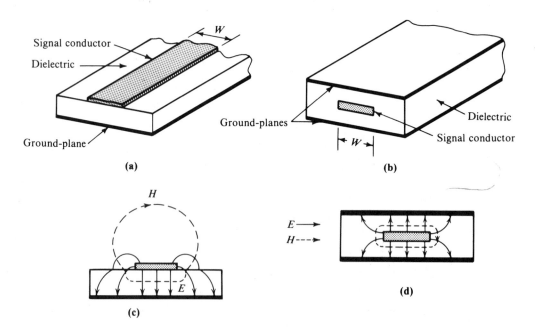

FIGURE 12.21. (a) Microstrip; (b) stripline; (c) electric field E and magnetic field H for microstrip; (d) E and H for stripline.

Although the conductor patterns for a given circuit configuration may be similar for both systems, the constructional details are very different. A typical substrate material for microstrip is Alumina, which is physically hard and can withstand the high temperatures and high vacuum conditions encountered during the deposition of the conductors. Both thick-film and thin-film deposition methods are used in practice. The nominal relative permittivity of Alumina is 9. An open view of a microstrip assembly is shown in Fig. 12.22, along with the equivalent circuit.

Stripline circuits are made using printed-circuit-board methods. However, special board material having low loss at microwave frequencies and very good mechanical stability must be used. Various board materials are available such as woven fiberglass and polyolefin, the relative permittivity for the latter being 2.32. Boards may come with copper cladding on one or both sides. The cladding is specified in ounces, which refers to the number of ounces of copper used for each square yard of surface. Common weights are 1 oz and 2 oz, corresponding to copper thicknesses of .0014 in. and .0028 in., respectively. Figure 12.23 shows an exploded view of a stripline assembly. Two printed circuit boards are used, each having a groundplane. The lower surface on the upper board is left blank, while the conductor pattern is etched on the upper surface of the lower board. When ready for assembly, the two substrates are tightly and uniformly clamped together between metal casings. The residual air gap around the conductor pattern may be filled using a silicone grease or a polyethylene laminate, or in some cases the clamping pressure may be relied upon to produce sufficient flow of the dielectric around the conductors.

The characteristic impedance of both types of line depends on the conductor line width, thickness, and separation from substrate, and also on the relative permittivity of the substrate. In the case of microstrip, an effective relative permittivity must be used which takes into account the combined substrate–air dielectric space. Design formulas and charts for determining characteristic impedance are available in specialist publications dealing with the design of these circuits. Practical values of characteristic impedance range from about 5 Ω to 110 Ω. The line impedance increases as the line width w decreases because a narrower line results in an increased series inductance and decreased shunt capacitance, and as shown by Eq. (12.14), both effects tend to increase Z_0.

Matching networks similar to those shown in Figs. 12.14 and 12.15 can be designed using microstrip and stripline, and the method of using the Smith chart to solve the matching problem is equally applicable, but of course the values must be normalized to the characteristic impedance of the line being used. A main advantage of stripline and microstrip is that various line sections having different characteristic impedances can be readily constructed together in the same circuit, resulting in versatile matching networks and other types of circuits. Because of the mixture of characteristic impedances encountered in such situations, the Smith chart does not provide a convenient method of

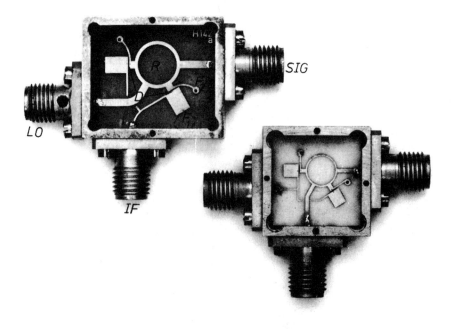

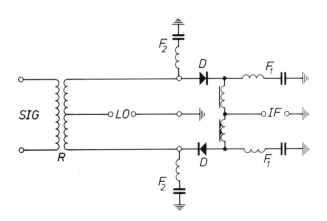

FIGURE 12.22. (a) Balanced mixer for 12 GHz, *left*: on quartz glass, *right*: on aluminum oxide. *R* hybrid ring. *D* Schottky-barrier diodes. F_1 band-stop filters for 12 GHz. F_2 band-stop filters for the second harmonic of the oscillator frequency. *LO* oscillator input. *SIG* aerial input. *IF* intermediate-frequency output. (b) Low-frequency equivalent of the mixer circuit (a). (Courtesy J. H. C. van Heuven and A. G. van Nie. Philips Tech. Rev **32**, 1971)

solution, but in fact the matching elements are easily calculated by direct means. Such networks make use of the circuit elements described in Section 12.11. Specifically, open- and short-circuited line lengths of $\frac{1}{8}\lambda$ and $\frac{3}{8}\lambda$ are commonly used, as these provide pure reactances. From Eq. (12.75), the impedance Z of a $\lambda/8$ short-circuited length of line is given by

$$Z = jZ \tan\left(\frac{2\pi}{\lambda} \frac{\lambda}{8}\right) = jZ_0 \tan\left(\frac{\pi}{4}\right) = jZ_0 \tag{12.99}$$

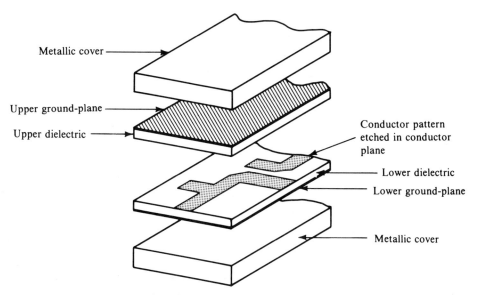

(a)   An exploded view of a stripline assembly.

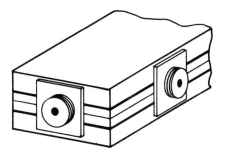

(b)   The assembled stripline with edge connectors.

FIGURE 12.23.

This shows that the input impedance of such a stub is purely reactive and numerically equal to the characteristic impedance of the stub. In a similar manner, a length of $3\lambda/8$ results in

$$Z = -jZ_0 \tag{12.100}$$

Usually, it is more convenient to work with admittance values, which are simply the reciprocals of the impedances:

$$\frac{\lambda}{8} \text{ short-circuited: } Y = -jY_0 \tag{12.101}$$

$$\frac{3\lambda}{8} \text{ short-circuited: } Y = jY_0 \tag{12.102}$$

Open-circuited stubs may also be used, and the admittances for the two cases are readily derived from Eq. (12.80) as

$$\frac{\lambda}{8} \text{ open-circuited: } Y = jY_0 \tag{12.103}$$

$$\frac{3\lambda}{8} \text{ open-circuited: } Y = -jY_0 \tag{12.104}$$

Along with stub sections, the $\lambda/4$ transformer section is also used, Eq. (12.81) being applicable in this case. The use of stripline (and microstrip) for construction of a matching network is illustrated in the following example.

Example 12.7   In order to optimize the low-noise performance of a microwave amplifier, the transistor must be fed from a source having an admittance $Y = 0.05 - j0.03$ siemens. The actual source impedance is 50 Ω. Design a suitable stripline matching network.

Solution   The schematic for the matching network is shown in Fig. 12.24(a). Writing the required source admittance as $Y = G + jB$, the $\lambda/4$ section has to transform the 50-Ω source into $1/G = 1/.05 = 20$ Ω. Using Eq. (12.81),

$$20 = \frac{Z_0^2}{50}$$

Therefore,

$$Z_0 = \sqrt{20 \times 50} = 31.62 \ \Omega$$

The stub must add a susceptance of $jB = -j.03$ siemens. Equation (12.101) shows that this can be accomplished through the use of a short-circuited stub of characteristic admittance Y_0, where

$$-j.03 = -jY_0$$

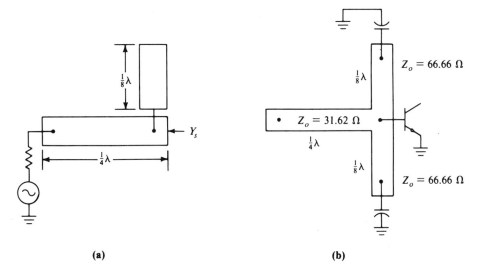

FIGURE 12.24. (a) Matching network schematic for Example 12.7; (b) the stripline layout.

Thus,

$$Z_0 = \frac{1}{Y_0} = 33.33 \ \Omega$$

In practice, to minimize the effect of the shunt to series connection, the shunt stub is usually balanced about the series line as shown in Fig. 12.24(b). This is equivalent to having two stubs in parallel, each of susceptance $-j.015$ siemens and, therefore, each having a characteristic impedance of 66.66 Ω. Capacitive short circuits are used at the ends of these stubs.

12.17 PROBLEMS

1. Explain what is meant by the characteristic impedance of a transmission line. Explain why you would expect an infinitely long uniform line to have an input impedance equal to the characteristic impedance.

2. The primary constants (which may be assumed independent of frequency) for a transmission line are $R = 0.5 \ \Omega/m$, $L = 250 \ \text{nH/m}$, $C = 100 \ \text{pF/m}$, and $G = 10^{-6} \ \text{S/m}$. Plot the variation of $|Z_0|$ against frequency for this line.

3. For the transmission line in Problem 2, calculate the attenuation coefficient and the phase-shift coefficient at a frequency of 20 MHz.

4. Explain what is meant by the propagation coefficient of a transmission line. The propagation coefficient for a transmission line is given as $(0.0005 + j\pi/10)$ m^{-1}. Determine, for a matched-line length of 50 m, (a) the total attenuation in nepers, (b) the total attenuation in decibels, and (c) the total phase shift.

5. The propagation coefficient γ for a transmission line is given by $\gamma = 0.0006 + j(\pi/10)$. The line is 32 m long and is correctly terminated to avoid reflections. Plot the phasor diagram showing the variation of current over the length of line (see, for example, Fig. 12.3) but plot the phasor length in nepers, using one-millineper steps.

6. For the transmission line in Problem 4, determine the attenuation, in nepers, of 15 km of line.

7. Explain what is meant by the *group delay time* and why this should be constant for a transmission line. Given that $\beta = \omega\sqrt{LC}$, $L = 1\ \mu\text{H/m}$, and $C = 11.11$ pF/m, calculate (a) the group delay time, and (b) the group velocity.

8. Explain what is meant by the *voltage reflection coefficient* for a transmission line. The voltage reflection coefficient on a 50-Ω line is 0.7 $\underline{/30°}$. Determine the load impedance terminating the line.

$0.632\underline{/108}$ ✓ 9. A 300-Ω line is terminated in a load impedance of $100 + j200\ \Omega$. Determine the voltage reflection coefficient.

10. Explain how standing waves may be set up on a transmission line, and state the relationship between voltage and current standing waves for a line whose characteristic impedance is purely resistive.

2.1
$3.65.1$

11. Calculate the voltage standing-wave ratio for a lossless line of characteristic impedance 50 Ω when it is terminated in (a) 100 Ω resistive, and (b) $(30 - j50)\ \Omega$.

12. Sketch the variation of input impedance with length for a lossless transmission line terminated in a short circuit. A section of 50-Ω line, 0.15λ in length, is terminated in a short circuit. Determine the effective input reactance of the section, showing clearly whether it is inductive or capacitive.

$/8\lambda$ $\frac{3\lambda}{8}$

13. Calculate the shortest stub length required to produce a reactance equal to (a) Z_0, and (b) $-Z_0$. Express the length as a fraction of a wavelength.

14. A 600-Ω lossless line is terminated in a 5-pF capacitor. The line is operated at a frequency which makes it $\frac{1}{4}\lambda$ long. Determine the nature of the input impedance of the line, giving the equivalent inductance or capacitance value.

2.0

15. Determine, using a Smith chart, the admittance value corresponding to an impedance of $150 + j80\ \Omega$. Give the admittance value in siemens, and also given the equivalent parallel component values in ohms.

16. A load impedance of $80 - j70$ Ω is connected to a 50-Ω transmission line. Find (a) the VSWR, (b) the distance, in terms of wavelength, to the first voltage minimum, and (c) the impedance at this minimum.

17. Measurements on a 50-Ω slotted line gave the following results: VSWR = 3.0; distance from load to first voltage minimum = 0.4λ. Determine, using a Smith chart, the value of the load impedance.

18. An impedance $100 - j50$ Ω is used to terminate a 50-Ω lossless line. The line is 0.6λ long. Determine, using a Smith chart, (a) the VSWR, (b) the input impedance, and (c) the impedance at a voltage maximum.

19. A load admittance of $4 - j10$ mS is to be matched to a 50-Ω line system using single stub matching. Determine the position for the stub to be connected in parallel with the main line, the length of the stub, and whether the stub is to be open- or short-circuited.

20. Repeat problem 19 for a series-connected stub.

21. A double stub matching section is arranged as shown in Fig. 12.15(a) with the distance $\ell = 0.1\lambda$. Determine which of the following normalized admittances can be matched: (a) $0.6 + j1.2$, (b) $0.6 - j1.2$, (c) $1.1 + j1.3$, (d) $3 - j2$.

22. A load impedance of $75 - j80$ Ω is to be matched to a 50-Ω line using the double stub section of Problem 21. Determine the required stub lengths, in wavelengths.

23. In a time-domain reflectometer measurement on a terminated transmission line, the reflected voltage was found to be a vertical downward step, of magnitude one-half the input step. If the line can be assumed lossless and has a characteristic impedance of 100 Ω, determine the value of load impedance.

24. Sketch the form of trace expected on a TDR measuring set when applied to a line terminated in a series RL circuit.

25. A transmission line is made of two parallel copper #18 AWG wires (1.024 mm in diameter) spaced 10 mm apart and embedded in a plastic spacer made of polyvinyl chloride (PVC). The wire has a resistance of 1.984 Ω/km and the insulation has a leakage conductivity of 2.5×10^{-10} siemen/km. Find the primary line constants. For PVC, $\varepsilon_r = 3.3$.

26. Repeat Problem 25 for a coaxial line made of the same wire and insulating material, with a copper sheath of inside diameter of 2 cm. Assume the leakage to have the same value.

27. A microwave amplifier requires to be fed from a source impedance of $12 + j8$ Ω. The actual source impedance is 50 Ω. Determine the impedance values for a suitable stripline matching section consisting of a λ/4 transformer section and a parallel-connected 3λ/8 stub, and state whether the stub should be open- or short-circuited.

0.25λ

0.335
0.25

.085

50

1
─
8

.125 λ

0.25
.125

.375

CHAPTER 13

Waveguides

13.1 INTRODUCTION

At frequencies higher than about 3000 MHz, transmission of electromagnetic waves along lines and cables becomes difficult, mainly because of the losses that occur both in the solid dielectric needed to support the conductors, and in the conductors themselves. It is possible to transmit an electromagnetic wave down a metallic tube called a *waveguide*. The most common form of waveguide is rectangular in cross section, as shown in Fig. 13.1(a). Induced currents in the walls of the waveguide give rise to power losses, and to minimize these losses the waveguide wall resistance is made as low as possible. Because of skin effect, the currents tend to concentrate near the inner surface of the guide walls, and these are sometimes specially plated to reduce resistance.

Apart from determination of losses, the walls of a waveguide may be considered to be perfect conductors. Two important boundary conditions result, which determine the mode of propagation of an electromagnetic wave along a guide: (1) electric fields must terminate normally on the conductor [i.e., the tangential component of the electric field must be zero (Fig. 13.1(b))]; (2) magnetic fields must lie entirely tangentially along the wall surface [i.e., the normal component of the magnetic field must be zero (Fig. 13.1(c))]. Knowing these boundary conditions provides a simple way of visualizing how the various modes of waveguide transmission occur, which will be discussed in the following sections.

Although the microwave frequency spectrum extends from about 300 MHz to 300 GHz (see Fig. 14.17), transmission lines can be utilized for the lower part of the range. Above about 3000 MHz, waveguides become necessary where large amounts of power have to be transmitted.

13.2 RECTANGULAR WAVEGUIDES

The boundary conditions already referred to preclude the possibility of a waveguide supporting *transverse electromagnetic* (TEM) wave (see Appendix B) propagation, since the magnetic field is at right angles to the direction of propagation (along the axis of the waveguide) and therefore would have to terminate normally to the sidewalls, which cannot happen. One possible solution is for the magnetic field to form loops along the direction of propagation lying parallel to the top and bottom walls and tangentially to the sidewalls (see Fig. 13.2(a)).

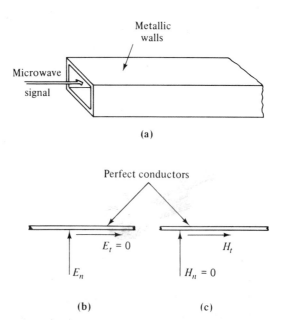

FIGURE 13.1. (a) Rectangular waveguide; (b) electric field boundary conditions; (c) magnetic field boundary conditions.

The variation of electric field as a function of distance along the direction of propagation is shown in Fig. 13.2(b) and along the cross section in Fig. 13.2(c). The propagation mode sketched in Fig. 13.2 is known as a *transverse electric* (TE) *mode* because the electric field is entirely transverse to the direction of propagation. (It is also known as an *H mode*, signifying that part of the magnetic field lies along the direction of propagation.)

Subscripts are used to denote the number of half-cycles of variation which occur along the a and b sides. As shown in Fig. 13.2, one half-cycle (i.e., one maximum) occurs along the a side and none along the b side; this mode is therefore referred to as the TE_{10} *mode*. The TE_{10} mode is the dominant mode in waveguide transmission, for, as will be shown, it supports the lowest-frequency waveguide mode.

13.2.1 Properties of the TE_{10} Mode

Figure 13.3(a) shows how a TE_{10} wave may be formed as the resultant of two TEM waves crossing each other. At the points of intersection shown, the individual waves add vectorially. At those points where the electric fields reinforce each other, shown by double crosses and double dots, the magnetic field is directly up and down; at the points where the electric fields cancel, shown by dots and crosses together, the magnetic field is directed left and right. Metallic walls may be placed along the L-R direction as shown in Fig. 13.3(b) without violating the boundary conditions, as it will be seen that the

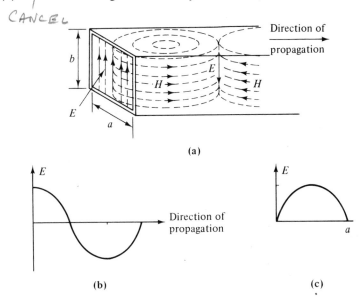

(a)

(b)

(c)

FIGURE 13.2. (a) One possible field configuration for a waveguide; (b) electric field amplitude along guide axis; (c) electric field amplitude along guide width.

electric field is zero and the magnetic field tangential at these walls. Metallic top and bottom walls can be put in position, as the electric field will terminate normally on these and the magnetic field lies parallel to them.

The direction of the magnetic field loops alternates as shown in Fig. 13.3(b), one loop occupying a distance of a half wavelength in the guide. This

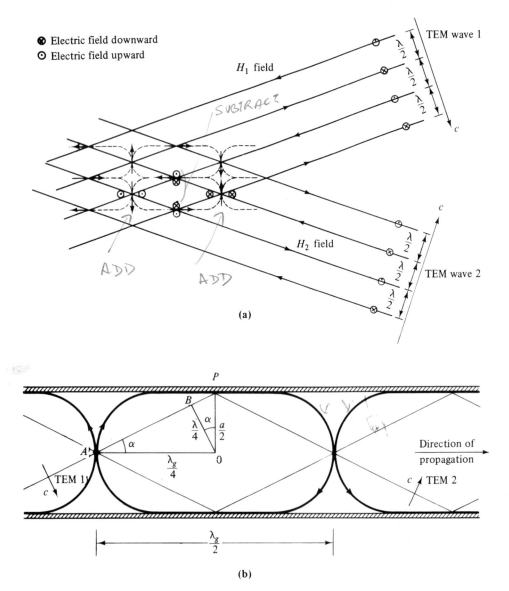

FIGURE 13.3. (a) Generating the waveguide mode from two TEM waves; (b) geometry used in determining TE_{10} properties.

will be different from a half wavelength of the individual TEM waves and is shown as $\frac{1}{2}\lambda_g$, where λ_g is termed the guide wavelength.

The frequency of the TE wave will be the same as that of the TEM waves (i.e., it will have the same number of cycles in a second). Let v_p be the phase velocity of the TE wave; then, using the general relationship $\lambda f = v_p$, derived in Appendix B, we have

TE wave: $$\lambda_g f = v_p \tag{13.1}$$

TEM wave: $$\lambda f = c \tag{13.2}$$
(in air)

from which it follows that

$$v_p = c\frac{\lambda_g}{\lambda} \qquad \qquad \tag{13.3}$$

From the geometry of Fig. 13.3(b), it can be seen that

$$\cos\alpha = \frac{\lambda}{2a} \quad \text{(from right-angled triangle OBP)} \tag{13.4}$$

$$\sin\alpha = \frac{\lambda}{\lambda_g} \quad \text{(from right-angled triangle ABO)} \tag{13.5}$$

and since

$$\sin^2\alpha + \cos^2\alpha = 1$$

it follows that

$$\left(\frac{\lambda}{\lambda_g}\right)^2 + \left(\frac{\lambda}{2a}\right)^2 = 1$$

or

$$\frac{1}{\lambda_g^2} = \frac{1}{\lambda^2} - \frac{1}{(2a)^2} \qquad \tag{13.6}$$

where λ_g is the guide wavelength of the TE_{10} mode, λ is the free-space wavelength of either TEM wave, and a is the broad dimension of the guide.

From Eq. (13.6), it can be seen that when $\lambda = 2a$, the guide wavelength becomes infinite; this corresponds to the individual TEM waves bouncing from side to side with no velocity component directed along the guide. Clearly, then, $\lambda = 2a$ represents the longest-wavelength TEM wave that can be induced into a guide, as a TE mode will not be generated for longer wavelengths. For shorter TEM wavelengths, Eq. (13.6) shows that the guide wavelength is real and positive, so that TE-mode propagation takes place.

The term $2a$ is referred to as the cutoff wavelength λ_c of the TE_{10} mode, and, rewriting Eq. (13.6), we have

$$\frac{1}{\lambda_g^2} = \frac{1}{\lambda^2} - \frac{1}{\lambda_c^2} \qquad (13.7)$$

From Eqs. (13.3) and (13.5), it follows that

$$v_p = \frac{c}{\sin \alpha} \qquad (13.8)$$

Since the sine of an angle is never greater than unity, v_p can never be less than c; it can, of course, be greater.

The individual waves have to travel a zigzag path, and therefore the velocity component along the guide, which is the velocity at which the energy in the waves is conveyed, is less than c. From Fig. 13.4 this component, known as the group velocity v_g, is

$$v_g = c \sin \alpha \qquad (13.9)$$

Combining Eqs. (13.8) and (13.9),

$$v_g v_p = c^2 \qquad (13.10)$$

Equation (13.9) can also be rewritten in terms of wavelengths, since these are

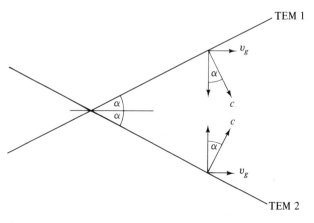

FIGURE 13.4. Group velocity as the horizontal components of c.

the quantities that are normally known:

$$v_g = c \frac{\lambda}{\lambda_g} \qquad (13.11)$$

Unless otherwise stated, it is assumed that the waveguide dielectric is dry air, for which free-space permittivity and permeability apply.

Example 13.1 A rectangular waveguide has a broad wall dimension of 0.900 inch, and is fed by a 10-GHz carrier from a coaxial cable as shown in Fig. 13.5. Determine whether a TE_{10} wave will be propagated, and, if so, find its guide wavelength, phase, and group velocities.

Solution
$$a = 0.900 \text{ inch}$$
$$= 2.286 \text{ cm}$$

Therefore,
$$\lambda_c = 2 \times 2.286$$
$$= 4.572 \text{ cm}$$
$$\lambda = \frac{3 \times 10^8}{10^{10}}$$
$$= 3 \text{ cm}$$

Note that it is the free-space wavelength, and not the wavelength along the coaxial cable, which is used. Therefore, $\lambda_c > \lambda$, and a TE_{10} wave will propagate. Hence,

$$\frac{1}{\lambda_g^2} = \frac{1}{\lambda^2} - \frac{1}{\lambda_c^2}$$

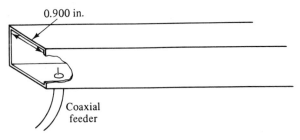

0.900 in.

Coaxial
feeder

FIGURE 13.5. Example 13.1.

Therefore,

$$\lambda_g = \frac{\lambda}{\sqrt{1 - (\lambda/\lambda_c)^2}}$$

$$= \frac{3}{\sqrt{1 - 0.431}}$$

$$= 3.975 \text{ cm}$$

$$v_p = c\frac{\lambda_g}{\lambda}$$

$$= 3 \times 10^8 \times 1.325$$

$$= 3.975 \times 10^8 \text{ m/s}$$

$$v_g = c\frac{\lambda}{\lambda_g}$$

$$= \frac{3 \times 10^8}{1.325}$$

$$= 2.264 \times 10^8 \text{ m/s}$$

The *wave impedance* is defined as the ratio of the transverse components of electric to magnetic fields. In synthesizing the TE_{10} mode, the two TEM waves are of equal amplitude and time phase; thus, $E_1 = E_2 = E$ and $H_1 = H_2 = H$. The entire electric field E is transverse, along dimension b, while from Fig. 13.3 it is easily ascertained that the transverse component of H lies along dimension a and is

$$H_a = H \sin \alpha \tag{13.12}$$

Hence the wave impedance for the TE_{10} mode is

$$Z_\omega = \frac{E_b}{H_a}$$

$$= \frac{E}{H \sin \alpha} \tag{13.13}$$

In Appendix B the wave impedance of a TEM wave in free space is shown to be (Eq. (B.10))

$$Z_0 = \frac{E}{H}$$

$$= \sqrt{\frac{\mu_0}{\epsilon_0}} \tag{13.14}$$

Thus, substituting Eqs. (13.14) and (13.5) in Eq. (13.13) gives

$$Z_\omega = Z_0 \frac{\lambda_g}{\lambda} \tag{13.15}$$

$$= 120\pi \frac{\lambda_g}{\lambda} \Omega \tag{13.16}$$

when the numerical values for μ_0 and ϵ_0 are substituted.

The importance of the wave impedance concept is that it can be used in an analogous manner to the characteristic impedance of a transmission line, so that the theory of reflections, standing waves, and Smith charts developed in Chapter 12 can also be applied to waveguides.

13.2.2 Standing Waves

Consider a section of waveguide closed by a perfectly conducting sheet at the end. The boundary conditions require that the fields adjust so that the electric field is zero and the magnetic field is entirely tangential at the closure (i.e., the field patterns are similar to those at the walls of the guide). The resultant wave pattern can be accounted for in terms of the incident TE wave and a reflected TE wave, the combination of which sets up a standing-wave pattern along the guide, similar to that described in Section 12.7 for transmission lines. It is important to realize that the resultant wave pattern is stationary in space and varies in time, whereas the traveling wave shown in Fig. 13.3(b) is time-invariant, but of course moves along the guide. Figure 13.6 emphasizes this point. On the left are shown the conditions in a short-circuited guide at time intervals of one-fourth the periodic time of the wave, the reference time being chosen at a maximum field condition. The sequence on the right shows how a single traveling wave would vary over the same time intervals. (For clarity, only the magnetic field loops are shown.)

It will be observed that at $\frac{1}{4}\lambda_g$ from the closed end the transverse magnetic field is zero, whereas the electric field is a maximum. Therefore, the wave impedance is infinite at this section across the guide. At the $\frac{1}{2}\lambda_g$ section, the electric field is zero and the transverse magnetic field component is a maximum, so that the wave impedance is zero. Thus the quarter-wavelength and half-wavelength sections of guide have the same transformation properties as those of the transmission-line sections described in Section 12.11.

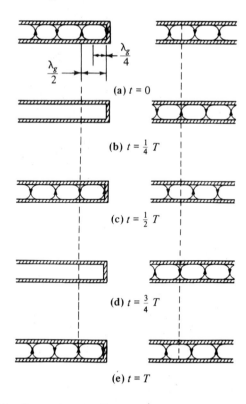

FIGURE 13.6. Comparing standing waves (left) to traveling waves (right).

13.2.3 Waveguide Terminations

In order to terminate a waveguide without causing reflections, the terminating impedance must equal the wave impedance of the incident wave. Consider first a thin sheet of resistive material closing the end of the waveguide (Fig. 13.7(a)). Between the top and bottom of the guide, the sheet presents a resistance of path length b and cross-sectional area ta, where t is the thickness of the sheet, a and b are the guide dimensions defined previously. Only the sheet resistance between top and bottom need be considered, since the current direction is at right angles to the H-field at the sheet. Let ρ be the resistivity of the sheet material; then the terminating resistance is

$$R_T = \rho \frac{b}{ta} \tag{13.17}$$

The *sheet resistivity* R_s is defined as

$$R_s = \frac{\rho}{t} \Omega \text{ per square} \tag{13.18}$$

The units of the square do not matter. R_s will have the same value for a square inch, a square centimeter, and so on. What does matter is that the units of ρ and t should be consistent. If ρ is in Ω-m, then t must be in meters and R_s will be in ohms per square. If ρ is in $\mu\Omega$-in, t must be in inches and R_s will be in $\mu\Omega$ per square.

Now, the current I in the sheet supports the tangential component of the magnetic field H_T, given by

$$H_T = \frac{I}{a} \tag{13.19}$$

Also, the voltage across the sheet gives rise to an electric-field component E_T:

$$E_T = \frac{V}{b} \tag{13.20}$$

Therefore,

$$\frac{E_T}{H_T} = \frac{Va}{Ib} \tag{13.21}$$

But $V/I = R_T = R_s(b/a)$, and substituting this into Eq. (13.21) gives

$$\frac{E_T}{H_T} = R_s \frac{b}{a} \frac{a}{b}$$

$$= R_s \tag{13.22}$$

E_T and H_T will consist of three components: incident, reflected, and transmitted waves, shown in Fig. 13.7(a) as W_i, W_r, and W_t, respectively. The transmitted wave can be eliminated by extending the guide $\frac{1}{4}\lambda_g$ beyond the termination sheet, as shown in Fig. 13.7(b), and short-circuiting the end. The $\frac{1}{4}\lambda_g$ shorted section presents an infinite impedance; therefore, no wave can be transmitted into it. Next, the sheet resistivity R_s is made equal to the

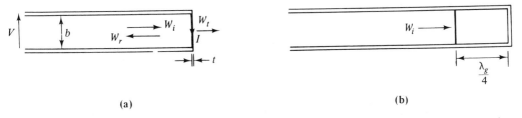

(a)   (b)

FIGURE 13.7. (a) Waveguide terminated with sheet resistance; (b) $\frac{1}{4}\lambda_g$ section added to termination.

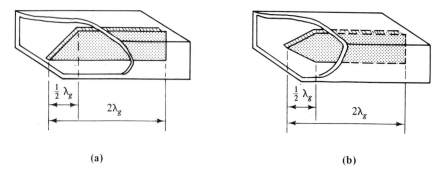

FIGURE 13.8. Practical waveguide terminations.

incident-wave impedance Z_ω:

$$Z_\omega = R_s \qquad (13.23)$$

This ensures that $E_T/H_T = E_i/H_i = E/H$; that is, the incident wave is absorbed by the sheet just as though it were continuing down an infinitely long waveguide in which the wave impedance was Z_ω (or R_s). Thus, the reflected wave is eliminated.

The arrangement discussed illustrates two important principles, that of sheet resistivity and the idea of using a $\frac{1}{4}\lambda_g$ shorted section to isolate the load from external impedances. The arrangement is difficult to implement in practice, however, largely because it requires an adjustable short circuit on the $\frac{1}{4}\lambda_g$ section which is rather critical to adjust. More practical arrangements for terminating a guide are shown in Fig. 13.8. The resistive strip is set in the plane of the maximum electric field, and the taper ensures a gradual change in electric field, which reduces reflections to a negligible level. Either form of taper shown in Fig. 13.8 is satisfactory, and the dimensions shown are typical of those used in practice. A carbon-coated strip may be used, or for more stable operation, a glass strip covered with a thin film of metal which, in turn, has a thin dielectric coating for protection. Surface resistivities of the order of 500 Ω per square are typical.

13.2.4 Attenuators

An arrangement similar to that shown in Fig. 13.8 can be used to provide attenuation in a waveguide. Two common methods are shown in Fig. 13.9. In Fig. 13.9(a), the thin resistive sheet can be moved from the sidewall, where it produces minimum attentuation, to the center of the guide, where it produces maximum attenuation. The mechanical drive for the sheet is often fitted with a micrometer control so that fine adjustment of attenuation can be made and accurately calibrated. The flap attenuator of Fig. 13.9(b) is simple to construct,

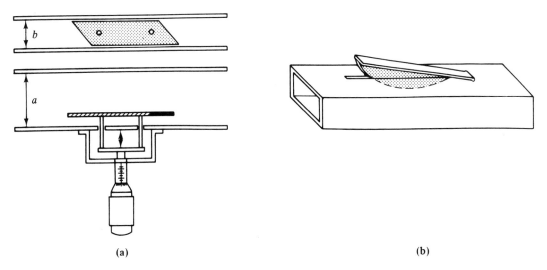

FIGURE 13.9. Waveguide attenuators.

and, as shown in the next section, the slot position is such that radiation is minimized. However, some radiation does occur, and this type is not used for accurate work.

13.2.5 Currents in Walls

As already stated, the traveling-wave pattern is time-invariant; that is, the pattern appears to keep its shape as it moves down the guide. At any given cross section of the guide, however, the magnetic field (and the electric field) appears to vary in time as the loops of alternate polarity sweep by. This gives rise to induced currents in the walls of the guide, at right angles to the magnetic field. The currents for the TE_{10} mode are as shown in Fig. 13.10(a). This pattern moves down the guide along with the field pattern at the phase velocity. It is emphasized that the pattern moves at the phase velocity, not the current (since that would imply electrons moving faster than the speed of light!). In effect, the current pattern builds up and decays as the TE wave sweeps by. An analogy can be drawn with a sea wave moving along a sea wall, creating a splash (Figure 13.10(b)). The splash travels at the phase velocity (which is faster than the velocity at which the wave approaches the wall), and people standing at the edge of the sea wall will move back as the splash passes them, so the ripple in the line also moves at the phase velocity.

Knowing the current pattern is important, as it enables slots to be correctly positioned in the walls, to serve various purposes. Slots that do not noticeably interrupt the current flow are termed *nonradiating*, since they result in minimum disturbance of the internal fields, and therefore little electromagnetic energy leaks through them. The positions of two nonradiating slots,

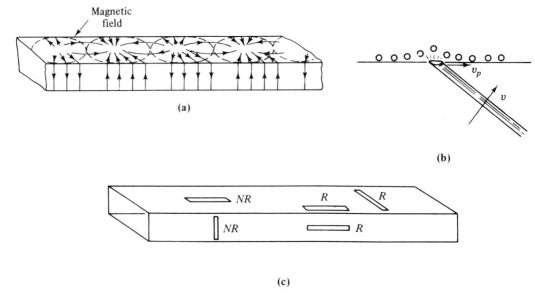

FIGURE 13.10. (a) Currents in walls for the TE_{10} mode; (b) sea wave analogy illustrating phase velocity; (c) nonradiating (NR) and radiating (R) slots.

labeled NR, are shown in Fig. 13.10(c). It will be seen that a nonradiating slot can be placed along the center of the broad wall of the guide, and use is made of this in the attenuator of Fig. 13.9(b). Another application is the slotted waveguide, which enables standing waves to be measured, the technique being similar to that described in Section 12.10 for transmission lines.

Slots that do interrupt the current flow, such as those labeled R in Fig. 13.10(c), are termed *radiating* slots. These produce maximum disturbance of the internal fields, which results in energy being radiated. Radiating slots form the basis of slot antennas.

13.2.6 Contacts and Joints

The properties of a short-circuited section can be used to provide an electrical short circuit without the necessity of providing a solid mechanical contact at the point of short circuit. This principle is incorporated into the design of some types of flanges used for coupling guide sections together and in the design of movable short-circuiting contacts. Two $\frac{1}{4}\lambda_g$ transformations take place, as shown in Fig. 13.11(a). The top $\frac{1}{4}\lambda_g$ section transforms the solid short circuit at the top to an open circuit at the junction of the two $\frac{1}{4}\lambda_g$ sections. Here a mechanical joint occurs, but since this is a high-impedance point the currents are small and the joint resistance is not important. The second $\frac{1}{4}\lambda_g$ section transforms the open circuit back to a short circuit at the entry point. Application of the principle to a coupling flange is illustrated in

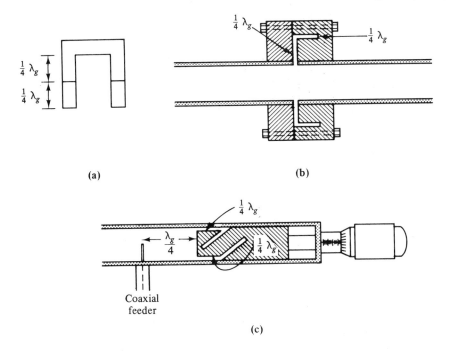

(a)　　　　　　　　　　　　　　　　**(b)**

Coaxial
feeder

(c)

FIGURE 13.11. Two $\frac{1}{4}\lambda_g$ transforming sections in series; (b) a coupling flange utilizing (a); (c) a movable short-circuiting contact utilizing (a), The contact illustrated provides a $\frac{1}{4}\lambda_g$ shorted section for the coaxial feeder.

Fig. 13.11(b) and to a movable short circuit in Fig. 13.11(c). In the latter figure, a coaxial feeder is shown by way of example, the $\frac{1}{4}\lambda_g$ short-circuited section performing the same function as that shown in Fig. 13.7(b).

13.2.7 Reactive Stubs

Equivalent reactive elements can be introduced into a waveguide in a variety of ways, a common method being to use an adjustable screw stub as shown in Fig. 13.12. When the stub is only a short way in, as shown in Fig. 13.12(a), it acts as a capacitor, as it produces an increase in the electric flux density of the wave in the vicinity of the stub.

With the screw all the way in, such that it forms a post between the top and bottom of the guide (Figure 13.12(b)), a path is provided for induced currents which set up a magnetic field; such a post therefore acts as an inductor. An equivalent series *LC* circuit is formed when the screw is sufficiently far in for both components to be significant but neither dominant, such as is shown in Fig. 13.12(c). Screw stubs are used singly and in groups of two and three, to provide matching devices between a waveguide and load.

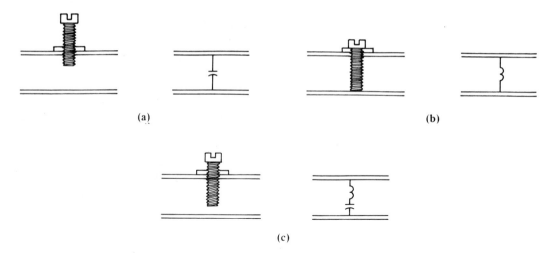

(a)   (b)

(c)

FIGURE 13.12. Reactive stubs: (a) capacitive stub; (b) inductive stub; (c) series LC stub.

13.3 OTHER MODES

Higher modes can occur in waveguides: an example of the TE_{20} mode is sketched in Fig. 13.13(a). *Transverse magnetic* (TM) *modes* can also occur, as the TM_{11} mode sketched in Fig. 13.13(b) illustrates. It can be shown that the

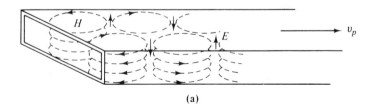

(a)

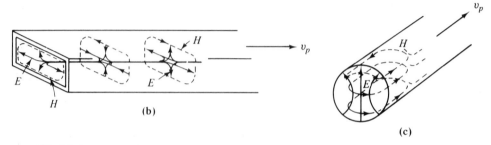

(b)   (c)

FIGURE 13.13. Other waveguide modes: (a) the TE_{20} mode; (b) the TM_{11} mode; (c) the TE_{11} circular waveguide mode.

cutoff wavelength for TE_{mn} and TM_{mn} modes in general is given by

$$\left(\frac{1}{\lambda_c}\right)^2 = \left(\frac{m}{2a}\right)^2 + \left(\frac{n}{2b}\right)^2 \tag{13.24}$$

where m and n are integers. Equation (13.24) reduces to $\lambda_c = 2a$ for $m = 1$ and $n = 0$, which is the condition already introduced in Eq. (13.7) for the TE_{10} mode.

The TM_{11} mode is the lowest TM mode that can occur, as the boundary conditions exclude the TM_{10} mode. For transmission purposes only the TE_{10} mode is used, and the guide dimensions are chosen, in conjunction with the input frequency, to cut off all but the dominant mode (TE_{10}). For example, for a standard guide WR 90, the dimensions are

Outside walls:	1.000×0.500 inch
Wall thickness:	0.050 inch

The inside wall dimensions are, therefore,

$$a = 2.286 \text{ cm}$$
$$b = 1.016 \text{ cm}$$

The cutoff wavelengths for some of the various modes are

$$E_{10}: \qquad \lambda_c = 2a$$
$$= 4.572 \text{ cm}$$
$$TE_{20}: \qquad \lambda_c = a$$
$$= 2.286 \text{ cm}$$
$$TE_{01}: \qquad \lambda_c = 2b$$
$$= 2.032 \text{ cm}$$
$$\left.\begin{array}{c} TE_{11} \\ TM_{11} \end{array}\right\}: \qquad \left(\frac{1}{\lambda_c}\right)^2 = \left(\frac{1}{2a}\right)^2 + \left(\frac{1}{2b}\right)^2$$

Therefore, $\qquad\qquad\qquad \lambda_c = 1.857 \text{ cm}$

These results are tabulated as follows:

Mode	λ_c(cm)	f_c(GHz)
TE_{10}	4.572	6.56
TE_{20}	2.286	13.1
TE_{01}	2.032	14.8
$\left.\begin{array}{c} TE_{11} \\ TM_{11} \end{array}\right\}$	1.857	16.2

The recommended operating frequency range for the WR90 guide is 8.20 to 12.40 GHz, and it will be seen that only the TE_{10} mode will be excited. Of course, if a higher frequency is fed into the guide, higher modes will be excited; for example, if a frequency of 15 GHz is used, the first three modes may exist simultaneously in the guide. For all these modes, including the circular mode described in the next paragraph, Eq. (13.7) applies.

Circular guides also support waveguide modes; the TE_{11} circular mode is sketched in Fig. 13.13(c). (The subscripts here denote circumferential and radial variations of the fields, which are much more complex than the modes in the rectangular guide; only the TE_{11} circular mode will be considered here.) The cutoff wavelength is related to the diameter of the guide, the value derived from theory being

$$\lambda_c = 1.71d \qquad (13.25)$$

where d is the diameter of the circular guide.

Circular guides have special properties which enable them to be used for rotating joints, and the TE_{10} mode has the unusual characteristic of attenuation becoming less as the frequency is increased, which is attractive for transmissions at the higher microwave frequencies. However, the mechanical problems of making and maintaining precise dimensions in a circular guide are much more formidable than in a rectangular guide, and the latter finds greater use for transmission purposes.

13.4 PROBLEMS

1. Determine λ_c, λ_g, λ, v_p, and v_g for standard waveguide WR62 for which the outside dimensions are 0.702×0.391 inch and the wall thickness is 0.040 inch, when the exciting frequency is 10 GHz.

2. Determine the TE_{10} wave impedance for the WR62 guide of Problem 1.

3. Explain how standing waves can be produced in a waveguide. What is the spacing, in terms of wavelength, between successive minima along the standing-wave pattern?

4. By analogy with the transmission-line theory presented in Section 12.8, determine the reflection coefficient for the electric field of a TE_{10} wave of wave impedance 680 Ω terminated as shown in Fig. 13.7(b), the sheet resistivity of the load being 500 Ω/square.

5. Also by analogy to the transmission theory presented in Section 12.9, determine the electric field standing-wave ratio for the guide conditions of Problem 4.

6. What is meant by sheet resistivity, and how does this differ from the resistivity of a material? What are the units for sheet resistivity?

7. Illustrate on a Smith chart the transformations that take place in the $\frac{1}{2}\lambda_g$ short-circuited section shown in Fig. 13.11(a).

8. Explain the difference between radiating and nonradiating slots in a TE_{10} mode waveguide.

9. A standard waveguide, WR187, has the following dimensions: outside walls 2.000×1.000 inches, wall thickness 0.064 inch. Determine the highest frequency than can be transmitted if only the TE_{10} mode is permitted. What modes can exist in it if it is excited by a wave of frequency 6.00 GHz?

CHAPTER 14

Radio-Wave Propagation

14.1 INTRODUCTION

Radio communications use electromagnetic waves propagated through the earth's atmosphere or space to carry information over long distances without the use of wires. Radio waves with frequencies ranging from about 100 Hz in the ELF band to well above 300 GHz in the EHF band have been used for communications purposes, and more recently radiation in and near the visible range (near 1000 THz, or 10^{15} Hz) have also been used. Figure 14.17 shows the frequency-band designations in common use.

Some of the basic properties of a *transverse electromagnetic* (TEM) wave are described in Appendix B. Although the electric and magnetic fields exist simultaneously, in practice, antennas are designated to work through one or other of these fields. Antennas are described in Chapter 15. Basically, in order to launch an electromagnetic wave into space, an electric charge has to be accelerated, which in practice means that the current in the radiator must change with time (e.g., be alternating). In this chapter, sinusoidal or cosinusoidal variations will be assumed unless stated otherwise.

14.2 PROPAGATION IN FREE SPACE

14.2.1 The Mode of Propagation

Consider first an average power P_T, assumed to be radiated equally in all directions (isotropically). This will spread out spherically as it travels away from the source, so that at distance d, the power density in the wave, which is the power per unit area of wavefront, will be

$$P_{Di} = \frac{P_T}{4\pi d^2} \qquad W/m^2 \qquad (14.1)$$

This is so because $4\pi d^2$ is the surface area of the sphere of radius d, centered on the source. P_{Di} stands for isotropic power density.

It is known that all practical antennas have directional characteristics; that is, they radiate more power in some directions at the expense of less in others. The directivity gain is the ratio of actual power density along the main axis of radiation of the antenna to that which would be produced by an isotropic antenna at the same distance fed with the same input power. Let G_T be the *maximum* directivity gain of the transmitting antenna; then the power density along the direction of maximum radiation will be

$$P_D = P_{Di}G_T$$
$$= \frac{P_T G_T}{4\pi d^2} \qquad (14.2)$$

A receiving antenna can be positioned so that it collects maximum power from the wave. When so positioned, let P_R be the power delivered by the antenna to the load (receiver) under matched conditions; then the antenna can be considered as having an effective area (or aperture) A_{eff}, where

$$P_R = P_D A_{eff}$$
$$= \frac{P_T G_T}{4\pi d^2} A_{eff} \qquad (14.3)$$

It can be shown that for any antenna, the ratio of maximum directivity gain to effective area is

$$\frac{A_{eff}}{G} = \frac{\lambda^2}{4\pi} \qquad (14.4)$$

Here, λ is the wavelength of the wave being radiated. Letting G_R be the maximum directivity gain of the receiving antenna, we have from Eqs. (14.3) and (14.4),

$$\frac{P_R}{P_T} = G_T G_R \left(\frac{\lambda}{4\pi d^2} \right)^2 \qquad (14.5)$$

This is the fundamental equation for free-space transmission. Usually it is expressed in terms of frequency f, in MHz, and distance d, in kilometers. As shown in Appendix B, $\lambda f = c$, and on substituting this in Eq. (14.5) and doing the arithmetic, which is left as an exercise for the reader, the result obtained is

$$\frac{P_R}{P_T} = G_T G_R \frac{0.57 \times 10^{-3}}{(df)^2} \tag{14.6}$$

By expressing the power ratios in decibels, Eq. (14.6) can be written as

$$\left(\frac{P_R}{P_T}\right)_{dB} = (G_T)_{dB} + (G_R)_{dB} - (32.5 + 20\log_{10}d + 20\log_{10}f) \tag{14.7}$$

The third term in parentheses on the right-hand side of Eq. (14.7) is the loss, in decibels, resulting from the spreading of the wave as it propagates outward from the source. It is known as the transmission path loss, L. Thus,

$$L = (32.5 + 20\log_{10}d + 20\log_{10}f)_{dB} \tag{14.8}$$

where d is in kilometers and f in megahertz.

Equation (14.7) then becomes

$$\left(\frac{P_R}{P_T}\right)_{dB} = (G_T)_{dB} + (G_R)_{dB} - (L)_{dB} \tag{14.9}$$

Example 14.1   In a satellite communications system, free-space conditions may be assumed. The satellite is at a height of 36,000 km above earth, the frequency used is 4000 MHz, the transmitting antenna gain is 15 dB, and the receiving antenna gain is 45 dB. Calculate (a) the free-space transmission loss, and (b) the received power when the transmitted power is 200 W.

Solution   (a)
$$L = 32.5 + 20\log_{10}36{,}000 + 20\log_{10}4000$$
$$= 196 \text{ dB}$$

(b)
$$\left(\frac{P_R}{P_T}\right)_{dB} = 15 + 45 - 196$$
$$= -136 \text{ dB}$$

This is a power ratio of 0.25×10^{-13}, and since $P_T = 200$ W,

$$P_R = 200 \times 0.25 \times 10^{-13}$$
$$= 5 \times 10^{-12}\text{W} = 5 \text{ pW}$$

Frequently, it is required to know the electric field strength of the wave at the receiving antenna. In Appendix B, E is given by Eq. (B.12) in terms of power density P_D and wave impedance Z_0 as

$$E = \sqrt{Z_0 P_D} \qquad (14.10)$$

Also, Z_0 is given by Eq. (B.10) as

$$Z_0 = \sqrt{\frac{\mu}{\varepsilon}} \qquad (14.11)$$

The free-space values are as follows: $\mu = \mu_0 = 4\pi \times 10^{-7}$ H/m; $\varepsilon = \varepsilon_0 = 8.854 \times 10^{-12}$ F/m. Substituting these in Eq. (14.11) gives

$$Z_0 = 120\pi \ \Omega \qquad (14.12)$$

The field strength may now be found by substituting Eqs. (14.2) and (14.12) in Eq. (14.10):

$$E = \frac{\sqrt{30 P_T G_T}}{d} \qquad \text{V} / \text{m} \qquad (14.13)$$

This is the fundamental equation which gives the field strength at the receiving antenna, for free-space propagation conditions. A receiving antenna has an effective length ℓ_{eff} (analogous to effective area) such that the open circuit EMF of the antenna E_s is given by

$$V_s = E\ell_{\text{eff}} \qquad (14.14)$$

Effective length is discussed in Section 15.9.

Example 14.2   Calculate the open-circuit voltage induced in a $\frac{1}{2}\lambda$ dipole when 10W at 150 MHz is radiated from another $\frac{1}{2}\lambda$ dipole 50 km distant. The antennas are positioned for optimum transmission and reception.

Solution
$$\lambda = \frac{3 \times 10^8}{150 \times 10^6}$$
$$= 2 \text{ m}$$

In Chapter 18 it is shown that, for a $\frac{1}{2}\lambda$ dipole, the maximum gain is 1.64:1, and the effective length is λ/π.
Therefore,

$$V_s = \frac{\sqrt{30 \times 10 \times 1.64}}{50 \times 10^3} \cdot \frac{2}{\pi}$$
$$= 282 \ \mu\text{V}$$

Equation (14.13) is sometimes expressed in terms of the field strength at unit distance, E_0. Thus, at $d = 1$ m,

$$E = E_0 = \frac{\sqrt{30P_TG_T}}{1} \quad \text{V/m} \tag{14.15}$$

and therefore Eq. (14.13) can be written as

$$E = \frac{E_0}{d} \quad \text{V/m} \tag{14.16}$$

Note that both E and E_0 are in units of V/m, although Eq. (14.16) may tend to suggest that E was in units of (V/m)/m, or V/m^2. Equation (14.16) really expresses a proportionality, and written in full it would be

$$E = E_0 \times \frac{1\,(\text{m})}{d\,(\text{m})} \quad \text{V/m} \tag{14.17}$$

14.2.2 Microwave Systems

Microwave radio systems operating at frequencies above 1 GHz propagate mainly in a line-of-sight or free-space mode, whether they are on the ground or in satellite systems. Since the 1950s, microwave radio systems have become the workhorses of long-distance telephone communications systems. These systems provide the needed transmission bandwidth and reliability to allow the transmission of many thousands of telephone channels as well as several television channels over the same route and using the same facilities. Carrier frequencies in the 3–12 GHz range are used. Since microwaves travel only on line-of-sight paths, it is necessary to provide repeater stations at about 50-km intervals. This makes the equipment costs for such a system very large, but this is more than made up for by the increased channel capacity. Transmitter output powers are low (they may be less than 1 watt), because highly directional high-gain antennas are used.

Figure 14.1(a) shows the equipment needed to provide one channel of a microwave system. It consists of two terminal stations and one or more repeater stations. At the sending terminal, the inputs comprising several hundred telephone channels and/or a television channel are frequency-multiplexed within the baseband bandpass of 6 MHz. The baseband frequency modulates a 70-MHz IF signal, which is then up-converted to the microwave output frequency f_1 within the 4-GHz band. This signal is amplified and fed through a directional antenna toward a repeater station some 50 km distant. At the repeater station, the signal at f_1 is received on one antenna pointed toward the originating station, down-converted to the IF, amplified, and up-converted to a new frequency f_2 for retransmission toward the receiving terminal station.

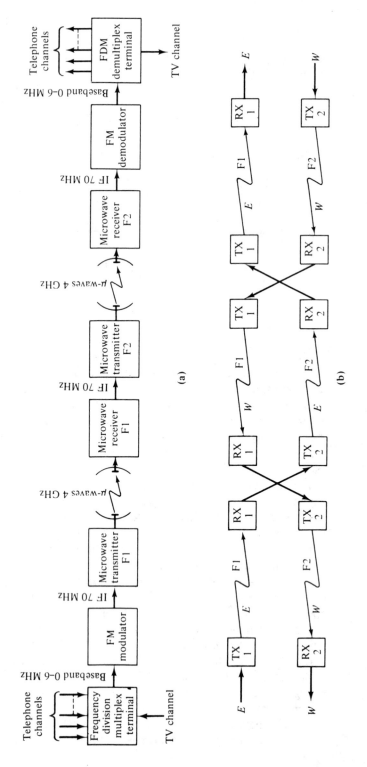

FIGURE 14.1. A microwave relay system: (a) a one-way channel showing the equipment used on the route; (b) a two-way channel pair showing frequency interchanges at intermediate repeater stations.

When the signal is passed through a chain of several repeaters, alternate links in the chain use alternate frequencies so that retransmitted energy at a repeater station does not feed back into its own receiver.

With two frequencies in use, alternating at each repeater, two channels, one in each direction, can be provided, as illustrated in Fig. 14.1(b). In some microwave systems several two-way channel pairs are provided, and a more complex system of frequency switching is used at the repeater stations.

At the receiving terminal station, the signal is down-converted to the IF and then demodulated to recover the baseband signal. This baseband signal is then demultiplexed to recover the individual telephone or television channel signals.

Terminal stations use two antennas, one for receiving and one for transmitting. The system may have several transmitters and receivers, but they all use the same antennas. Repeater stations are provided with two antennas pointing in each direction, for a total of four. Repeater station sites are chosen at high points, such as on hilltops or tall buildings, and sturdy towers provide additional elevation to maximize the distance between stations.

One-way single-channel microwave systems using portable terminal and repeater stations are frequently used for remote television pickup of special events and for other temporary installations such as the testing of new microwave routes. This equipment is often mounted in mobile vans outfitted with telescopic towers that can be quickly erected.

14.3 TROPOSPHERIC PROPAGATION

14.3.1 The Mode of Propagation

The troposphere is the region of the earth's atmosphere immediately adjacent to the earth's surface and extending upward for some tens of kilometers (i.e., the region in which we normally live and travel, including high-flying jet planes). In this region, the free-space conditions are modified by (1) the surface of the earth, and (2) the earth's atmosphere.

Considering first the effect of the earth's surface, a much-simplified model has been developed which successfully describes electromagnetic propagation over a wide range of practical circumstances and is illustrated in Fig. 14.2(a). In this simplified picture the earth is assumed to be flat, and the space wave reaching the receiver has two components: the direct wave, which follows a ray path s_d, and a ground-reflected wave, which follows a ray path s_i. The reflected wave travels a greater distance than the direct wave, and although this has negligible effect on the amplitude, it does introduce a phase difference which is highly significant. Let Δs be the path difference; then, since a phase angle of 2π radians corresponds to a path length of one wavelength (λ), the phase angle corresponding to Δs is

$$\phi_s = \frac{2\pi}{\lambda}\Delta s \tag{14.18}$$

(Note that $2\pi/\lambda$ is simply the phase-shift coefficient introduced in Eq. (12.23)).

The problem now is to find Δs, and this can be solved from the geometry of Fig. 14.2(b). h_T is the height of the transmitting antenna above ground, h_R is the height of the receiving antenna, and d is the distance along the flat earth between the two. From triangle FBD (Fig. 14.2(b)),

$$s_i^2 = (h_T + h_R)^2 + d^2$$

From triangle ABC,

$$s_d^2 = (h_T - h_R)^2 + d^2$$

Therefore,

$$s_i^2 - s_d^2 = (h_T + h_R)^2 - (h_T - h_R)^2$$
$$= 4h_T h_R \tag{14.19}$$

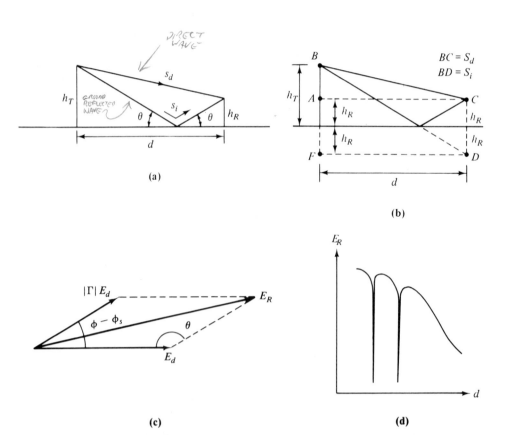

FIGURE 14.2. (a) Simplified model for tropospheric propagation ray paths; (b) geometry used to determine phase difference; (c) the field strength phasor diagram at the receiver; (d) sketch of field strength modulus versus distance, Eq. (14.26).

Also, applying the formula for the difference of two squares,

$$s_i^2 - s_d^2 = (s_i + s_d)(s_i - s_d)$$
$$= (s_i + s_d) \Delta s$$

For most practical purposes, $s_i \cong s_d \cong d$. Therefore,

$$s_i^2 - s_d^2 \cong (2d) \Delta s \qquad (14.20)$$

Equating Eqs. (14.19) and (14.20), the result is

$$2d \, \Delta s = 4h_T h_R$$

Therefore,

$$\Delta s = \frac{2h_T h_R}{d} \qquad (14.21)$$

Substituting this in Eq. (14.18) gives

$$\phi_s = \frac{4\pi h_T h_R}{\lambda d} \qquad (14.22)$$

This is not the only phase change which takes place. Reflection at the earth's surface also affects the amplitude and phase of the reflected wave relative to the direct wave. The nature of the reflection depends in a complicated way on the constitution of the reflection surface, the angle of incidence of the wave, and whether the wave is horizontally or vertically polarized (meaning whether the electric field of the wave is horizontal or vertical, respectively, with respect to the earth's surface).

Let the electric field reflection coefficient at the point of reflection be represented by $\Gamma = |\Gamma|/\underline{\phi}$, analogously to the voltage reflection coefficient introduced in Section 12.8. The amplitude of the reflected wave at the receiving point will be $|\Gamma| E_d$, where E_d is the amplitude of the direct wave. The phase of the reflected wave relative to the direct wave is $(\phi - \phi_s)$, since it is known that ϕ_s is a phase lag because of the longer path length for the reflected wave, and a positive value of ϕ is assumed to be a phase lead. The phasor diagram for the field strength at the receiving point is therefore as shown in Figure 14.2(c). The angle θ is equal to $180° - (\phi - \phi_s)$, and applying the cosine rule to the triangle to find the resultant field strength gives

$$E_R^2 = E_d^2 + |\Gamma|^2 E_d^2 - 2|\Gamma| E_d^2 \cos \theta$$
$$= E_d^2 (1 + |\Gamma|^2 + 2|\Gamma| \cos(\phi - \phi_s))$$

Therefore,

$$E_R = E_d \sqrt{1 + |\Gamma|^2 + 2|\Gamma| \cos(\phi - \phi_s)} \qquad (14.23)$$

For a wide range of conditions it is found that the reflection coefficient is equal to -1, that is, the reflection occurs with negligible amplitude change but $180°$

phase change. For this particular condition, eq. (14.23) reduces to

$$E_R = E_d\sqrt{1 + 1 + 2\cos(180 - \phi_s)}$$
$$= E_d\sqrt{2(1 - \cos\phi_s)} \tag{14.24}$$

Using the identity $2\sin^2 A = 1 - \cos 2A$ allows this to be written as

$$E_R = 2E_d\sin\left(\frac{\phi_s}{2}\right) \tag{14.25}$$

Equation (14.16) gives E_d as

$$E_d = \frac{E_0}{d}$$

Thus, substituting Eqs. (14.16) and (14.22) in Eq. (14.25) results in

$$E_R = \frac{2E_0}{d}\sin\left(\frac{2\pi h_T h_R}{\lambda d}\right) \tag{14.26}$$

The variation of E_R with d is sketched in Fig. 14.2(d). The sharp dips toward zero occur where the sine term goes to zero. When d is large, so that the angle within the brackets of Eq. (14.25) is small (less than about 0.5 rad), the approximate form

$$E_R \cong \frac{2E_0}{d}\left(\frac{2\pi h_T h_R}{\lambda d}\right)$$

or

$$E_R = E_0\frac{4\pi h_T h_R}{\lambda d^2} \tag{14.27}$$

may be used.

Example 14.3 In a VHF mobile radio system, the base station transmits 100 W at 150 MHz, and the antenna is 20 m above ground. The transmitting antenna is a $\frac{1}{2}\lambda$ dipole for which the gain is 1.64. Calculate the field strength at a receiving antenna of height 2 m at a distance of 40 km.

Solution
$$\lambda = \frac{300 \times 10^6}{150 \times 10^6}$$
$$= 2\text{ m}$$
$$E_0 = \sqrt{30 \times 100 \times 1.64} \qquad (\text{Eq. (14.15)})$$
$$= 70\text{ V/m}$$

The angle $2\pi h_T h_R/\lambda d$ is very much less than 0.5 rad, as an approximate calculation will show. Therefore, Eq. (14.27) can be used:

$$E_R = \frac{70 \times 4 \times \pi \times 20 \times 2}{2 \times (40 \times 10^3)^2}$$

$$= 11 \ \mu V/m$$

Equation (14.27) shows the importance of antenna heights. For example, doubling the height of the base-station antenna in the previous example will double the field strength at the receiving point. Note also that the field strength is proportional to E_0, which, in turn, is proportional to the square root of the transmitted power. Doubling the power results in only a $\sqrt{2}$ increase in field strength.

Equations (14.26) and (14.27) also apply when a transmission takes place from the h_R antenna and is received at the h_T antenna: it is only necessary to change E_0 to the new value determined by the gain of the antenna at h_R and the power transmitted from there.

As the distance d increases, it becomes necessary to take into account the curvature of the earth. Reflection from the curved surface reduces both the amplitude of the reflected wave and the phase difference. These two effects tend to offset each other and the resultant amplitude does not vary rapidly as a result.

14.3.2 Radio Horizon

The curvature of the earth has a more important effect in that it presents a horizon that limits the range of transmission. This range is greater than the optical range because the effect of the earth's atmosphere is to cause a bending of the radio wave, which carries it beyond the optical horizon. Figure 14.3(a) shows a typical radio-wave ray path, and Fig. 14.3(b) shows how the path can be considered straight by assigning a greater radius to the earth than it actually has. For standard atmospheric conditions the increase in radius has been worked out at $\frac{4}{3}$, so that

$$a' = \tfrac{4}{3}a \tag{14.28}$$

where a is the earth's actual radius and a' is the fictitious radius which accounts for refraction. From Fig. 14.3(b),

$$(a')^2 + d_1^2 = (a' + h_T)^2$$

Therefore,

$$d_1^2 = 2a'h_T + h_T^2 \tag{14.29}$$

But since $a' \gg h_T$

$$d_1^2 \approx 2a'h_T \qquad (14.30)$$

Similarly,

$$d_2^2 \approx 2a'h_R \qquad (14.31)$$

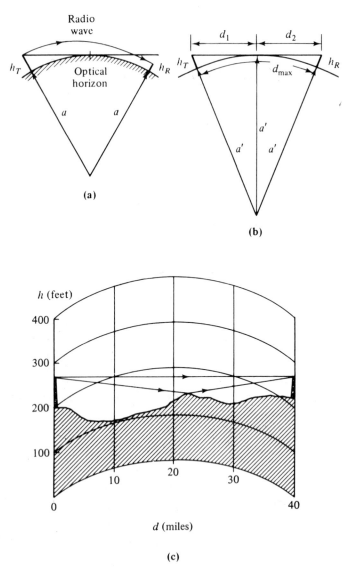

FIGURE 14.3. (a) Curvature of ray path resulting from change of the refractive index of air; (b) equivalent straight-line ray path for fictitious earth's radius a'; (c) example of a contour map for radio path planning.

sions. These effects are discussed in connection with satellite communications in Section 19.6.

14.3.6 VHF / UHF Radio Systems

Propagation in the VHF and UHF bands between 30 MHz and 3 GHz takes place in the tropospheric mode.

The major use of two-way radio communications in the VHF and UHF bands is communications between a fixed base station and several mobile units, located on vehicles, ships, or aircraft in the frequency band 30–470 MHz. Typical applications are in control-tower-to-aircraft communication at airports, fire departments, ship control within harbors, police departments, armed-forces field operations, pipeline and transmission line maintenance, highway maintenance, taxicab and delivery vehicle dispatch, and personnel paging systems. Since these systems operate in frequencies above 30 MHz, their range of operation is limited to within the line-of-sight horizon of the base station (see Section 14.3.2), or that much further again if a repeater station is used. Large obstacles such as hills or tall buildings in an urban zone create shadows and odd reflection patterns which make complete coverage of the zone from a single base station difficult. For this reason, and to increase the horizon somewhat, it is usual practice to locate the base-station antenna on top of a high hill or building to gain additional height.

A limited number of channel assignments is available within the spectrum, mostly within the bands 148 to 174 MHz and 450 to 470 MHz. FM operation is preferred, and the maximum permissible channel spacing for this service has been progressively reduced from 120 kHz to the 15 kHz presently allowed, so that more channels can be assigned. Because of the narrow bandwidths used, the transmitters and receivers must be very stable, and must maintain their operating frequency within ± 5 parts per million. Crystal control is a must if this type of stability is to be realized.

Dispatch systems for automobiles are usually required to cover as much area as possible, and omnidirectional vertically polarized antennas are usually used to accomplish this, both at the base station and in the mobile units. In some applications, such as pipeline and highway maintenance systems, the field of operations is strung out in a line over many miles, and for these systems vertically polarized multi-element Yagi antennas aimed along the path are frequently used. This provides little coverage off the sides, but does provide better coverage along the line up to the horizon. The antennas used on the vehicles are nearly always short ground-whip antennas mounted on top of the vehicle. The longer whips used for the 50-MHz VHF band are not as popular.

Transmitter power in both the mobile units and the base-station units is usually limited to about 150 W, mainly because of the limited power available from the vehicle system. Voltage supplies for mobile equipment range from

12 V nominal for automobiles, 28 V and 48 V for aircraft, and 48 V for railway locomotives. The base stations are usually operated directly from the 110-V, 60-Hz power mains, although for some applications backup battery power is also provided in case of power failures.

The transceivers are designed to alternately transmit and receive on the same frequency. For aircraft and ship control use, and such systems as police and fire operation, the units may be designed to operate on one of several channels, with manual switching between channels provided. Each mobile unit is provided with a control head which is usually separate from the main chassis, conveniently located near the operator. The control head provides a power on/off switch, audio volume control, muting threshold control, and a handset containing a microphone, a telephone receiver, and a push-to-talk switch. The base station may be self-contained and directly connected to an antenna near the operator's location, but usually the base-station transceiver unit is located with the antenna at a convenient high point and the operator is provided with a remote-control console. Connection between the base station and the control console is made by means of a pair of wires if the distance is short, or a leased telephone line if the distance is considerable. A typical dispatch system is shown in Fig. 14.5(a).

Often it is found that it is impossible to cover the desired area from the single base-station location. In this case one or more additional base stations can be established. These may be connected back to the operator's console over wire lines and operated independently of each other by the operator. Alternatively, a radio repeater link can be established. This requires the use of a second frequency for the link between the main base station and the repeater station. Figure 14.5(b) shows an arrangement that might be used. Under normal operation, the base operator can communicate over the local base station on Frequency 1 with any mobile unit within coverage area 1. When it is necessary to reach a vehicle in coverage area 2, he may turn off the local-area base station and turn on the repeater link on Frequency 2. Now when the base operator transmits, he transmits on F2 toward the repeater. The repeater receives control on F2 and retransmits on F1 to the extended-coverage area. When a mobile unit in the extended area transmits, it is received at the repeater on F1 and retransmitted on F2 toward the base station.

When a system with several fixed stations is being operated in a repeater mode, the base operator must continuously monitor all the incoming signals from all the repeaters. If a mobile unit should be in an overlap area between two repeater stations, he may "key" both repeaters, causing interference at the control console. The operator must be able to purposely disable all but one repeater during a conversation. This can be accomplished by sending a coded tone signal to turn off the desired repeater stations, and then turn them back on when the conversation is complete. The detection characteristics of FM are such that a receiver located in the presence of two co-channel (same frequency) transmissions (such as f_1 base and f_1 repeater in Fig. 14.5(b)) will suppress the weaker signal. This is known as the *capture effect*.

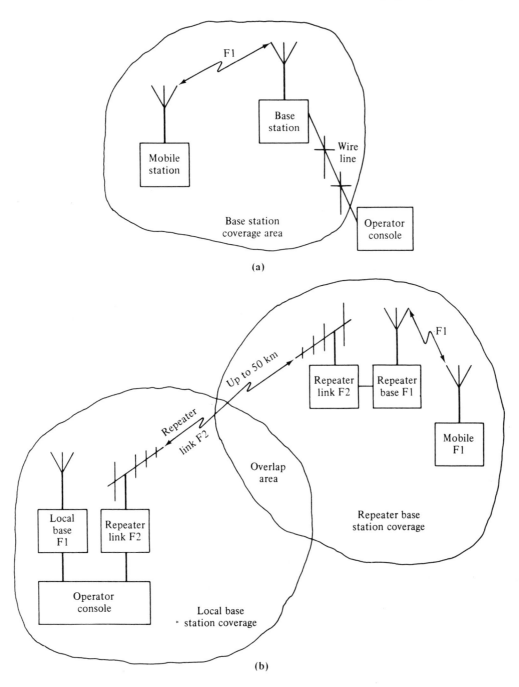

FIGURE 14.5. (a) A simple mobile dispatch system; (b) a dispatch system with a repeater station to extend coverage area.

14.4 IONOSPHERIC PROPAGATION

14.4.1 Ionospheric Layers

The upper reaches of the earth's atmosphere are ionized (i.e., electrons are detached from atmospheric gas atoms), mainly as a result of receiving ultraviolet radiation from the sun, although other sources, such as cosmic rays, also contribute.

A state is reached where the free-electron density is maintained almost constant, the ionization rate being balanced by the recombination rate of electrons with positive ions. Clearly, the electron density will vary between day and night conditions, and will also show a variation between winter and summer, as the ionization rate is dependent on solar radiation.

Various peaks are observed in electron density corresponding to the heights at which various gases settle in the upper atmosphere. The layers follow a meteorological classification, being known as the *C layer*, *D layer*, *E layer*, and *F₁ and F₂ layers*. Figure 14.6(a) shows some recently published results for typical electron-density distribution with height. It will be seen that for nighttime conditions, only the F₂ layer remains. This is because in the lower layers, collision processes in the denser gas atmosphere cause a more rapid recombination rate compared to the less dense F₂ region.

Figure 14.6(a) also shows the main ionizing radiations, and along the top of the graph is shown the *plasma frequency f_N*, an important parameter in radio communications via the ionosphere.

14.4.2 Plasma Frequency and Critical Frequency

When an electromagnetic wave enters an ionized region at vertical incidence, as shown in Fig. 14.6(b), the electric field acts as a force on the charged particles (electrons and ions), resulting in charge movement and hence current flow. Although a positive ion will carry the same magnitude of charge as an electron, it is more than 1000 times as massive and, therefore, its velocity will be correspondingly smaller and the ionic contribution to the current can be neglected.

The electron cloud will oscillate in the electric field of the wave, but with a phase retardation of 90° (for a sinusoidal wave) because of the electron mass inertia. This motion of the electron cloud produces a space current; in addition, the electric field has its own capacitive displacement current which leads the field by 90°. The space current is therefore in phase opposition to the displacement current, and it appears to reduce the relative permittivity (dielectric constant) of the ionized medium, which is then given by

$$\varepsilon_r = \left(1 - \frac{Nq_e^2}{\varepsilon_o m \omega^2}\right) \qquad (14.35)$$

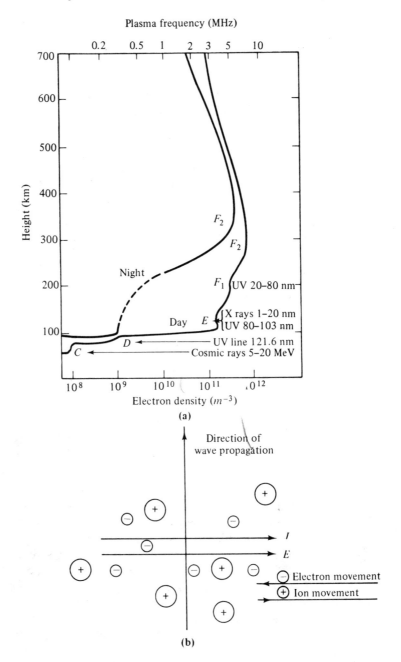

FIGURE 14.6. (a) Typical electron density distributions for summer noon and midnight conditions at mid-latitudes, and the principal ionizing radiations at each level — plasma frequency f_N (Hz) is related to electron density (m^{-3}) by $f_N = 9\sqrt{N}$; the plasma frequency at the peak of a layer is called the critical frequency (f_0) (Reprinted by permission of the IERE); (b) Electric field E and electric current density J vectors in an ionospheric layer.

where

$$N = \text{Electron density, } m^{-3}$$

$$m = \text{Electron rest mass} = 9.11 \times 10^{-31} \text{ kg}$$

$$q_e = \text{Electron charge magnitude} = 1.6 \times 10^{-19} \text{ C}$$

$$\omega = \text{Angular frequency of wave, rad/sec}$$

$$\varepsilon_o = \text{Permittivity of free space} = 8.854 \times 10^{-12} \text{ F/m}$$

The angular velocity of the wave can have a value that makes ε_r equal to zero, and this is termed the *plasma angular velocity*, ω_N. From Eq. (14.35), ω_N is seen to be

$$\omega_N^2 = \frac{Nq_e^2}{m\varepsilon_o} \tag{14.36}$$

or

$$f_N^2 = \frac{Nq_e^2}{(2\pi)^2 m\varepsilon_o} \tag{14.37}$$

Putting in the numerical values for the constants gives

$$f_N = 9\sqrt{N} \tag{14.38}$$

Equation (14.35) can now be rewritten as

$$\varepsilon_r = 1 - \frac{f_N^2}{f^2} \tag{14.39}$$

The significance of f_N is that when a wave of this frequency reaches a region of electron density N, as given by Eq. (14.38), the relative permittivity, as given by Eq. (14.39), is seen to be zero. This, in turn, means that the total displacement current density is zero, and hence the *effective* electric field is zero. This can be accounted for in terms of a reflected wave which exactly cancels the incident wave at the point of reflection. Of course, it should be possible to receive this reflected wave, which is exactly what happens in short-wave radio communications via the ionosphere.

The highest frequency wave that will be reflected from a given layer will be determined by the maximum electron density of that layer, and will be given by

$$f_0 = 9\sqrt{N_{\text{max}}} \tag{14.40}$$

Here f_0 is known as the *critical frequency*.

14.4.3 Phase and Group Velocities

An interesting point arises in connection with the velocities associated with the wave. The phase velocity has already been shown (Eq. (12.6)) to be

$$v_p = \frac{c}{\sqrt{\varepsilon_r}} \qquad (14.41)$$

Hence, in the ionosphere, when the wave reaches a height such that ε_r is zero, v_p becomes infinite! Now, as stated in Section 12.6, the energy in a wave travels at the group velocity v_g, and it can be shown that for an ionized layer

$$v_p v_g = c^2 \qquad (14.42)$$

(This was shown in Eq. (13.10) to hold for the special case of a waveguide.) The wavefront in the ionosphere is a step function that will propagate energy at the group velocity, and hence, from Eq. (14.42), when the phase velocity is infinite, the group velocity is zero, so that the energy ceases to be propagated upward.

14.4.4 Secant Law and Maximum Usable Frequency

When a wave enters an ionized layer at an oblique angle of incidence i, it follows a curved ray path (Fig. 14.7(a)). The phase velocity v_p at any given height can be determined by application of Snell's law of refraction for optics, the details of which will be omitted here. The result is that

$$\frac{\sin i}{c} = \frac{\sin r}{v_p} \qquad (14.43)$$

Here r is the angle of refraction at the height where v_p occurs, as shown in Fig. 14.7(a).

At the apex of the path, $r = 90°$, and therefore

$$v_p = \frac{c}{\sin i} \qquad (14.44)$$

It can be seen that as the angle of incidence i approaches zero, the phase velocity approaches infinity, in agreement with the results in the previous section.

As already shown,

$$v_p = \frac{c}{\sqrt{\varepsilon_r}}$$

It follows, therefore, that

$$\sqrt{\varepsilon_r} = \sin i$$

Substituting Eq. (14.39) for ε_r,

$$\left(1 - \frac{f_N^2}{f^2}\right) = \sin^2 i$$

from which the reader should be able to derive that

$$f = f_N \sec i \qquad (14.45)$$

This is known as the *secant law*. The highest frequency that can be used will be determined by N_{max}, and hence f_0; this is known as the *maximum usable frequency* (MUF):

$$\text{MUF} = f_0 \sec i \qquad (14.46)$$

It is possible for the wave to travel a considerable distance in the horizontal direction in the layer, as shown by the dotted ray path in Fig. 14.7(a), but irregularities in the ionized layer will eventually deflect the wave.

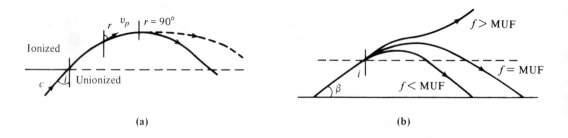

(a)

(b)

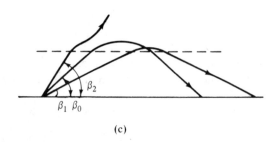

(c)

FIGURE 14.7. (a) Curved path of a wave in the ionosphere entering at oblique incidence; (b) ray paths for fixed angle of incidence and varying frequencies; (c) ray paths for fixed frequencies and varying angles of incidence.

Figure 14.7(b) shows what happens when the angle of incidence is kept fixed and the frequency is varied. At frequencies lower than the MUF, the wave is reflected from a lower point in the layer. At higher frequencies than MUF, refraction is insufficient to return the wave to earth and it escapes through the layer (it may, of course, be reflected from a more dense layer higher up).

Figure 14.8(c) shows what happens when the frequency is held constant while the angle of incidence is varied. Since in practice it would be the angle of elevation of the antenna beam (sometimes referred to as the takeoff angle) that would be varied, this is shown. The critical angle of elevation is shown as β_0, and at this angle, f becomes the MUF. At angles less than critical, the wave is reflected from a lower region than N_{max}, and at angles greater than critical, the wave escapes.

In the discussion so far, the curvature of the earth, and the ionosphere, have been ignored. This introduces little error where the ground distance between transmitter and receiver is less than about 1000 km. Beyond this, a correction factor has to be introduced in order to apply the secant law and the MUF equation.

14.4.5 Optimum Working Frequency

The frequency normally used for ionospheric transmissions is known as the *optimum working frequency* (OWF) and is chosen to be about 15% less than the MUF. It is desirable to use as high a frequency as possible, as the attenuation of the wave as it passes through the lower ionospheric layers is inversely proportional to the square of the frequency. This arises because as the electrons collide with the gas molecules, they lose kinetic energy which they gained from the passing wave. (Electrons that do not collide return the energy periodically through reradiation in correct phase.) The kinetic energy of an electron of mass m is $\frac{1}{2}mv^2$, where v is the velocity. The velocity will be proportional to the time which the electric field acts in any given direction, and this will be proportional to the periodic time of the wave. But the periodic time is equal to $1/f$ (Eq. (B.3)), where f is the frequency. Therefore, the kinetic energy lost is proportional to $(1/f^2)$.

The argument then is to use the highest possible frequency, which of course is the MUF. Irregularities in the ionosphere may, however, result in a MUF wave being occasionally deflected upward to escape through the layer. Practical experience has shown that frequencies about 15% lower than the MUF should be used.

14.4.6 Virtual Height

A wave traveling in a curved path has a horizontal component of group velocity v_h given by (see Fig. 14.8(a))

$$v_h = v_g \sin r \tag{14.47}$$

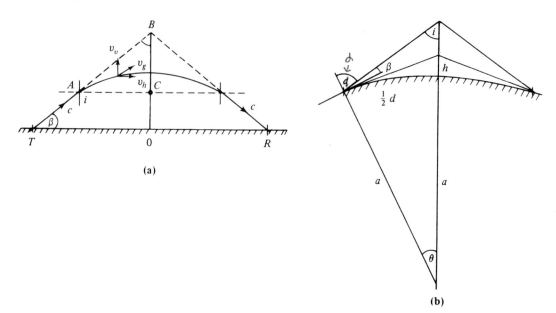

FIGURE 14.8. (a) Determination of virtual height; (b) determination of path distance.

From Eq. (14.42),

$$v_g = \frac{c^2}{v_p}$$

Therefore,

$$v_h = c^2 \frac{\sin r}{v_p}$$

Substituting for $(\sin r)/v_p$ from Eq. (14.43),

$$v_h = c^2 \frac{\sin i}{c}$$

$$= c \sin i \qquad (14.48)$$

This shows that the horizontal component of group velocity is constant and independent of height in the ionized layer. Therefore, the time t required for the wavefront to reach the highest point in the path will be

$$t = \frac{AC}{v_h} \qquad (14.49)$$

where AC is the horizontal distance from point of entry to the vertical dropped from the highest point (Fig. 14.8(a)).

Substituting from Eq. (14.48) for v_h, Eq. (14.49) becomes

$$t = \frac{AC}{c \sin i}$$

But from Fig. 14.8(a), $AC = AB \sin i$, and therefore

$$t = \frac{AB}{c} \tag{14.50}$$

Thus, the wave can be considered to have traveled distance AB at constant velcotity c in time t. The virtual height h of a layer is then OB (Fig. 14.8(a)), which is the height the wave appears to be reflected from had it been traveling at constant velocity c. The virtual height has the great advantage of being easily measured, and it is very useful in transmission-path calculations.

For flat-earth assumptions, and assuming that the ionospheric conditions are symmetrical for the incident and reflected waves, the transmission-path distance TR, Fig. 14.8(a), is

$$TR = \frac{2h}{\tan \beta} \tag{14.51}$$

Where the curvature of the earth is taken into account, the transmission-path distance can be determined from the geometry of Fig. 14.8(b):

$$\frac{\sin i}{a} = \frac{\sin(180 - \alpha)}{a + h}$$

$$= \frac{\sin \alpha}{a + h} \tag{14.52}$$

Also,

$$180 - \alpha = 180 - (i + \theta)$$

Therefore,

$$i = \alpha - \theta$$

Substituting Eq. (14.53) in Eq. (14.52),

$$\frac{\sin(\alpha - \theta)}{a} = \frac{\sin \alpha}{a + h}$$

Therefore,

$$\theta = \alpha - \sin^{-1}\left(\frac{a}{a + h} \sin \alpha\right) \tag{14.54}$$

In terms of the angle of elevation, Eq. (14.55) becomes

$$\theta = (90 - \beta) - \sin^{-1}\left(\frac{a}{a + h} \cos \beta\right) \tag{14.55}$$

Also, from Fig. 14.7(b), the arc length

$$\frac{d}{2} = a\theta \tag{14.56}$$

where the angle θ is in radians. Substituting for θ from Eq. (14.55) and using radian measure for all angles,

$$d = 2a\left[\left(\frac{\pi}{2} - \beta\right) - \sin^{-1}\left(\frac{a}{a+h}\cos\beta\right)\right] \tag{14.57}$$

Example 14.5   Calculate the transmission-path distance for an ionospheric transmission that utilizes a layer of virtual height 200 km. The angle of elevation of the antenna beam is 20°.

Solution   The flat-earth approximation (Eq. (14.51)), gives

$$d = \frac{2 \times 200}{\tan 20°}$$
$$= 1100 \text{ km}$$

Using Eq. (14.57),

$$d = 2 \times 6370\left[(1.57 - 0.349) - \sin^{-1}\left(-\frac{6370}{6570}\cos 20°\right)\right]$$
$$= 966 \text{ km}$$

Measurement of virtual height is usually carried out by means of an instrument known as an *ionosonde*. The basic method is to transmit vertically upward a pulse-modulated radio wave with a pulse duration of about 150 μs. The reflected signal is received close to the transmission point, and the time T required for the round trip is measured. The virtual height is then

$$h = \frac{cT}{2} \quad \text{where } c = \text{speed of light} \tag{14.58}$$

The ionosonde will have facilities for sweeping over the radio-frequency range; typically, it will sweep from 1 MHz to 20 MHz in 3 minutes. It will also have facilities for automatic plotting of virtual height against frequency, the resultant graph being known as an *ionogram* (Figure 14.9). The ionogram shows two critical frequencies, f_0F_2 and f_xF_2, which will be explained in the next section.

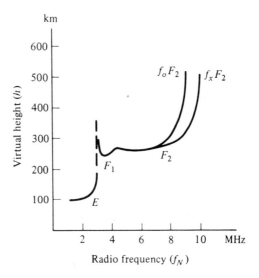

FIGURE 14.9. Sketch of an ionogram.

14.4.7 Effects of Earth's Magnetic Field

When a charged particle is displaced in a magnetic field, it experiences a force that causes it to move in a curved path. The magnetic field of the earth exerts such a force on the electrons in an ionized layer which are displaced by the electric field of a radio wave. (There is also the magnetic field of the radio wave, but the force exerted by it is negligible in this situation.) In general, the electron paths will be helixes, as sketched in Fig. 14.10(a). At one particular wave frequency, known as *the gyrofrequency*, where the periodic time of the wave is equal to the time required for one complete revolution about the magnetic field axis, the electron path becomes a very wide single loop (Fig. 14.10(b)). The gyrofrequency can be shown to be equal to

$$f_g = \frac{q_e}{2\pi m} B \qquad (14.59)$$

where q_e is the charge of the electron, m is its mass, and B is the magnetic field strength. Substituting values for q_e and m, and using an average value of 0.5×10^{-4} T for the earth's magnetic field, results in a gyrofrequency of 1.4 MHz.

The significance of the gyrofrequency is that because of the wide path followed by the electrons, the number of collisions between electrons and molecules in the D layer is increased, resulting in increased attenuation in the reflected wave at frequencies in the vicinity of f_g. Thus, most of the medium-wave broadcast band suffers high attenuation of the reflected-wave component during the day, when the D layer is present.

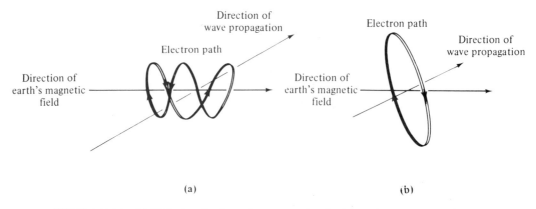

FIGURE 14.10. (a) Helical path of an electron moving in the earth's magnetic field; (b) electron path at gyrofrequency.

Another effect of the earth's magnetic field is that the relative permitivity of the ionized layer develops two components. The details are complicated and will not be shown here, but the result is that two critical frequencies occur for an ionized layer. In practice this only shows up for the F_2 layer, the two critical frequencies being known as the critical frequency for the ordinary ray, denoted by f_0F_2, and the critical frequency for the extraordinary ray, denoted by f_xF_2. These are shown in Fig. 14.9.

14.4.8 Service Range

NB
SERVICE
RANGE

The service range of a transmission of a given frequency is determined by the critical ray at the nearest point and the glancing ray at the farthest point, as shown in Fig. 14.11(a). Rays from the transmitting antenna at angles greater than the critical angle of elevation (or what is equivalent, at angles of incidence less than the critical value) will escape, creating a skip distance in which no signal is received.

The maximum possible range is reached when the critical ray coincides with the glancing ray (Fig. 14.11(b)), i.e., when the glancing ray is returned from the greatest virtual height h_m. The geometry of Fig. 14.11(b) is similar to that of Fig. 14.8(b) with β equal to zero and h being replaced by h_m. The angle θ is

$$\theta = \cos^{-1}\left(\frac{a}{a + h_m}\right)$$

and, from Eq. (14.56),

$$d = 2a\theta$$

$$= 2a\cos^{-1}\left(\frac{a}{a + h_m}\right) \tag{14.60}$$

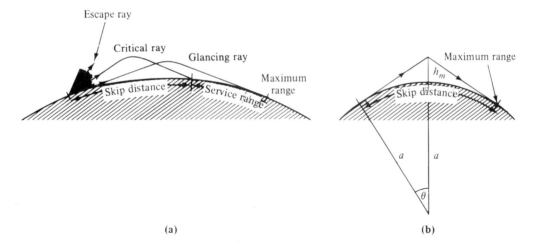

FIGURE 14.11. (a) Service range, showing skip distance; (b) maximum range.

Equation (14.60) gives the maximum possible distance for a "single hop," that is, one reflection involving the ionosphere. The range can be increased by using multiple hops, in which the wave is reflected from the earth after the first hop, to be reflected from the ionosphere once again farther on.

Of course, the ionospheric conditions at the point of reflection must be used in calculations to determine the transmission path. With multiple-hop transmissions the situation becomes more complicated, since two or more points must be taken into consideration. This is particularly so for paths following lines of latitude, since conditions may change from day to night over the path.

Irregularities in the ionosphere have been used successfully for *scatter propagation* similar to that which occurs in the troposphere. With ionospheric scatter, distances on the order of 1000 miles are achieved at frequencies in the low end of the VHF band (i.e., between about 20 and 60 MHz). Fading effects are more troublesome in ionospheric scatter systems than in tropospheric scatter systems, which tends to limit *ionoscatter* propagation to narrow-band (e.g., 3-kHz) signals.

14.4.9 Ionospheric Irregularities, Ionic Disturbances, and Fading

In previous sections, a very simple model of the ionosphere was assumed in which electron densities varied smoothly and uniformly, and changes resulting from diurnal and seasonal variations were assumed to be predictable. Calculations of MUF, virtual height, and so on, based on this model provide only an estimate for average conditions. In practice, the ionosphere exhibits irregularities in electron densities in the various layers, the extent of which ranges from less than 100 meters to many hundreds of kilometers. It is known that some irregularities travel through the ionosphere with a horizontal component of velocity ranging from a few meters per second to greater than 1 kilometer per second. These are known as *traveling ionospheric disturbances* (TID's), which can seriously affect the accuracy of high-frequency direction finders.

Many of the causes of ionospheric disturbances are not well understood. Some of the factors thought to be involved are large-scale gravity waves in the atmosphere, electric currents and plasma instabilities in the ionosphere, and, in particular, solar activity. It has been observed that severe attenuation, or even complete loss of radio signals, accompanies solar flares. Intense x-ray radiation from the flares increases the ionization of the D layer, resulting in increased absorption, as described in Section 14.4.5. The fadeouts occur very suddenly and are known as *sudden ionospheric disturbances* (SID's). They are also known as *Dellinger fadeouts* and *Mogel-Dellinger fadeouts*, having been reported by Mogel in Germany in 1930 and by Dellinger in the United States in 1937.

Protons are also discharged from the sun during solar flares. They affect the outer reaches of the earth's magnetic field about 30 hours after a flare has been observed and give rise to what is termed a *magnetic storm*. These storms adversely affect radio communications, particularly at high latitudes (near the magnetic poles). Magnetic storms not connected with solar flares, but associated with sunspots, have been recorded; yet others have been observed apparently not connected with solar flares or sunspots, but which have a periodicity of recurrence of about 27 days, the rotational period of the sun. These appear to originate in well-defined regions of the sun (labeled M regions) and, in fact, are more noticeable at sunspot minima. They rise much more gradually than those associated with sunspots or solar flares, and although they are weaker, they persist much longer, sometimes up to about 10 days. Weak sporadic magnetic storms apparently not directly related to solar activity have also been observed.

Another form of ionospheric disturbance which can seriously affect radio communications is known as *sporadic E*. Thin, highly ionized layers occur in the E layer from time to time; the extent, position, and timing of these are all irregular, hence the term "sporadic E layer." Because of its high electron density, sporadic E can often take over a transmission normally beamed on the F layer; VHF reflections have also been observed from sporadic E, resulting in interference in the VHF television channels (e.g., the "freak" reception of distant stations, which sometimes occurs).

Fluctuations in electron densities occur continuously in the ionosphere, giving rise to fluctuating phase differences in the various ray paths of a signal. Since the waves following these ray paths combine as phasors at the receiver, fluctuations in the received signal strength, called *interference fading*, will be observed. Another form of fading, known as *polarization fading*, occurs where the ordinary and extraordinary rays combine phasorally. The ordinary and extraordinary waves are always perpendicularly polarized with reference to each other, and because of the random variations in amplitude and phase of each, when they combine, the net polarization also varies in a random manner. The receiving antenna will of course be arranged for reception of a fixed-polarization wave (e.g., one that is vertically polarized).

With modulated signals, fading can affect a very narrow range of the frequency spectrum independently of other parts of the spectrum, and this gives rise to what is termed *selective fading*. Selective fading comes about essentially because the ray paths in the ionosphere will be different for different frequencies, and they will not necessarily all experience a disturbance in a given region. Selective fading limits ionospheric transmissions to narrow-band signals (e.g., 3-kHz bandwidth).

14.4.10 Summary of Layers

In spite of the difficulties and uncertainties inherent in ionospheric transmission, it does provide an inexpensive and relatively quick means for setting up medium- to long-distance radio communications. Of course, it has been superseded on many circuits by satellite systems or submarine cables, both of which provide much better service, but at increased cost.

Methods are being developed for predicting ionospheric conditions which may be utilized in the planning of radio circuits. Equipment is also being developed that will automatically select optimum frequency for transmission in an effort to combat fading.

The main features of the various layers are summarized here, with values obtained from Fig. 14.6(a). Frequencies, heights, and so on, will of course, vary considerably, as already discussed, and the values are presented simply to give an order-of-magnitude comparison for the various layers.

C and D layers. Virtual height, 60 to 80 km. Reflect low and very low frequencies (see Section 14.6), but for HF communications the D layer, in particular, introduces attenuation.

E layer. Virtual height $\sim$ 110 km. Critical frequency, $\sim$ 4 MHz. Maximum single-hop range, $\sim$ 2350 km.

F_1 layer. Virtual height, $\sim$ 180 km. Critical frequency, $\sim$ 5 MHz. Maximum single-hop range, $\sim$ 3000 km.

F_2 layer. Virtual height, $\sim$ 300 km daytime, $\sim$ 350 km nighttime. Critical frequency, $\sim$ 8 MHz daytime, $\sim$ 6 MHz nighttime. Maximum single-hop range, $\sim$ 3840 km daytime, $\sim$ 4130 km nighttime.

14.4.11 HF Radio Systems

High-frequency radio systems using the bands from 1.6 to 30 MHz have been popular since the early days of radio in applications where relatively long distances must be covered at a low cost and under relatively light traffic conditions. Recent advances in satellite communications have usurped many of the traditional applications, but communications by satellite tend to be costly in terms of equipment, and HF radio will continue to be used for many applications.

The typical HF radio system is a two-point system with a transmitter and a receiver located at each end of the link. The transmitter and receiver used is usually a general-purpose communications set, although for some fixed applications single-frequency units can be used. Transceiver units are popular for temporary systems where one of the terminal points may be frequently moved to different locations.

The mode of transmission most often used for HF systems is single-sideband, either with or without pilot carrier operated in the simplex mode. Independent sideband suppressed carrier transmission is useful where full duplex operation is desired with only one frequency assignment available. Separate transmitting and receiving antennas separated by some distance are required to prevent overloading of the local receiver by its associated transmitter. Dipoles are favored because they give some directional gain while being structurally simple. Fixed telephone stations may use more complex arrays, such as broadsides, to provide greater directivity. Transmitter powers ranging from a few watts to several kilowatts are commonly used.

The main advantages of HF systems lie in their ability to provide communication over great distances at relatively low cost. This makes them ideal for communications between remote settlements or camps, where the density of calls is low. HF systems do have several severe disadvantages which have resulted in a search for other means of communications, stemming mainly from the overall unreliability of the systems resulting from variable propagation conditions (see Section 14.4.9). The problem of variable propagation conditions can be partially overcome by using frequency diversity, in which a given system is provided with several frequency assignments spanning the HF band of frequencies, so that the operator may choose the channel that gives the best results at any given time. However, this gives no protection against the total blackouts which periodically occur because of magnetic storms and other solar-induced phenomena.

Skip is also a problem. If HF stations are located closely enough together to use ground-wave propagation, distant stations frequently interfere through sky-wave propagation. Fading and distortion because of the changing ionospheric conditions cause annoying interference, especially in the longer circuits. Sometimes space diversity can be used to ease the problem, but since the diversity stations must be separated by several wavelengths, this becomes very expensive. In regions above the artic circle, auroral phenomena frequently make HF communications impossible by disturbing the necessary ionospheric layers. HF systems in equatorial regions are much more reliable.

Perhaps the most limiting disadvantage of the HF system is the fact that the bandwidth allowed for a channel is very narrow, sufficient for only one voice channel. This means that for every voice circuit, a separate pair of radio circuits is required. Further, since transmission is unreliable, the use of HF in public telephone systems is limited. In the past, HF radio was used extensively for overseas telephone communications. Recent advances in cable telephony and in satellite communications has just about eliminated this application except for some remote areas. It is still being used extensively in the South Pacific area.

14.5   SURFACE WAVE

14.5.1   The Mode of Propagation

A radio wave propagating close to the surface of the earth will follow the curvature of the earth as a result of the phenomenon of *diffraction*. This is the same phenomenon which causes sound waves, for example, to travel around an obstacle. Diffraction effects depend on the wavelength in relation to the size of the obstacle, and are greater the longer the wavelength. In the case of radio waves, the "obstacle" around which the wave must travel is the earth itself, and the surface wave is of importance at frequencies below about 2 MHz.

The conductivity and permittivity of the surface play an important part in the propagation of the surface wave, as the wave will introduce both displacement and conduction currents in the surface. These currents may penetrate to depths of from about 1 m at the highest frequency to tens of meters at the lowest, so the actual conditions on top of the surface are relatively unimportant.

The energy lost in the surface comes from the radio wave, which is therefore attenuated as it passes over the surface. The attenuation increases with increase in frequency, which is another factor which limits the usefulness of the surface wave to frequencies below about 2 MHz.

The surface wave in practice is always vertically polarized, as the conductivity of the ground would effectively short-circuit any horizontal electric field component.

The electric field strength as given by Eq. (14.13) can be modified to take into account the various factors affecting propagation by introducing an attenuating factor A, such that

$$E = \frac{A\sqrt{30P_T G_T}}{d} \quad \text{V/m} \tag{14.61}$$

It is normal practice to evaluate E for standard conditions of 1 kW radiated from a short unipole antenna, which has a power gain of 3. Equation (14.61) then becomes

$$E = 300\frac{A}{d} \quad \text{V/m} \tag{14.62}$$

Graphs are available which present the factor A as a function of distance for various values of permittivity, conductivity, and frequency. At large distances (e.g., $d \geq 100\lambda$) the factor A becomes inversely proportional to distance; the field strength plotted against distance is shown in Fig. 14.12(a) for a frequency of 500 kHz.

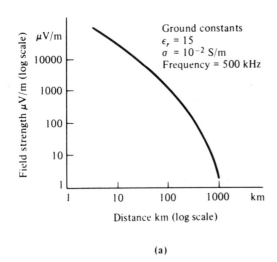

(a)

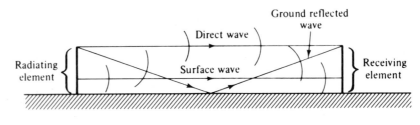

(b)

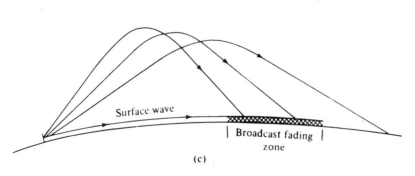

(c)

FIGURE 14.12. (a) Example of variation of surface-wave field strength with distance; (b) the ground wave; (c) broadcast fading zone.

The change of refractive index with height of the atmosphere also causes diffraction of the surface wave, which can be allowed for by adopting an earth's radius of $\frac{4}{3}$ times the actual radius a. The main low-frequency diffraction is then determined for an obstacle of radius $4a/3$.

14.5.2 Ground Wave

At low enough frequencies that the height of the transmitting antenna above ground, in terms of wavelength, is small, the direct wave and the ground-reflected wave (both of which go to make up the space wave—see Section 14.3) effectively cancel each other, leaving only the surface wave. At higher frequencies the height of the antenna may be such that the space wave is comparable in magnitude to the surface wave, the resultant wave being the phasor sum. The resultant wave is termed the *ground wave* (not to be confused with surface wave alone). The situation is shown in Fig. 14.12(b). Normally, the ground-wave field strength is greater than that of the surface wave alone, and this is taken into account by introducing a multiplying factor, known as the *height-gain* factor in Eq. (14.60). The height-gain factor depends on the physical heights of the transmitting and receiving antennas and also on the factors that are used in the determination of the attenuation factor A. Like A, the height-gain factor is available in graphical form for a wide range of practical conditions.

14.5.3 Broadcast Fading Zone

The medium-wave (550 to 1600 kHz) broadcast service normally utilizes the surface wave. Some energy will, however, be transmitted into the ionosphere, where during daytime conditions it is almost completely absorbed in the D layer.

During nighttime conditions, an appreciable portion of the ionospheric wave is returned to earth, extending the service area of the station well beyond that covered by the surface wave. There will also be a zone in which both the surface wave and the ionospheric wave are of the same order of magnitude and the resultant signal strength is the phasor sum of the two. Unfortunately, fluctuations in the ionosphere produce fluctuations in the phase of the reflected wave relative to the surface wave, resulting in severe fading of the combined wave. The area in which this occurs is known as the *broadcast fading zone* and is shown in Fig. 14.12(c).

14.6 LOW-FREQUENCY PROPAGATION AND VERY-LOW-FREQUENCY PROPAGATION

At frequencies below about 300 kHz it is convenient to consider propagation in two bands: 3 to 30 kHz, called the very-low-frequency (VLF) band, and 30 to 300 kHz, called the low-frequency (LF) band. It is an interesting point that the transatlantic radio links engineered by Marconi were at these low frequencies in the year 1901.

At frequencies in these bands, propagation is by means of surface waves up to distances of about 1000 km; beyond this, the sky wave plays an increasingly important part. The sky wave is propagated by means of multiple hops between the ionosphere and the earth. In the LF part of the band, theoretical treatment based on multiple hops of a TEM wave gives satisfactory results, while in the VLF band, the propagation is best considered to take place as a TM waveguide mode (see Chapter 13), the earth and the ionosphere forming the walls of a spherical waveguide.

Because of the low carrier frequencies and consequent narrow band-widths available, communications channels are limited to slow data rates. There are broadcast services of standard frequencies and time transmitted from various countries. The main characteristic of these low and very low frequencies is that they provide highly reliable radio links. Of course, the antenna structures have to be very large, and they are unfortunately inherently inefficient.

Radio navigational systems make extensive use of the low and very low frequency bands. The highly stable phase characteristics of the propagation allow the phase delay in a wave as a function of distance to be accurately known in advance, so that it can be measured by receiving equipment on the ship, aircraft, and so on, to determine position. The most common method is to use two or more transmissions to generate a pattern of *lines of position*, which can then be superimposed upon a map. Referring to Fig. 14.13(a), the signals received at position X, as given by Eq. (B.1), will be

$$A \sin(\omega_a t - \beta_a |s_1|) \qquad \text{from transmitter } A \qquad (14.63)$$

$$B \sin(\omega_b t - \beta_b |s_2|) \qquad \text{from transmitter } B \qquad (14.64)$$

By making ω_a and ω_b multiples of a common generating frequency ω_0 to which both are synchronized, it becomes possible to compare the phase of the two signals. The phase angles are $\beta_a |s_1|$ and $\beta_b |s_2|$, where the phase-shift coefficients β are as defined by Eq. (12.23). Let $\omega_a = 2\pi a f_0$ and $\omega_b = 2\pi b f_0$; then, by putting both signals through frequency dividers at the receiver, as shown in Fig. 14.13(b), and comparing phases, a measure of $|s_1| - |s_2|$ is obtained. The phase of the signals after frequency division will be

$$\text{phase of signal } A = \frac{\beta_a |s_1|}{a} \qquad (14.65)$$

$$\text{phase of signal } B = \frac{\beta_b |s_2|}{b} \qquad (14.66)$$

Now, $\beta_a = 2\pi/\lambda_a$ and $\lambda_a = \lambda_0/a$, so the phase angles of signals A and B, after frequency division, become

$$\text{phase of signal } A = \frac{2\pi}{\lambda_0}|s_1| \tag{14.67}$$

$$\text{phase of signal } B = \frac{2\pi}{\lambda_0}|s_2| \tag{14.68}$$

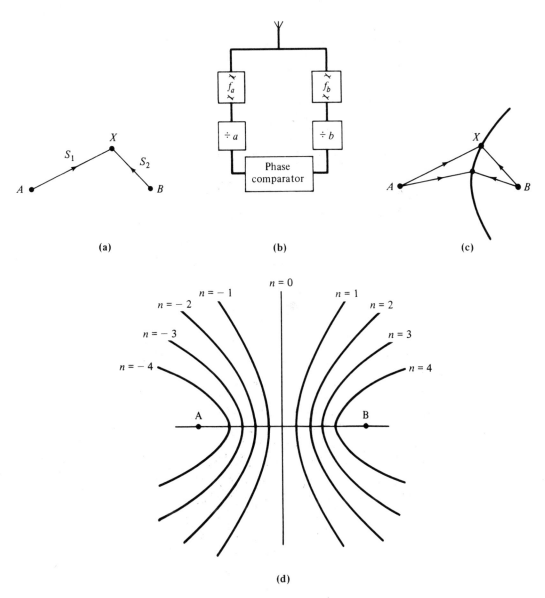

FIGURE 14.13. (a) Ray paths; (b) comparison of signal phases; (c) hyperbolic line of position; (d) family of lines of position.

Therefore, the phase difference Δ is

$$\Delta = \frac{2\pi}{\lambda_0}(|s_1| - |s_2|) \tag{14.69}$$

In effect, the phase difference goes through a series of maxima and minima with change in distance, just like that which occurs for standing waves on transmission lines (see Section 12.7). The two signals will be in phase when $\Delta = n2\pi$, where n is an integer. It follows, therefore, that

$$|s_1| - |s_2| = n\lambda_0 \tag{14.70}$$

For a given value of n, $|s_1| - |s_2|$ is a constant (since λ_0 is also fixed). Now, this is a property of a hyperbola (Fig. 14.13(c)), so that for a given value of n, a hyperbolic curve can be drawn along which $|s_1| - |s_2|$ is constant, and hence along which the signals are in phase. A whole family of curves can be constructed, one curve for each value of n, as shown in Fig. 14.13(d). These curves are called *lines of position*. The space between lines of position are known as *lanes*.

Along the base line joining A and B, the spacing is narrowest. Let L be the width of a lane on the base line. Then, on moving from line n to $n + 1$ along the base line, the total change in the path difference will be $2L$, since $|s_1|$ increases by L and $|s_2|$ decreases by L. In terms of wavelength, the change will be $(n + 1)\lambda_0 - n\lambda_0 = \lambda_0$. Hence,

$$2L = \lambda_0$$
$$L = \tfrac{1}{2}\lambda_0 \tag{14.71}$$

Equation (14.71) applies to the phase-comparison system shown in Fig. 14.13(b). In other systems the frequencies may be multiplied up to their lowest common multiple of f_0, or the carrier may first be divided down to f_0 and then multiplied up to the slave frequency to which it is being compared. The term *master* is used to denote the transmitter to which the others are synchronized, while the others are known as *slaves*. In the *Decca navigational system*, the master is at $6f_0$, the red slave at $8f_0$, the green slave at $9f_0$, and the purple slave at $5f_0$. (The colors refer to the color of the grids superimposed on the navigation map.) By using more than one slave, two or more families of hyperbolas are made to intersect, enabling a position to be pinpointed, as in Fig. 14.14(a). The receiving equipment has to be able to integrate the phase change measured—in effect, to count the number of lines of position crossed as well as to indicate the position within a lane.

The Decca System works very well with surface-wave coverage, but phase variations resulting from the ionosphere reduce its accuracy where the sky wave has to be used. A worldwide radionavigational system named *Omega*,

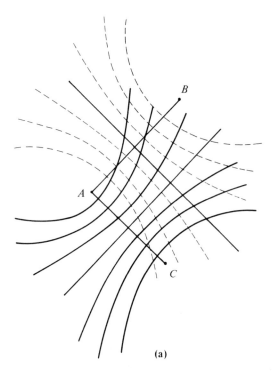

(a)

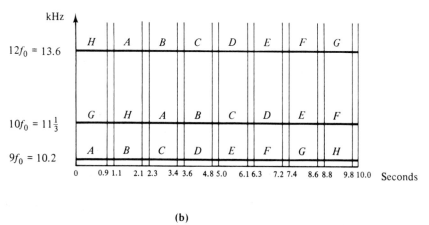

(b)

FIGURE 14.14. (a) Intersecting lines of position; (b) chart showing the transmission schedule for the *Omega* navigational system.

being developed by the U.S. Navy, operates in the VLF band. In this band, phase variations resulting from ionospheric changes are tolerable. Eight stations are involved, labeled $A, B, \ldots, H$, and some of these are presently in operation—for example, A in Norway, B in Trinidad, and D in New York. The schedule of transmissions for the Omega System is shown in Fig. 14.14(b).

14.8   EXTREMELY LOW-FREQUENCY PROPAGATION

Extremely low-frequency (ELF) propagation is used to penetrate to great depths into the ground and the oceans. The impetus to its development has been the need for communications to submarines, especially those which form part of the nuclear deterrent force.

The ELF band extends from 30 Hz to 300 Hz, the corresponding wavelength range being 10,000 km (6210 miles) to 1000 km (621 miles). Now, as pointed out in the introduction to Chapter 8, efficient radiation of electromagnetic waves requires that antenna dimensions be of the same order as the wavelength being radiated, and this conditions cannot be met in the ELF range. Also, as seen in the studies of modulation, a modulated carrier wave requires a certain frequency bandwidth which is typically on the order of 1% of the carrier frequency. Thus, at ELF this means that only 1 Hz of bandwidth is available for the information content. As shown in Chapter 17, rate of information transmittal is directly proportional to the bandwidth used, and therefore ELF does not offer the capacity for high rate of information transmission. The question then arises, why are ELF waves used for transmission purposes at all? The answer is that they provide the only practical means presently known of communications with submarines. Since this is primarily a matter of national defense, the problem of low efficiency antennas can be overcome by spending more money to increase power transmitted, and the problem of low communication rate can be overcome by just transmitting the code letters for a given message which is part of a list of standard messages aboard the submarine. For example, a bandwidth of 1 Hz allows a transmission rate of 1 bit/sec. Ten bits allows $2^{10} = 1024$ messages to be stored, and the time required to transmit the necessary ten bits in this example would be 10 seconds.

As has been pointed out in the case of VLF propagation, the propagation around the earth may be considered to take place in a waveguide mode, the earth and the ionosphere forming the waveguide boundaries. Because of the very long wavelengths, the transmitting antenna has to be mounted horizontally. The length of the antenna is very much less than a wavelength. The situation is sketched in Fig. 14.15. It is found that the far field consists of a horizontal magnetic field and a vertical electric field, and propagation in the earth-ionosphere waveguide is by means of a quasi-TEM mode. A leakage field also occurs, directed into the surface of the earth, and subsurface communications utilize this leakage field. The leakage field is a plane TEM wave which follows a propagation law similar to that of Eq. (12.20) for the voltage on a transmission line; that is, the electric field at depth z is related to $E(0)$, the electric field at the surface, by

$$E(z) = E(0)e^{-\gamma z} \tag{14.72}$$

The propagation coefficient γ is made up of the two components as given by Eq. (12.22), the attenuation coefficient α, and the phase shift coefficient β. These depend on the conductivity, permittivity, and permeability of the medium, and on the frequency of the wave. The ratio $\sigma/\omega\varepsilon$ is known as the loss tangent of the medium, and for $\sigma/\omega\varepsilon \gg 1$, the attenuation coefficient is numerically equal to the phase shift coefficient, each being given by

$$\alpha = \beta = \sqrt{\frac{\omega\sigma\mu}{2}} \tag{14.73}$$

The permeability μ may be assumed to be equal to $4\pi \times 10^{-7}$ H/m.

Example 14.6   Sea water has an average conductivity of 4 S/m and a relative permittivity of 80. Calculate the attenuation coefficient in dB/m for a signal of (a) 100 Hz, and (b) 1 MHz.

Solution   (a) First it is necessary to evaluate the ratio $\sigma/\omega\varepsilon$. At 100 Hz this is

$$\frac{\sigma}{\omega\varepsilon} = \frac{4}{2\pi \times 100 \times 80 \times 8.854 \times 10^{-12}}$$
$$= 8.9 \times 10^{7}$$

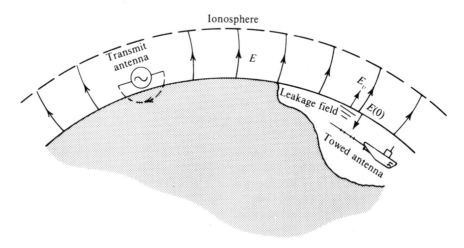

FIGURE 14.15.   ELF Propagation. (After Bernstein et al., "Long Range Communication at Extremely Low Frequencies." *Proc. I.E.E.E.* 62(3) 1974, pp. 292–312.)

Since this is very much greater than unity, Eq. (14.73) may be used to calculate the attenuation coefficient as

$$\alpha = \sqrt{\frac{2\pi \times 100 \times 4 \times 4\pi \times 10^{-7}}{2}}$$

$$= 3.974 \times 10^{-2} \, \text{N/m}$$

Using the conversion factor (see Eq. (A.7)) 1 N = 8.686 dB, we have

$$[\alpha] = 3.974 \times 10 \times 8.686 = 0.345 \, \text{dB/m}$$

(b) At $f = 1$ MHz, the loss tangent is $8.9 \times 10^7 \times 100/10^6 = 8.9 \times 10^3$, and since this is also very much greater than unity, Eq. (14.73) can be used in this case also. Thus,

$$\alpha = 3.974 \times 10^{-2} \times \sqrt{\frac{10^6}{100}} = 3.974 \, \text{N/m}$$

and

$$[\alpha] = 3.974 \times 8.686 = 34.5 \, \text{dB/m}$$

An interesting graph of attenuation vs. frequency for a plane TEM wave propagating in sea water is given in the book, *ELF Communications Antennas*, by Michael L. Burrows, (Peter Peregrinus Ltd., 1978). The graph is reproduced here in Fig. 14.16. It shows that the attenuation at ELF is a fraction of a decibel per meter, but it rises to the enormous attenuation of 1000 dB/m at about 1 GHz. This means, for example, that 3 mm of sea water will attenuate a signal at 1 GHz by 3 dB, or a factor of 2 : 1. At the visible end of the electromagnetic spectrum the attenuation decreases to a comparatively low value again, as shown in Fig. 14.16.

An important feature of ELF propagation is that noise at the surface will also be attenuated as it penetrates into the earth, so that the signal-to-noise ratio, as transmitted, is virtually independent of depth. Also, atmospheric noise is mostly lightning-induced and consists of sharp spikes which are readily removed by simple peak limiters at the receiver. Sea water flattens out the peaks, and in order for such noise limiters to be fully effective it is necessary to include an "inverse ocean filter" in the receiver ahead of the limiters. Of course, in addition to atmospheric noise, noise will be generated in the receiver itself, and this will ultimately set a limit on receiver sensitivity.

As already mentioned, the transmitting antenna has to be very large, and such antennas have been built which cover many square miles of ground. It is found that the antenna wires, which are horizontally mounted, may be buried as deep as 10 feet without noticeably affecting the radiation efficiency. The electric field increases with decreasing conductivity, and a low-conductivity ground is chosen for the antenna site. The signal strength is also influenced by the ionosphere height, and it must be kept in mind that the latter is only a small fraction of a wavelength at ELF. Transmitting antenna currents are of the order of 100 A, and some figures quoted in the paper by Bernstein et al. (Fig. 14.15) highlight the low efficiency of radiation. In one experiment quoted, the power supplied to the antenna was 3.88 MW. Of this, the percentages dissipated as heat were 43% in the conductors, 11% in the end-grounds for the antenna, and 46% in the ground return wire, the remainder, amounting to 69 watts, being the power radiated.

At the frequency of operation (100 Hz), the attenuation was 1.5 dB per 1000 km, and at a depth of 10,000 km, a transmission rate of 1 bit/sec was achieved with the 69 watts radiated. Special coding and modulation techniques were used.

Although transmitting antennas have to be large, the same is not necessarily true for receiving antennas. Towed antennas consisting of two electrodes spaced 300 m along a towed cable have been used with submarines. The electrodes respond to the electric field gradient through the water. The cable itself is 600 m long, and the furthest electrode is 75 m from the end. Other types of receiving antennas are being experimented with, based on magnetometer principles. These are sensors which respond to the magnetic field of the wave. One of these has the acronym SQUID, for *s*uperconducting *qu*antum

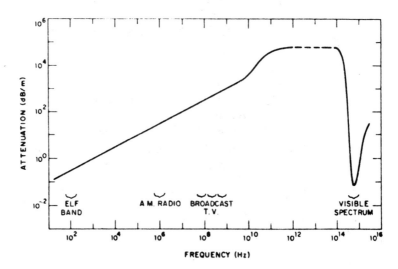

FIGURE 14.16. Attenuation of a Plane electromagnetic wave in sea water as a function of frequency. [From Michael L. Burrows, *ELF Communications Antennas* (Peter Perrigrine Ltd., 1978). Institution of Electrical Engineers, with permission.]

*i*nterference *d*evice. The essential element in the SQUID is a small (2-mm diameter) loop of superconducting material. Any change in the incident magnetic field at the loop changes a superconducting current flowing in the loop, and in this way detection of the signal is achieved. Such an antenna requires a large amount of support equipment.

Because of the factors outlined, ELF transmission is really suitable only for one-way signalling, from a high-power ground station to the subsurface receiving station. The practical difficulties of employing large transmitting antennas and high-power transmitters at the subsurface station prevent transmissions being sent up to the surface. Also, atmospheric noise, which is attenuated in the downlink, increases towards the surface, adding to the difficulties of establishing a two-way communications link.

14.9 SUMMARY OF RADIO-WAVE PROPAGATION

The chart in Figure 14.17 summarizes the classification of the various radio bands in terms of frequency and wavelength. It should be realized, of course, that sharp boundaries do not exist between the bands.

14.10 PROBLEMS

1. Part of a microwave link can be approximated by free-space conditions. The antenna gains are each 40 dB, the frequency is 10 GHz, and the path length is 80 km. Calculate (a) the transmission path loss, and (b) the received power for a transmitted power of 10 W.

2. What feature of the microwave relay system makes it attractive for providing telephone trunk circuits? How many repeater stations would be required on a 6000-km route?

3. A VHF radio link is set up between a shore station and an island in a lake 10 miles offshore. The antenna site on the island is atop a 100-ft cliff. Calculate the minimum height of the shore-station antenna if the minimum acceptable signal strength at either station is 10 μV/m. The frequency is 150 MHz and the transmit power is 1 W from a $\frac{1}{2}\lambda$ dipole from each station.

4. Derive Eq. (14.34).

5. Why are VHF systems popular for dispatch work?

6. Why must a radio repeater station not retransmit on the same frequency on which it receives?

7. Explain the significance of the plasma frequency in connection with ionospheric radio communications. How is the critical frequency for an ionized layer related to the plasma frequency?

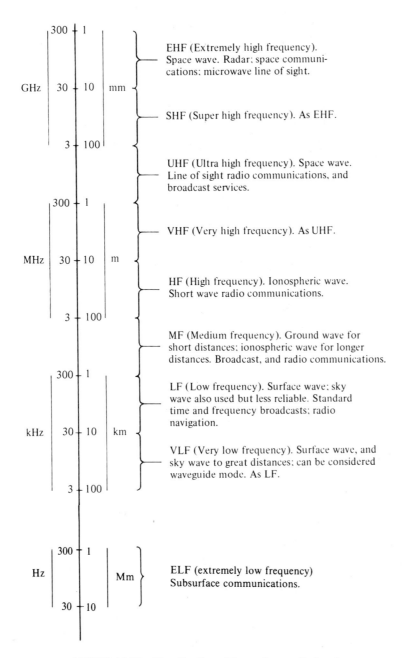

FIGURE 14.17. Classification of the various radio bands.

8. A radio communications link is to be established via the ionosphere. The maximum virtual height of the layer is 100 km at the midpoint of the path, and the critical frequency is 2 MHz. The distance between stations is 600 km. Determine suitable values for the optimum working frequency and the angle of elevation for the antenna main beam.

9. Explain what is meant by the gyro frequency and why frequencies in the region of the gyro frequency are not suitable for ionospheric transmissions.

10. Calculate the maximum range obtainable in a single-hop transmission utilizing the F_2 layer for which h_{max} is 400 km. The earth's radius may be taken as 6370 km.

11. Discuss briefly the factors that give rise to fading in ionospheric radio transmissions.

12. List the advantages and disadvantages of HF radiotelephone circuits.

13. Explain the difference between the surface wave and the ground wave for radio transmissions in the frequency range 300 kHz to 2 MHz.

14. Given that the free-space velocity of light c is 299,792.5 km/s and the ratio of phase velocity in the atmosphere to the free space values v_p/c is 0.9974, calculate the wavelength of the generating frequency f_0 in the Omega navigation system.

466

CHAPTER 15

Antennas

15.1 INTRODUCTION

In a radio system, an electromagnetic wave travels from the transmitter to the receiver through space, and antennas (or aerials) are required at both ends for the purpose of coupling the transmitter and the receiver to the space link. Many of the important characteristics of a given antenna are identical for both transmitting and receiving functions, and the same antenna is often used for both.

Antennas may be constructed from conducting wires or rods, as, for example, the ordinary domestic TV antenna. At microwave frequencies apertures coupled to waveguides may be used, such antennas are naturally called *aperture antennas*. A horn antenna is an example of an aperture antenna. Antennas may be further classified as *resonant antennas*, in which the current distribution exists as a standing-wave pattern, and *nonresonant* antennas, in which the current exists as a travelling wave. Again, the ordinary TV antenna is an example of a resonant antenna, usually cut to one-half wavelength, which gives it its resonant properties. Nonresonant antennas are used mainly for short-wave communications links, and will be described later.

The types of structures used for antennas are many and varied, ranging from a simple length of wire suspended above the ground to the curtain arrays used for very low-frequency (VLF) broadcasting, from the insignificant-looking lens antenna on a traffic policeman's radar apparatus to the huge parabolic dish antennas of the astronomer's radio telescope. Several of these will be discussed in detail in this chapter.

15.2 ANTENNA EQUIVALENT CIRCUITS

In a radio communications link, the transmitting antenna is coupled to the receiving antenna through the electromagnetic wave. The arrangement is somewhat similar to the transformer coupling described in Section 1.12, except that with antennas the coupling is normally very weak, and an electromagnetic wave is involved rather than just the magnetic field, as in the case of transformers. Further, the finite propagation time required for the wave to travel from the transmitting antenna to the receiving antenna can be significant. The antenna coupling system can, however, be represented as a four-terminal network as illustrated in Fig. 15.1(a). The network representation is useful largely because it allows the well-known network theorems to be applied, and important general results can be obtained which are valid for any antenna.

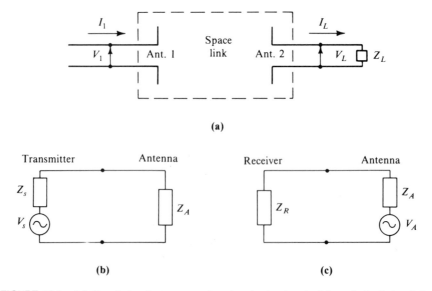

FIGURE 15.1. (a) Coupled antennas as a four-terminal network; (b) equivalent circuit for the transmit antenna; and (c) equivalent circuit for the receive antenna.

Figure 15.1(a) shows antenna 1 transmitting. The input current is I_1 and the input voltage is V_1. Thus, the antenna impedance for transmitting is

$$Z_A = \frac{V_1}{I_1} \tag{15.1}$$

For simplicity, antenna terminals are shown. In the case of an aperture antenna fed by a waveguide terminals do not of course exist in this sense, but impedance can still be measured in terms of the reflection coefficient as given by Eq. 12.74:

$$Z_A = Z_0 \frac{1 + \Gamma_A}{1 - \Gamma_A} \tag{15.2}$$

Here, Z_0 is the wave impedance of the waveguide, and the same equation can be used with transmission lines, with Z_0 the characteristic impedance of the line.

In order to find the impedance of antenna 1 in the receiving mode, Thévenin's theorem can be applied, and the Thévenin voltage equivalent generator found. The Thévenin voltage is the open-circuit voltage at antenna 1 terminals when antenna 2 is transmitting. The Thévenin impedance is found by shorting the EMF source at antenna 2, applying a voltage at antenna 1's terminals, and measuring the resulting current. The Thévenin impedance is then the ratio of this voltage to current. Now, assuming that the coupling between the antennas is sufficiently weak that the short-circuited antenna 2 has no effect on the current in antenna 1, then the ratio of voltage to current at antenna 1 terminals in this case will be the same as given by Eq. (15.1). The antenna impedance is therefore the same for both transmitting and receiving. The equivalent circuit for transmitting is shown in Fig. 15.1(b), and that for receiving is shown in Fig. 15.1(c). In (b), the transmitter is shown as an equivalent voltage generator feeding an antenna impedance of Z_A, and in (c) the antenna is represented by its Thévenin voltage generator equivalent circuit, feeding a receiver load impedance Z_R. The connection between antenna and transmitting or receiving equipment will be by means of feeder lines or guide, and the effect of mismatch on this is considered later.

A network theorem known as the *reciprocity theorem* can also be applied to the antenna system shown in Fig. 15.1(a). This theorem states, in essence, that if an EMF E applied to the terminals of antenna 1 gives rise to a terminal current I at antenna 2, then applying E to the terminals of antenna 2 will give rise to the current I at the terminals of antenna 1. Now it is known that all practical antennas are directive; that is, they radiate better in some directions than others and receive better from some directions than others. A consequence of the reciprocity theorem is that the directive pattern for a given antenna will be the same for both the transmitting and the receiving modes of operating. Directivity is discussed in more detail in Section 15.7.

The antenna impedance Z_A is a complex quantity:

$$Z_A = R_A + jX_A \tag{15.3}$$

The reactive part X_A results from the reactive fields surrounding the antenna. As with any reactance, energy is stored in these fields and returned to source. Wherever possible, the reactance will be tuned out, so that the antenna presents a purely resistive load to the transmission line. The resistive part R_A is given by

$$R_A = R_{\text{loss}} + R_{\text{rad}} \tag{15.4}$$

The resistance R_{rad} is a fictitious resistance termed the *radiation resistance*, which, if it carried the same rms terminal current as the antenna, on transmission would dissipate the same amount of power as was radiated. A certain amount of power will be dissipated in the antenna as heat, and the power dissipated in R_{loss}, when carrying the same current as R_{rad}, gives the power lost in this way. The resistance R_{loss} therefore represents the losses in the antenna. The concepts of loss resistance and radiation resistance are most useful with wire antennas, for which the terminal currents are easily identified and the loss resistance is mainly the resistance of the antenna wire. For this type of antenna, let I be the rms terminal current. Then the total power supplied to the antenna is $I^2 R_A$, and the power radiated is $I^2 R_{\text{rad}}$. The antenna efficiency is therefore

$$\eta_A = \frac{I^2 R_{\text{rad}}}{I^2 R_A} = \frac{R_{\text{rad}}}{R_A} \tag{15.5}$$

In the receiving mode, the efficiency is defined as the ratio of power delivered to a matched load from the actual antenna to the power delivered to a matched load from the antenna with R_{loss} assumed equal to zero. Applying the maximum power transfer theorem (Section 1.14) to the receiving antenna circuit of Fig. 15.1(c), for the real antenna the maximum power is $V_s^2 / 4 R_A$, and for the lossless antenna it is $V_s^2 / 4 R_{\text{rad}}$. Thus, the receiving efficiency is also given by Eq. (15.5).

Antenna matching to the feeder line is important for eliminating reflected waves and obtaining maximum power transfer. Examples of matching circuits are given in Sections 5.8 and 12.12, and in general, the matching network has to provide both reflectionless matching and matching for maximum power transfer. A matching network designed to meet one of these conditions automatically satisfies the other condition. Thus, as shown in Fig. 15.2(a), the impedance seen looking into the network at the feeder side is Z_0, and at the antenna side Z_A^*, where $Z_A^* = R_A - jX_A$ is the complex conjugate of Z_A. This is required for maximum power transfer as described in Section 1.14.

In effect, the matching network tunes the antenna to resonance and then transforms the resistive part to Z_0 and can therefore be represented by the

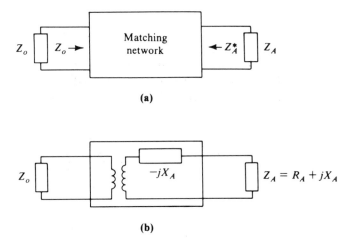

(a)

(b)

FIGURE 15.2. (a) Terminal impedances of a matching network providing reflectionless matching and maximum power transfer; (b) a circuit that meets the general impedance transformations required by (a).

general arrangement shown in Fig. 15.2(b). Working from antenna to transmission line, the reactance $-jX_A$ tunes out the $+jX_A$ component of antenna impedance, and the transformer transforms the remaining R_A component to Z_0. Now, assuming that the transmission line is matched at its far end, the matching network sees an impedance Z_0, and working from line to antenna, the transformer transforms this Z_0 back into R_A. This R_A is in series with the $-jX_A$ element, and therefore the output impedance is $R_A - jX_A$, or Z_A^*, as required for maximum power transfer to Z_A.

The antenna feeder should be matched at both ends as shown in Fig. 15.3. This provides maximum power transfer from the transmitter or into the receiver. It also provides reflectionless matching for the line, which prevents multiple reflections occurring should the line be mismatched at the antenna end.

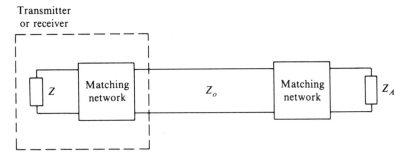

FIGURE 15.3. The antenna feeder matched at both ends.

Consider now the effect of a mismatch at the antenna end. Under transmitting conditions the transmitter and line appear as an EMF source of internal resistance Z_0 feeding a load Z_A as shown in Fig. 15.4(a). The current flowing in this circuit is $V_0^2/|Z_0 + Z_A|^2$ and the power delivered to Z_A is $R_A V_0^2/|Z_0 + Z_A|^2$. The power delivered under matched conditions would be $V_0^2/4Z_0$, and therefore the matching efficiency can be written as

$$\eta_\Gamma = \frac{R_A V_0^2}{|Z_0 + Z_A|^2} \frac{4Z_0}{V_0^2}$$

$$= \frac{4R_A Z_0}{|Z_0 + Z_A|^2} \tag{15.6}$$

Under receiving conditions (Fig. 15.4(b)), the power delivered to Z_0 is $Z_0 V_A^2/|Z_0 + Z_A|^2$, and the power that would have been delivered under matched conditions is $V_A^2/4R_A$. Hence, the matching efficiency in the receiving case is also

$$\eta_\Gamma = \frac{Z_0 V_A^2}{|Z_0 + Z_A|^2} \frac{4R_A}{V_A^2}$$

$$= \frac{4R_A Z_0}{|Z_0 + Z_A|^2}$$

Thus, the matching efficiency is the same for both transmitting and receiving conditions. It is left as an exercise for the student to show that, using the relationship given by Eq. (15.2) and the fact that $R_A = \frac{1}{2}(Z_A + Z_A^*)$, the matching efficiency is given by

$$\eta_\Gamma = 1 - |\Gamma_A|^2 \tag{15.7}$$

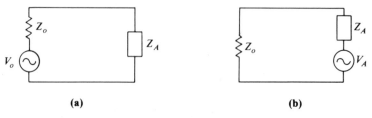

(a) (b)

FIGURE 15.4. Mismatch conditions: (a) transmit; (b) receive.

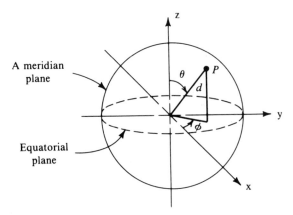

FIGURE 15.5. The spherical coordinates for point P on the surface of the sphere are radius d and angles θ and φ.

15.3 COORDINATE SYSTEM

The directional characteristics of an antenna are usually described in terms of spherical coordinates, shown in Fig. 15.5. The antenna is imagined to be at the center of a sphere, and any point P on the surface of the sphere can be defined in relation to the antenna by the radius d and the angles θ and ϕ. These are shown with reference to the rectangular coordinates x, y, and z. Also shown in Fig. 15.5 is the *equatorial plane*, which is the xy plane. Once the equatorial plane has been defined, any plane at right angles to it that contains the center of the sphere is known as a *meridian plane*. In practice, the equatorial plane and one of the meridian planes will be defined by the planes of symmetry for the antenna. Examples will be given later.

15.4 RADIATION FIELDS

An electric current in a wire is always surrounded by a magnetic field. When the current is alternating, the free electric charges in the wire are accelerated, which gives rise to an alternating electromagnetic field which travels away from the wire in the form of an electromagnetic wave. (An analogy may be made by considering a semirigid sheet being moved at constant velocity, which results in a steady air displacement. If the sheet is vibrated, which in effect imparts acceleration to some areas of it, a sound wave will be generated which travels through the air away from the sheet.)

The total field originating from an alternating current in a wire is complicated, consisting of (*i*) an electric field component which lags the current by 90° and which decreases in amplitude as the cube of the distance; (*ii*) an electromagnetic field (i.e., a combined electric and magnetic field) which is in phase with the current and which decreases in amplitude as the square of the distance; and (*iii*) an electromagnetic field which leads the current by 90° and which decreases in amplitude directly as the distance increases. Only the latter electromagnetic field reaches the receiver in a normal radio communications system, where it appears to the receiving antenna as a plane transverse electromagnetic (TEM) wave. The basic properties of a TEM wave are discussed in Appendix B. A useful rule of thumb is that for antennas for which the largest dimension D is very much greater than the wavelength being radiated ($D \gg \lambda$), the far-field zone becomes the only significant one for distances d greater than $2D^2/\lambda$:

$$d \geq 2D^2/\lambda \tag{15.8}$$

15.5 POLARIZATION

In the far-field zone, the *polarization* of the wave is defined by the direction of the electric field vector in relation to the direction of propagation. *Linear* polarization is when the electric vector remains in the same plane, as shown in Fig. 15.6(a). A linear polarized wave that is propagated across the earth's surface is said to be *vertically polarized* when the electric field vector is vertical, and *horizontally polarized* when it is parallel, to the earth's surface. For example, in North America, television transmissions are horizontally polarized, and it will be observed that receiving antennas are also horizontally mounted, whereas in the United Kingdom vertical polarization is used, and there, antennas are mounted vertically.

In certain situations the electric vector may rotate about the line of propagation. This can be caused, for example, by the interaction of the wave with the earth's magnetic field in the F_2 layer of the ionosphere. Rotation of the electric vector can also be produced by the type of antenna used, and this effect is put to good use in satellite communications, as described in Chapter

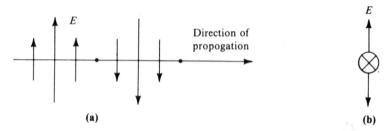

FIGURE 15.6. Linear polarization: (a) as viewed on axis of propagation; and (b) as viewed along direction of propagation.

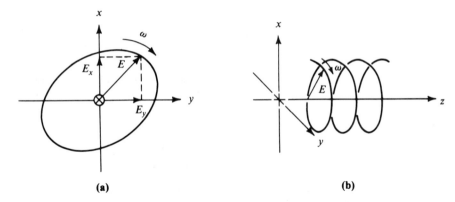

FIGURE 15.7. Elliptical polarization: (a) as viewed along direction of propagation, and (b) as viewed on axis of propagation.

19. The path traced out by the tip of the electric vector may be an ellipse, as illustrated in Fig. 15.7, in which case it is referred to as *elliptical polarization*. If the rotation is in a clockwise direction when looking along the direction of propagation, the polarization is referred to as *right-handed*; if it is anticlockwise, it is called *left-handed*. In Fig. 15.7(a) the direction of propagation is into the paper, so the polarization is right-handed. A special case of elliptical polarization is *circular polarization*, as illustrated in Fig. 15.8, and both right-handed and left-handed circular polarization are used in satellite communications systems as described later. Linear polarization can also be considered to be a special case of elliptical polarization. As shown in Fig. 15.7, elliptical polarization can be resolved into two linear vectors, E_x and E_y. Linear polarization results when one of these components is zero.

In order to receive a maximum signal, the polarization of the receiving antenna must be the same as that of the transmitting antenna, which is defined

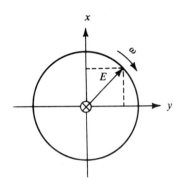

FIGURE 15.8. Circular polarization.

to be the same as that of the transmitted wave. For example, a wire dipole antenna, illustrated in Fig. 15.9(a), will radiate a linear polarized wave. A similar receiving dipole must be oriented parallel to the electric vector for maximum reception. If it is at some angle ψ, as illustrated in Fig. 15.9(b), then only the component of the electric field parallel to the receiving antenna will induce a signal in it. This component is $E \cos \psi$, and therefore the polarization loss factor is

$$\text{plf} = \cos \psi \tag{15.9}$$

A similar situation can exist with an aperture antenna as illustrated in Fig. 15.9(c). The angle ψ is the angle between the induced field in the aperture and the incoming electric field, and the polarization loss factor is also $\cos \psi$ in this case. In both cases the direction of propagation is normal to the plane of the antenna.

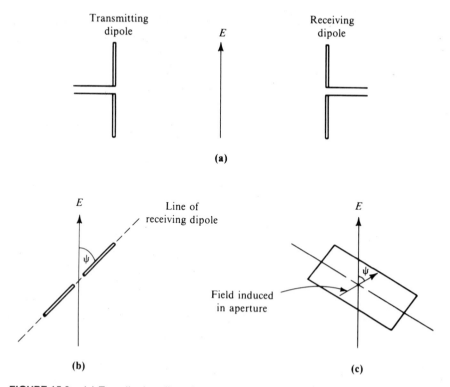

FIGURE 15.9. (a) Two dipoles aligned with same polarization; (b) receiving dipole in same plane as E but polarization misaligned; and (c) incoming wave E in same plane as aperture, but aperture polarization misaligned.

15.6 THE ISOTROPIC RADIATOR

The word *isotropic* means "equally in all directions," so that an isotropic radiator is one which radiates equally in all directions. A star is an example of an isotropic radiator of electromagnetic energy, but on a more practical level, all real antennas radiate better in some directions than others and cannot be isotropic. However, the concept of an isotropic radiator is a very useful one and provides a standard to which real antennas can be compared. Furthermore, since this is a hypothetical radiator, it may be assumed lossless, that is, its efficiency is unity. Let P_s represent the power input to a lossless isotropic radiator. Then since its efficiency is unity, this is also the power radiated. Consider this antenna at the center of the sphere shown in Fig. 15.5. Then, since any sphere has a solid angle of 4π steradians at its center, the power per unit solid angle is

$$P_i = \frac{P_s}{4\pi} \qquad \text{W/sr} \qquad (15.10)$$

This quantity is used as a standard to which real antennas can be compared. Another useful quantity is the power density. The surface area of a sphere of radius d is $4\pi d^2$, and therefore the power density for the lossless isotropic radiator is

$$P_{Di} = \frac{P_s}{4\pi d^2} \qquad \text{W/m}^2 \qquad (15.11)$$

It can be seen that the power density and the power per unit solid angle are related by

$$P_{Di} = \frac{P_i}{d^2} \qquad (15.12)$$

15.7 THE POWER GAIN OF AN ANTENNA

For any practical antenna, the power per unit solid angle will vary depending on the direction in which it is measured, and therefore it may be written generally as a function of the angular coordinates θ and ϕ as $P(\theta, \phi)$. The power gain of the antenna is then defined as the ratio of $P(\theta, \phi)$ to the power per unit solid angle radiated by a lossless isotropic radiator. The gain function, denoted by $G(\theta, \phi)$, is

$$G(\theta, \phi) = \frac{P(\theta, \phi)}{P_i}$$

$$= \frac{4\pi P(\theta, \phi)}{P_s} \qquad (15.13)$$

The gain function is a very important antenna characteristic which can be measured or, in some cases, calculated, and some examples will be given later.

For most antennas, the gain function shows a well defined maximum, which will be denoted by G_M, and the radiation pattern of the antenna is

$$g(\theta, \phi) = \frac{G(\theta, \phi)}{G_M} \qquad (15.14)$$

The radiation pattern is seen to be simply the gain function normalized to its maximum value. The maximum value G_M is referred to as the *gain* of the antenna, but this is only a gain in the sense that the antenna concentrates or focuses the power in the maximum direction. It does not increase the total power radiated.

Closely associated with the power gain is the *directive gain* of the antenna. This is the ratio of $P(\theta, \phi)$ to the average power per unit solid angle radiated by the *actual* antenna, and is denoted by $D(\theta, \phi)$. The average power per unit solid angle is $\eta_A P_s / 4\pi$, where η_A is the antenna efficiency and P_s is the power input, as before. Thus, the average is seen to be equal to $\eta_A P_i$, and therefore, the directivity is related to power gain by

$$D(\theta, \phi) = \frac{G(\theta, \phi)}{\eta_A} \qquad (15.15)$$

In particular, the maximum value of $D(\theta, \phi)$ is termed the *directivity*, or *directive gain*, given by

$$D_M = \frac{G_M}{\eta_A} \qquad (15.16)$$

When the gain function is plotted, a three-dimensional plot results, as sketched in Fig. 15.10(a). The length of the line from the origin to any point on the surface of the figure gives the gain in the direction of the point. The maximum gain G_M is shown, as well as the gain $G(\theta_1, \phi_1)$ in the direction (θ_1, ϕ_1). Minor lobes, as indicated by G_2 and G_3, also occur in general.

In practice, two-dimensional plots are often used, one for the equatorial plane, and one for the meridian plane, the function $g(\theta, \phi)$ usually being the one that is plotted. In the equatorial plane this is denoted by $g(\phi)$ since θ is constant, and in the meridian plane by $g(\theta)$ since ϕ is constant. This is illustrated in the following example.

Example 15.1   For the Hertzian dipole (to be described later), the radiation pattern is described by $g(\theta) = \sin^2 \theta$ and $g(\phi) = 1$. Plot the polar diagrams.

Solution   The polar diagrams are plotted in Fig. 15.10(b) and (c).

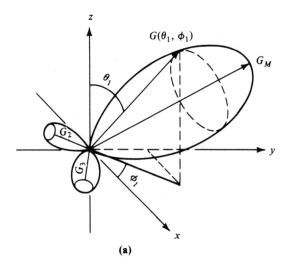

(a)

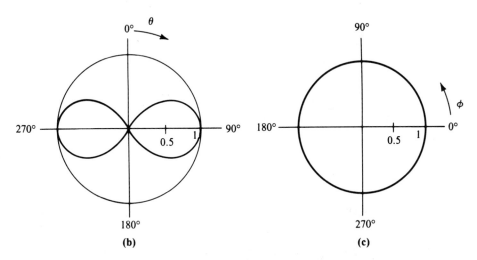

(b) **(c)**

FIGURE 15.10. (a) The gain function $G(\theta, \phi)$; (b) the polar diagram function $g(\theta)$; and (c) the polar diagram function $g(\phi)$, for example 15.1.

The -3 *dB beamwidth* of an antenna is the angle subtended at the center of the polar diagram by the -3 dB gain lines. This is illustrated in Fig. 15.11, where the beamwidth is

$$\theta_3 = \theta_2 - \theta_1 \qquad (15.17)$$

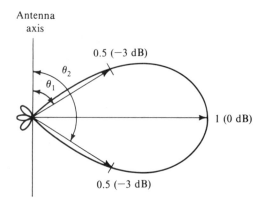

FIGURE 15.11. The -3 dB beamwidth.

Example 15.2   Determine the -3 dB beamwidth for the Hertzian dipole of Example 15.

Solution $g(\theta_1) = \sin^2 \theta_1 = .5$, giving $\theta_1 = 45°$. Also, $g(\theta_2) = \sin^2 \theta_2 = .5$, giving $\theta_2 = 135°$. Therefore,

$$\theta_3 = \theta_2 - \theta_1 = 135° - 45° = 90°$$

It will be observed from the above example that the beamwidth applies only to the meridian plane for this antenna, since the equatorial plane polar diagram is a circle.

In certain cases the beamwidth may be specified for levels other than -3 dB, other common values being -10 dB and -60 dB. Obviously the beamwidth level must be specified along with the beamwidth.

15.8 EFFECTIVE AREA OF AN ANTENNA

A receiving antenna may be thought of as having an effective area which collects electromagnetic energy from the incident wave, rather as a solar collector collects energy from sunlight. Assuming that the antenna is in the far-field zone of the radiated wave, the wave incident on it will be a plane TEM wave having a power density of P_D W/m^2 of wavefront. Let the receiving antenna be at the center of a spherical coordinate system, and let the incoming wave direction be specified by the angular coordinates (θ, ϕ) with reference to the antenna. The power delivered to a matched load (receiver) will be a function of direction, and this is taken into account by making the effective area a function of direction, i.e., $A = A(\theta, \phi)$. Thus, if P_R is the power delivered to a matched load,

$$P_R = P_D A(\theta, \phi) \tag{15.18}$$

Equation (15.18) serves as a defining equation for effective area. The effective area will have some maximum value A_{eff}, called the *effective area of the antenna*, just as the maximum power gain is called the gain of the antenna. As a result of the reciprocity theorem, the effective area normalized to its maximum value has the same functional form as the normalized power gain (Eq. (15.14)), i.e.,

$$\frac{A(\theta, \phi)}{A_{\text{eff}}} = g(\theta, \phi) \tag{15.19}$$

It follows that A_{eff} is proportional to G_M. It can also be shown from the reciprocity theorem that the constant of proportionality is the *same for all antennas*, namely, $\lambda^2/4\pi$. Thus,

$$\frac{A_{\text{eff}}}{G_M} = \frac{\lambda^2}{4\pi} \tag{15.20}$$

Thus, if the gain of an antenna is measured under transmitting conditions as G_T, its effective area under receiving conditions can be found from Eq. (15.20) to be

$$A_{\text{eff}} = \frac{\lambda^2 G_T}{4\pi} \tag{15.21}$$

This relationship was used, for example, in deriving Eq. (14.5) for free-space transmission. Note that G_T takes into account antenna efficiency, and therefore, so does the effective area. Often, in theoretical calculations, the directivity D_T is used in Eq. (15.21) instead of G_T, and this will give a higher value of effective area since directivity excludes the antenna efficiency.

Another factor which can reduce the effective area is the mismatch factor, given by Eq. (15.7). The effective area is reduced directly by this factor, and, of course, if the antenna is matched to the line, no reduction occurs.

Previously, the effective area was shown to be a function of the angular coordinates θ and ϕ. Defined in this way, $A(\theta, \phi)$ automatically takes into account any loss resulting from polarization misalignment. However, it is usually the maximum value A_{eff} of $A(\theta, \phi)$ that is known, as well as the polarization loss factor plf, as given by Eq. (15.9). Since the plf is defined for electric field strength, and A_{eff} for power, the reduction in A_{eff} as a result of polarization misalignment is $(\text{plf})^2$.

15.9 EFFECTIVE LENGTH OF AN ANTENNA

Although the concept of effective area can be used with any antenna, it is particularily useful with microwave antennas. At lower frequencies, where the physical structure of the antenna is of the form of a linear conductor or an array of conductors, an analogous concept, the *effective length*, proves to be more useful.

For a receiving antenna, the open circuit EMF appearing at the terminals is V_A, the Thévenin equivalent voltage, as shown in Fig. 15.1(c). Now, this is produced by a wave having an electric field strength of E V/m sweeping over the antenna, and therefore an effective length ℓ_{eff} can be defined by

$$V_A = E\ell_{\text{eff}} \tag{15.22}$$

The effective length ℓ_{eff} as defined by Eq. (15.22) is the maximum value. The effective length will in general be a function of θ and ϕ, just as the effective area is in general, and with ℓ_{eff} it is to be understood that the antenna is oriented for maximum induced EMF, so, for example, the polarization loss factor would be unity.

For the transmitting antenna, effective length is defined in a different manner, but it can be shown by the use of the reciprocity theorem that the effective length for transmitting is the same as that for receiving. The definition of ℓ_{eff} under transmitting conditions is in terms of the current distribution. The antenna current will vary as a function of physical length along the antenna. The effective length is defined such that the product of input terminal current and effective length is equal to the area under the actual current–length curve. Let I_0 represent the input terminal current; then

$$I_0\ell_{\text{eff}} = \text{Area under current–length curve} \tag{15.23}$$

Example 15.3   For the $\frac{1}{2}\lambda$ dipole described in Section 15.11, the current–length curve may be assumed to be $I = I_0 \cos \beta\ell$, where $\ell = 0$ at the input terminals. Find the effective length.

Solution   The physical length of the antenna is $\lambda/2$, and the average current for the cosine distribution is

$$I_{\text{av}} = \frac{2}{\pi} I_0$$

Thus,

$$\text{Area} = I_{\text{av}} \times \text{physical length}$$

$$= \frac{2}{\pi} I_0 \frac{\lambda}{2} = \frac{I_0\lambda}{\pi}$$

Hence, from Eq. (15.23),

$$\ell_{eff} = \frac{\lambda}{\pi}$$

For low- and medium-frequency antennas which are mounted vertically from the earth's surface, the effective length is usually referred to as the *effective height* h_{eff}. This is directly related to the physical height. Effective height must not be confused with the physical heights h_T and h_R introduced in Section 14.3. For example, a $\frac{1}{2}\lambda$ dipole may be mounted at some mast height h above the ground, but its effective length is *always* λ/π.

15.10   THE HERTZIAN DIPOLE

The Hertzian dipole is a short linear antenna, which, when radiating, is assumed to carry uniform current along its length. Such an antenna cannot be realized in practice, but longer antennas can be assumed to be made up of a number of Hertzian dipoles connected in series. The radiation properties of the Hertzian dipole are readily calculated. This is useful in itself in that it helps to illustrate the general properties discussed in the previous sections, but also, the properties of longer antennas can often be deduced by superimposing the results of the chain of Hertzian dipoles making up the longer antenna.

An approximation to a Hertzian dipole can be achieved by capacitively loading the ends of a short center-fed dipole as shown in Fig. 15.12(a). The capacitive ends allow a nearly uniform charging current to be maintained in the wire if its length $\delta\ell$ is much shorter than a wavelength ($\delta\ell \ll \lambda$). Thus, over the entire length of the dipole, the current is assumed to be

$$i = I_0 \sin \omega t \tag{15.24}$$

The electric field in the far-field zone is directly proportional to the component of current which is parallel to the electric field, which, as shown in Fig. 15.12(b), is $I_0 \sin \theta$. From the physics of radiation, it is also found that the instantaneous electric field is proportional to the rate of change of current and inversely proportional to the distance d, the full expression being

$$e_\theta = \frac{60\pi \, \delta\ell \, I_0 \sin \theta}{\lambda d} \cos \omega(t - d/c) \tag{15.25}$$

The subscript θ is used to show that the field is in the direction of the coordinate θ and at right angles to the direction of propagation. A delay time d/c is included, which is the time taken for a change in current to be effective at the point in the far-field zone.

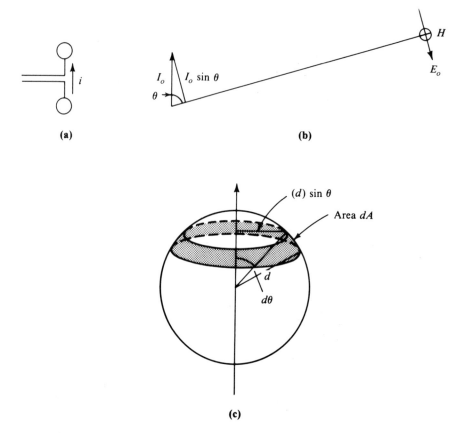

FIGURE 15.12. (a) The Hertzian dipole; (b) the far-zone field E_0 is proportional to the current component $I_0 \sin \theta$; (c) sphere used for determining radiated power.

The maximum value of the electric field, denoted by E_0, is

$$E_0 = \frac{60\pi\, \delta\ell\, I_0 \sin\theta}{\lambda d} \qquad (15.26)$$

The rms current I may be substituted for the maximum current I_0 to give the rms field strength E. In Appendix B it is shown that the power density in the far-field zone is given by $E^2/120\pi$, and using the rms field obtained from Eq. (15.26) gives, for the power density,

$$P_D = P_{DM} \sin^2 \theta \qquad (15.27)$$

where

$$P_{DM} = 30\pi \left(\frac{\delta \ell \, I}{\lambda d} \right)^2 \tag{15.28}$$

The total power flow through a strip (Fig. 15.12(c)) is given by $dP = P_D \, dA$, where dA is the area of the strip. From the geometry of the figure, $dA = 2\pi d^2 \sin\theta \, d\theta$. The total power is the summation of all such elemental amounts dP, which in the limit becomes the integral

$$P_T = \int_0^\pi (P_{DM} \sin^2\theta)(2\pi d^2 \sin\theta \, d\theta)$$

$$= 2\pi d^2 P_{DM} \int_0^\pi \sin^3\theta \, d\theta$$

$$= \frac{8\pi d^2 P_{DM}}{3} \tag{15.29}$$

Substituting for P_{DM} and simplifying,

$$P_T = I^2 80 \left(\frac{\pi \delta \ell}{\lambda} \right)^2 \tag{15.30}$$

Now, the radiation resistance is defined by the relationship $P_T = I^2 R_{\text{rad}}$, and therefore, from Eq. (15.30),

$$R_{\text{rad}} = 80 \left(\frac{\pi \delta \ell}{\lambda} \right)^2 \tag{15.31}$$

The directive gain can also be determined. This is defined as the ratio of the power per unit solid angle to the average power per unit solid angle, as in the derivation of Eq. (15.16). The relationship between power per unit solid angle and power density is $P(\theta, \phi) = d^2 P_D$ (see, for example, Eq. (15.12)). The average power per unit solid angle is $P_T/4\pi$, and therefore,

$$D(\theta, \phi) = \frac{4\pi P(\theta, \phi)}{P_T}$$

$$= \frac{4\pi d^2 P_{DM} \sin^2\theta}{8\pi d^2 P_{DM}/3}$$

$$= 1.5 \sin^2\theta \tag{15.32}$$

From Eq. (15.32), the directivity is seen to be $D_M = 1.5$, and the normalized gain is $g(\theta, \phi) = \sin^2\theta$. Because of the symmetry of the antenna, there is no variation of gain in the equatorial plane (i.e., $g(\phi) = 1$), and the gain variation in the meridian plane is $g(\theta) = \sin^2\theta$. These are the functions which were plotted in example 15.1.

I love A man!

Lastly, the effective area of the Hertzian dipole is found, using Eq. (15.21), to be

$$A_{\text{eff}} = 1.5 \frac{\lambda^2}{4\pi} = .119\lambda^2 \tag{15.33}$$

This is the effective area for unity efficiency.

15.11 HALF-WAVE DIPOLE

The *half-wave dipole* is a resonant antenna, the total length of which is nominally $\frac{1}{2}\lambda$ at the carrier frequency. Standing waves of voltage and current exist along the antenna, a good approximation to the distribution being obtained by assuming the antenna to be an opened-out $\frac{1}{4}\lambda$ section of an open-circuited transmission line. As shown in Section 12.7, the spacing between a standing-wave maximum and minimum is $\frac{1}{4}\lambda$, and since the current must be zero at the open circuit, it will be a maximum $\frac{1}{4}\lambda$ in from the end, while the voltage is a maximum at the end, going to a minimum at the $\frac{1}{4}\lambda$ point. Figure 15.13(a) shows how the magnitudes of voltage and current vary as a function of distance from the end, while Fig. 15.13(b) shows the line opened out to form the $\frac{1}{2}\lambda$ dipole, along with the voltage and current distributions, which are assumed unchanged. Because a 180° phase shift also occurs along the $\frac{1}{4}\lambda$ section ($\frac{1}{2}\pi$ radians for the incident wave, and $\frac{1}{2}\pi$ radians for the reflected wave in the opposite direction), it is convenient to show the voltage and current as in Fig. 15.13(b). The voltage is assumed to go through zero at the feed point. Of course, the voltage must be finite at the feed point, and the amplitude distribution cannot be identical to that of the transmission line section since the geometry of the antenna is different; however, results based on these assumptions agree very well with the measured results. The $\frac{1}{2}\lambda$ dipole may then be considered as consisting of a large number of Hertzian dipoles connected in series, the current in each being determined by the current distribution shown in Fig. 15.13(b).

There will also be a phase difference between radiation from different elements on the half-wave dipole as a result of the difference in distance $(d_0 - d)$, as shown in Fig. 15.14(b). Applying the cosine rule to the triangle formed by ℓ, d_0, and d results in $d^2 = d_0^2 + \ell^2 - 2d_0\ell\cos\theta$, and for $d_0 \gg \ell$, this gives $d_0^2 - d^2 \cong 2d_0\ell\cos\theta$. It is left as an exercise for the student to show that, applying the same mathematics as was used in deriving Eq. (14.21),

$$(d_0 - d) \cong \ell\cos\theta \tag{15.34}$$

The phase shift resulting from this is $(d_0 - d)2\pi/\lambda$, and this is seen to depend on both ℓ and θ. Thus, the response at some distant point P resulting from all

Hertzian dipole elements making up the $\frac{1}{2}\lambda$ dipole must take into account this phase difference, as well as the current distribution on the $\frac{1}{2}\lambda$ dipole. The total response can be found by integrating the individual fields for the full length of the $\frac{1}{2}\lambda$ dipole. This is quite a difficult integration to perform, and only the results will be given here. Corresponding to Eq. (15.26), the peak field strength is

$$E_0 = \frac{60I_0}{d}F(\theta) \tag{15.35}$$

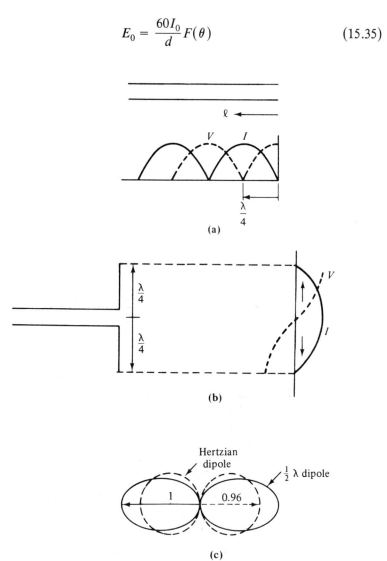

(a)

(b)

(c)

FIGURE 15.13. The half-wave dipole: (a) current and voltage standing waves on an open-circuit line; (b) current and voltage standing waves on a $\frac{1}{2}\lambda$ dipole; (c) radiation of $\frac{1}{2}\lambda$ dipole compared to that of a Hertzian dipole.

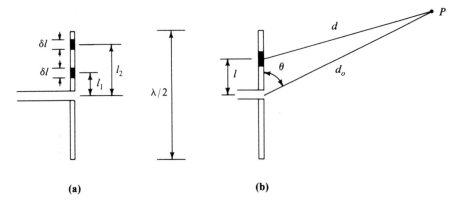

FIGURE 15.14. (a) Half-wave dipole made up of Hertzian dipoles; (b) path difference $(d - d_o)$ introduces a phase difference.

where

$$F(\theta) = \frac{\cos\left(\frac{\pi}{2}\cos\theta\right)}{\sin\theta} \tag{15.36}$$

The normalized power gain function is therefore

$$g(\theta) = F^2(\theta) \tag{15.37}$$

As with the Hertzian dipole, $g(\phi) = 1$ because of symmetry. It is left as an exercise for the student to show that the -3 dB beamwidth for the $\frac{1}{2}\lambda$ dipole is

$$\theta_3 = 78° \tag{15.38a}$$

The *field strength* polar diagram $F(\theta)$ is shown in Fig. 15.13(c), along with that for the Hertzian dipole ($F(\theta) = \sin\theta$)) for comparison, both normalized to unity for the $\frac{1}{2}\lambda$ dipole. Other important results for the $\frac{1}{2}\lambda$ dipole are

$$\ell_{\text{eff}} = \lambda/\pi \tag{15.38b}$$

$$D_M = 1.64 \tag{15.39}$$

$$A_{\text{eff}} = .13\lambda^2 \tag{15.40}$$

$$R_{\text{rad}} = 73 \; \Omega \tag{15.41}$$

The total impedance will be a function of frequency, being capacitive for frequencies just below the resonant value and inductive for frequencies above the resonant value, up to the next resonant value, which occurs when the physical length is approximately one wavelength. Because the velocity on the wire is slightly slower than that in free space, resonance does not occur at exactly the $\frac{1}{2}\lambda$ point, but at a slightly shorter length, in practice about 95% of the $\frac{1}{2}\lambda$ value. At $\ell = \frac{1}{2}\lambda$, the antenna impedance is $73 + j42.5 \; \Omega$, and at the 5% lower length it is $73 \; \Omega$, as illustrated in Fig. 15.15.

Transmitted waves are rarely single-frequency sinusoids, but are modulated. All modulated waves are made up of a carrier and a number of sideband frequencies spread out on either side. Since the half-wave dipole (or any other tuned antenna) is resonant at only one frequency, and since it behaves like a frequency-dependent reactance at other frequencies, the sideband frequencies will be distorted somewhat. For narrow-band transmissions this distortion is not significant, but at higher bandwidths it can cause trouble. The characteristics can be improved slightly by spoiling the effective Q of the antenna by making the radiating conductors large in diameter. This has the effect of increasing the capacitance and lowering the inductance, thus lowering the Q of the antenna. The result is a wider-frequency deviation between the 3-dB frequency points, or a wider bandwidth.

Resonance in the dipole is not limited to the half-wave frequency, but occurs also at all integer multiples of the half-wave frequency. The current distributions are different for each case, and the result is a different radiation pattern for each resonant frequency. This is illustrated in Fig. 15.16 for the cases of effective length equal to 1, 2, 3, and 4 times the half-wavelength. The drawings show that the number of lobes on each side of the radiator is equal to the multiple of half-wavelengths used. Current phasing changes by $180°$ from one $\frac{1}{2}\lambda$ section to another and thus is shown by the current arrows. Again, these patterns are the ones that would result if the antenna were mounted free in space, away from reflections. It must also be noted that the current node occurs at different positions on the antenna length, and if a resonant feed point is to be maintained, it must be located at one of the current nodes that occur; that is, it must be located at $\frac{1}{4}\lambda, \frac{3}{4}\lambda, \frac{5}{4}\lambda,\dots$ away from either end of the radiator.

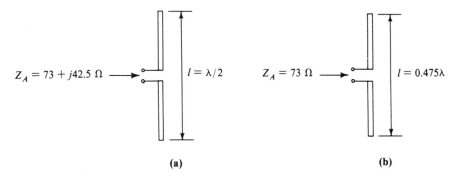

$Z_A = 73 + j42.5 \; \Omega$ $l = \lambda/2$ $Z_A = 73 \; \Omega$ $l = 0.475\lambda$

(a)   (b)

FIGURE 15.15. (a) Input impedance of half-wave dipole cut exactly to $\frac{1}{2}\lambda$ is $73 + j42.5$ ohms; (b) cutting the dipole about 5% shorter reduces reactive component to zero.

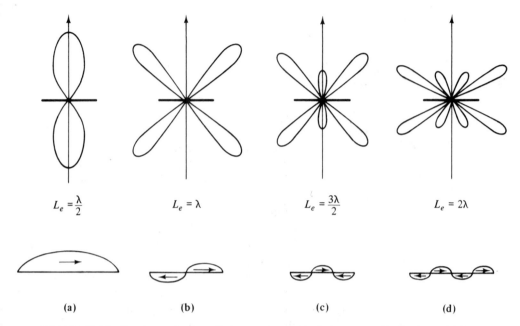

$$L_e = \frac{\lambda}{2}$$

$$L_e = \lambda$$

$$L_e = \frac{3\lambda}{2}$$

$$L_e = 2\lambda$$

(a)　　　　　　　(b)　　　　　　　(c)　　　　　　　(d)

FIGURE 15.16. Resonant dipole radiation patterns with their current distributions: (a) for $L_e = 1 \times \lambda/2$; (b) for $L_e = 2 \times \lambda/2$; (c) for $L_e = 3 \times \lambda/2$; (d) for $L_e = 4 \times \lambda/2$.

With the increase in the number of lobes, the lobes nearest the antenna axis will always be larger than the others. These are the major lobes, and they get progressively closer to the axis with increasing numbers of lobes.

15.12 VERTICAL ANTENNAS

15.12.1 Ground Reflections

The ground will act as an almost-perfect reflecting plane for any antenna placed near its surface, and an apparent mirror image of the antenna will appear to be located immediately beneath the surface below the antenna. This is illustrated by Figure 15.17, which shows a distant observer receiving a direct wave from a point on the antenna, and a reflected wave which appears to come from the corresponding point on the image antenna.

Because of the interaction between the direct and reflected waves, the radiation polar pattern is drastically modified and appears to be the vector sum of radiation from two separate antennas, the real one and the image. The amount of interaction is dependent on how far above the ground the antenna is placed. If the effective height is several wavelengths, then practically no interaction will occur, and the antenna may be considered to be mounted in free space. For heights up to a few wavelengths, the reflections must be

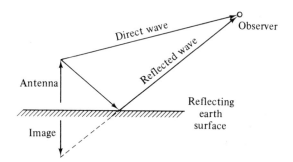

FIGURE 15.17. A vertical antenna and its reflected image.

considered, and the antenna and its image act as a phased array of two antennas.

The image antenna is an exact mirror image of the real antenna, and the apparent currents within it are at each point the same as those within the real antenna, except for reversed polarity. This is true whether the antenna is grounded or not grounded, horizontally or vertically polarized.

15.12.2 Grounded Vertical Antennas

Most of the medium-frequency (MF) broadcast antennas fall into this category, as do the VHF mobile-whip antennas. This type of antenna, known as the *Marconi antenna*, is made up of a vertical mast, pole, or rod which forms the main radiating conductor. It may be free-standing or supported by insulated guy wires, and is placed in a location where good electrical ground is available. Good locations for MF antennas include marshy fields and seacoast flats. If poor soil conditions exist, an artificial ground plane may be created by burying a mat of heavy conductors extending radially from the mast for up to at least a quarter-wavelength, and preferably at least a half-wavelength, from the mast. The ends of the buried radials are usually connected together, with deep grounding stakes driven at their extremities. The base of the mast is electrically insulated from ground, and the feed line from the transmitter is connected between the mast base and the ground. (In some special cases the feed point may be located at a current node farther up the mast.)

Since high-power transmitters may generate potentials of several hundred kilovolts on the antenna structure, high-quality insulators are a must. At lower frequencies special plastics such as polystyrene and Teflon are used.

The grounded vertical antenna combines with its image to act in exactly the same manner as a doublet or dipole, with the radiator vertical to the ground surface. Figure 15.18 shows grounded verticals of various lengths with the current distributions on the antenna and its image, and the resulting radiation pattern in the vertical plane. The radiation pattern in the horizontal plane is circular. For vertical radiators of $\frac{1}{4}\lambda$ or less in effective height, the current distributions on the combined antenna-image structure is identical to a doublet of dipole of twice the length of the radiator, and produces a radiation pattern that is the same. However, half of the radiation pattern appears to lie below the ground surface, and in fact does not exist. All the power from the radiator is contained in that portion of the pattern which is above the surface. The $\frac{1}{4}\lambda$ vertical acts similarly to the half-wave dipole. The current distribution and radiation pattern are shown in Fig. 15.18(a). The structure behaves like two antennas being fed in parallel; the radiation resistance is reduced to half that of the dipole, or about 36.5 Ω.

Figure 15.18(b) shows the current distributions and pattern for a $\frac{1}{2}\lambda$ vertical radiator. The structure no longer behaves like a combined dipole, but more like an array of two separate antennas of the same length fed 180° out of phase with each other. Each half is a $\frac{1}{2}\lambda$ antenna, and the two halves aligned so that their radiation patterns are in phase along the ground surface. This makes the antenna appear to have the same pattern shape as the $\frac{1}{2}\lambda$ dipole, but with an additional gain of 3 dB.

When the length of the radiator is increased above $\frac{1}{2}\lambda$, side lobes begin to appear, and the main lobe lifts off the ground. Figure 15.18(c) shows the result

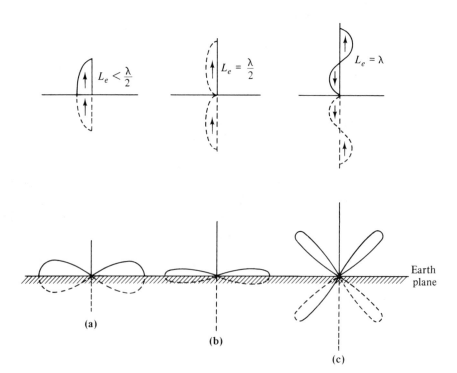

FIGURE 15.18. Grounded vertical radiators. Current distributions and vertical radiations patterns are shown for effective lengths of (a) less than $\lambda/2$; (b) $\lambda/2$; (c) λ.

for the 1λ radiator. This is identical in shape to the pattern for the 1λ dipole shown in Fig. 15.16(b), except that the bottom half of the pattern has been folded over and added to the top half. It is obvious from this that if a vertical is to radiate along the surface of the ground, it must not be higher than $\frac{1}{2}\lambda$, or too much power will be radiated into the sky. If sky waves are to be used, a longer radiator must be used to get the main lobe off the ground.

The $\frac{1}{4}\lambda$ verticals are favored because they may be fed directly with a cable at the low-impedance current node which occurs at the base of the mast. At low frequencies it is often not physically feasible to build a mast which is a full $\frac{1}{4}\lambda$ in height, so a form of "top-hat" loading is employed to make the radiator look longer than it physically is. A horizontal wire is connected to the top of the mast to make up the missing length. This may be a simple wire, as in the inverted L and T antennas, or it may be in the form of a disk. The length of the horizontal or the disk radius is adjusted so that the current node occurs at the base of the mast, and from the feed point the antenna appears to be a $\frac{1}{4}\lambda$ tuned vertical. The current density in the horizontal portion is low and does not radiate nearly as much power as does the vertical portion, with the result that little change occurs in the radiation pattern. Some minor lobes in the upward direction do occur, but the major lobe is caused to lie closer to the ground because of more even current distribution within the vertical mast. This type of antenna is sometimes referred to as a *unipole antenna*, because only one radiating element is fed.

Grounded verticals are frequently used at VHF for mobile service because of their simple structure. In this case the radiator length varies from a few feet down to a few inches, and a real earth ground cannot be used. For fixed antennas a series of radial ground-plane rods are frequently used, while for the mobile antennas, the flat metal top of the vehicle is used to provide the ground plane. Figure 15.19 illustrates these grounds.

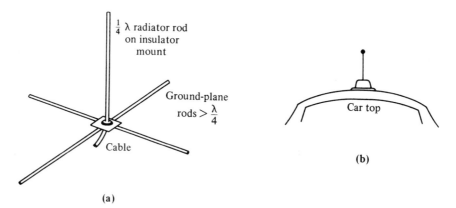

FIGURE 15.19. Whip antennas: (a) VHF vertical with simulated ground plane; (b) UHF mobile whip mounted on car top.

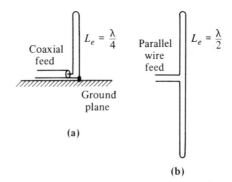

FIGURE 15.20. Folded dipoles: (a) folded quarter-wave vertical; (b) folded half-wave dipole.

15.13 FOLDED ELEMENTS

The antennas discussed so far all use single-conductor radiators. However, if each conductor is twinned with a second conductor, insulated from, but closely parallel to it, and connected together at the voltage node points, then a similar current pattern will be induced in the second conductor. The radiation pattern will be exactly the same as for the single-conductor antenna, but the radiation resistance will be different, namely, just four times that produced by the single radiator. Thus, a folded vertical antenna as shown in Fig. 15.20(a) will have a radiation resistance four times that of the single vertical (4 × 36.5), or 146 Ω. The folded dipole shown in Fig. 15.20(b) has a radiation resistance of 4 × 73, or 292 Ω.

The folded dipole antenna is favored as the driving element of VHF dipole arrays because it can be made very inexpensively with self-supporting tubing and provides a higher terminal impedance, which tends to offset the reduction of impedance resulting from the loading of the parasitic elements. Since the center of the second conductor is at a current node, it may be electrically connected to a grounded support rod without affecting the characteristics, making a very sturdy mechanical structure.

15.14 LOOP AND FERRITE-ROD RECEIVING ANTENNAS

15.14.1 Loop Antenna

The *loop antenna* is made up of one or more turns of wire on a frame, which may be rectangular or circular, and is very much smaller than one wavelength across. This antenna is popular for two reasons: (1) it is relatively compact, lending itself to use with portable receivers; and (2) it is quite directive, lending itself to use with direction-finding equipment.

A loop antenna is shown in Fig. 15.21(a), with its radiation pattern. The radiation pattern is the doughnut shape of the doublet antenna, except that the plane of the doughnut corresponds to the plane of the loop, so that the loop will radiate equally well in all directions within its own plane. A distinct null response occurs along the axis of the loop, and it is this null which is used in direction finding. The physical shape of the loop does not drastically affect the radiation pattern except when the length and width of the loop are unequal, in which case a squashed doughnut shape occurs.

Loop antennas made of several turns of wire around a rectangular frame were popular for earlier model broadcast receivers, with the loop being mounted in the back of the cabinet. Recently, these have been almost entirely replaced by the smaller ferrite-rod antennas.

When the loop is aligned for maximum signal strength, the magnetic flux linkages are BAN, where B is the rms magnetic flux density in teslas, A is the physical loop area in m², and N is the number of turns in the loop. The

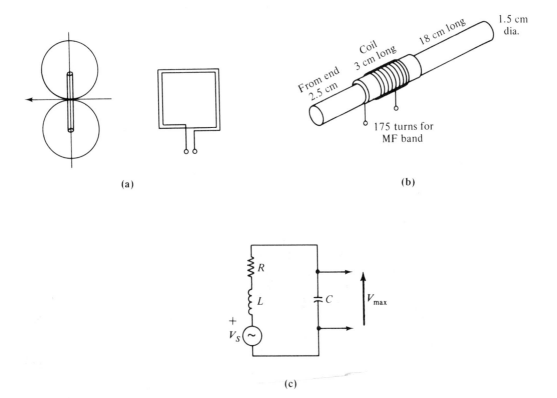

(a) (b)

(c)

FIGURE 15.21. Loop antennas: (a) square-loop antenna with its radiation pattern; (b) ferrite-rod antenna; (c) equivalent circuit for the ferrite-rod antenna.

induced EMF is given by Faraday's law as the rate of change of flux linkages, which for a wave of angular frequency ω gives

$$V_s = \omega BAN \tag{15.42}$$

When the loop is tuned by means of an external capacitor to the received frequency, the voltage at the capacitor terminal is magnified by the circuit Q to give

$$V_{max} = V_s Q = \omega BANQ \tag{15.43}$$

Since the loop is usually much smaller than the received wavelength, the induced voltage may be quite small. It can be increased by increasing any one of the factors in Eq. (15.43). The Q is determined by the desired selectivity. The area must be kept small: increasing the number of turns increases the coil inductance and changes the Q, and even changing the flux density B affects the Q. However, changing the flux density by using a magnetic core can be achieved with a minimal change of Q using ferrite cores. This alternative is now in widespread use.

15.14.2 Ferrite-Rod Antenna

The *ferrite-rod antenna* is made by winding a coil of wire on a ferrite rod similar to the one illustrated in Fig. 15.21(b). Ferrites are materials that exhibit the properties of ferrimagnetism. The materials exhibit a high relative permeability in the same manner as magnetic metals do, but unlike the ferromagnetic metals, they also have a high bulk resistivity. This means that at high frequencies, eddy currents induced within the materials are practically nonexistent, and high-Q coils can be made. Typical values for μ are around 100 and for resistivity 10,000 Ω-cm. A high length-to-diameter ratio for the rod gives a high permeability, which is desirable.

The size of coil is a compromise among several factors. If the coil is too long compared to the rod length, the change of permeability with temperature will cause a noticeable change in the inductance. If it is too short, the Q will be low. Positioning the coil on the core is critical as well, since the effective permeability is a function of position on the rod, ranging from a maximum at the center to a minimum at either end. The coil is usually placed near the quarter-point, allowing adjustment in either direction to trim the coil inductance. When more than one coil is mounted on the same rod, they must be placed at opposite ends to minimize interaction between them.

The coil of wire on the ferrite rod is basically a modified loop antenna, so the induced maximum EMF appearing at its terminals is given by

$$V_s = \omega BANF\mu_r \tag{15.44}$$

where

F = modifying factor accounting for coil length, ranging from unity for short coils to about 0.7 for one that extends the full length of the rod

μ_r = effective relative permeability of the rod, as measured for the actual coil position

A = rod cross-sectional area.

An expression for the effective length of a ferrite rod antenna can be derived by combining Eqs. (B.4), (B.8), (15.22), and (15.44) to give

$$\ell_{\text{eff}} = \frac{2\pi A N F \mu_r}{\lambda} \tag{15.45}$$

Since the voltage appearing at the terminals is of more importance in a receiving antenna, the factor $Q\ell_{\text{eff}}$ is often given as a figure of merit for rod antennas. The directional properties of the ferrite-rod antenna are similar to those of the loop antenna, although the null may not be quite so pronounced.

15.15 NONRESONANT ANTENNAS

15.15.1 Long-Wire Antenna

The *long-wire antenna* is just that, a wire several wavelengths in length that is suspended at some height above the earth. The wire is driven at one end and has a resistive termination at the remote end which is matched to the characteristic impedance of the line at that end. This forms a transmission line with a ground return and a matched termination. When an alternating current wave is transmitted down this line toward the terminated end, about half of the energy is radiated into space. Since there is no reflection at the far end, no return wave exists, and no standing waves appear on the wire, regardless of its length-to-wavelength ratio. Of the energy not radiated, a small amount is dissipated in the wire, and the remainder is dissipated in the terminating resistance.

The long wire is illustrated in Fig. 15.22(a), with its horizontal radiation pattern. The radiation pattern shown would be true for any direction at right angles to the wire if the wire were mounted in free space. Usually it is a fraction of a wavelength above the ground, and ground reflections cause most of the energy to be radiated upward so that the vertical pattern would be a single lobe of twice the strength of the horizontal lobes.

This antenna is not often used because it is not very efficient, has a comparatively low gain, and takes up a lot of space. Also, matching the transmitter to the line can be a problem. However, since no standing waves exist, the antenna has no resonances, and as long as the length of the wire lies in the range 2λ to 10λ, its characteristics remain relatively constant for all frequencies in that range. It is thus used as a broadband antenna for low-cost point-to-point communications, especially in the HF band from 3 to 30 MHz. The upward tilt of the pattern lends itself to skywave propagation in this band.

15.15.2 Rhombic Antenna

The *rhombic antenna* takes its name from its diamond-shaped layout. It is an array of four interconnected long-wire antennas, laid out in the manner shown in Fig. 15.22(b). Each of the four legs has the same length and lies in the range 2λ to 10λ. The transmission line feeds one end and transmits an unreflected current wave down each side toward the resistive termination at the far end. The lengths of the sides and the angle ϕ are interrelated and must be

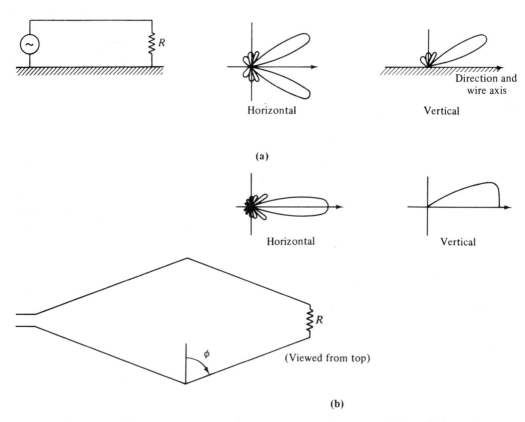

FIGURE 15.22. Nonresonant antennas: (a) the long-wire antenna, with its radiation patterns in the horizontal and vertical planes along the wire axis; (b) the rhombic antenna, with its radiation patterns in the horizontal and vertical planes along the rhombic major axis.

carefully chosen so that the side lobes cancel properly, leaving only a single main lobe lying along the main axis of the rhombus. Again, ground reflections cause the lobe to be tilted upward into the sky, and the amount of tilt is a function of the length of the legs.

The resistive termination is chosen so that no reflections occur, and the antenna is untuned, as is the long-wire antenna. Its frequency range is broad, almost 10 to 1, allowing a single structure to be used over most of the HF bands. It is highly directional and, if the tilt is chosen properly, is ideal for point-to-point skywave propagation.

The feed-point impedance falls in the range 600 to 800 Ω, allowing direct feed with an open-wire parallel line, and the terminating resistor is in the same range. The angle ϕ falls between $40°$ and $75°$ and the leg length between 2λ and 10λ. The resulting directive gain obtained with the rhombic ranges from 15 to 60. The physical structure is relatively simple and inexpensive and usually takes the form of four poles placed at the apexes of the rhombus and wires supported on tension insulators for the radiators. The feed line is a parallel line and can be any length within reason, although it is usual practice to place the transmitter house near the feed end of the rhombus. For example, a 2λ rhombic for 3 MHz would have diagonals of about 320 m long by about 250 m wide. This antenna is not practical in an urban environment.

15.16   DRIVEN ARRAYS

A *linear array* of antennas consists of a number of basic antenna elements, usually $\frac{1}{2}\lambda$ dipoles, equispaced out along a line referred to as the array axis. By suitably phasing the radiation from each element, the directivity can be altered in a number of ways. To illustrate this, arrays of identical elements operating in the transmitting mode will be considered, as sketched in Fig. 15.23(a). It is assumed that each element is fed with currents of equal amplitudes, but with successive phase shifts α. Thus, the current to the first element, is $I\underline{/0}$, to the second, $I\underline{/\alpha}$, to the third, $I\underline{/2\alpha}$, and to the nth element, $I\underline{/(n-1)\alpha}$. Figure 15.23(b) shows the plan view of the array, and for convenience this is taken to be the equatorial plane. The distance between successive wavefronts is $s\cos\phi$, and therefore the phase lead of element n with respect to element $(n-1)$ is $(2\pi/\lambda)s\cos\phi$. Thus, the total phase lead of element n with respect to element $(n-1)$ is

$$\psi = \frac{2\pi}{\lambda}s\cos\phi + \alpha \tag{15.46}$$

Figure 15.23(c) shows the phasor diagram for the individual field strengths at a point in the far-field zone. The individual field strengths are assumed to have equal amplitudes E_1 and to undergo successive phase shifts ψ. The resultant

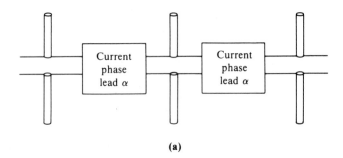

(a)

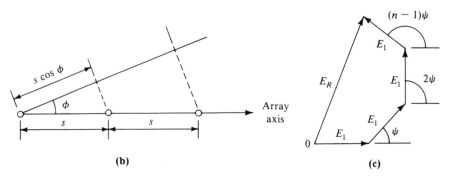

(b)

(c)

FIGURE 15.23. (a) A linear array of dipoles; (b) plan view of (a); (c) phasor diagram for the field strength in the far-field zone.

field strength amplitude E_R is the closing side of the polygon, and a geometrical result for such a polygon is

$$E_R = E_1 \frac{\sin(n\psi/2)}{\sin(\psi/2)} \qquad (15.47)$$

Now, if the *same total power* had been radiated from a single element, the field strength would have been $\sqrt{n}\, E_1$. The *array factor* AF is the ratio of E_R to $\sqrt{n}\, E_1$, and therefore,

$$\text{AF} = \left| \frac{\sin(n\psi/2)}{\sqrt{n}\sin(\psi/2)} \right| \qquad (15.48)$$

This is the factor by which the array increases the field strength over that of a single element *radiating the same total power*.

The array factor has a maximum which occurs at $\psi = 0$. As ψ approaches zero, $\sin(n\psi/2) \to (n\psi/2)$ and $\sin(\psi/2) \to (\psi/2)$, so that, from Eq. (15.48),

$$\text{AF}_{\text{max}} = \frac{(n\psi/2)}{\sqrt{n}\,(\psi/2)} = \sqrt{n} \qquad (15.49)$$

The direction of the maximum can be selected by an appropriate choice of element spacing and current phasing, as illustrated in the following sections.

15.16.1 The Broadside Array

The *broadside array*, as the name suggests, has its maximum directed along the normal to the plane of the array, i.e., when ϕ is $90°$ in Fig. 15.23(b). This requires ψ to equal zero as shown in Eq. (15.49), and from Eq. (15.46), this in turn requires α to equal zero, or the currents to be in phase. This can be achieved very conveniently by spacing the elements $\frac{1}{2}\lambda$ apart, and alternately crossing the feed points as shown in Fig. 15.24(a). The field strength polar diagram in the equatorial plane is given by the normalized array factor. The diagram is sketched in Fig. 15.24(a). Increasing the number of elements sharpens the beam in the broadside direction, but will also introduce small sidelobes as sketched in Fig. 15.24(a). The polar diagram in the meridian plane will be that for a single element (normalized to the maximum value $\sqrt{n}\,E_1$), and for an array of half-wave dipoles, it would be similar to that shown in Fig. 15.13.

Quite often the broadside array is used in conjunction with a second array of reflectors, which is made the same size and mounted a half-wavelength behind the main array. The back lobe is now reflected forward and adds directly to the forward lobe, improving its gain and directivity and making the structure unidirectional. The reflectors may be driven, or they may be parasitic reflectors.

A variation of this array, called a *collinear broadside array*, is formed when a number of dipoles driven in phase are spaced in-line along the same axis. This array radiates equally well in all directions within the plane normal to the axis, but produces very little radiation off this plane and none along the array axis.

15.16.2 The End-Fire Array

The end-fire array, as the name suggests, has the main beam directed along the axis of the array, i.e., when ϕ is 0 in Fig. 15.23(b). As shown in Eq. (15.49), a maximum requires ψ to be zero, and hence, from Eq. (15.46),

$$0 = \frac{2\pi}{\lambda} s \cos(0) + \alpha$$

Thus,

$$\alpha = -\frac{2\pi}{\lambda} s \tag{15.50}$$

The negative sign indicates that successive phase lags are required, propor-

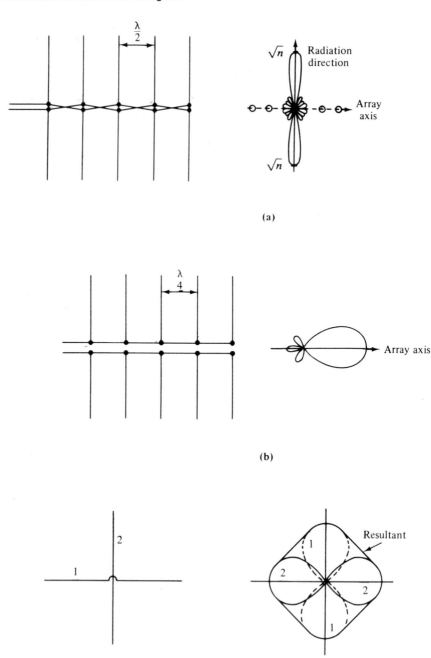

FIGURE 15.24. Driven arrays: (a) broadside array with radiation pattern in the equatorial plane; (b) end fire array with its radiation pattern in the equatorial plane; (c) the turnstile antenna, showing its radiation pattern as the sum of two dipole patterns at right angles to each other.

tional to the spacing s. Thus, the phase angle ψ for the end-fire array becomes

$$\psi = \frac{2\pi}{\lambda} s (\cos \phi - 1) \tag{15.51}$$

This shows that with ϕ equal to zero, ψ is also zero, and therefore the array factor becomes the maximum value as given by Eq. (15.49). When $\phi = 180°$, the angle ψ becomes $-4\pi s/\lambda$, which, for $s = \lambda/2$, yields $\psi = -2\pi$. This would also give a maximum, and for proper end-fire operation with only a single beam, this spacing of $\lambda/2$ must be avoided. A common arrangement used in practice is to space the elements by $\lambda/4$ and directly feed them in parallel as shown in Fig. 15.24(b). Depending on the number of elements, some minor backlobes may exist, but these are generally much smaller than the forward lobe. For a given number of elements, the end-fire array does not produce as narrow a beam as the broadside array.

Both the broadside and the end-fire arrays may be used for any frequency band. Because of their physical size, they are usually limited to the HF bands and higher, but they have been used in the LF bands for point-to-point communications. Overseas broadcasting stations in the MF and HF bands frequently use them as well, for repeated broadcasting to the same distant area.

15.16.3 Turnstile Antenna

Figure 15.24(c) shows a simple *turnstile antenna*, consisting of two half-wave dipoles placed at right angles to each other and fed 90° out of phase with each other. This results in the two dipole patterns combining in the manner shown in the figure, producing an almost circular pattern in the plane of the turnstile. The pattern also has the feature that it is polarized in that same plane, so that if it were mounted in the horizontal plane, the antenna would radiate horizontally polarized waves about equally well in all directions along the ground. Several of these turnstiles may be stacked along a vertical axis and phased so as to improve the radiation directivity along the ground (i.e., in the plane of polarization). This type of antenna is frequently used for television broadcasting in the VHF-UHF bands.

15.17 PARASITIC ARRAYS

15.17.1 Parasitic Reflectors

Parasitic elements are secondary antennas which are placed in close proximity to the main or driven antenna. They are not directly fed, but have currents induced in them from the main element (or from the received wave in the case of a receiving antenna). The secondary antennas are tuned so as to cause a lagging or leading phase shift in energy which is reradiated from them, and this changes the radiation pattern of the main antenna, as shown in Fig. 15.25(a) and (b).

The reflector element is placed behind the driven dipole and is made about 5% longer than the driven ($\lambda/2$) dipole so that it is inductive. Spacing is adjusted until maximum radiation occurs along the normal in front of the dipole (array axis) as shown in Fig. 15.25(a). Optimum spacing, which is found by experiment, is usually about 0.15λ.

15.17.2 Parasitic Directors

The director element, which is placed in front of the driven dipole, is made about 5% shorter than the dipole so that it is capacitive. It is spaced to provide maximum radiation in the forward direction, and optimum spacing is again found experimentally to be about 0.15λ. The pattern is shown in Fig. 15.25(b).

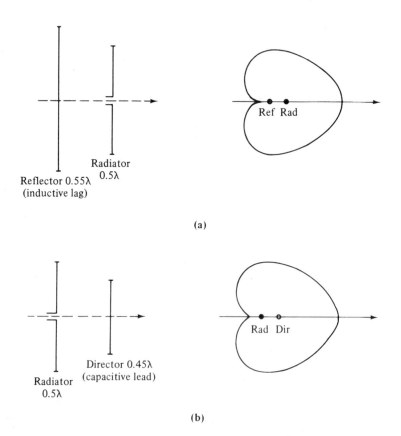

(a)

(b)

FIGURE 15.25. The effects of parasitic elements on the radiation pattern of a dipole: (a) reflector; (b) director.

15.17.3 Yagi-Uda Array

The *Yagi-Uda* (or simply the *Yagi*) *antenna* is a parasitic array comprising a driven half-wave dipole antenna which is usually a folded dipole, a single parasitic reflector, and one or more (up to 13) director elements with each director cut to act as if the previous one were the driven element, so that the whole structure tapers in the direction of propagation. The structure is illustrated in Fig. 15.26. All the elements are electrically fastened to the conducting, grounded central support rod. This has no effect on the currents since the support point in the center of each element is at a current node.

Only one reflector need be used, since the addition of a second or third reflector adds practically nothing to the directivity of the structure. The directive gain is improved considerably by the addition of more directors to give directive gains from about 7 dB for a three-element Yagi to about 15 dB for a five-element Yagi. The pattern consists of one main lobe lying in the forward direction along the axis of the array, with several very minor lobes in other directions. Polarization is in the direction of the element axes.

A folded dipole is often used as the driven element to raise the antenna terminal impedance. The parasitic elements are quite closely coupled to the driven element, which results in a lowering of the effective radiation resistance. The fourfold multiplication of the folded dipole brings the impedance level back up to reasonable levels. A Yagi with a folded dipole would have a terminal resistance of 200 to 300 Ω.

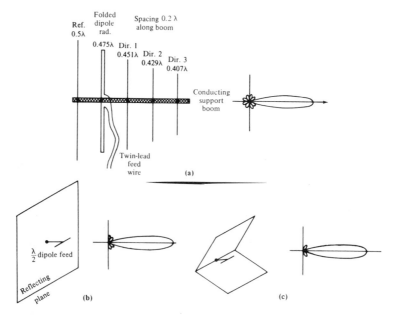

FIGURE 15.26. (a) A five-element Yagi-Uda array; the structure of the antenna is shown with its dimensions and the radiation pattern along its major axis. (b) A plane reflector with a dipole feed. (c) A corner reflector with a dipole feed.

This type of antenna would be very bulky at low frequencies; hence, it is used most often in the VHF range. Radio amateurs have built Yagis for the 20-m band, but the structure is large and cumbersome. Its high directional gain makes it ideal for point-to-point fixed-frequency communications networks, either at terminal or at repeater stations. They have also been used for base stations on mobile communications systems where the working area is strung out along a line, such as railroads, highways, or pipelines.

15.17.4 Plane Reflector Arrays

At UHF it is common practice to use a *plane reflecting surface*, either a flat surface or a corner of two surfaces, in place of the single reflector element of the Yagi. The reflecting surface must be at least one wavelength across in each direction, and can be much larger. It may be a solid metal surface, or it may be wire mesh or a network of interconnected metal rods. Somewhat sharper directivity is obtained with the corner reflector, shown in Fig. 15.26(b), but only in the plane across the fold.

The plane reflector is arranged so that the driven dipole (or directional antenna of any type) is mounted a quarter-wavelength ahead of the reflector surface. A mirror image of the driven antenna occurs a quarter-wavelength behind the surface and appears to radiate a wave which arrives at the driven antenna in phase with the driving antenna radiation, but delayed one period. The resulting radiation pattern appears to be the vector sum of the radiation from two in-phase antennas, one of which is the image.

15.18 VHF-UHF ANTENNAS

15.18.1 Discone Omni

The *discone antenna* is designed to radiate an omnidirectional pattern in the horizontal plane, with vertical polarization. It is a broadband antenna with usable characteristics over a frequency range of nearly $10:1$. It is usually designed to be fed directly from a 50-Ω coaxial line and is mounted directly on the end of that line. The discone is illustrated in Fig. 15.27(a).

This type of antenna is ideal for base-station operation for urban mobile communication systems, since it gives a good omnidirectional pattern, is physically very compact and rugged, and is quite inexpensive to construct. Its directional gain along the horizontal plane is comparable to that of the dipole antenna.

15.18.2 Helical Antenna

The radiator element of a *helical antenna* is basically a coil of wire. If the helix diameter is much less than one wavelength, its length less than one wavelength, and it is center-fed, the whole structure will behave very much like

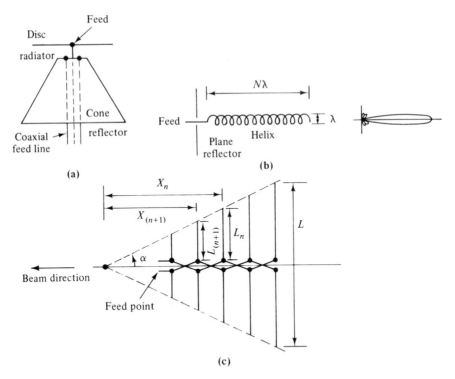

FIGURE 15.27. VHF-UHF antennas: (a) discone omnidirectional antenna; (b) end-fire helix antenna; (c) log periodic dipole array.

a compact dipole antenna, radiating in the "normal" mode. Since the current wave must propagate along the helix conductor, its actual velocity along the axis will be much less than free-space velocity. The velocity will be determined by the diameter and pitch of the helix coil, and the half-wave resonant length of the helix will physically be very much less than the free-space half-wavelength. For most combinations of helix, the polarization is elliptical. This type of antenna is sometimes used in locations where it is not possible to mount a full-sized dipole, such as in urban areas or on rooftops.

If the helix diameter is made approximately one wavelength, and the helix made several wavelengths long and end-fed, the helix radiates in an end-fire mode, producing a narrow beam of circularly polarized waves. The 3-dB beam width obtainable with single helixes is on the order of $15°$ to $30°$. When used in this mode, the radiator is usually end-fed and a plane reflector is placed behind the feed end. The result is a highly directional antenna which is physically compact and which can be easily mounted on a movable table for tracking moving sources. Arrays of end-fire helixes are often used for tracking satellites. These structures are much less expensive than the large parabolas used for radio astronomy and have been very popular with amateur satellite trackers. The end-fire helix is illustrated in Fig. 15.27(b).

15.18.3 Log Periodic Antenna

The *log periodic antenna* is basically an array of dipoles, fed with alternating phase, lined up along the axis of radiation. The structure is illustrated in Fig. 15.27(c). The element lengths and their spacing all conform to a ratio, given as

$$\tau = \frac{L_{(n+1)}}{L_n} = \frac{X_{(n+1)}}{X_n} \tag{15.52}$$

Also, the angle of divergence is given as

$$\alpha = \tan^{-1}\left(\frac{L_n}{X_n}\right) \tag{15.53}$$

The open-end length L must be larger than $\frac{1}{2}\lambda$ if high efficiency (90%) is to be obtained.

This antenna has the unique feature that its impedance is a periodic function of the logarithm of the frequency—hence its name. The antenna characteristics are broadband, and it has the directional characteristics of a dipole array. This type of antenna is often used for mobile-base-station operations, where many channels must be handled over a single antenna system with good directive characteristics.

15.19 MICROWAVE ANTENNAS

15.19.1 Horns

Radio waves can be radiated directly from the end of a waveguide in the same way as from the end of an open transmission line. The end of the waveguide represents an abrupt transition from the characteristic impedance of the waveguide into that of free space, and the radiation resulting is neither efficient nor very directive. This state of affairs can be improved considerably by flaring out the end of the waveguide to form a hornlike structure. A gradual transition can thus take place as the wave passes from the mouth of the horn.

Narrow-mouthed horns with long flare sections produce sharper beams than shallow, wide-mouthed ones. Also, the wider-mouthed horns tend to produce a wavefront with a distinct curvature, which is undesirable. The ideal would be for the waves to leave the horn with a completely planar wavefront, and to accomplish this a focusing mechanism, such as a curved reflector or a lens, may be used with the horn.

Three types of horns are shown in Fig. 15.28. The first is the sectoral horn, which is flared in only one plane (Fig. 15.28(a)); the second is the pyramidal horn, which is flared in both planes (Fig. 15.28(b)). Both of these are used with rectangular waveguides. The third type is conical (Fig. 15.28(c)) and is used with a circular waveguide to produce a circularly polarized beam. Horn-type antennas do not provide very high directivity but are of simple, rugged construction. This makes them ideal as primary feed antennas for parabolic reflectors and lenses.

The choice of horn dimensions is dependent on the desired beam angle and directive gain and involves specification of the ratio of flare length to wavelength L/λ and flare angle ϕ.

15.19.2 The Paraboloidal Reflector Antenna

The most widely used antenna for microwaves is the paraboloidal reflector antenna, which consists of a primary antenna such as a dipole or horn situated at the focal point of a paraboloidal reflector, as shown in Fig. 15.29. The mouth, or physical aperture, of the reflector is circular, and the reflector contour, when projected onto any plane containing the focal point F and the vertex V, forms a parabola as shown in Fig. 15.29(b). The path length $FAB = FA'B'$ for this curve, where the line BB' is perpendicular to the reflector axis. The important practical implication of this property is that the reflector can focus parallel rays onto the focal point, and conversely, it can produce a parallel beam from radiation emanating from the focal point. Figure 15.29(c) illustrates this. An isotropic point source is assumed to be situated at the focal point. In addition to the desired parallel beam being shown, it can be seen that some of the rays are not captured by the reflector, and these constitute "spillover." In the receive mode, spillover increases noise pickup, which can be particularly troublesome in satellite ground stations. Also, some radiation from the primary radiator occurs in the forward direction in addition to the desired parallel beam. This is termed "blacklobe" radiation since it is

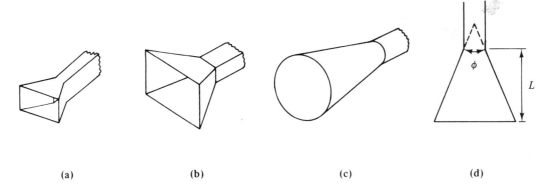

(a)   (b)   (c)   (d)

FIGURE 15.28. Microwave horn antennas: (a) sectoral horn; (b) pyramidal horn; (c) conical horn; (d) horn flare dimensions.

from the backlobe of the primary radiator. Backlobe radiation is undesirable because it can intefere destructively with the reflected beam, and practical radiators are designed to eliminate or minimize this. The isotropic radiator at the focal point will radiate spherical waves, and the paraboloidal reflector converts these to plane waves. Thus, over the aperture of an ideal reflector, the wavefront is of constant amplitude and constant phase.

The directivity of the paraboloidal reflector is a function of the primary antenna directivity and the ratio of focal length to reflector diameter, f/D. This ratio, known as the aperture number, determines the angular aperture of the reflector, 2Ψ (Fig. 15.30(a)), which in turn determines how much of the primary radiation is intercepted by the reflector. Assuming that radiation from

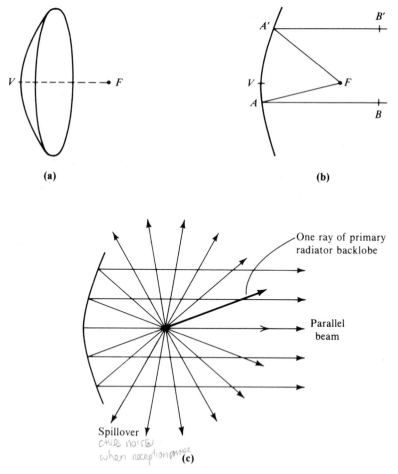

(a)

(b)

One ray of primary radiator backlobe

Parallel beam

Spillover

(c)

FIGURE 15.29. (a) A paraboloidal reflector; (b) the parabola; and (c) radiation from the paraboloidal reflector and primary radiator.

the primary antenna is circularily symmetric about the reflector axis $(F - V)$ and is confined to angles ψ in the range $-\pi/2 < \psi < \pi/2$, it is found that the effective area is given by

$$A_{eff} = AI(\theta) \tag{15.54}$$

where $A = \pi D^2/4$ is the physical area of the reflector aperture, and $I(\theta)$ is a function, termed the *aperture efficiency* (or *illumination efficiency*), which takes into account both the radiation pattern of the primary radiator and the effect of the angular aperture. With the focal point outside the reflector, as shown in Fig. 15.30(a) (which requires $f/D > 1/4$), the primary radiation at the perimeter of the reflector will not be much reduced from that at the center, and the reflector illumination approaches a uniform value. This increases the aperture efficiency, but at the expense of spillover occurring. Making f/D too large increases spillover to the extent that aperture efficiency then decreases. Reducing f/D to less than $1/4$ places the focal point inside the reflector, as shown in Fig. 15.30(b). Here, no spillover occurs, but the illumination of the reflector tapers from a maximum at the center to zero within the reflector region. This nonuniform illumination tends to reduce aperture efficiency. Also, placing the primary antenna too close to the reflector results in the reflector affecting the primary antenna impedance and radiation pattern, which is difficult to take into account. It can be shown that the aperture efficiency peaks at about 80%, with the angular aperture ranging from about $40°$ to $70°$ depending on the primary radiation pattern. The relationship between aperture number and angular aperture is

$$\frac{f}{D} = .25 \cot\left(\frac{\Psi}{2}\right) \tag{15.55}$$

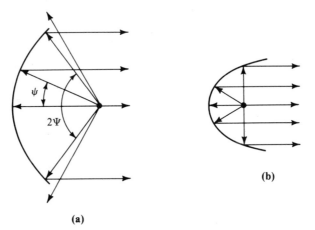

FIGURE 15.30. (a) Focal point outside reflector; (b) focal point inside reflector.

Typically, for an angular aperture of $55°$, the aperture number is

$$\frac{f}{D} = .25 \times 1.92 = .48$$

This shows that the focal point should lie outside the mouth of the reflector, since f/D is then greater than $1/4$. Satisfactory results are obtained in practice if the main lobe of the primary antenna intercepts the perimeter of the reflector at the -9 to -10 dB level as shown in Fig. 15.31.

On substituting $\pi D^2/4$ for A in Eq. (15.54) and using Eq. (15.20) for gain, we get

$$G = I(\theta)\left(\frac{\pi D^2}{4}\right) \tag{15.56}$$

The beamwidth also depends on the primary radiator and its position. In practice, it is found that for most types of feed the -3 dB beamwidth is given approximately by

$$\text{BW}_{(-3\ \text{dB})} \cong \frac{70\lambda}{D} \text{ degrees} \tag{15.57}$$

and the beamwidth between nulls by

$$\text{Nulls BW} = 2\left(\text{BW}_{(-3\ \text{dB})}\right)$$

$$= \frac{140\lambda}{D} \text{ degrees} \tag{15.58}$$

Example 15.4   Find the directivity, beamwidth, and effective area for a paraboloidal reflector antenna for which the reflector diameter is 6 m and the illumination efficiency is .65. The frequency of operation is 10 GHz.

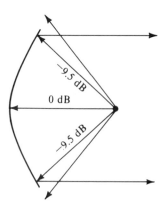

FIGURE 15.31.   Edge illumination from primary antenna is between 9 and 10 dB below that at vertex.

Solution

$$\lambda = \frac{c}{f} = \frac{300 \times 10^6}{10 \times 10^9} = 0.03 \text{ m} = 3 \text{ cm}$$

$$A = \frac{\pi D^2}{4} = \frac{3.14 \times 6^2}{4} = 28.26 \text{ m}^2$$

$$A_{\text{eff}} = 0.65A = 18.4 \text{ m}^2$$

$$D_0 = \frac{4\pi}{\lambda^2} A_{\text{eff}} = 257{,}000 \ (54.1 \text{ dB})$$

$$\text{BW}_{(-3 \text{ dB})} = \frac{70\lambda}{D} = \frac{70 \times 0.03}{6} = 0.35°$$

$$\text{BW}_{(\text{null})} = 2 \times 0.35 = 0.70°$$

15.19.3   Variations on the Parabolic Reflector

The *parabolic reflector* is a favorite antenna for fixed point-to-point microwave communications systems. It is relatively simple in construction, and unless large in size, it is quite inexpensive. Huge steerable parabolic dishes have been built for use with the radio telescopes, up to 200 ft in diameter, and mounted on a movable turret which allows rotation in both the horizontal and vertical directions to allow the tracking of moving targets such as satellites and radio stars.

Antennas used for radioastronomy must utilize all their area for reception to get the highest efficiency and the lowest noise figure. Special feed systems are used so that the feed antenna is reduced in size or physically located out of the path of the incoming radiation. Two types of feed are shown in Fig. 15.32. The first of these uses a dipole antenna, which normally radiates outside the parabola as well as onto it, but has a spherical reflector placed directly behind the dipole to prevent direct radiation. The backlobe radiation is reflected back at the parabola and is added to the main portion of the radiation. Some tuning is necessary since the reflector position is different for different frequencies.

The second method is known as the *Cassegrain feed system* (Fig. 15.32(b)). The horn feed antenna, the paraboloid reflector, and the hyperboloid subreflector have a common axis of symmetry as shown, and the virtual focal point of the hyperboloid is coincident with the focal point of the paraboloid. Radiation reflected off the subreflector illuminates the main reflector approximately uniformly, and equally important, spillover at the edges is low. This is of particular advantage in low-noise receiving systems, where large spillover results in high noise levels.

Two types of modified horns also use parabolic surfaces. The first of these is the *Cass horn* (Fig. 15.32(c)). Incoming radiation is reflected from a large lower parabolic section which has a rectangular sectional front and forms the bottom of the horn, up to a smaller parabolic surface which forms the top of the horn, and then to a horn antenna located at the focal point of the combined parabolas. This horn has the advantage that incoming radiation is not blocked by the feed structure, but it is much more complex to build.

The second horn is the *hog horn* (Figure 15.32(d)). This is basically a large horn antenna with a parabolic reflector mounted directly in front of it and oriented to radiate at right angles to the axis of the horn. It is often used for microwave communication links, such as the Bell TD-2 system, because it is compact and simple in construction while giving the high directivity of the parabola and also high efficiency. It also has the advantage that it may be rotated at the neck on a sliding joint, allowing steering of the beam without undue manipulation.

15.19.4   Dielectric Lens Antennas

Electromagnetic radiation is refracted when it passes through a surface separating a zone of lower dielectric constant from one of higher dielectric constant in exactly the same manner that light is refracted. The angles of incidence and refraction are related by the modified version of Snell's law, which states (referring to Figure 15.33(a)) that

$$\frac{\sin \phi_r}{\sin \phi_i} = \sqrt{\frac{\varepsilon_{ri}}{\varepsilon_{rr}}} = \frac{1}{n} \tag{15.59}$$

for waves entering the region of high dielectric constant, where n is the refractive index of the material. The refractive property is reciprocal, and the same relationships obtain when the radiation passes from the higher dielectric

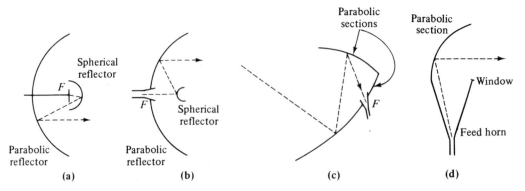

FIGURE 15.32. Methods of feeding microwave antennas: (a) parabolic reflector fed from a dipole and a small spherical reflector; (b) the Cassegrain feed: parabola fed from a horn and a small hyperboloid reflector; (c) Cass horn; (d) hog horn.

medium into the lower dielectric medium, except that the subscripts i and r are interchanged.

The material used for the lens is usually one of the high-dielectric plastics, such as polystyrene or Teflon.

Figure 15.33(b) illustrates the principle of the collimating lens, which is used to make a diverging beam of radiation into one traveling in only one direction with a planar wavefront. The lens in this case is a convex one. Radiation along the axis of the lens passes through both surfaces at right angles, so no refraction takes place. Radiation at an angle from the axis is incident at the curved interface at an angle other than normal and is refracted toward the normal as it passes into the lens, at A'. The curvature of the lens is such that after refraction the rays are all parallel to the axis.

Radiation at the angle along FA' takes slightly longer to reach the refracting surface, arriving with a slight time lag compared to that along the axis FA. However, the velocity of propagation within the dielectric is slower than that in air, and a compensating delay occurs in the radiation along the axis AB as compared to that along $A'B'$, so that all the radiation arrives at the flat interface BB' in phase. The result is a planar wavefront leaving the lens.

Many types of lens systems can be used for the collimating or paralleling function. Two of these are shown in Fig. 15.33(c). These are *zoned lenses*,

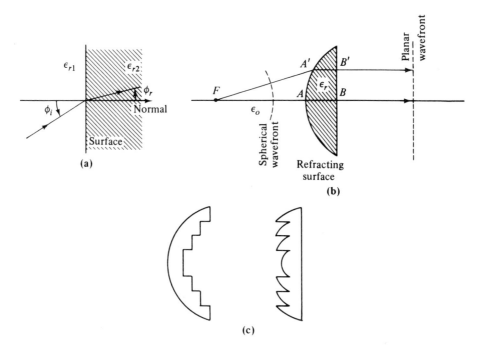

FIGURE 15.33. Dielectric lens antennas: (a) illustration of Snell's law of refraction; (b) the principle of the collimating lens; (c) cross sections of two types of zoned lenses.

which are basically the same in function as the convex lens illustrated in Fig. 15.33(b). However, the zone structure allows a large reduction in the volume of dielectric material which must be used to make the lens, with a corresponding savings in cost and weight. This saving is made at a slight sacrifice of directivity.

Again, ideally, the lens should be fed with even illumination over its entire surface to achieve maximum efficiency and gain. The horn antenna is the most popular method of feed and most closely approximates the even-illumination requirement. For the higher microwave frequencies, the lens makes a very compact, highly directive narrow-beam antenna which is popular for applications such as portable communication links and mobile radar systems of the type used for vehicle-speed monitoring.

15.19.5 Slot Antennas

When a slot in a large metallic plane is coupled to an RF source, it behaves like a dipole antenna mounted over a reflecting surface. Figure 15.34(a) shows a rectangular slot in a large plane. The slot is coupled to a line or feed waveguide in such a manner that the E field lies along the short axis of the slot, as indicated. At microwave frequencies, the slot may be energized as described in Section 13.2.5. At lower frequencies, a transmission line may be connected directly across the slot as shown in Fig. 15.34(a).

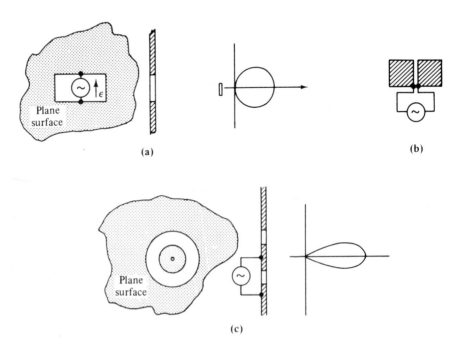

FIGURE 15.34. Slot antennas: (a) rectangular slot; (b) complementary dipole equivalent of a rectangular slot; (c) annular slot.

The behavior of this *slot antenna* can be shown to be equivalent to a complementary antenna in which the slot is replaced by a sheet dipole with a reflector behind it in free space, as shown in Fig. 15.34(b). The dimensions of the slot are usually such that the long axis is approximately a half-wavelength at the operating frequency. The resulting radiation pattern is similar to that for the dipole with a reflector.

The feed impedance of the slot can be calculated by first calculating the impedance for the complementary dipole. The actual slot impedance is related to the dipole impedance by the complementary characteristic, which states that

$$Z_r \times Z_c = \left(\frac{Z_{0\,\text{space}}}{2} \right)^2 = (60\pi)^2 \tag{15.60}$$

where Z_r is the radiation impedance of the slot and Z_c is the radiation impedance of the complementary (dipole) antenna. The shape of the slot does not have to be rectangular, but can be any convenient shape. Rectangular and circular (annular) slots are favored because they are easy to make and relatively easy to analyze. Figure 15.34(c) shows an annular slot, which produces a narrow beam of radiation.

Arrays of slots are ideal for use in aircraft. The slots can be formed directly in the metal skin of the aircraft and then windowed with a dielectric material such as polystyrene. The smooth surface produced does not interfere with the streamlining of the aircraft. Phasing the feed to the slots allows production of a beam which may be swept through a wide angle without physically moving the structure, thus allowing its use for mobile radar systems.

A variation of the slot antenna is the *notch antenna*, where an appropriately shaped notch is cut out of the edge of a large metal surface and connected to an RF source. Again, the chief use is on aircraft, where the notches can be made in the edges of the wing surfaces and filled with dielectric material to make them aerodynamically smooth.

15.20 PROBLEMS

1. (a) The terminal input current to an antenna is $2\underline{/11°}$ A when the terminal voltage is $100\underline{/0°}$ V. Determine the antenna impedance. (b) The voltage reflection coefficient measured on a 50-Ω transmission line feeding an antenna is $0.1\underline{/5°}$. Determine the antenna impedance.

2. Assuming 100% efficiency for the antennas in Problem 1, determine the radiation resistance in each of cases (a) and (b), and the power transmitted in (a).

3. Derive Eq. (15.7) of the text.

4. An antenna of impedance $Z_A = 45 - j10\ \Omega$ is connected directly to a 50-Ω transmission line. Determine the matching efficiency.

5. The (x, y, z) coordinates for a point in space are given as $(10, 5, -2)$. Determine the polar spherical coordinates (d, θ, ϕ) as shown in Fig. 15.5.

6. A paraboloidal reflector antenna has a diameter equal to 100λ. Determine the distance at which the far-field zone is the only effective component of the total radiated field. The frequency of operation is 10 GHz.

7. Two $\frac{1}{2}\lambda$-dipoles are arranged for transmission and reception as shown in Fig. 15.9. The receiving antenna is in the plane of the incoming wavefront, and the maximum induced EMF is measured as 10 μV. Calculate the induced EMF when the receiving antenna is rotated (a) 30°, (b) 60°, and (c) 90° in the plane of the wavefront.

8. A power of 100 W is radiated from an isotropic radiator. Calculate the power radiated per unit solid angle and the power density at a distance of 5 km from the antenna.

9. An isotropic antenna radiates energy equally in all directions. The total power delivered to the radiator is 100,000 W. Calculate the power density and electric field intensity at distances of (a) 100 m, (b) 1 km, (c) 100 km, and (d) 1000 km. Plot the power density against distance on log-log paper.

10. Calculate the directivity for the antennas for which the following specifications apply: (a) power gain $10^3 : 1$, efficiency 90%; (b) power gain 45 dB, efficiency 90%.

11. The radiation pattern for an antenna in the meridian plane is given by $\sin^4 \theta$. Plot this function (a) on polar graph paper, and (b) on linear graph paper. (c) Determine the -3 dB and the -10 dB beamwidths.

12. An antenna has a gain of 35 dB at a frequency of 300 MHz. Calculate the effective area.

13. Calculate the effective length of an antenna which has a directive gain over an isotropic antenna of 17 dB and a radiation resistance of 350 Ω at a frequency of 144 MHz. Use the relationship shown in Problem 14.

14. By making use of Eq. (B.12) and the maximum power transfer theorem, show that the effective area and effective length of an antenna are related by

$$A_{\mathrm{eff}} = \frac{30\pi \ell_{\mathrm{eff}}^2}{R_{\mathrm{rad}}}$$

15. An elementary doublet has an electrical length of 0.0625λ and carries an RF current of 2.5 A rms. Calculate the field intensity at a point located 40 km from the doublet and at an angle of 25° from the main lobe of radiation.

16. A directional antenna has an effective radiated power of 1.1 kW when it is fed with a terminal input power of 90 W. The radiation resistance is found

to be 74 Ω at resonance, and the measured antenna current is 1.088 A rms. Find (a) the antenna efficiency; (b) the terminal resistance; (c) the antenna power loss; and (d) the antenna directive gain, in dB, over an isotropic radiator.

17. Calculate the current induced in the terminals of a vertical receiving antenna if it has a gain of 6 dB over an isotropic antenna, a terminal impedance of 35 Ω, a load impedance of 35 Ω, and it is in an electric field with an intensity of 10 μV/m, at a frequency of 7 MHz.

18. Calculate the capture area of the antenna in Problem 17.

19. Plot the normalized radiation pattern for a $\frac{1}{2}\lambda$ dipole antenna, using Eq. (15.37), on a polar graph, for $0° < \theta < 360°$. Calculate values for each $10°$ displacement of θ.

20. Calculate the 3-dB beam width of the $\frac{1}{2}\lambda$-antenna in Problem 19.

21. Using the relationships given by Eqs. (15.40), and (15.41) and the results of Problem 14, calculate the effective length of a $\frac{1}{2}\lambda$-dipole. Compare the result with that given in Example 15.2.

22. A loop antenna is made by winding 10 turns of wire on a 1-m² frame. It is located in a magnetic field of 0.015 μteslas, at 10 MHz and oriented for maximum signal strength. Find (a) the induced EMF in the antenna; and (b) the terminal voltage if the antenna is tuned to resonance at 10 MHz, with a total resistance of 65 Ω in series with a 25-pF capacitor.

23. A direction finder using a loop antenna is used to locate an illegal transmitter. The operator goes to location A, where he receives a good signal. He mounts his receiver so that the antenna faces due north and then rotates the antenna from $0°$ north clockwise through $48°$ to obtain a null. He moves the direction finder to a new location B, which he determines is exactly 2550 m due east of A, and obtains a new bearing by rotating the antenna $15°$ counterclockwise from $0°$ north. Compute the distance from point A and from point B to the transmitter.

24. Calculate the effective length of a ferrite-rod receiving antenna which has 120 turns wound on a 1.40-cm-diameter ferrite rod which has a relative permeability of 160. Assume the length factor to be 0.75 and the frequency to be 1 MHz.

25. Plot the array factor as a function of angle ϕ for a four-element broadside array for which the elements are spaced by $\frac{1}{2}\lambda$.

26. Determine the current phasing required for an end-fire array for which the elements are spaced $\frac{1}{4}\lambda$. Plot the array factor as a function of angle ϕ for a four-element array with this spacing.

27. Calculate the lengths of the elements and spacing for a five-element Yagi antenna for Channel 4 television (66 to 72 MHz). The effective length factor is to be 5%, and each element is to be 95% of the previous one in length.

28. Calculate the element lengths for a 10-element log-periodic array if the smallest element is not less than 10% of the largest element, the angle of divergence $\alpha = 15°$, and the longest element is cut for a frequency of 50 MHz.

29. Calculate the angular aperture for a paraboloidal reflector antenna for which the aperture number is (a) 0.25, (b) 0.5, and (c) 0.6. Given that the diameter of the reflector mouth is 10 m, calculate the position of the focal point with reference to the reflector mouth in each case.

30. (a) Specify the diameter of a parabolic reflector required to provide a gain of 75 dB at a frequency of 15 GHz. The area factor of the feed is 0.65. (b) Calculate the capture area of the antenna and its 3-dB beamwidth.

CHAPTER 16

Telephone Systems

16.1 WIRE TELEPHONY

16.1.1 Telephone Circuits

The word *telephone* is derived from the Greek words *tele*, meaning far, and *phone*, meaning sound. In today's context, *telephony* involves the conversion of the sound signals into an audio-frequency electrical signal which can then be transmitted over an electric transmission system and then reconverted to sound pressure signals at the receiver end. The electrical signals may be transmitted by radio or by wire, and a system may well use both means to establish any given circuit. The wire telephone system was the earliest of such systems and still forms the backbone of modern communications.

Telephone systems may be categorized by the mode of transmission used. The simplest is a single one-way voice system which connects a single transmitter in one location to a single receiver in another location. Such one-way systems are rare, and most systems provide two-way communications in one of the following two ways.

The *simplex* system allows transmission in only one direction at a time, but provides communication in either direction alternately on a push-to-talk basis. The operator at the end originating the transmission closes a switch to turn on the transmitter and turn off the receiver, so that he or she may talk toward the far-end receiver. As long as the receiving operator does not close his transmitting switch, he will receive the signals from the far end. If both should attempt to transmit at the same time, the message will be lost and must be repeated. Simplex operation is most often used with radio systems, where transmission in both directions must take place on the same frequency.

Duplex systems allow simultaneous transmission in both directions. This may be accomplished by simply providing two separate circuits, one for each direction, but this is not economical because of duplication of facilities. Duplex in most telephone systems means simultaneous transmission over the same pair of wires without the need for switching. Since most telephone systems today use full duplex operation, this mode will be described in detail.

A telephone system must be able to transmit voice signals in both directions and it must provide a means of signaling from each terminal toward the other. This signal originally only summoned the person at the far receiver, but this was rapidly changed to the process of signaling a central operator who connected the desired line and who then signaled the far-end receiver. The same signal channel is used in automatic telephone systems to operate automatic line-switching apparatus at the central location, so that an intermediate operator is not needed.

A simple two-point telephone circuit uses a single pair of wires (or one wire with ground return) and two telephone sets of the self-contained type used on manually operated rural multiparty lines. Each set is self-contained and does not use any central office equipment. The telephone set is similar to that used in the subscriber loop circuit of Fig. 16.1, except that the dial pulsing unit is left out, and a battery and a source of ac such as a hand-cranked alternator for signaling are included. With all of the handsets hung on the hook switches, the signaling circuit is complete, and turning the alternator crank at one set will cause the bells at all other sets to ring. Several sets may be included on one line in the multiparty configuration. When the handsets are picked up, the battery in each set is connected to the carbon transmitter and the conversation may proceed. Only the ac component of the voice is transmitted on the line.

Telephone systems that require switching at a central location to connect the calling party line to the called party line operate on a "loop" basis. The telephone "subscriber's loop" consists of a pair of wires between the subscriber's location and the telephone switching center, a telephone terminal set, and a circuit at the switching location to supply battery current, signaling current, and a means of connecting to the switching machine.

Figure 16.1 shows a typical subscriber's loop circuit which might be used with a step-by-step or cross-bar switching office. Signaling inward is done by dc switching, while signaling outward is done by an ac source to ring the subscriber's bell. The subscriber's terminal consists of a handset containing a carbon transmitter and an earphone receiver. The handset is connected to the line through an autotransformer (induction coil) which provides the necessary impedance matching between the line and the handset. The impedance ratios are adjusted so that the "sidetone" or signal feedback from the local transmitter into the local receiver is adjusted to an acceptable level, while still providing a good match to the line itself. If sidetone is extremely objectionable, a special transformer is included which cancels out part of the sidetone but does not interfere with normal operation. A signal bell is connected across the line through the hook-switch contact.

Current to operate the transmitter is supplied from the central office battery over the loop pair. This current also passes through the pulsing contact B in the dial pulsing unit, which interrupts current to transmit the coded information to select the proper "address" in the automatic switching machine. Contact A closes to short-circuit the receiver while dialing is in progress, so that the caller is not subjected to the clicks of the pulse train. It releases as the dial returns to its rest position.

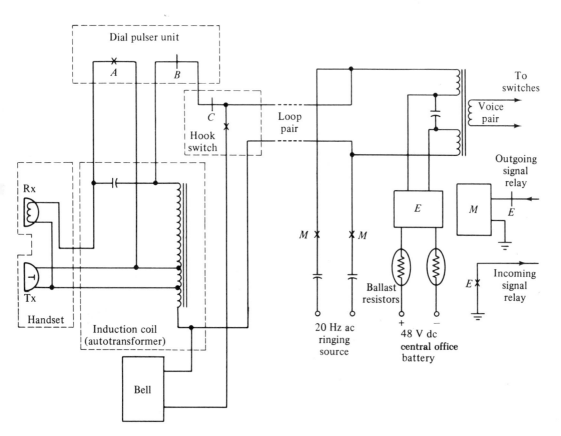

FIGURE 16.1. A telephone subscriber's loop circuit.

The loop circuit at the central office location contains a bridging transformer which allows separation of the voice signal from dc signaling current. The direct current is supplied from the common central office battery through the coils of the sensitive E relay, which relays the dial pulses and the fact that the hook switch is open (busy), and two ballast resistors which are used to compensate for different loop resistances and provide about the same value of loop current regardless of loop length. A second M relay operates on a demand signal from the switching machine to connect the ac bell-ringing supply to the line. Signaling operation is now described.

On originating a call, the subscriber picks up the handset, which closes the hook switch to connect the handset to the line and disconnect the bells. This action closes the dc path in the loop, and the flow of direct current from the central battery picks up the E relay. The E relay contacts send a signal to the switching machine, indicating that the line is busy, and also disengage the M or ringing relay. The talking circuit is now energized. A slow-operating relay also operates to "hold" the busy condition during dialing. Operating the dial sends several trains of 1 to 10 pulses at a rate of about 10 pulses per second through the E relay to select the proper switch combination in the switching machine. If the called line indicates a busy condition, a special busy tone is sent back. If it is not busy, the M relay of the called loop is energized and a special ringing tone is sent back to the originator's receiver to indicate that signaling is proceeding.

When the demand signal from the switching machine energizes the M relay to the called loop, its contacts connect the ac ringing source to the line. This signal is an ac supply of several hundred volts at 20 Hz frequency which is interrupted for about 1 out of every 2 seconds and which rings the bell at the called station. When the called party lifts his receiver, his hook switch closes the dc path in the called loop, operates the E relay, and disconnects the M relay and ringing source. The conversation can then proceed.

Loop lengths are limited primarily by the dc resistance in the loop. If the dc loop resistance becomes greater than about 1200 Ω, dc signaling becomes unreliable. Depending on the type of cable used for the loop pair, this value of resistance allows loops from as little as 5 km using 26 AWG copper to as much as 45 km using 16 AWG copper. Open wire lines using heavy-gauge copper-clad steel conductors are frequently used for long rural loops.

A further limitation is placed on the loop length by the attenuation factor of the cable. Loops with losses in excess of 10 dB total are not acceptable, and repeater amplifiers must be included in the loop to make up for these losses. Table 16.1 shows typical exchange cable characteristics. Uncompensated or unloaded cable exhibits an increasing attenuation factor as frequency increases. It is necessary to correct this in long loops. The correction or compensation is known as *loading* and consists of placing series inductances at intervals along the cable. These inductances act with the cable capacitance to create a distributed filter with a cutoff frequency of about 3.5 kHz and a flat

TABLE 16.1. Characteristics of Cables Used for Customer Loops in Telephone Exchanges

Cable AWG Gauge No.	Loop Resistance (Ω/Loop-Mile) Round Trip	Attenuation Factor (dB/Mile) Unloaded	Attenuation Factor (dB/Mile) M88 Loading*	Length for 1200-Ω Loop Resistance (Miles)	Total 1200-Ω Loop Loss (dB) Unloaded	Total 1200-Ω Loop Loss (dB) Loaded
26	431	2.67	—	2.8	7.5	—
24	271	2.15	1.31	4.5	9.7	5.9
22	170	1.80	0.92	7.1	12.8	6.5
19	85	1.12	0.44	14.1	15.8	6.2
16	42.4	0.76	0.24	28	21.3	6.7

*M88 loading: 88-mH loading coils placed at 9000-ft intervals.
Source: *IT & T Reference Data for Radio Engineers*, 5th ed., Howard W. Sams & Co., Inc., Indianapolis, Ind., 1969.

attentuation function over the frequency range from 300 to 3000 Hz. The attenuation factor tends to be lower than in the case of the unloaded cable, as indicated in Table 16.1. The total loop attenuation for unloaded cable cut for 1200-Ω dc resistance rises as the physical length increases (lower gauge number or larger-diameter wire), while that for the loaded cable works out about the same in each case at about 6 dB.

Touch-tone signaling has become popular recently and will probably completely replace the familiar dial signaling in the next few years. Touch-tone signaling uses a two-of-seven code to represent the 10 decimal digits and two other symbols. Each of the seven states is designated by the presence of a separate tone, each of which falls in the voice-frequency band. The dial unit on the telephone terminal set is replaced by a touch pad or keyboard with 12 pushbutton switches. When one of the switches is depressed, two of the seven tone generators are energized, and the two tones are sent out audibly in the

TABLE 16.2. Touch-Tone Dial Signaling Frequency Assignments

Low Group (HZ)	Code Designation*		
697	1	2	3
770	4	5	6
852	7	8	9
941	Spare	0	Spare
	1209	1336	1477 (HZ) High Group

*Each code consists of one tone from the high group, and one from the low group, of frequencies.
Source: *IT & T Reference Data for Radio Engineers*, 5th ed., Howard W. Sams & Co., Inc., Indianapolis, Ind., 1969.

voice channel. Tone-recognition equipment at the switching center decodes the dial information and sets up the switches. The E relay is still present, but it is used only to provide indication of whether or not the loop is busy. Table 16.2 lists the tones used in the North American telephone system for tone signaling. The actual frequencies used have been carefully selected to minimize duplication by the normal voice signals.

16.1.2 Transmission Bridges

The term *transmission bridge* applies specifically to the circuit used at the central office end of the loop to separate the voice path from the signaling path and pass it on to another loop. This circuit must isolate the voice signals from the dc signaling circuit and from the central battery supply, and pass these signals into the switching network for cross-connection to another loop.

Figure 16.2(a) shows a simple bridge circuit. The loops are operated in an unbalanced (one-side-grounded) mode, and the dc from the battery is injected through the coil of a relay and an inductor or choke coil. The choke provides the large impedance necessary to prevent voice signals from entering the signal and battery circuit. Capacitors are used to couple the voice signals into the switching network, while keeping direct currents in the loop. Such systems are not often used because of transmission and noise problems created by the unbalanced transmission mode, even though the switching system is simpler because it only has to switch one wire.

The unbalance problems are not present in the Stoneman bridge shown in Fig. 16.2(b). In this circuit, each loop has two inductors, one for each side of the loop line. No connection is made to earth or ground, so noise due to ground currents is reduced. The inductors have large reactance, and most of the voice currents are passed through the blocking capacitors to the switching network. The switching network in this case must switch two wires instead of one so that the entire circuit, including both loops, may be operated in a balanced mode. Ballast resistors are included in the battery leads to compensate for various loop lengths (loop resistances). The signaling relays shown are simple two-coil relays which operate when current flows. In practice, special multicoil "polarized" relays of the type used in wire telegraph loops are used. These relays operate in a differential mode and provide much more sensitivity than ordinary relays. They also provide a bias adjustment which allows compensation for any pulse distortion due to the limited frequency response of the loop cables.

Figure 16.2(c) shows a variation of the Stoneman bridge. The only difference between this circuit and that of Figure 16.2(b) is that transformer coupling is used to connect the loop into the switched network. Transformer coupling reduces the insertion loss of the coupling and also eliminates frequency distortion due to reactance. Moreover, the transformer allows impedance matching between the loop and the switching system, which may be operating

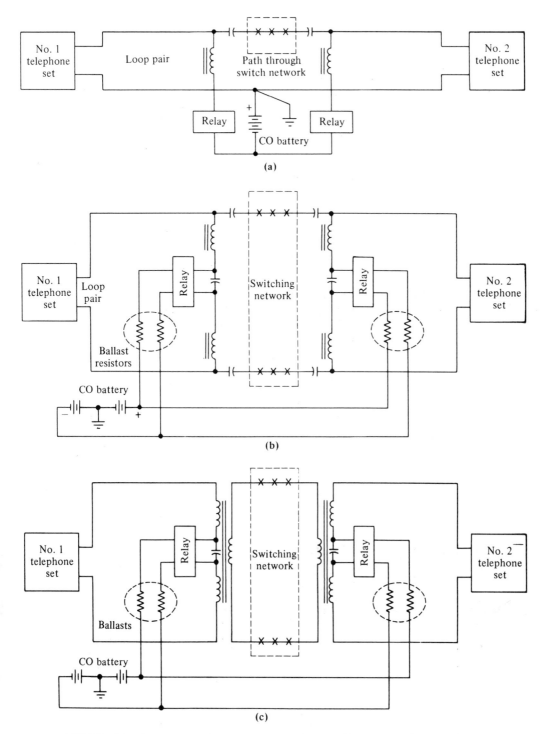

FIGURE 16.2. Transmission bridges: (a) simple bridge; (b) Stoneman bridge; (c) transformer bridge.

at a different characteristic impedance than that of the loop circuit. Typically, the characteristic impedance of loop circuits is about 900 Ω, while that of interoffice trunk circuits is 600 Ω.

16.1.3 The Four-Wire Terminating Set

The *four-wire terminating set* shown in Fig. 16.3(a) is used to provide two-wire to four-wire connection, as required by repeater stations. This circuit uses two separate transformers, each with four identical windings. The two-wire

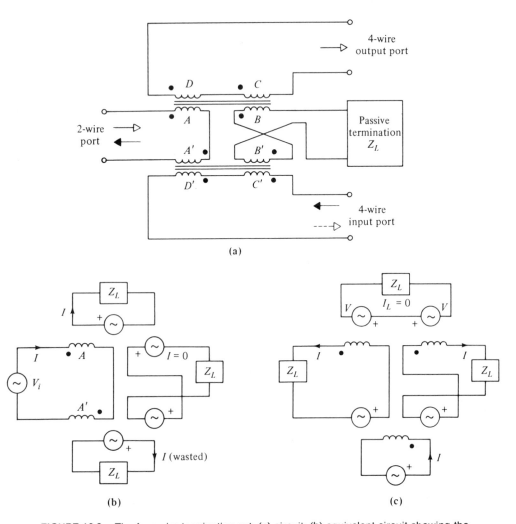

(a)

(b)   (c)

FIGURE 16.3. The four-wire terminating set: (a) circuit; (b) equivalent circuit showing the currents and EMF's due to a two-wire signal; (c) equivalent circuit showing the currents and EMF's due to a received four-wire signal; the input may be applied to either of the four-wire ports.

line is connected to the two *A* windings connected in series aiding. The passive termination is connected to the two *B* windings connected in series opposing. The transmitting pair of the four-wire line is connected to the *C, D* windings of one transformer in series aiding, and the receiving pair to the *C, D* windings of the other transformer. The passive termination is built to have an impedance identical to that of the two-wire line.

Figure 16.3(b) shows the equivalent circuit for currents applied from the two-wire line. Currents in windings *A* induce identical EMF's in windings *B* in both transformers. But these windings are connected series opposing, so that the EMF's cancel and no current flows in the passive termination. Currents are induced in the *C, D* windings of both transformers. Those in the output port are transmitted on one side of the four-wire line, but those applied to the input port are blocked because of unilateral amplifiers in the input line. Half of the power sent in is sent out the output pair, so that the set creates a 3-dB *insertion loss* in the line.

When current is applied to the input port, as in Fig. 16.3(c), equal currents are induced in the *A, B* windings. These currents flow from the *A* windings down the two-wire line and the *B* windings into the passive termination, and are returned through the *A, B* windings of the other transformer. But these windings are connected series opposing, so that no current is generated in the output port. In practice, a small amount (typically less than -50 dB) of the signal applied to the input port leaks through to the output port, this being termed the *feedthrough loss*. Again, the input power is split two ways, causing a 3-dB insertion loss.

16.1.4 Two-Wire Repeaters

Repeaters are amplifiers which are inserted in transmission lines at intervals to amplify the signal and compensate for transmission loss on the cable. On two-wire transmission lines, transmission takes place in both directions, and the amplifier must also be bilateral. "Negative impedance amplifiers" employing a principle similar to the negative resistance oscillators described in Section 6.8 are used. Negative impedance amplifiers are two-terminal bilateral devices. Before these became available, only unidirectional amplifiers were available, and it was necessary to split the two-way transmission into two separate paths before unilateral amplifiers could be applied.

Special terminal sets like the 4-wire terminating set of Section 16.1.3 are required to divide the two-way path into 2 one-way paths or to convert from two-wire to four-wire transmission.

Figure 16.4(a) shows the typical arrangement of a two-wire repeating amplifier. Two unilateral amplifiers provide the necessary gain to overcome the insertion losses of the terminating sets and the transmission losses over a section of the line. Two terminating sets split the two-wire circuit into two one-way circuits containing the amplifiers, connected back to back. If ordinary transformers were used to accomplish this split, there would be a loop-trans-

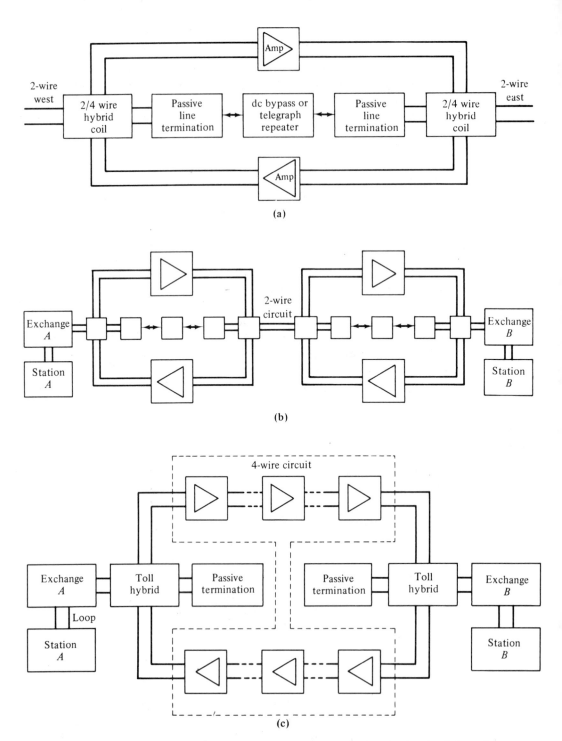

FIGURE 16.4. (a) A two-wire repeater station; (b) a two-wire circuit with 2 two-wire repeaters incorporated; (c) a four-wire toll trunk circuit.

mission path around the two transformers and the two amplifiers with a net loss of only 6 dB. Any amplifiers would have to have a gain of less than this if oscillations were to be avoided. The terminating set prevents signals from the output of an amplifier from reaching the other amplifier around this short loop, and the gain of the amplifiers can be made much higher. The total gain is limited by the degree of matching on the lines necessary to reduce reflected waves, which can cause oscillations. For the repeater to be stable, it is necessary for the net repeater loop-gain, including the two amplifier gains, and the hybrid feedthrough losses, to be less than unity. For the system to be stable, it is also necessary for the net line loop-gain, including the two amplifier gains, the terminating set insertion losses, and the reflection losses of the two cables, to be less than unity. Here, *reflection loss* is the ratio (in dB) of the reflected power to the transmitted power at the transmitting point on a cable and includes both return losses at reflection points and line attenuations.

Example 16.1 (a) If the repeater of Fig. 16.1(a) uses hybrid transformers with a direct feed-through loss of − 50 dB, and the lines are matched so that reflections are negligible, calculate the maximum gain in each direction if the gains are to be equal and if a gain margin against singing (oscillation) of 12 dB is to be provided.

(b) If the line reflection loss measured on the west line were − 20 dB and that on the east line were − 40 dB, find the maximum gain in each direction for a 12-dB singing margin.

Solution (a) The net loss around the loop is twice the feed-through loss of the hybrids used:

$$\text{Net loss} = -50 \times 2 = -100 \text{ dB}$$

The net allowed loop gain if stability is to be maintained is

$$\text{gain} = -\text{loss} - \text{margin} = 100 - 12 = 88 \text{ dB}$$

The net repeater amplifier gain is one-half of the net loop gain, or 44 dB.

(b) The net loop loss is twice the insertion loss of the hybrid (since the signal passes through each hybrid twice, once going and once on return) times two again because there are two transformers, plus the two reflection losses of the lines:

$$\text{net loss} = (-3 \text{ dB} \times 2 \times 2) + (-20) + (-40) = -72 \text{ dB}$$

$$\text{allowed net loop gain} = 72 - 12 = 60 \text{ dB}$$

$$\text{Maximum repeater amplifier gain} = 60 \div 2 = 30 \text{ dB}$$

Figure 16.4(b) shows a two-wire circuit incorporating 2 two-wire repeater stations.

Negative-resistance amplifiers may also be used for loop gain. The use of such repeaters is dependent on how well the lines can be matched to prevent reflections from occurring, because the amplifier is simply connected in parallel with the two-wire line, and there is no local feedback loop. The only way oscillations can occur is if the total inserted gain of the negative-resistance amplifier is greater than the sum of the two reflection losses on the lines. The better the lines are matched, the higher these reflection losses are, and the more the allowed gain. Since there is no break in the line itself, no dc bypass circuit is needed, and only a blocking capacitor to the amplifier need be provided. Transformer coupling can also be used, to insert the negative resistance in series with the line. The Bell System E6 exchange loop repeater is a good example of this. It is a transistorized amplifier which operates directly from the dc talk battery current in the line, and allows the reduction of losses on long loops. Separate signaling repeaters may also be required to regenerate the dc loop signals on very long loops.

16.1.5 Four-Wire Transmission

Two-wire circuits are difficult to balance accurately, and the maximum gain that can be included in repeaters is limited. For this reason, long distance and interoffice trunk circuits are nearly always done on a four-wire basis, with two wires providing the transmission path in one direction and the other two providing the transmission path in the other direction. When carrier circuits are used, four-wire or two-path transmission must be used because the carrier amplifiers along the route only allow transmission in one direction. Figure 16.4(c) shows the typical arrangement of equipment along the path of a long-distance circuit. The customer's loop at each end is switched on a two-wire basis into a toll (long-distance) switching center. A four-wire terminating set located at the toll center converts the circuit to the two-path transmission mode. The long-distance facilities between this center and the far toll center are all four-wire and may include cable circuits, repeaters, cable carrier circuits, and radio circuits. At the far toll center the transmission is converted back to two-wire before being connected to the far customer's loop.

16.2 PUBLIC TELEPHONE NETWORK

16.2.1 The System

The basic purpose of the public telephone system is to provide two-way voice communication between any pair of subscribers within the system. A wide variety of subsidiary services has been added to the list in recent years, such as communication with subscribers in other systems, telegraph, teletype-writer and facsimile service, computer communications, conference calls, and private network service. The basic components of this system are a network of cable and radio channels linking many centers, switching devices at each center to allow interconnection, and local subscriber stations.

The basic component of the system is the exchange switching machine, or "office." This is a switching system that allows direct interconnection of up to 10,000 subscriber stations located mostly within a radius of about 10 km of the office, each of which is assigned a separate four-digit number. The office also provides facilities for connection through exchange interoffice trunk circuits to several other nearby exchanges, or into toll connecting trunks to a nearby toll or long-distance switching center. Early telephone switching offices were manually operated by operators who physically made the connections necessary to complete a call by means of plug leads on a switchboard. Manual switchboards are still used to serve private company systems with a few hundred lines, but nearly all public system switching is now done automatically.

In the 1950's the Bell System introduced the concept of continent-wide automatic toll switching, or *direct distance dialing* (DDD). Before this time, all long-distance connections were done by means of manual switching by long-distance operators. Each exchange could automatically switch the connections between a subscriber and the nearest toll office, but then the toll operator had to manually complete the call to the far end. This often involved switching by several operators at intermediate centers along the route of the call. With DDD, the connections at the toll center between the toll connecting trunks and the intertoll long-distance trunk circuits are made automatically, and billing information is automatically recorded by a computer. Backup manual facilities are maintained on a limited basis in case of difficulty or for the establishment of special service calls such as person-to-person calls and conference calls.

For the purpose of making long-distance calling more efficient with respect to the use of toll-trunk circuit facilities, a hierarchy of toll offices was established. Figure 16.5 illustrates the interconnections possible within the DDD system. A specific location was chosen within each major population center for the establishment of the local toll-connecting offices. These toll-connecting offices might serve anywhere from 10 to 100 exchange offices, depending on the local population density. Toll-connecting trunk circuits link each exchange office to the toll-connecting office. Within each exchange calls are made mostly on a flat-rate basis, so that call-connection recording is not required, such as (A) to (B) in Fig. 16.5. Exchange interoffice trunk circuits allow flat-rate calls to be made to the other nearby exchange locations within a designated flat-rate calling area, such as (C) to (D) in the figure. Again, no record is kept of the calls.

Any call that requires a specific charge must be routed through the nearest toll office, where the record of the call is made. At the toll-connecting level, three levels of long distance calls are possible. First, the call may be to a second exchange which is served by the same toll center, in which case a direct connection between two toll-connecting trunks is made. Second, the call may be to an exchange served by a nearby toll center, in which case the call may be routed over a direct toll trunk between the two centers. Third, the call may be between centers in two different regions, separated by a considerable distance. Each region is served by a regional toll switching center, and any calls between regions are routed through these centers. The facilities linking these regional

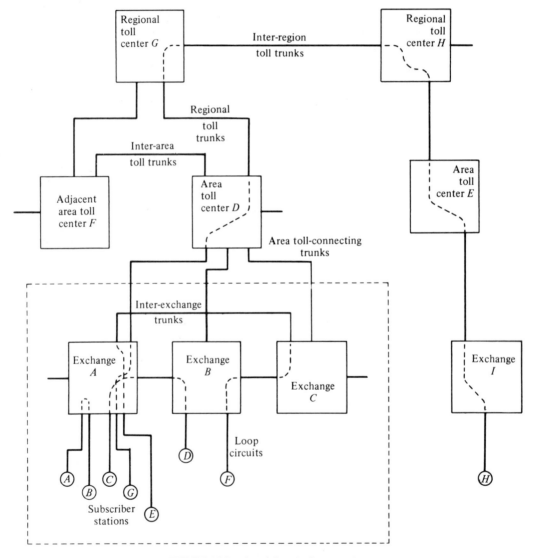

FIGURE 16.5. A public telephone system.

centers are mostly heavy-route microwave relay systems, or the satellite relay system.

A limited number of trunk circuits is provided between any two nearby switching machines, be they exchange or toll. The number provided is based on long-range forecasts of area population growth and future demand for service, done on a statistical basis to provide a certain maximum probability of not being able to complete a call during maximum "busy hours." Calls between each pair of exchanges within the system are assigned first-, second-, and third-choice routes if they are available, and the switching machine automatically selects these in sequence before indicating a "trunk-busy" condition. An example of this is provided by Ⓒ to Ⓓ and Ⓔ to Ⓕ in Fig. 16.5. Ⓒ was able to establish a direct trunk connection to Ⓓ through offices A_x and B_x, a first-choice selection. When Ⓔ tried to make a call to Ⓕ between the same two offices, all the A_x to B_x trunks were busy, and the alternate route through office C_x was chosen. In a similar manner, toll calls within a region may be made on first choice over direct trunks, and on second choice through one or more intermediate toll offices or through the regional toll office.

The automatic switching machines for each exchange are designated by a three-digit address code. Any local call within an area must be established by means of seven address digits, three prefix digits for the office address, and four for the particular subscriber's station being called. The digit "1" is reserved as an access code within each exchange to make connection to a toll-connecting trunk circuit. A second three-digit code immediately after the "1" addresses the toll area in which the destination subscriber is located. This is followed by the seven-digit number which completes the call in the destination area.

16.2.2 Step-by-Step Switching

The heart of the automatic telephone system is the switching machine. The first of these systems used step-by-step switching by means of the *Strowger stepping switch*, whose basic mechanism is illustrated in Fig. 16.6(a). The switch is constructed with a single pair of contacts attached to an arm at the end of the shaft. These contacts can make with any of 100 pairs of contacts in a stator assembly laid out in 10 rows of 10 sets of contacts in a semicylindrical array around the shaft axis. A rachet and pawl operated by a magnet allows the armature to be stepped up one level of contacts for each of 10 pulses received, from a rest position below the bottom row of contacts, and to one side of the array. A second rachet and pawl allows the armature to be stepped along any previously selected row, one step for each received pulse. Variation of the switch allows the armature to step continuously along each horizontal row until it reaches an unoccupied set of contacts in the row, or until it reaches the end of the row and drops out completely to the rest position.

Within a step-by-step office, these switches are used for two different purposes, line finding and line selecting. Figure 16.6(b) shows the typical chain of switching which takes place in making an exchange call with step-by-step equipment. Each subscriber station is equipped with a dial pulsing unit which sends out trains of 1 to 10 pulses representing the 10 decimal digits by interrupting the loop dc current. These pulses are relayed to the switching train

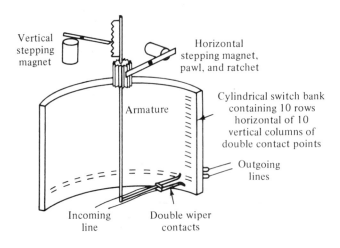

(a)

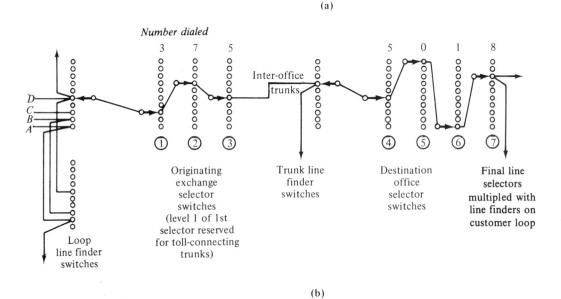

(b)

FIGURE 16.6. Dial step-by-step switching: (a) the stepping switch; (b) connections for a call made through a step-by-step switching machine.

to activate the switches completing the call. Each subscriber loop is connected to one pair of points on each of several *line-finder* switches. The number of line finders to which each line is connected is chosen to minimize the possibility of finding all the line finders within the group busy. When a subscriber attempts to make a call, he engages the hook switch at his set by lifting his receiver. This closes the loop circuit, and a relay energizes one of the line-finder switches in the group which is not already engaged with another call. The line finder then automatically steps up until it finds the level on which the demanding subscriber is located, and then across until it connects to that point. As long as loop current is maintained, the line finder will remain in that position. When the loop is broken for a period exceeding the dial pulse period (about half a second), the line finder drops out and returns to the idle position.

Each line finder has its armature connected directly to the armature of the first-group selector switch, which is now ready to receive the dial pulses for the first digit of the number being called. Each level of the selector corresponds to the digits 1 through 0. If the first number dialed is a 3, the first selector will step up three levels as the three dial pulses are received. When loop current is maintained for a period longer than the dial pulse, the switch automatically steps along the third row until it finds an unoccupied point. Each point in the third row may be connected to the armature of a selector switch in the second group, level 3 subgroup. Each of these second-group selectors is connected to several first-group selectors in corresponding points, so that each subscriber has access to several group 2-3 selectors, and so several different subscribers may use each selector at different times. Again, the number of selectors in the group is chosen to minimize uncompleted calls.

If the second digit dialed is a 7, the second-group selector steps up to the seventh level and searches for a free group 3-7 selector, as the pulses are received. Dialing the third digit operates this switch to step up to the appropriate level and search for a free point. The points on one row of one group of these final selectors are connected back into the local office, while all the others are connected to interoffice trunks so that connections can be made to them.

The far ends of the trunk circuits appear on the banks of line-finder switches which operate in the same manner as the subscriber line-finder switches. Two succeeding stages of group selector switches follow and respond to the first two of the last four digits in the number being called. The outputs of these switches are connected to the wipers of several multipled final selector switches. These final switches are arranged to step both up and across according to the last two digits. The point arrived at is connected to corresponding points of each of the other multipled final selectors, to several initial line finders, and to the loop circuit of the called subscriber. If the line is already busy, the final selector will drop out and send back a busy-tone signal. If it is free, it will send ringing current out to ring the bells, and when the subscriber answers, it will disconnect ringing tone and hold the chain of switches until the calling subscriber breaks the loop.

The step-by-step office contains many hundreds of mechanical switches, each with moving parts and wiping contacts. As a result, maintenance on this type of system is costly and difficult. The switching of currents in magnetic circuits also generates a high level of impulse-type switching noise within the system, which, while tolerable in voice communications, is completely unacceptable in data communications. As a result, very few new step-by-step offices are being installed. The immediate successor of the step-by-step office is the cross-bar switching office, which is discussed in the next section.

16.2.3 Cross-Bar Switching

The *cross-bar switching system* derives its name from the switching device used, the cross-bar switch. Figure 16.7(a) shows the mechanical layout of a 200-point cross-bar switch unit. This unit is laid out in 10 horizontal rows and 20 vertical columns, with a set of four contacts located at each of the 200 cross-points. Each cross-point contact set can be actuated by energizing one horizontal magnet and one vertical magnet, first causing a horizontal bar to rotate and then a vertical holding bar to rotate. There are only five horizontal bars, each of which can rotate in two directions, so that each bar can control cross-point switches in one of two rows at any one time. There are 20 vertical holding bars which rotate in only one direction, so that only 20 simultaneous independent paths can be held through the switch.

Figure 16.7(b) and (c) shows one of the cross-points in more detail. Each horizontal bar carries 20 selecting fingers or wires, which in the idle state lie in the gap between two sets of cross-point contacts (Fig. 16/7(b)). When one of the horizontal magnets is energized, the horizontal selecting bar rotates, causing all 20 selecting fingers to move over on top of the contact sets on one side. Now when one of the vertical holding magnets is energized, its vertical holding bar pushes down on the selecting spring at the desired cross-point and closes the four contact sets under it through the push rod. When the horizontal selecting bar is released, all springs except the held spring move back into the idle position. The operated spring remains trapped under the vertical holding armature to hold the contacts down until the vertical holding magnet is released. Each of the 200-point cross-bar switch units is capable of making and holding a total of 20 independent contact closures at one time. Switch units are used in pairs, and several pairs are multipled within a frame to provide the required fanout of the connections in the switching system.

Figure 16.8 shows the major pieces of equipment in a typical cross-bar exchange office and the path of a call established through it. The line link, district link, office link, line connecting link, and incoming trunk link are crossbar switch banks used to establish signal paths through the machine. The sender link is a crossbar switch used to connect the district junctor to a sender during call establishment. The district junctor, sender, marker, and line link controller are each special-purpose computers made of electromechanical relay circuits that control the action of the various switches as described below. An incoming subscriber's loop is connected to several vertical points within one or more cross-bar switches, so that it has access to several horizontals. When the calling station loop is energized, a relay network searches for and engages an idle line-link controller. This, in turn, controls the action of the line-link cross-bar switches. It searches to find a primary cross-point to engage on the incoming line, and a secondary cross-point to engage an idle district junctor

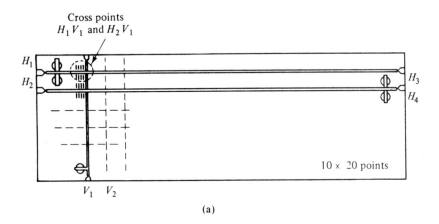

(a)

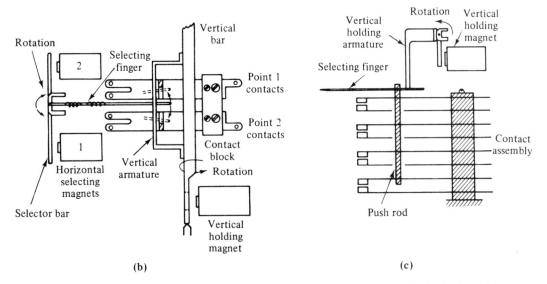

(b)   (c)

FIGURE 16.7. (a) Illustration of the structure of a cross-bar switch unit; the horizontal selecting bars and the vertical holding bars are shown with their activation magnets; (b) two contact assemblies showing the action of the selection wire; (c) contact assembly showing the holding armature action.

unit. Once these have been found, the two cross-bar switchpoints necessary to complete the connection are closed, and holding control is passed to the district junctor unit. At this point the line-link controller disconnects. The district junctor unit searches for an idle sender unit and operates the two switches in the sender link to connect the junctor to the sender. The sender, in turn, connects itself to an idle marker unit. Now the district junctor provides a dial tone, indicating that it is ready to receive dial pulses.

The first three digits of the number being dialed are passed directly to the marker unit, which decodes and stores them as the address of the office in which the called station is located. It then searches for and makes crosspoint connections in the district link and office link frames necessary to connect the district junctor through to an idle trunk circuit out to the destination office. At this point, the marker and sender have completed their part of the job, and the

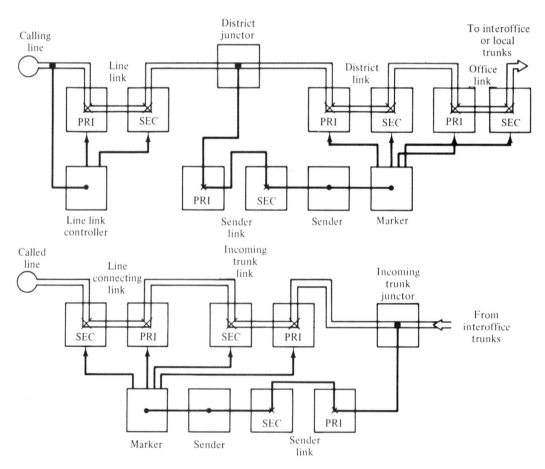

FIGURE 16.8. Exchange cross-bar switching system; the route of a local call is shown.

marker, sender, and sender link are all released, leaving the district junctor to hold the connections in the district and office link switches. At the far end of the trunk circuit, which may terminate in a remote office or in the originating office itself, an incoming trunk junctor circuit similar to the district junctor takes over control. It engages a new sender and marker unit, and waits for the final four digits of the number. These are received, decoded, and stored, and the called line is searched for. A path between the incoming trunk junctor and the called line is searched for through the incoming link switches and the line-link switches, and, if the line is not already engaged, the connection is made. At this point the marker and sender are dropped, leaving control to the incoming trunk junctor. If the line is already busy, the connection is dropped and a busy tone is transmitted back to the originating station. If it is free, the connections are held and ringing current is sent out to the called station. Operating the hook switch in the called loop to close the dc path cuts off the ringing current, and the conversation can proceed. Control is maintained by the calling party's loop current, and the connection is not dropped until the calling party hangs up. When it does, the two junctor circuits release all the cross-bar switch points in the circuit, which now become available to establish other connections.

The number of units of each type provided in an office depends on the statistical probability that a call cannot be completed because all the units are busy. The line link must provide at least one vertical on its primary for each line in the office, and each is multipled over several verticals. The number of other switch units in the district and office links and the incoming links is considerably smaller and depends on the expected maximum traffic. The sender and marker units are important, but since they do their job in a very short period of time, only a few are needed in a given office. Generally, for a 10,000-line office, only six markers are necessary. Only two junctor circuits are necessary for each call in process at a given time. One or two hundred of these would be sufficient to operate the office.

16.2.4 Electronic Switching

In the 1950s electronic switching systems were developed for use in telephone exchanges. The earliest of these simply replaced the cross-bar switch with its electronic equivalent and replaced the wire spring relays used in the sender, marker, and link control units with electronic circuits. Of course, such an approach is not the most efficient way to realize such a system, and an alternative was found in the general-purpose time-shared computer. Future electronic switching systems will use such digital computers to realize most of the functions involved in such switching.

Figure 16.9 illustrates how a digital computer may be used for message switching in a telephone system. Each subscriber loop and interoffice trunk circuit terminates on an electronic cross-point switch which performs the same

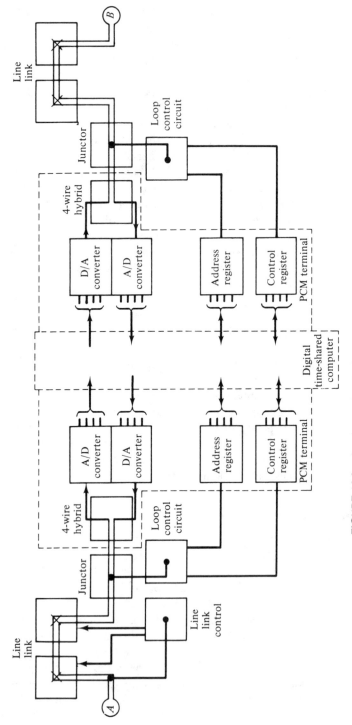

FIGURE 16.9. An electronic telephone exchange switching system.

function as the line link in a cross-bar system. The switch points are provided by using such devices as light-coupled silicon-controlled rectifiers. A unit providing 200 cross-points, as in the cross-bar unit, can be mounted on a single printed circuit card and can occupy a smaller volume than does the mechanical version.

In attempting to make a call, loop A energizes the link control circuit, which searches for a free computer terminal circuit and closes the link points to connect it. A junctor circuit separates the control signals and supplies the loop currents. The loop control circuit receives the dialed or pulsed number, which is the address of the distant terminal, and enters it in the address register. The proper command number is entered into the control register to tell the computer to complete the connection. Once the terminal circuit has been engaged, the computer periodically samples the address and control registers, and when the command is picked up, it acts to find a second free terminal in the group which can access the destination loop. The control and address numbers are entered into the second unit registers, and the loop control circuit picks up the free link circuit to connect the called loop. Busy or ring-down signals are sent, and once the distant party answers, the two loop control circuits hold the connections until such time as either loop is broken, terminating the call. The voice signals are separated by four-wire hybrid terminating transformers, with transmitted signals being encoded by the analog-to-digital (A/D) converter into a number which the computer can handle, and the numbers transferred from the far terminal being converted back to an analog signal by the digital-to-analog (D/A) converter. The computer periodically connects to each busy terminal in turn, picks up the number from the A/D converter along with the two addresses of the terminal points and the command status, and transfers the data to the proper D/A converter. Each register is sampled at least 8000 times per second so that no frequency distortion of the 4-kHz audio signals occurs. In a 10,000-line office, several hundred terminal units may be provided, but at any one time only a few of these would be in use. The number provided would depend on the maximum busy-period call rate, so that the chance of having an uncompleted call is minimized.

This is only one of several possible equipment arrangements, and new versions of the electronic office are being developed continually by several companies as new technology becomes available. The availability of single-chip micro-computers opens up a whole new field of possibilities for such switching centers. The British Post Office (BPO) has recently revealed its plans to modernize its entire telephone system by the early 1990s. The heart of this modernization is a flexible computer-based digital switching system, which is illustrated in Fig. 16.10. Analog telephone circuits are pulse-code-modulated and time-division-multiplexed onto high-speed data transmission lines (at 2 Mbits/second) before being connected to the switching system. Switching is done in both a spatial mode and a temporal mode within the switching network. All of these switches will be electronic in nature.

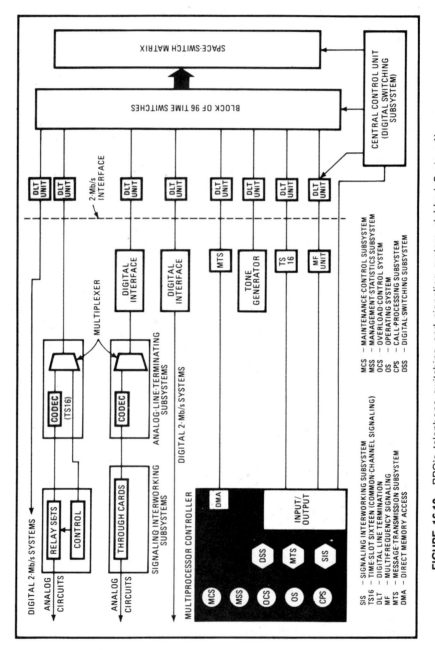

FIGURE 16.10. BPO's telephone switching and signaling system, dubbed System X, uses a modular approach to ensure compatibility and interchangability. Circles in multiprocessor indicate software; hexagons indicate software handlers. Reprinted from *Electronics*, May 24, 1979; Copyright © McGraw-Hill, Inc., 1979 all rights reserved.

SIS — SIGNALING INTERWORKING SUBSYSTEM
TS16 — TIME-SLOT SIXTEEN (COMMON-CHANNEL SIGNALING)
DLT — DIGITAL LINE TERMINATION
MF — MULTIFREQUENCY SIGNALING
MTS — MESSAGE-TRANSMISSION SUBSYSTEM
DMA — DIRECT MEMORY ACCESS

MCS — MAINTENANCE-CONTROL SUBSYSTEM
MSS — MANAGEMENT-STATISTICS SUBSYSTEM
OCS — OVERLOAD-CONTROL SYSTEM
OS — OPERATING SYSTEM
CPS — CALL-PROCESSING SUBSYSTEM
DSS — DIGITAL-SWITCHING SUBSYSTEM

A multiprocessor unit controls the action of the switching network. Within this network are several microcomputer units such as the Intel 8080 series or the General Instruments 1650. These microcomputers are each assigned a specific task within the system, such as the Maintenance Control Subsystem (MCS), the Call Processing Subsystem (CPS), and the Digital Switching Subsystem (DSS). All of these microcomputers share a large common memory system, and one of the microcomputers acts as the Operating System (OS) to provide the central control in passing instructions to the other microcomputers and taking care of the housekeeping chores.

The PCM encoding circuits (CODEC) and multiplexers may be adjacent to the switching unit, but are more likely to take the form of line concentrator units in the neighborhood centers surrounding the switching office. Most of the interconnecting transmission will be at 2 Mbits/second, either on coaxial cables or on optic fiber cables.

System flexibility is maintained by using similar equipment, subsystems, and operating programs, whether the switching unit is to be used as a large central office or as a small PBX.

16.2.5 Trunk Circuits

Trunk circuits are transmission circuits interconnecting two different switching centers. These centers may be two adjacent exchanges, or they may be separated by a considerable distance and the connection will involve one or more toll (long-distance) circuits.

Exchange interoffice circuits are merely extensions of the loop circuits connected at either end. They must provide a two-way voice channel, which can be done on either a two- or four-wire basis, and they must provide a two-way telegraph channel for supervisory control signals such as dial information and busy signals. Although two-wire circuits operating in the same manner as loops may be used, they are not favored for interoffice connections. Carrier systems are more commonly used, so that fewer cable pairs between offices may be used. The transmission objective on such an interoffice trunk circuit is that the total transmission loss between the points where the two loops are connected should approach 0 dB. Each loop is allowed a 3- to 5-dB loss maximum so that the net connection loss measured at 1000 Hz is no more than 6 to 10 dB.

Toll connecting and intertoll trunks are always designed on a four-wire basis. Ideally, the total transmission loss from end to end of any toll connection should be 0 dB. However, since each end is terminated in a two-wire system which does reflect energy, it is possible to get energy reflected from the far end returning on the return channel to the source. This can result in one of two things happening. If the loss is too low and the phase is right, oscillations, or singing, may occur. More likely, there will be sufficient loss in the loops to prevent singing, but there will be an echo. Since the velocity of propagation over long circuits is finite, it is possible to have return delays in excess of about 45 ms, with levels that make the echo intolerable to the user. The echo on circuits with shorter delays can also be unacceptable if the return level is too high. Toll-trunk circuits are designed on the *via net loss* (VNL) *principle*, so each circuit has a minimum amount of return loss built into it to keep the echo within acceptable levels. The VNL required for a toll circuit is calculated from the equation

$$\text{VNL} = \frac{0.2 L}{V} + 0.4 \text{ dB} \tag{16.1}$$

where

V = velocity of propagation on the facility, km/ms
L = circuit length, kilometers
0.2 = No. of dB loss that must be inserted for each millisecond of transmission delay, to maintain acceptable echo levels
$\dfrac{0.2}{V}$ = VNLF (via net loss factor) for a given facility, dB/km of length

Tables of VNLF's for various facilities are available. (See "Reference Data for Radio Engineers," 1972, Howard W. Sams.)

If the total VNL of a given trunk circuit exceeds about 2.5 dB, devices called *echo suppressors* are inserted in the channel. These are voice-operated relays which sense the high level of a near-end talker on the outgoing channel and insert a high loss in the return channel, as long as the speech continues. They disconnect in about 100 ms after speech ceases, so that the far-end party can start talking. A second circuit on the return channel does the same for transmission in the other direction. Because the velocity of propagation varies widely depending on the type of transmission facilities, spacing of echo suppressors may vary from a few hundred kilometers on cable circuits to a thousand kilometers on microwave circuits.

Noise on trunk circuits must also be kept to acceptable levels. Noise on long circuits must be kept below 44 dBrn, and those on short and medium-length circuits must be kept below 38 dBrn. The reference level for the noise measurements is derived from an arbitrary minimum level of output from a white-noise source which has first been passed through a weighted filter network with a frequency response similar to a standard telephone circuit. In terms of the Western Electric 3A telephone set (C message weighting, or the 500-type telephone receiver), 0 dBrn of noise corresponds to -90 dBm (ref. 1 mW) of 1000-Hz pure tone, or to -88 dBm of white noise (equivalent to thermal noise) in the frequency passband of 0 to 3 kHz. Special noise-measuring sets are used in practice which contain a white-noise source, a weighting filter, a calibrated attenuator, and a meter for measuring and comparing the channel noise to the reference-source noise. The abbreviation dBrn simply means decibels relative to reference noise level.

16.2.6 Private Telephone Networks

Many companies and institutions use private telephone networks to facilitate communication among their own members. These systems can be anything from two or three handsets on a single loop to an international network involving several centers, each with hundreds of stations. Private networks may be operated on a fully manual basis with all control and switching performed by operators, or they may be fully automatic switching systems similar to the public telephone network, or, as is more probable, they may be a combination of both manual and automatic operations.

The small manually operated system with several telephone stations near a single switchboard is known as a *private branch exchange* (PBX). Internal switching is performed by the company telephone operator, who often doubles as the office receptionist. One or more dial trunk circuits connect the PBX to a nearby public exchange office where they terminate on loop circuits corresponding to the number assigned to the company. Usually this is a single number, with access to several trunk circuits and a special switch which searches to find an idle trunk circuit.

The small automatic system is known as a *private automatic branch exchange* (PABX). This system is usually set up so that calls between stations within the system are done on an automatic switching basis by equipment similar to that used in the dial exchange office of the public system. One or more dial trunks connect the switching system with a nearby public exchange and may be arranged for automatic dial-out operation with manual answering or may be completely handled by the operator. An operator is usually provided with a manual switchboard to allow handling of outside calls and also special-service internal calls.

Large companies may have several PBX's or PABX's at different locations many kilometers apart. Communication between these PBX's may be done by means of long-distance calls over the public network, but if the volume of traffic is high, the company can get preferred rates by renting the use of private trunk circuits or tie lines between the various PBX's, reserved only for their use. Some large companies may even install their own transmission facilities over considerable distances.

Although a company may purchase, install, and maintain its own private telephone system, it has generally been more convenient to rent such systems from the public telephone companies. All the station and switching equipment then remains the property of the public company, but its use is reserved specifically for one customer. The service provided under these conditions must be at least as good as that provided by the public system and in some instances must meet even higher requirements because of special services, such as computer interconnections.

16.3 PROBLEMS

1. Calculate the dc-loop resistance of a telephone subscriber's loop comprised of 5.6 km of AWG22 cable and a further 10 km of 19 AWG cable. If M88 loading is used, calculate the overall signal attentuation at 1000 Hz.

2. Explain the purpose of the induction coil in a telephone station set.

3. Explain the term "full duplex" as applied to telephony.

4. Explain the principle of operation of the transformer-type transmission bridge shown in Fig. 16.2(c).

5. A repeater circuit similar to that of Fig. 16.4(a) is to be included in a two-wire trunk circuit. The maximum return loss on the east circuit is -40 dB and that on the west circuit is -55 dB. The two hybrid coils each have a feed-through loss of -50 dB and a transmission loss of -3 dB in either direction. If a 10-dB gain margin against singing is to be provided, calculate the maximum loop gain in the two amplifiers and the net gain of the repeater under these conditions between the two two-wire ports.

6. Explain, with the aid of equivalent circuits, the operation of the four-wire terminating set.

7. Explain the function of a line-finder switch in a step-by-step switching system. Explain the function of a line-selector switch.

8. What is the main advantage of the cross-bar switching system over the step-by-step system? What advantage does the electronic system have over the cross bar?

9. Calculate the via net loss of a trunk circuit which involves 2000 km of microwave relay circuits with an average velocity of propagation of 299×10^6 m/s. How long would this circuit have to be before an echo suppressor must be included?

10. Explain the term dBrn as used in connection with telephone-circuit noise measurement. Give the definition of 0 dBrn in reference to a pure tone of 1000 Hz.

11. Find the via net loss required for a 5000-km cable circuit on which the average propagation velocity is 80% of the speed of light. Does the circuit require an echo suppressor?

CHAPTER 17

Digital
Communications

17.1 INTRODUCTION

When a finite set of discrete values is used to represent information, the resulting signal is known as a *digital signal*. This differs from pulse modulation, described in Chapter 11, where although discrete values of the information signal were used, they could take on *any value* within the range of the continuous signal.

Digital data may arise in one of two ways. They may be sampled data from a continuous signal, as in pulse modulation, but with the sample quantized to the nearest value within the finite set of numbers. Or they may originate in digital form, as for example, a sequence of numbers giving the quantities of parts in a stockroom.

In the decimal system any one of the numbers from 0 to 9 inclusive is a digit. In the binary system, only two digits, 0 and 1, are used. The majority of digital electronic equipment utilizes the binary number system because of the ease with which the 0 to 1 states can be generated and manipulated. A binary digit is sometimes referred to as a *binit*, but more usually it is known as a *bit*. There is a distinction between these two abbreviated forms which can be important in certain circumstances. The *bit* is a measure of information content, and in a binary message where the 0's and 1's are *equiprobable*, one binary digit (binit) contains one bit of information, by definition. However, if the 0's and 1's do not occur with equal probability, a binit (which is still either one of the binary digits) will contain more or less than one bit of information. The distinction is not always observed simply because in many cases the equiprobable condition applies (or is assumed), and *bit* has become a synonym for binit. Following popular usage, the term *bit* will be used here.

17.2 BIT TRANSMISSION

Regardless of the origin of the data, a binary digital signal will consist of a sequence of bits which must be represented by a voltage or current waveform. Figure 17.1(a) shows the sequence 01000111011 as it would originate from a teletypewriter sending the character **G** in ASCII (*American Standard Code for Information Interchange*). Figure 17.1(b) shows the waveform that results for *unipolar synchronous transmission*, in which the 1 bits are sent as positive voltage pulses ($+V$), and the 0 bits as zero voltage. In ACSII, the first 0 bit is used for *start synchronization*, the next seven digits contain the character being transmitted (see Fig. 17.13 for an ASCII listing), the eighth bit is for a form of error detection, and the last two 1 bits are for *stop synchronization*. Coding and error detection are described in more detail in Section 17.10. For the moment, the point being made is that the digital message appears as a pulse-type waveform.

Unipolar transmission is wasteful of power in that a dc component (approximately $V/2$) is transmitted which does not add to the information

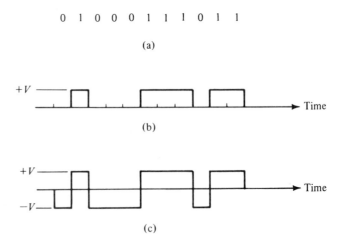

FIGURE 17.1. Binary digital sequence: (a) in digital form; (b) as a unipolar synchronous wave (c) as a polar synchronous wave.

content of the signal. This dc component can be eliminated by using *polar transmission* in which the 1 bits are sent as $+V$ pulses and the 0 bits as $-V$ pulses, The polar synchronous waveform for the sequence 01000111011 is shown in Fig. 17.1(c).

Other types of waveforms, including multilevel waveforms, are used to represent binary data. However, many of the important properties of digital systems can be illustrated using the polar synchronous waveform, and this is used in the following sections. Note that these pulsed waveforms are continuous in time and provide an analog representation of the binary data. Two of the most important parameters in a digital transmission are *signalling rate* and *error probability*, and these are discussed next.

17.3 SIGNALLING RATE

For a binary waveform, the bit rate is equal to the signalling rate and is measured in bits per second. Let τ represent the time it takes for one bit to be transmitted; then the signalling rate is

$$r = \frac{1}{\tau} \tag{17.1}$$

When the signal is transmitted through a baseband channel, the channel bandwidth sets a limit on the signalling rate. The limit is reached for the signal with the greatest number of changes per second, which in polar synchronous transmission is a square wave representing the digital signal...101010.... The period for this square wave is 2τ, and from the results of Section 2.2, the fundamental frequency spectrum component is $f_0 = 1/2\tau = r/2$. Now, a baseband channel acts as a low-pass filter which passes all frequencies from zero up to some cutoff value. Assuming that the frequency response is zero above some frequency limit B, it is clear that f_0 must be no greater than B for the fundamental component of the square wave to be transmitted; that is, $B \geq f_0$, or

$$B \geq r/2 \tag{17.2}$$

This is known as the *Nyquist criterion*. It states that for a given signalling rate r, the smallest bandwidth that can be used is $B = r/2$.

Figure 17.2(a) illustrates what happens when the square wave is applied to the LPF system. The output consists of the fundamental spectrum component only, and is therefore a sinusoid. In order for the original digital information to be recovered from this, it should be sampled at a rate r such that the samples occur at the peaks of the sinusoid, shown by the dashed lines in Fig. 17.2(a). Figure 17.2(b) illustrates the situation for a periodic signal of the form ...100100100.... The signalling rate is still r, and the bandwidth of the system is $1/2\tau$. Again, from the results of Section 2.2, the fundamental spectrum component of the input wave is $f_0 = 1/3\tau$, which, since the second harmonic is $2f_0$, lies above B. Thus, only the fundamental component is transmitted by the LPF system. Again, therefore, the output is a sinusoid, and if this is sampled at rate r at the positions shown by the dashed lines in Fig. 17.2(c), the original digital sequence is recovered. Note that the positive samples have a greater amplitude than the negative samples. These examples illustrate the minimum bandwidth criterion and show that accurate timing of the sampling process at the receiver is needed.

Signalling rate does not set an upper limit to bandwidth, for, as shown in Section 11.2, the square wave (in this case representing ...101010...) will be transmitted with little distortion when $B \gg 1/\tau$. In practice it is usually desirable to use the smallest practical bandwidth, as this reduces noise (see Chapter 4) and of course allows more channels to be fitted in a given frequency band. However, as shown in Section 11.2, bandwidth restrictions lead to intersymbol interference (ISI), and therefore some acceptable compromise must be reached between these two conflicting requirements in practice. Nyquist showed that in theory it is possible to eliminate ISI altogether while restricting the bandwidth of the transmission channel. This requires a special output pulse shape which has one main peak, and zero crossings on the time axis which are spaced by the sampling period. Thus, where the pulses overlap in time, the zero crossings of the interfering pulses coincide with the peak of the pulse being sampled at that instant, and ISI is avoided. The output pulse shape is controlled by controlling the overall frequency response of the system. Let $V_i(f)$ represent the frequency spectrum of the input pulse. For example, if

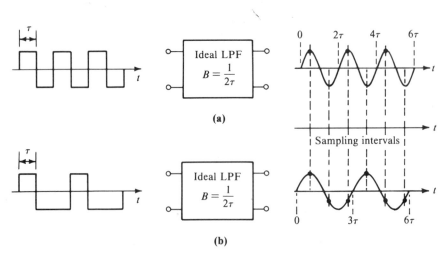

FIGURE 17.2. The bandwidth limitation B of the LPF converts: (a) the input square wave; and (b) the input pulsed wave, both to output sinusoids.

this is a flat-topped pulse, $V_i(f)$ will have the shape shown in Fig. 2.3(d). The spectrum is further shaped by three blocks—the transmitting filter $H_T(f)$, the channel filtering, $H_{CH}(f)$, and the receiving filter $H_R(f)$,—and the output spectrum for a single pulse, from the results of Section 1.15.1, is

$$V_0(f) = V_i(f)H_T(f)H_{CH}(f)H_R(f) \tag{17.3}$$

The system is illustrated in Fig. 17.3.

The transmitting filter is used to limit the average power transmitted. The receiving filter is adjustable and controls the final shape of the output spectrum. The channel frequency response $H_{CH}(f)$ is seldom within the control of the system designer, being a function of the particular routing used, which may include a variety of land lines, radio links, etc. The receiving filter is termed an *equalizer filter*. In some systems it can be adjusted automatically through feedback techniques to suit the particular channel in use, in which case it is referred to as an *adaptive filter*.

One theoretical function which is widely used in practice as a model for $V_0(f)$ is the *raised cosine rolloff function* shown in Fig. 17.4(a). This function is chosen because it can be approximated very closely in practice. As seen from the figure, the output spectrum is constant up to some frequency f_1 and is zero for all frequencies above a bandwidth limit B. Between the limits f_1 and B the normalized curve follows the equation

$$V_0(f) = \frac{1}{2}\left[1 + \cos\frac{\pi(f - f_1)}{(B - f_1)}\right] \tag{17.4}$$

The frequency f_0 at which the normalized response drops to 0.5 must be made equal to $r/2$ for the avoidance of ISI. The bandwidth B is chosen to lie between f_0 and $2f_0$, and the *rolloff factor* ρ for the response is defined as

$$\rho = \frac{B - f_0}{f_0} \tag{17.5}$$

Since the rate r will normally be known, this can be rewritten as

$$\rho = \frac{2B - r}{r} \tag{17.6}$$

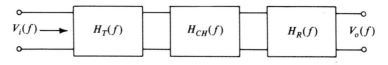

FIGURE 17.3. The three filter blocks that convert the input spectrum $V_i(f)$ to the output spectrum $V_0(f)$ as given by Eq. (17.3).

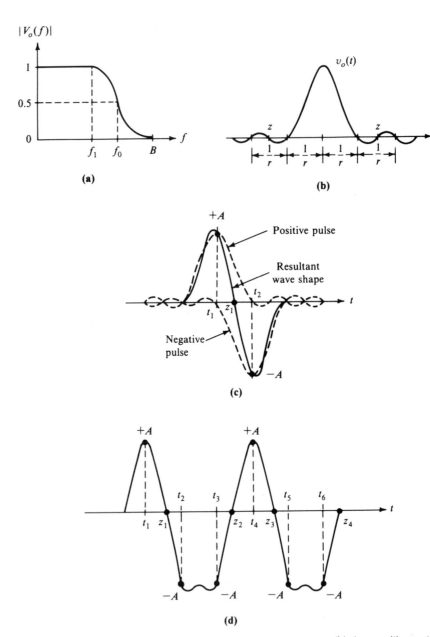

FIGURE 17.4. (a) The raised cosine roll-off frequency response; (b) the resulting output pulse waveform; (c) a positive and negative pulse added; (d) the output waveshape for the binary signal 100100.

The frequency f_0 falls midway between f_1 and B, i.e., $f_0 = (f_1 + B)/2$. Thus, from Eqs. (17.5) and (17.6), the frequencies B and f_1 can be expressed in terms of the rolloff factor ρ and signalling rate r as

$$B = (1 + \rho)\frac{r}{2} \tag{17.7}$$

and

$$f_1 = (1 - \rho)\frac{r}{2} \tag{17.8}$$

The upper limit for ρ is 1, and this results in $f_1 = 0$ and $B = r$. The lower limit for ρ is zero, which results in $B = r/2$, or $r = 2B$. This is the maximum signalling rate as determined by the Nyquist criterion. It will be seen that when $\rho = 0$, $B = f_1 = f_0$; that is, the output frequency response is rectangular in shape.

When the overall frequency response for a single pulse has the raised cosine rolloff shape and $f_0 = r/2$, the output voltage–time waveform has the shape shown in Fig. 17.4(b). Periodic zero crossings occur at time intervals of $1/2f_0$. Other zero crossings may occur, but those spaced at $1/2f_0$ are important because, by making $f_0 = r/2$, they occur at intervals of $1/r$, which is the reciprocal of the signalling rate. Thus, when a train of pulses is transmitted, although there will be some overlap in time, the zero crossovers of the interfering pulses will coincide with the peak of the pulse being sampled, and if this is sampled at the peak, the ISI will be zero. Figure 17.4(c) shows the resulting waveshape when a positive pulse is followed by a negative pulse. The sampling time t_1 and t_2 are spaced exactly $1/r$ apart.

Other zeroes shown by z on Fig. 17.4(b) may occur in the output pulse. In particular, when the rolloff factor is equal to unity, the additional zero crossings fall exactly midway between the $1/r$ sampling points. This has the further advantage that when a positive and negative pair of pulses occur, the crossover, shown as z_1 on Fig. 17.4(c), falls exactly midway between the sampling instants t_1 and t_2. From this information, the receiver timing can be accurately synchronized. This requires $\rho = 1$, and from Eq. (17.7), the bandwidth is $B = r$. This is twice the theoretical minimum bandwidth required to handle a signalling rate r and is the price paid for acquiring the timing information in this way.

The waveshape for the 100100 sequence is shown in Fig. 17.4(d) for $\rho = 1$. This should be compared with the output waveshape for this sequence shown in Fig. 17.2(b), which is the output obtained when $\rho = 0$.

It should be kept in mind that the raised cosine rolloff function is a theoretical curve, and the closer the rolloff factor is to zero, the more difficult it becomes to approximate the curve in practice. Also, as already noted, with $\rho = 1$, valuable timing information is obtained, and in many cases this is the best practical choice for the overall frequency response.

Signalling rate and bandwidth are discussed further in connection with pulse code modulation of speech signals in Section 17.6, where it is shown that the signalling rate for a single speech channel is typically about 64 kbits/s. Some other values encountered in practice are 100 bits/s for manually operated teletypewriters at the low end, to about 100 Mbits/s for pulse-code-modulated color video at the high end.

17.4 ERROR PROBABILITY

Errors can occur in a digital message as a result of noise. As described in Chapter 4, there are many sources of noise, some fundamental, some a result of improperly designed equipment. In binary digital systems it is possible for a pulse in one part of the system to induce a pulse in other parts, giving rise to an error. Even when all precautions are taken to minimize unwanted coupling, thermal noise (see Section 4.2) will be a limiting factor.

For baseband polar synchronous transmission in the presence of thermal noise, the probability of error depends on the probability that the noise voltage plus signal will cross a "decision level" of zero volts. This means that when a 0 bit $(-V)$ is being transmitted, an error will occur if the noise drives the voltage positive, and when a 1 bit $(+V)$ is being transmitted, an error will occur if the noise drives the voltage negative. This is illustrated for the 01000111011 bit sequence in Fig. 17.5(a). Note that zero volts is termed a decision level because the system is designated so that any positive voltage received is interpreted by the receiver as a 1 bit, and any negative voltage as a 0 bit.

The probability of error is a function of the ratio of signal amplitude $|V|$ to the rms noise voltage V_n. Figure 17.5(b) shows in graphical form the probability of error versus the ratio $|V|/V_n$. Acceptable error limits depend on application. For example, telemetry data such as temperature and pressure measurements which in analog form may be measured to an accuracy of 10% would probably be satisfactory in digital form with an error probability of 10^{-2}. Other data, such as fault monitoring on power transmission lines, may require an error probability of 10^{-6} or less.

The error probability known, the average number of errors in a message m bits long may be calculated from

$$\text{Avg. No. of Errors} = m \times \text{Error Probability} \qquad (17.9)$$

For $m \gg 1$ and (error probability) $\ll 1$, the *standard deviation* of the error distribution is approximately equal to the square root of the mean (for the Poisson probability distribution assumed for this situation). Thus,

$$\text{Std. Dev.} = \sqrt{\text{Avg. No. of Errors}} \qquad (17.10)$$

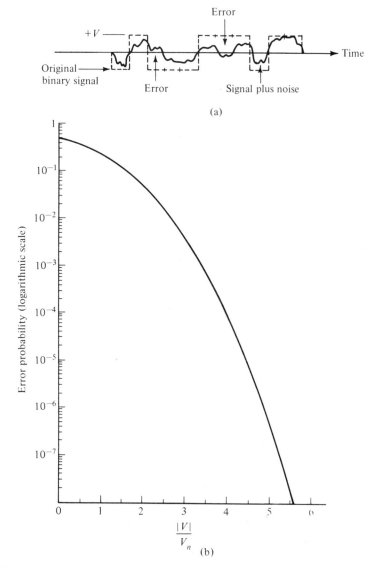

FIGURE 17.5. (a) Errors in a binary signal resulting from noise; (b) graph of error probability vs. signal-to-noise ratio for polar baseband transmission.

There is about a 95.5% probability that the number of errors will neither be more than Mean + 2(Std. Dev.) nor less than Mean − 2(Std. Dev.).

Example 17.1 A digital message is 10^7 bits long, and the error probability has been calculated as 10^{-5}. Calculate the average number of errors and the likely range in the number of errors.

Solution

$$\text{Avg. No. of Errors} = 10^7 \times 10^{-5}$$
$$= 100$$
$$\text{Std. Dev.} = \sqrt{100} = 10$$
$$\text{Range} = \text{Mean} \pm 2 \times \text{Std. Dev.}$$
$$= 100 \pm 20$$

Thus, one can expect anywhere from 80 to 120 errors.

17.5 DIGITAL FILTERING

As shown in Chapter 2, an analog signal may be filtered such that the output time function is some modified version of the input time function. In effect, the filter removes, or attenuates, certain frequencies relative to the others in the spectrum. For example, Fig. 2.10 shows how a low-pass filter (LPF) can convert a square wave to a triangular wave.

A digital filter performs a similar operation on discrete input data. These data can be samples of an analog signal, or they can originate as discrete data. The symbol x_n is often used for input data, and if these are samples of an analog signal $x(t)$, then x_n stands for $x(nT_s)$, where T_s is the sampling interval and n is the sample number in the sequence. Thus, x_n may represent a sequence of input values, and y_n the corresponding sequence of output values, as shown in Fig. 17.6.

Digital filters are classified as *recursive* and *nonrecursive*. With the recursive type, past and present values of input and past values of output are used to determine the present output value. With the nonrecursive type, only present and past values of the input are used to determine the present value of output. Here we are considering only *causal* filters, which respond only to present and past values, and not to future values. Figure 17.6(b) shows in block diagram form a simple nonrecursive digital filter, illustrating the basic elements used therein, viz., the adder, the unit delay, and the multiplier. For Fig. 17.6(b), the output is related to the input by

$$y_n = a_1 x_n + a_0 x_{n-1} \tag{17.11}$$

Here, a_0 and a_1 are constants. When x_n exists at the input to the unit delay, x_{n-1} exists at the output. The inputs to the top multiplier are a_0 and x_{n-1}, and the output is $a_0 x_{n-1}$. The inputs to the other multiplier are x_n and a_1, and its output is $a_1 x_n$. Both multiplier outputs are fed as inputs to the adder, and the output from this is y_n, as given by Eq. (17.11). Figure 17.6(c) shows a simple recursive filter. The output y_n exists at the input to the unit delay, and therefore y_{n-1} exists at the output of the unit delay. This is multiplied by the coefficient

b_1 and fed as an input to the adder, together with x_n. The output of the adder is the sum $x_n + b_1 y_{n-1}$, and this is multiplied by coefficient a_0, giving the output

$$y_n = a_0(x_n + b_1 y_{n-1}) \tag{17.12}$$

Equations (17.11) and (17.12) describe the operation of these simple filters on the input sequence x_n, and if x_n represents sampled values of a continuous time function, the equations describe the filtering operation in the time domain. The digital filter can also be characterized by its transfer function

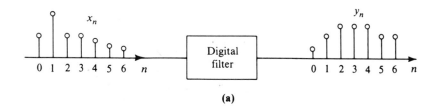

(a)

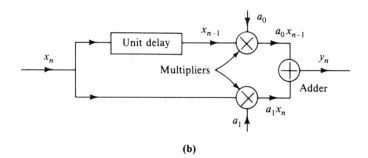

(b)

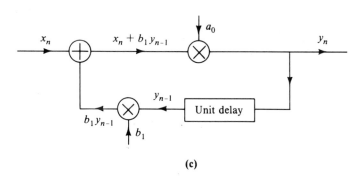

(c)

FIGURE 17.6. (a) Digital filtering of an input sequence x_n to produce an output sequence y_n; (b) a simple nonrecursive digital filter; (c) a simple recursive digital filter.

in the frequency domain, and parallels exist for all the classes of analog filters, such as low-pass, etc. Often the digital filter is designed to simulate one of the standard forms, for example, the Butterworth type, but new digital filter forms are also possible which cannot be realized in analog circuits.

The digital filter may be realized in software; that is, the filter is in the form of a computer program whose input is the x_n sequence, and whose output sequence y_n is the result of the computation. The digital filter may also be designed in hardware using binary digital logic circuits. The input sequence x_n must then be in binary form, and if it consists of real-time samples of an analog waveform, the samples can be converted to the required binary digital numbers through an analog-to-digital (A/D) converter.

Improved accuracy and stability can be achieved with digital filters compared with analog filters, and the digital filter characteristics can be made programmable. Thus, adaptive filters can be made which are automatically adjusted through feedback signals to give the desired filtering, as required in certain communications links. (See, for example, Section 17.3.) However, digital filters tend to be more expensive than their analog counterparts.

Any filter can be completely characterized by its response in the time domain to an impulse input. The impulse response of a nonrecursive filter is finite, meaning that the output sequence contains a finite number of values when the input is a single impulse. The nonrecursive filter is therefore also referred to as a *finite impulse response* (FIR) filter.

The impulse response of a recursive filter continues indefinitely, so that it is necessary to truncate the output sequence at a point determined by the accuracy required. The recursive filter is therefore also known as an *infinite impulse response* (IIR) filter. In comparison, the recursive filter requires fewer components than the nonrecursive filter to provide a given amplitude–frequency response, but its phase–frequency response cannot be exactly linear. In some applications a linear phase–frequency response is essential, and it can be achieved using a nonrecursive filter.

As an example of the operation of the recursive filter of Fig. 17.6(c), consider applying a step input to it. The step input consists of a sequence of unit values as shown in Fig. 17.7(a) defined by

$$step\ input: \quad x_n = 0 \quad for \quad n < 0$$

$$= 1 \quad for \quad n \geq 0 \tag{17.13}$$

For a causal filter, the output y_i is zero for $i < 0$. Thus, when $n = 0$, $x_n = 1$, and $y_{n-1} = y_{-1} = 0$. Substituting these values in Eq. (17.12) gives

$$y_0 = a_0 \tag{17.14}$$

Proceeding in this way, for $n = 1$,

$$y_1 = a_0(1 + b_1 y_0) = a_0(1 + b_1 a_0) \tag{17.15}$$

For $n = 2$,

$$y_2 = a_0(1 + b_1 y_1) = a_0\left(1 + b_1 a_0 + (b_1 a_0)^2\right) \tag{17.16}$$

And in general,

$$y_n = a_0\left(1 + b_1 a_0 + \cdots + (b_1 a_0)^n\right) \tag{17.17}$$

The recursive operation is evident in the buildup of these equations. The right-hand side of Eq. (17.17) is a geometric progression whose sum is

$$y_n = a_0 \frac{1 - (b_1 a_0)^{n+1}}{1 - b_1 a_0} \tag{17.18}$$

Thus, the output for a step input has been found in terms of the index n and the filter constants a_0 and b_1. When the product $b_1 a_0$ is less than unity, the filter can be designed to have a response similar to that for a simple RC low-pass analog filter. To see this more clearly, Eq. (17.18) can be rewritten as

$$y_n = A(1 - Be^{-na}) \tag{17.19}$$

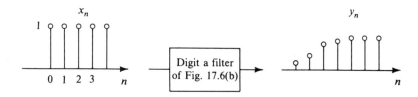

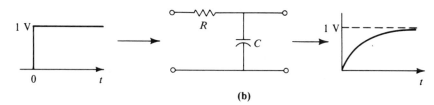

(b)

FIGURE 17.7. (a) A step-input sequence applied to the digital filter of Figure 17.6(c); (b) a step input applied to an RC low-pass filter, and the corresponding output.

where

$$A = \frac{a_0}{1 - b_1 a_0} \qquad (17.20)$$

and

$$B = b_1 a_0 = e^{-a} \qquad (17.21)$$

The charging curve for the RC low-pass filter, (Fig. 17.7(b)) is

$$y(t) = 1 - e^{-t/RC} \qquad (17.22)$$

At sampling time $t = nT_s$, this becomes

$$y(nT_s) = 1 - e^{-nT_s/RC} \qquad (17.23)$$

Comparing Eqs. (17.23) and (17.19), it can be seen that the forms are similar, although not identical, because of the factor B in Eq. (17.19). If the filters are designed to have the same time constant value, then a in Eq. (17.19) can be equated to T_s/RC in Eq. (17.23), and for the same final value, A in Eq. (17.19) can be equated to unity. This results in

$$a_0 = 1 - e^{-T_s/RC} \qquad (17.24)$$

$$b_1 = \frac{1 - a_0}{a_0} \qquad (17.25)$$

Thus, for the ratio T_s/RC for the analog system, Eqs. (17.24) and (17.25) allow the digital filter constants a_0 and b_1 to be evaluated.

Example 17.2 For the LPF of Fig. 17.7(b), $T_s/RC = 0.1054$. Determine the digital filter constants a_0 and b_1, and plot the output curves for each filter, for step inputs.

Solution From Eq. (17.24),

$$a_0 = 1 - e^{-0.1054} = 0.1$$

From Eq. (17.25),

$$b_1 = \frac{1 - 0.1}{0.1} = 9$$

From Eq. (17.20),

$$A = \frac{0.1}{1 - 9 \times 0.1} = 1$$

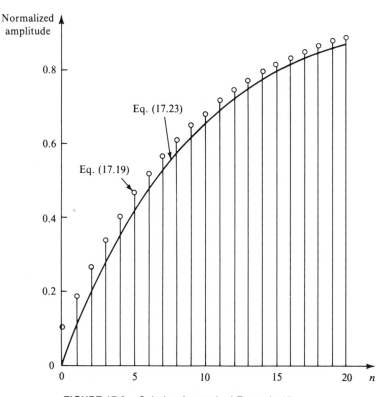

FIGURE 17.8. Solution for worked Example 17.

From Eq. (17.21),

$$B = 9 \times 0.1 = 9$$

Since the time constants are chosen to be equal, $a = 0.1054$. Thus, for the continuous time function, Eq. (17.23) gives

$$y(n) = 1 - e^{-0.1054n}$$

Also, for the discrete-valued output from the digital filter, Eq. (17.19) gives

$$y_n = 1 - 0.9e^{-0.1054n}$$

Both of these functions are plotted for the first 15 values of n in Fig. 17.8.

17.6 PULSE-CODE MODULATION

Pulse-code modulation (PCM) is used to convert analog signals into binary digital form. As already mentioned, the pulse modulation techniques described in Chapter 11 are analog methods. They suffer from transmission distortion and noise to an appreciable extent, and are therefore not used that often for direct transmission, but more often as intermediate steps in the generation of PCM. In the PCM system, groups of pulses or codes are transmitted which represent binary numbers corresponding to modulating voltage levels. Recovery of the transmitted information does not depend on the height, width, or energy content of the individual pulses, but only on their presence or absence. Since it is relatively easy to recover pulses under these conditions, even in the presence of large amounts of noise and distortion, PCM systems tend to be very immune to interference and noise. Regeneration of the pulses en route is also relatively easy, resulting in a system that produces excellent results for long-distance communication.

17.6.1 Quantization

The first step in the PCM system is to *quantize* the modulating signal. The modulating signal can assume an infinite number of different levels between the two limit values which define the *range* of the signal. In PCM, a coded number is transmitted for each level sampled in the modulating signal. If the exact number corresponding to the exact voltage were to be transmitted for every sample, an infinitely large number of different code symbols would be needed. Quantization has the effect of reducing this infinite number of levels to a relatively small number which can be coded without difficulty.

In the quantization process, the total range of the modulating signal is divided up into a number of small subranges. The number will depend on the nature of the modulating signal and will be from as few as eight to as many as 128 levels. A number that is an integer power of two is usually chosen because of the ease of generating binary codes. A new signal is generated by producing, for each sample, a voltage level corresponding to the midpoint level of the subrange in which the sample falls. Thus, if a range of 0 to 4 V were divided into four 1-V subranges, and the signal were sampled when it was 2.8 V, the quantizer would put out a voltage of 2.5 V and hold that level until the next sampling time. 2.5 V corresponds to the midpoint of the third range. The result is a stepped waveform which follows the contour of the original modulating signal, with each step synchronized to the sampling period. Figure 17.9 illustrates the quantization process.

17.6.2 Quantization Noise

The quantized *staircase* waveform is an approximation to the original waveform. The difference between the two waveforms amounts to "noise" added to the signal by the quantizing circuit. The mean square quantization noise voltage has a value of

$$E_{nq}^2 = \frac{S^2}{12} \tag{17.26}$$

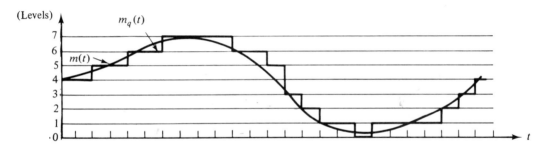

FIGURE 17.9. Quantization of a signal. $m(t)$ is the signal to be quantized. $m_q(t)$ is the quantized result. The difference $m(t) - m_q(t)$ is the quantization error.

where S is the voltage of each step, or the subrange voltage span. As a result, the number of quantization levels must be kept high in order to keep the quantization noise below some acceptable limit, given by the power signal-to-noise ratio, which is the ratio of average signal power to average noise power. For a sinusoidal signal which occupies the full range, the mean square signal voltage is

$$E_s^2 = \frac{1}{2} E_{\text{peak}}^2 = \frac{1}{2} \left(\frac{MS}{2} \right)^2 = \frac{(MS)^2}{8} \quad (17.27)$$

where M is the number of steps and S is step height voltage. The signal-to-noise ratio is now given by

$$\frac{\text{Signal}}{\text{Noise}} = \frac{E_s^2}{E_{nq}^2} = \frac{(MS)^2}{8} \times \frac{12}{S^2} = \frac{3}{2} M^2. \quad (17.28)$$

The number of levels M is related to the number n of bits per level by

$$M = 2^n \quad (17.29)$$

Substituting this in Eq. (17.28) gives, for the signal-to-noise ratio,

$$S/N = \tfrac{3}{2} \times 2^{2n} \quad (17.30)$$

In decibels this becomes

$$(S/N)_{\text{dB}} = 10 \log\left(\tfrac{3}{2} \times 2^{2n} \right)$$
$$= 1.761 + 6.02n \quad \text{dB} \quad (17.31)$$

A relationship between signal-to-noise and PCM bandwidth is developed in Section 17.6.5.

Example 17.3   In a PCM system the signal-to-noise (quantization-noise) ratio is to be held to a minimum of 40 dB (a power ratio of 10,000). Determine the number of levels needed.

Solution   From Eq. (17.28),

$$M = \sqrt{\frac{2}{3} \frac{\text{Signal}}{\text{Noise}}} = \sqrt{\frac{2}{3} \cdot 10,000} = 81.7$$

The number of binary bits required is found from Eq. (17.29) to be

$$M = \text{Log}_2(81.7) = 6.35$$

The next higher integer (7) would be used, giving an actual value of $M = 128$ and an actual signal-to-noise ratio of $1.5 \times 128^2 = 24,576$ (43.9 dB).

17.6.3   Companding

The signal-to-noise ratio derived above was related to a maximum-amplitude signal, one with a peak-to-peak voltage equal to the full signal range. In practice, the signal may be many times smaller than this, as much as 30 dB less. Since the noise level is dependent on the step size, which is a constant, with small signals a much lower signal-to-noise ratio will result. The process of *companding* is used to overcome this degradation of the S/N ratio. Companding is a compound process of volume compression before transmission combined with volume expansion after transmission. Companding itself does not produce any distortion in the recovered signal, but it does reduce the level of quantization noise during low-signal-level periods. The compressor amplifier amplifies low-level signals more than it does high-level signals, thus compressing the input voltage range into a smaller span. The steps transmitted have equal amplitudes, but are equivalent to smaller steps used in the low-level signals and larger steps used in the high-level signals. The result is lower-amplitude quantization noise during the low-level periods.

17.6.4   PCM Encoding

The encoding process generates a binary code number corresponding to the quantization level number to be transmitted for each sampling interval. Any one of the codes mentioned in the following sections (such as the ASCII, or biquinary) may be used as long as it provides a sufficient number of different symbols to represent all of the levels to be transmitted. Ordinary binary coding, in which the binary number corresponding to the decimal number of the level in question is transmitted, is most often used. This binary

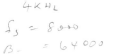
number will contain a train of 1 and 0 pulses with a total of $\log_2 N$ pulses in each number. (N is the number of levels in the full range.) This system is very economical to realize, because it corresponds exactly to the process of analog-to-digital (A/D) conversion. Microcircuits containing complete converters are available at very low cost.

17.6.5 PCM Bandwidth and Signal-to-Noise Ratio

Let each quantized level be converted to n bits, and let f_s represent the sampling frequency; then the signalling rate is $r = nf_s$ bits/s. If the highest frequency in the analog signal spectrum is represented by W, then $f_s = 2W$, and the signalling rate becomes $r = 2Wn$ bits/s. If the bit stream is coded as a polar synchronous wave (see Fig. 17.1), then, from the results of Section 17.3, the bandwidth required for transmission of the signal is

$$B = (1 + \rho)\frac{r}{2}$$
$$= (1 + \rho)Wn \qquad (17.32)$$

For example, for telephone speech signals, the highest frequency is about 3400 Hz, and to allow for the transition band of the audio frequency filter, the bandwidth is standardized in most cases at $W = 4$ kHz. If this is coded for $n = 8$, and a rolloff factor $\rho = 1$ is used, the bandwidth required by the PCM system is $B = (1 + 1) \times 4 \times 8 = 64$ kHz. This is considerably greater than the baseband width of 4 kHz, but results in a very high signal-to-noise ratio, as shown next.

Assuming that regenerators are used frequently enough so that the signal-to-noise ratio is determined by the quantization signal-to-noise ratio given by Eq. (17.31), Eq. (17.32) can be used to substitute for n to get

$$(S/N)_{dB} = 1.761 + \frac{6.02}{(1 + \rho)} \cdot \frac{B}{W} \qquad (17.33)$$

This shows that the signal-to-noise ratio, in decibels, is directly proportional to the bandwidth expansion factor B/W. In the above example, where $B = 64$ kHz, $W = 4$ kHz, and $\rho = 1$, the signal-to-noise ratio is 49.92 dB.

17.6.6 PCM System

The block diagram of a PCM system is shown in Fig. 17.10. The modulating signal $m(t)$ is applied to the input of the volume compressor unit. A sampling circuit generates a PAM signal from the compressed signal, which is then multiplexed with signals from other input channels. An analog-to-digital converter performs the two functions of quantization and encoding, producing a binary-coded number for each channel-sampling period. A commutator circuit transmits the code bits in serial fashion.

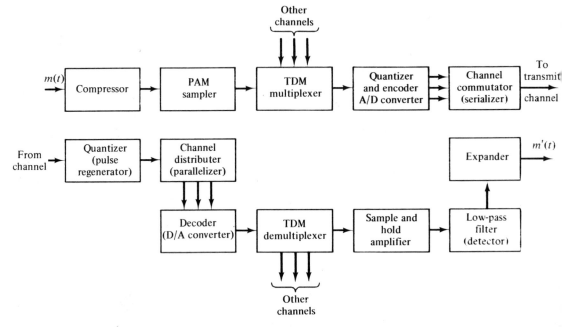

FIGURE 17.10. Pulse-code-modulation (PCM) transmission system.

At the receiver, a two-level quantizing circuit reshapes the incoming pulses and eliminates most of the transmission noise. A distributor circuit decommutates the pulses and passes the bits in parallel groups to a digital-to-analog (D/A) converter for decoding. Another distributor demultiplexes the several PAM signals and routes them to the proper output channels. Each channel has a sample-and-hold amplifier which maintains the pulse level for the duration of the sampling period, recreating the staircase waveform approximation of the compressed signal. A low-pass filter may be used to reduce the quantization noise, and finally, an expander circuit removes the amplitude distortion which was intentionally introduced in the compression of the signal, to yield the output signal $m'(t)$.

The Bell T1 PCM system is an example of an operational system. In this system, 24 voice channels are multiplexed, the sampling frequency being 8 kHz. The channeling scheme is shown in Fig. 17.11. The frame period is the time interval between successive samples of a given channel, and as seen from Fig. 17.11, the frame period is equal to the sampling period, which is

$$T_s = \frac{1}{f_s}$$

$$= \frac{1}{8 \text{ kHz}} = 125 \mu s \tag{17.34}$$

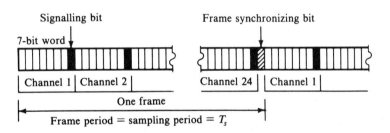

FIGURE 17.11. Channeling scheme for the Bell TI PCM 24-channel system.

Each channel is coded into 8 bits, which includes signalling information (dialling tones, etc.), and therefore, in one frame period the 24 channels require $24 \times 8 = 192$ bits. One frame-synchronizing bit is also required in each frame period; therefore, the total number of bits in a frame period is $192 + 1 = 193$. The signalling rate is therefore

$$r = \frac{193}{125} = 1.544 \text{ Mb/s}$$

Assuming that the system uses a filter with a rolloff factor $\rho = 1$, the bandwidth required is, from Eq. (17.7),

$$B = (1 + 1) \times \frac{1.544}{2} = 1.544 \text{ MHz}$$

17.7 DELTA MODULATION

Delta modulation (DM) is a process of modulation in which a train of fixed-width pulses is transmitted whose polarity indicates whether the demodulator output should rise or fall at each pulse. The input is caused to rise or fall by a fixed step height at each pulse. Figure 17.12(a) shows the block diagram of a DM system.

The modulating signal $m(t)$ is applied to the noninverting input of a high-gain differential comparator. A reconstructed version of this signal $m'(t)$ is applied to the inverting input. The differential comparator will be in saturation, either positive or negative, depending on the polarity of the *difference* voltage between the input signals. Thus, the output will be ± 1, ignoring the ambiguous range in the middle.

The modulator receives a train of unipolar pulses $p_i(t)$ recurring at the desired sampling rate and either transmits them directly for a $+1$ input or inverts their polarity for a -1 input. This signal is transmitted as the output signal $p_0(t)$ and also is passed to a local integrating circuit. This integrator causes $m'(t)$ to rise or fall by a fixed step height for each $+$ or $-$ pulse applied to its input. The height of the pulse can be adjusted by changing the gain

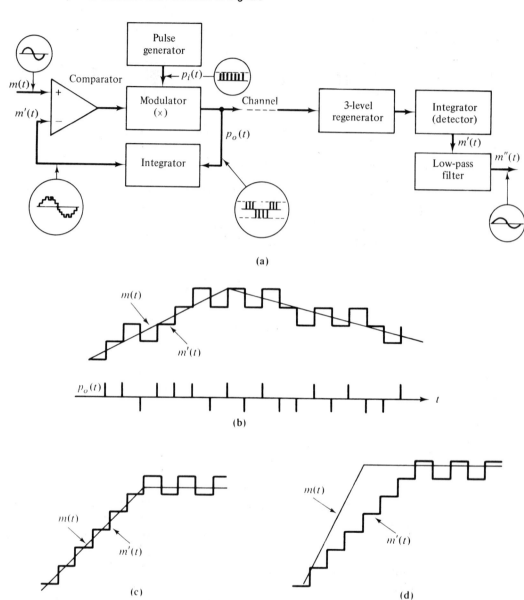

FIGURE 17.12. (a) A delta modulation (DM) transmission system; (b) input, output, and regenerated signal for input slope less than R_{max}; (c) input and regenerated signal for slope equal to R_{max}; (d) input and regenerated signal slope greater than R_{max}. R_{max} is the maximum rate of rise of the step waveform.

factor of the integrator. An increase will cause the step waveform to have a higher maximum rate of rise R_{max}.

At the receiver, a regenerator reshapes the received signal and removes most of the noise. The signal is then fed to another integrator, which reconstructs $m'(t)$, the step waveform. This is then passed through a low-pass filter to remove the quantization noise, leaving a replica $m''(t)$ of the original signal.

17.7.1 Slope Overload

The integrator will not be able to follow a signal with a very fast rate of rise. Figure 17.12(b) shows how the pulse train and the reconstructed step waveform follow a slowly rising or falling input. Figure 17.12(c) shows the case where the input is rising at the same rate as the staircase (i.e., it is rising at its maximum rate). At this rate, the output signal $p_0(t)$ will consist only of positive pulses. When the signal levels off, the step waveform will alternate above and below it, or "hunt." The output will be alternating positive and negative pulses. Figure 17.12(d) shows the case where the input is rising faster than the step waveform. The step waveform will continue to rise constantly at its maximum rate until it catches up, but in the meantime, it will be less than the input, causing an error in the demodulated signal. This condition is known as *slope overload*.

17.7.2 Adaptive Delta Modulation

To counteract slope overload, a variation called *adaptive delta modulation* has been developed. This system works in exactly the same manner as ordinary DM, except that a variable-gain amplifier is included in the input to the integrator. The gain of this amplifier is controlled by the average rate of pulses, either positive or negative, being transmitted. Thus, if more positive pulses are transmitted than negative pulses, the gain will increase. Also, if more negative pulses are transmitted than positive pulses, the gain will increase. The gain will be lowest when the number of positive and negative pulses are equal, or at constant input signal level. The effect of this is to cause the step height to be larger for faster-rising signals, so the step waveform can follow faster waveforms than the ordinary DM system.

Finally, comparing DM and PCM systems, DM requires a higher sampling rate and larger bandwidth than PCM to yield comparable fidelity, but DM requires much simpler and less costly hardware than PCM.

17.8 CODING

17.8.1 Information Content: The Bit

Information theory was developed by Shannon and others to provide a means of quantitatively measuring the rate and efficiency of information flow. According to this theory, the most basic piece of information possible is the choice of one of two equiprobable events—the familiar true or false, or, in the case of binary signals, one or zero. This basic unit of information is called the *binary* dig*it*, or *bit*, and is the basis of the binary number system. In a communication system, the basic bit of information may be conveyed as the presence or absence of signal during a specified time period (i.e., by a pulse). The PCM system is a case in point. Each pulse carries one bit of information. Several of these bits must be combined to extract enough data to reconstruct the signal level again.

If we consider a system of N equiprobable events (such as the N quantization levels in the PCM system), and we wish to specify *one* of those equiprobable events so that it cannot be confused with any other of that set of events, a certain minimum amount of information has to be provided. The number of bits required to define 1 out of N events in the binary system is

$$I = \log_2 N \tag{17.35}$$

Since a fraction of a bit is an impossible situation, the next higher integer number of bits must be used for the message, which specifies the 1-out-of-N event.

Coding can be done in higher-base number systems as well, although these cannot be so readily transmitted as the binary codes. Suppose that each pulse in a train of pulses can be amplitude-modulated to 10 levels. Then each pulse can carry the information to choose 1 out of 10 equiprobable events. Trains of pulses of this nature would allow coding in terms of ordinary decimal numbers. Each unit is a *d*ecimal dig*it*, or *dit*, and to encode a single event out of N events would require

$$I = (\log_{10} N) \text{dits of information} \tag{17.36}$$

One dit is equivalent to $\log_2 10$, or 3.32 bits, but 4 (the next higher integer) bits must be used to represent it.

17.8.2 Code Efficiency

Code efficiency is defined as the ratio of the exact number of bits necessary to define 1 of N events divided by the total number of bits that are used to transmit information.

Example 17.4   In the BCD code, 4 bits of information are transmitted in each *character* or message to define a single event. The total number of events out of which each is chosen is $N = 10$. However, 4 bits can allow the choice of 1 out of 16 events.

The efficiency is calculated as

$$\text{code efficiency} = \frac{\log_2 N}{B} \quad (\text{where } B = \text{bits used})$$

$$= \frac{\log_2 10}{\log_2 16}$$

$$= \frac{3.32}{4.00} = 0.83 \tag{17.37}$$

17.8.3 Telegraph Speed

Several types of binary codes, notably the international Morse code and the five-bit Baudot teletype code, use unequal length pulses in their coding. A special measure of message transmission speed was developed to describe this situation. The *baud* is defined as the inverse of the period of the shortest information unit (which corresponds to 1 bit) used in the code transmission. The baud is thus the maximum information rate of the code transmission, i.e.,

$$\text{telegraphic speed} = \frac{1}{T_{p\,\text{min}}} \quad \text{baud} \tag{17.38}$$

where $T_{p\,\text{min}}$ is the duration of the shortest information unit. This measure is used primarily in connection with teletypewriter codes.

In transmitting English text, the average word length has been found to be six characters, including the space separating it from the next word. The speed of transmission on a teletype or telegraph channel is quite often referred to in terms of words per minute (wpm). This can be readily converted to a baud rate if the character structure of the code is known.

Example 17.5 Suppose that a teletypewriter circuit uses the ASCII code and transmits at a speed of 100 wpm. Each character contains 11 bits, all of equal length. Eight bits carry message information, one is used for start synhronization, and two are used for stop synchronization. The code has an efficiency of 8/11, or 0.727. The baud rate is the same as the bit rate and is calculated as

$$\text{speed} = \frac{100 \text{ wpm}}{60 \text{ s/min}} \times 6 \text{ char/word} \times 11 \text{ bits/char}$$

$$= 110.0 \text{ baud (bits/s)}$$

The period of each bit is $1/110.0 = 9.1$ ms.

17.8.4 Parallel Transmission

In a given code, each character of the code is represented by I bits of information. In the teletype codes this information is usually transmitted in the bit-serial mode, with extra bits added for synchronization. Message transmission may be speeded up if a separate transmission channel is provided for each bit of the code character, and the message is sent out in bit-parallel (character-serial) mode with synchronization characters included after a block of characters has been transmitted. For the same channel bandwidth, information may now be transmitted at I times the speed, but this increase in speed is at the expense of providing the additional parallel channels, i.e., at the expense of increased total bandwidth.

17.9 CODES

One of the earliest binary codes was the code developed by Samuel Morse for use on his telegraph system. This system used alternate on and off periods of current on the telegraph wire to code the information. The current caused the armature of a solenoid coil much like a relay to rise if current was on, and fall when current was off. Each time the armature rose or fell, an audible click was produced. Two different lengths of pulses were used, and various combinations of up to six long and short pulses made up each character. The code was deciphered aurally by the operator, and a good telegrapher could read the code at speeds of up to 150 wpm. The Canadian National Railways in Canada discontinued the last of their Morse telegraph circuits in 1970.

A variation of Morse's code was adopted for use by radiotelegraphers and is still being used by amateur radio operators all over the world. This code uses combinations of long and short bursts of tone to define each character. The code may be transmitted by turning on and off a radio transmitter to produce the dots and dashes. The character symbols have been modified from Morse's original, and many new ones have been added. A partial listing of the code is shown in Fig. 17.13(a).

The advent of the teletypewriter machine produced the need for codes with a fixed character length and structure. The earliest of these, developed in the early 1900s, was the Baudot code. The character structure of this code is shown in Fig. 17.14(a). Each character contains five equal-length bits of information, a start bit, and a longer stop synchronizing pulse of 1.5-bit units, making a total of 7.5 bits per character. The coding allows a basic block of $2^5 = 32$ characters, but two of these characters are used as uppercase shift and lowercase shift. In uppercase, the same block of 30 characters has a different meaning than it does in lowercase. Thus, a code containing 60 useful characters is generated. The machines will remain in uppercase until a lowercase shift is

Character	Symbol
A	·—
B	—···
C	—·—·
D	—··
E	·
F	··—·
G	——·
H	····
I	··
J	·———
K	—·—
L	·—··
M	——
N	—·
O	———
P	·——·
Q	——·—
R	·—·
S	···
T	—
U	··—
V	···—
W	·——
X	—··—
Y	—·——
Z	——··
.	·—·—·—
,	——··——
?	··——··
EOM	·—·—·
EOT	····—·—
ERROR	······
Ø	——————
1	·————
2	··———
3	···——
4	····—
5	·····
6	—····
7	——···
8	———··
9	————·

(a)

	Character		12345	1234567
1	A	—	11000	0011010
2	B	?	10011	0011001
3	C	:	01110	1001100
4	D	WRU	10010	0011100
5	E	3	10000	0111000
6	F	%	10110	0010011
7	G	@	01011	1100001
8	H	£	00101	1010010
9	I	8	01100	1110000
10	J	BELL	11010	0100011
11	K	(	11110	0001011
12	L	)	01001	1100010
13	M	.	00111	1010001
14	N	,	00110	1010100
15	O	9	00011	1000110
16	P	Ø	01101	1001010
17	Q	1	11101	0001101
18	R	4	01010	1100100
19	S	'	10100	0101010
20	T	5	00001	1000101
21	U	7	11100	0110010
22	V	=	01111	1001001
23	W	2	11001	0100010
24	X	/	10111	0010110
25	Y	6	10101	0010101
26	Z	+	10001	0110001
27	CARR. RETURN		00010	1000011
28	LINE FEED		01000	1011000
29	FIGS. SHIFT		11011	0100110
30	LETT. SHIFT		11111	0001110
31	SPACE		00100	1101000
32	BLANK		00000	0000111

(b)   (c)

Bit position

76 01	76 10	8	76	54321
SPACE	@			00000
!	A			00001
"	B			00010
#	C			00011
$	D			00100
%	E			00101
&	F			00110
'	G			00111
(	H			01000
)	I			01001
*	J			01010
+	K	P		01011
,	L	A		01100
—	M	R		01101
.	N	I		01110
/	O	T		01111
0	P	Y		10000
1	Q	or		10001
2	R			10010
3	S	1		10011
4	T			10100
5	U			10101
6	V			10110
7	W			10111
8	X			11000
9	Y			11001
:	Z			11010
;	[			11011
<	\			11100
=	]			11101
>	∧			11110
?	—			11111

76 = 00 group-control codes
76 = 11 group-lower case alpha
and special marks

(d)

FIGURE 17.13. Partial listings of some commonly used codes: (a) international Morse code; (b) CCITT-2 teleprinter code; (c) ARQ teleprinter code; (d) US-ASCII teleprinter code. (Courtesy Howard W. Sams and Company, Inc.)

received. The listing for one variation of this code, the CCITT-2 code, is shown in Fig. 17.13(b). Most of the teletypewriters using this code operate at a speed of 60 wpm.

The ASCII code has recently been developed to provide a standard code for transmitting data between computers and for teletypewriters. The structure of this code is shown in Fig. 17.14(b), and a partial listing is shown in Fig.

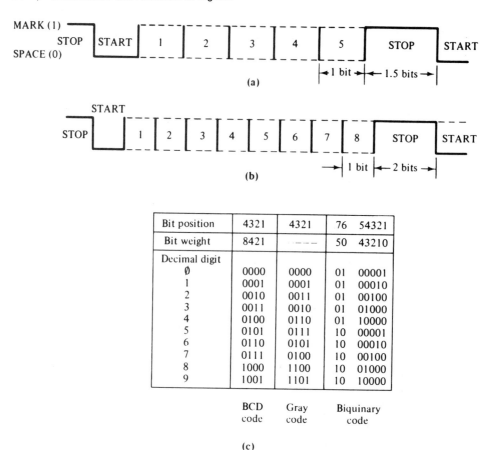

FIGURE 17.14. (a) Character structure of the five-level teletype code; (b) character structure of the eight-level teletype code; (c) some binary numeric codes.

17.13(d). It uses 11 equal-length bits per character: one start, two stop, and eight information bits. The eighth bit is usually reserved for parity, leaving a total character complement of $2^7 = 128$. This allows plenty of room in the code for all the common alphanumerics and punctuations, as well as a number of special functions. Only about 80 of the characters available are used in the eight-level teletypewriter machine coding, although the meaning of each character may vary considerably depending on the use the machine is put to. It should be noted from Fig. 17.13(d) that the 10 decimal digits are assigned within one block of characters, so that the last four bits of the character form the binary equivalent of the digit, resulting in very easy decoding of numerical data for computer entry.

Ordinary binary coding is quite often used for the interchange of numerical data between computers, especially on high-speed links. Such data are usually transmitted in blocks of characters, each block containing some form of error detection and correction mechanism. Numerical data are also quite often transmitted in the form of *binary-coded decimal* (BCD). This code uses four-bit binary characters to represent the 10 decimal digits; each one is the direct binary equivalent of the digit it represents. The code is shown in Fig. 17.14(c) and is known as a binary-weighted code. Several different weighted and unweighted codes are derived by choosing a different group of 10 characters out of the 16 available with four-bit coding.

17.10 ERROR DETECTION AND CORRECTION CODES

17.10.1 Parity

The simplest form of error detection is *parity*. Single parity is established as follows. First, the information is coded in the normal manner using one of the standard binary codes. Each character is then examined to determine whether it contains an even or an odd number of 1 bits. If even parity is to be established, a 1 bit is added to each odd character, and a 0 bit is added to each even character. The result is that all the characters now contain an even number of 1 bits. After transmission, each character is examined to see if it still contains an even number of 1 bits. If it does not, the presence of an error is indicated. If it does, the parity bit is removed and the data passed to the user. This form of parity will detect errors only if an odd number of bits is disturbed. An even number of errors within the same character will compensate for each other and go undetected. The total error security for this type of detection is quite low.

Multiple forms of parity provide a more secure system. In this case, a group of several characters are combined to form a block of information. Within a block, each character forms a row and has simple parity applied to it. In addition, each column in the block may have parity established for it by adding a parity word to the end of the block. Now if a single error occurs within the block, a parity error will be indicated for the word in which the error occurs and also for the column in which it occurs. The intersection of the row and column is the location of the faulted bit, and *correction* can be achieved by merely inverting that bit. The system will also detect multiple errors, but if these occur in more than one word or in more than one row of the block, correction cannot be accomplished. The coordinates for two errors will intersect at four locations, and unless a third parity (such as spiral) is added, there is no way to distinguish which of the four locations are in error. In that event, retransmission is requested.

17.10.2 Redundant Codes

A highly redundant code with natural built-in parity was developed in the 1940s for radio telegraphy. This is the ARQ code. Each character contains seven bits, of which only three are 1 bits, which gives the code its error-detection capability. There are 35 symbols of this type out of the 128 symbols possible with 7 bits, and only 32 of these are used. They are designated on the same basis as the CCITT-2 teletype code. The efficiency of the code is only 5/9.5 as compared to 5/7.5 for the CCITT-2, but this is more than compensated for by the increased security against errors. The ARQ code is listed in Fig. 17.13(c).

Another code which is similar to the ARQ code is the biquinary code, used for transmitting numerical data. It is also a seven-bit code, of which only two 1 bits are transmitted in each character. The bits are weighted from left to right as 50-43210, so the character 1000100 represents the decimal digit 7. The efficiency is 3.32/7, but it is very unlikely that an error would go undetected. This code is listed in Fig. 17.14(c).

The Gray code is an unweighted reflected binary code which is often used for transmitting telemetry data. It has the advantage that in progressing up or down through the code in counting, adjacent characters differ by only one bit. Telemetry data are usually varying slowly, so that the same character may be transmitted for several successive sampling periods before a change occurs. When that change does occur, it will be to the next higher or the next lower count. If a greater change is received, the character may be rejected, or an error indicated. Such a scheme also eliminates the ambiguity that occurs between two characters. Thus, if the sequence 8 8 8 4 7 7 7... were received, the receiver could correctly interpret that the 4 was a spurious signal and that a transition from the 8 to the 7 level occurred at the time that the 4 was encoded. The Gray code for the 10 decimal digits is shown in Fig. 17.14(c). Its similarity to the binary number sequence should be noted.

17.10.3 Hamming Codes

A somewhat different form of error detection and coding correction is developed by adding three or more check bits to each character, usually less than the number of bits in the original coding. These check bits are chosen so that the expanded characters used differ from each other by at least three bits. The number of bits difference is known as the *Hamming distance*, and a code of this type is known as a *Hamming code*. Obviously, many variations can be derived.

All the preceding codes treat the data on a block basis. The data are divided up into blocks of bits of predetermined length, and the error-detecting redundancy is applied to each block independently of the other. With a Hamming code, a continuous form of error checking is applied to form a convolution code. In this case, m code bits are derived by a logical combination of n successive bits of the original bit stream. The original code is shifted through a shift register serially, and at each shifting the combinational circuit produces a group of m bits of the coded stream. Thus the information for each input bit is contained in mn of the coded bits, and the likelihood of errors disturbing all the *mn* bits is very low. At the receiving end, another shifting and combinational operation decodes the original bit stream. A coding system of this type would be ideal for application to a PCM system or to any process that produces a continuous stream of code bits.

17.11   DIGITAL CARRIER SYSTEMS

Slow-speed data systems such as teletypewriters are usually modulated on a tone carrier, which is then frequency-multiplexed into a slot in a voice channel.

17.11.1   Frequency-Shift Keying

Two-tone frequency-shift keying is commonly used. In this system, two tones within the voice channel, which can be separated easily with bandpass filters, are assigned as carriers. One tone is transmitted for a 1 bit, and the other for a 0 bit. Two bandpass filters, rectifiers, and a differential amplifier demodulate the signal. Figure 17.15(a) shows a possible FSK system. Teletype carrier systems used on the telephone system divide the 4-kHz voice band up into as many as 20 channels, with a channel separation of about 120 Hz for 60 wpm, or as few as six 170-Hz channels for 100-wpm eight-level teletypes.

Radio-telegraph systems quite often frequency-modulate the RF carrier directly, transmitting the nominal carrier frequency for the 0-bit condition and transmitting a frequency about 70 Hz lower for the 1 bit. Alternatively, the two-tone FSK system can be either amplitude-modulated or frequency-modulated onto the RF carrier.

17.11.2   Phase-Shift Keying

Phase-shift keying is also used. For simple phase-shift keying, the unshifted carrier $V \cos \omega_0 t$ is transmitted to indicate a 1 condition, and the carrier shifted by $180°$, or π radians $[V \cos(\omega_0 t + \pi) = -V \cos \omega_0 t]$, is transmitted to indicate a 0 condition. The modulating circuitry is quite simple for this, since it is only necessary to provide two switches and an inverter. Figure 17.15(b) shows a possible phase-shift keyer arrangement. Demodulation is accomplished by subtracting the received carrier from a derived synchronous reference carrier of constant phase.

17.11.3   Differential Phase-Shift Keying

Phase-shift keying requires a local oscillator at the receiver which is accurately synchronized in phase with the unmodulated transmitted carrier, and in practice this can be difficult to achieve. Differential phase-shift keying (DPSK) overcomes the difficulty by utilizing the phase shift between successive

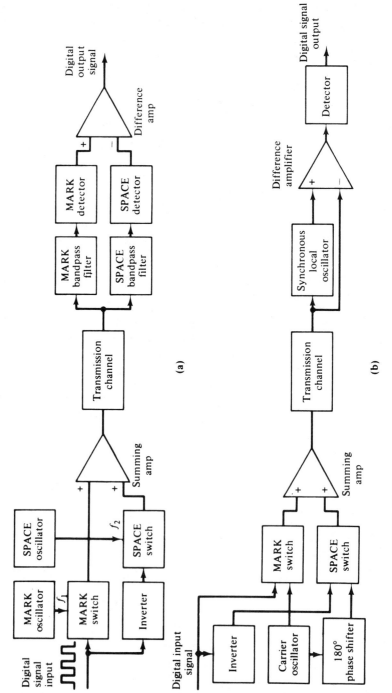

FIGURE 17.15. (a) FSK (frequency-shift keying) digital transmission system; (b) PSK (phase-shift keying) digital transmission system.

bits, hence the name *differential* PSK. The phase of the first message bit has to be compared to a reference bit, which may be arbitrarily chosen as 0 or 1. Figure 17.16(a) shows a 0 reference bit. If the compared bits do not differ, this is encoded as a 1, or zero phase shift of carrier. If they do differ, this is encoded as a 0, or π radians shift of carrier. The figure shows the sequence of events for the message 01100. Comparing the first message bit with the reference, they are seen to be the same, so this is encoded as a 1, and

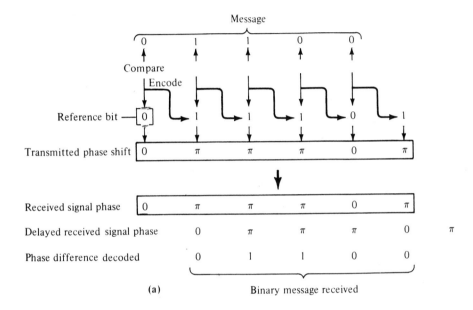

(a)

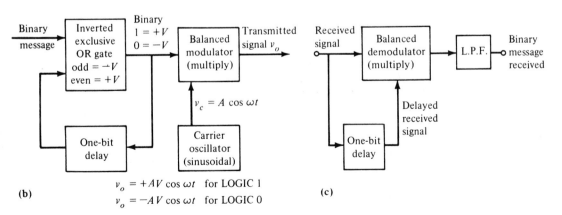

(b)

$v_o = +AV \cos \omega t$ for LOGIC 1
$v_o = -AV \cos \omega t$ for LOGIC 0

(c)

FIGURE 17.16. Differential phase-shift keying: (a) encoding and decoding a message sequence; (b) an encoding arrangement; (c) the receiver.

transmitted as zero phase shift. (Note that the first transmitted bit is the reference bit, a 0 in this case.) The second message bit is compared with the second encoded bit, both are seen to be 1, so this information is also encoded as a 1, and the carrier phase shift remains at zero. Continuing in this way, the third message bit is seen to be the same as the third encoded bit, so again this is encoded as a 1, or zero phase shift of carrier. The fourth message bit is compared with the fourth encoded bit. They are seen to be different, so this is encoded as a 0, and the carrier phase shift is π radians. The last message bit is compared with the fifth encoded bit; they are the same, so this is encoded as a 1, or again, a carrier phase shift of zero. Figure 17.16(b) shows how encoding might be achieved by digital hardware.

At the receiver, the signal is multiplied by a one-bit-delayed version of itself. This is shown schematically in Fig. 17.16(c). If the delayed signal is in phase with the direct signal, the output following the low-pass filter will have a positive dc component and will be decoded as a 1. If the two signals differ in phase by π radians, the dc output will be negative and interpreted as a binary 0. This sequence of events at the receiver is also illustrated in Fig. 17.16(a).

17.12 TELEPRINTERS AND TELEGRAPH CIRCUITS

17.12.1 Teleprinters

Teleprinters were originally designed for the transmission and reception of printed text, including both letters and numbers. They were a logical extension of the ordinary typewriter, and the early models were simply modified typewriters. The basic operation of a teleprinter unit can be followed from Fig. 17.17(a). The figure shows the various functional blocks involved in a teleprinter circuit connecting two keyboard-send-receive (KSR) units, which are used for direct communication between two operators. Each KSR consists of a keyboard with one key for each pair of symbols to be transmitted. A "shift-up" character in the five-level teletypewriter system shifts the printer up into the second group in the same manner as the uppercase shift on a typewriter, where it remains until a "shift-down" character is received. Each code character is used for two different symbols. In eight-level systems a shift key and another control key allow the generation of four different characters by each of the other keys, and the printer simply prints the corresponding symbol.

The keyboard and the printer are both mechanical. Both units are driven from a common motor through a set of shafts. Depression of a key on the keyboard sets a series of mechanical latches in the encoding section and releases the transmission clutch. The transmitter shaft makes a single revolution, during which a set of cams causes a contact to open and close according to the selected code. This contact breaks the current in the electrical transmis-

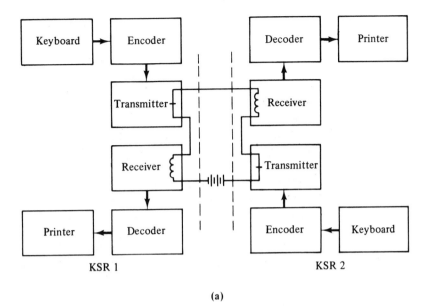

(a)

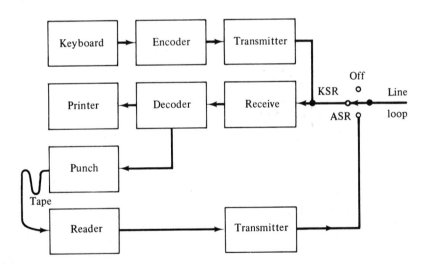

(b)

FIGURE 17.17. (a) Teletypewriter keyboard-send-receive (KSR) circuit; (b) automatic send-receive (ASR) station.

sion loop whenever a 0 is transmitted and makes the current whenever a 1 is transmitted. This transmitter is called a *commutator* and serves the function of time-division multiplexing of the bits of information that make up each character.

The current pulses are transmitted over telegraph circuits to the distant station, where a relay in the receiver section reconstructs the pulses and passes them to an electromagnet which controls the receiver operation. When this electromagnet is deenergized, it trips the printer clutch in the machine, which then makes one revolution. During this revolution a series of cams set or leave unset a series of mechanical latches in the decoder unit, depending on whether or not the receiver magnet is energized. Once the code has been interpreted by setting the latches, a final cam operates the printer to strike the selected character.

The receiving unit and printer at the sending location also monitors the line and prints the character as it is transmitted. Some machines are arranged so that the mechanical functions of the transmitter and receiver are combined, and the printer is operated directly from the keyboard by mechanical linkages. The keyboard must be locked out while the receiver is receiving a character from the line in order to prevent interruption errors. Special send-only machines are arranged in this manner but have the receiver-decoder functions omitted. These are known as *keyboard-send-only* (KSO) *machines. Receive-only* (RO) *machines* contain only the receiver-decoder-printer mechanisms and are often used for monitoring purposes.

Paper-tape punches and readers are also available, either as separate units or as integral parts of send-receive machines. Figure 17.17(b) shows the equipment features of an *automatic send-receive* (ASR) *station*. It is comprised of a keyboard with its encoder and transmitter, and a printer with its receiver and decoder, which can be either connected and used as a KSR unit on-line or used off-line on a local loop for tape preparation as a keyboard with a local printer. The tape punch is attached to, and operates as part of, the printer unit but can be disabled by a panel switch. Operated off-line, the station can be used to prepare a punched-tape message which can later be sent on-line by a paper tape reader with its own encoder and transmitter. The reader may also be used off-line to feed the punch so that duplicate tapes may be prepared, or so that the taped message may be typed for reading locally. This type of machine is particularly useful in applications on busy telegraph circuits, where the manual operation of a keyboard at speeds less than continuous maximum would be wasteful of transmission capacity. A tape of the message is prepared off-line at the operator's leisure and then transmitted at maximum speed when the telegraph circuit becomes available.

All the preceding machines have printers which produce a single copy of text, with up to five carbon copies, using fanfold or roll paper. A special form of the receive-only machine is frequently used in stock-exchange applications, in which the message is printed out on a narrow paper tape. This machine is called a *ticker*, because of the ticking noise the early machines made. It is enclosed in a soundproofed enclosure to reduce annoyance in quiet office surroundings. Similar machines are sometimes used in commercial telegraph offices, where an incoming message is typed on ticker tape and then cut up into sections and glued to a standard telegram form. It is doubtful that such practices will survive for long with computers available that can use a standard printer to automatically format a message on a form.

Teletypewriter machines are used at continuous speeds in the range from 60 wpm to about 200 wpm, corresponding to bit rates of 45 baud for 60-wpm five-level, and 220 baud for 200-wpm eight-level, machines. Several telegraph channels of this nature may be frequency-division-multiplexed onto a single voice channel.

High-speed paper-tape readers and punches are also available which operate at speeds up to 2000 wpm eight-level. These machines need a channel bandwidth of at least 1100 Hz, and only two or three channels may be carried on a single voice channel. They are useful in speeding up service on heavy-traffic circuits or for use with computer terminals. Transmission is always at the maximum rate, so line utilization is maintained at a high level.

Special versions of the keyboard, printer, tape punch, and tape reader are available which transmit and receive data in bit-parallel form. These machines require an eight-channel parallel transmission facility and are typically used in computer terminal applications.

Special very high-speed page-printing units have been developed for use as computer terminals. These may be classed as teleprinters, and could be used as such if broadband communications circuits are available. These machines feature printing on a line-at-a-time basis rather than a character at a time, usually print only a single copy on fanfold or roll paper, and can print at speeds up to 100 lines per minute, with as many as 200 characters per line. A small control computer with facilities for storing two or more lines of data at a time is needed with such a machine to control its formatting and operation. Data are transmitted to it and operated on in blocks which are usually one line in length. Transmission bandwidths for such applications are in the megahertz region, and long distance use is not common, except on intercomputer links. Such machines are usually used as the hard-copy output machines associated with local computer systems.

17.12.2   Telegraph and Teletype Transmission

Telegraph and teletype circuits operate by interrupting or reversing the polarity of a direct current in a pair of wires, in pulse fashion. The current level used is between 20 and 200 mA and is provided by batteries usually located at a central switching office. Local batteries located at the terminal machine are sometimes used to aid the central battery to guarantee enough current in long loop circuits. The loop circuits are usually telephone subscriber loop pairs especially assigned for the telegraph circuit, although in some cases the telegraph circuit may be multiplexed on a voice circuit which does not use dc signaling.

The polar relay has two opposing windings and a double magnetic circuit which is arranged so that if the difference between the currents in the two coils is positive, it will operate in one direction; and if the difference is negative, it will operate in the other direction. It is extremely sensitive, and a very small difference is sufficient to cause it to operate.

The polar relay is used to square up pulses in telegraph circuits and to provide an adjustable threshold. Polar relays located at intervals along a long telegraph line sense pulses from the sending end and apply fresh battery current to send on toward the receiver. The result is to square up pulses that have been skewed by the line, and to correct for bias distortion, but it will not correct timing distortion once it has been introduced by characteristic distortion or by a misadjusted circuit.

Figure 17.18(a) shows a simplified half-duplex (alternate transmission in either direction similar to simplex mode) teletype circuit. This circuit operates on unipolar current pulses in both the loop and the line, with the presence of current being a mark, or 1, condition, and the absence of current being a space, or zero. When the circuit is idle, mark current flows in both the loops and the trunk circuit. Mark current for the sending station loop is provided by two batteries, one positively connected to one side of the loop pair, and one negatively connected through the contacts of the receiver relay and the coil of the sending relay. This battery also supplies bias current for the sending relay, at one-half of the line current level. The net flux generated by the two coils is such as to hold the sending relay in its mark position.

When the sending contact of the teletype machine opens, signaling a zero or space bit, the line current is interrupted, and the net flux in the sending relay coils reverses polarity so that the relay operates to the space position. This provides battery of opposite polarity to the loop, which is opposed by the battery of that same polarity at the far end, and the loop current drops to zero. The bias current is reversed, so that the receiving relay remains in the mark position. Transmission toward the teletype machine proceeds in a similar manner. A third break relay (which is not shown) is included to prevent false operation if the trunk circuit should attempt to send into the loop when the loop is open-circuited, by forcing the trunk circuit current to remain marking.

Half-duplex circuits of this nature may be quite long and can span several exchange offices. If they are long, it may be necessary to include a regenerative-type repeater circuit somewhere in the middle of the circuit. The repeater is either a mechanical device similar to the receiver and transmitter of a teletype machine or an electronic circuit. This circuit acts to receive and interpret a distorted signal from one side, reconstruct it so that the transitions from mark to space and space to mark all occur at the proper time intervals, and retransmit it on the other side with fresh battery.

Dc trunk circuits of great length are not used. Instead, the teletype signals are received at the local exchange office and applied to a *modem*, or

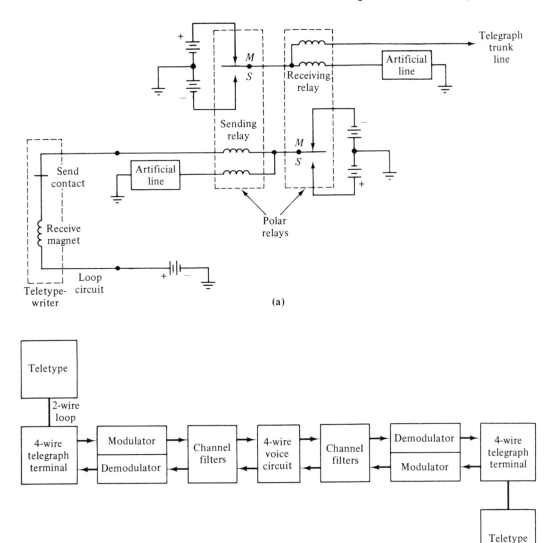

FIGURE 17.18. (a) Simplified half-duplex telegraph circuit; the circuit operates on unipolar or neutral pulses; (b) a carrier telegraph circuit; a teletypewriter system uses a tone carrier telegraph circuit superimposed on a telephone voice circuit.

data terminal set, which modulates the pulses onto a tone carrier which can be transmitted over an ordinary telephone circuit. The modulation method may be one of several types, but is frequently FSK. The *modem* (derived from *mo*dulator-*dem*odulator) also contains narrow bandpass filters to separate the desired carrier signals from others which may be on the same voice circuit, or from interference from supervisory signals. A single voice channel can accommodate as many as twenty 60-wpm five-level teletype channels or twelve 100-wpm eight-level channels, with FSK modulation.

Figure 17.18(b) illustrates a carrier teletype circuit. The loops of either end are arranged in the same manner as those of the dc circuit previously discussed. Relay terminal circuits separate the east- and west-bound signals. Outgoing signals are applied to the modulator input of the data modem circuit, and incoming signals are picked off the demodulator output. The outputs from the modems are connected to the voice trunk circuit through channel bandpass filters.

Quite often the data modem is included directly at the teletype machine. This has the advantage that the signal may be placed directly on an ordinary telephone loop circuit without the necessity of modifying any line equipment. Dial switching networks similar to the telephone switching networks have been developed by several companies. In some of these, special interoffice trunk circuits have been constructed and assigned just for teletype service, but some telephone companies use the same switching network for both telephone and teletype service. The terminal machines for such service include a data modem, dial-pulse generator, monitor handset, and calling and answering circuits, as well as the teleprinter machinery. The TWX service offered by Bell System companies and AT & T in North America operate such a service using the telephone network, while several railway companies and Western Union Telegraph operate a system called Telex.

Slow-speed data with rates less than about 200 baud can be transmitted over standard telegraph circuits such as the ones discussed in the previous section. Higher-speed data transmission needs special facilities. Data transmission with rates up to about 2000 baud can be transmitted on tone carriers over standard telephone channels. Rates up to about 4000 baud can be accommodated on compensated broadband voice circuits of the type used to feed radio network signals, while higher rates require special circuits.

Modified telephone carrier systems with special broadband data channels can be set up to accommodate data rates up to about 500,000 baud. Terminal loops for such systems require the use of coaxial cables to provide the necessary bandwidth and require special installations. Yet higher bandwidths can be provided over television channels on the radio systems. If such facilities are leased from telephone companies, the circuit is reserved specifically for the one user or customer. Most facilities of this type are used to interconnect computer systems and tend to be very costly.

In some cases of large systems interconnecting many computers, it may be more economical for the companies involved to construct their own microwave relay system. The costs of such a system are very high, and unless the system is very large and has a heavy volume of traffic, it would not be economical.

Standard data modems using a variety of modes of transmission and modulation have been developed. All of these have been designed so that the terminal connections and overall mode of operation are the same for all the various types of data sets. Thus, any data user with a standard data interface circuit can connect to a standard circuit and communicate with any other data user with similar terminal equipment. Interconnection of various types of equipment becomes a simple matter.

17.13 RADIOTELEGRAPH TRANSMITTERS

Figure 17.19(a) shows the block diagram of an interrupted continuous wave (ICW) transmitter. Such transmitters are rarely used these days, except for emergency communications and by radio amateurs and the military. These communications are for the most part concentrated in the HF band from 1.6 MHz up to 30 MHz, although some are used in the LF band below 0.5 MHz. Most modern telegraph communications use an audio-tone subcarrier which is then modulated on a continuous RF carrier by one of the standard voice-modulation techniques. However, many sections of the ICW transmitter are common to all types of transmitters, and these are briefly described here.

The circuit in Fig. 17.19(a) shows that the first and main component of the CW telegraph transmitter is a stable oscillator circuit at the assigned frequency. The oscillator is nearly always followed by a buffer amplifier, which is a tuned class A amplifier. This amplifier is used to prevent variations of load in subsequent amplifiers due to tuning from affecting the frequency of the main oscillator circuit. It also acts to provide any impedance transformation that may be required. All the tuned circuits in the CW transmitter are constructed with as high Q as possible, consistently with circuit stability, to provide as much rejection of harmonics as possible.

The buffer amplifier is followed by one or more tuned class C amplifiers which amplify the signal to the level necessary to drive the final amplifier. The final amplifier may have an output of several kilowatts or more, and the drive power may well be several hundred watts. Thus, the driver amplifier itself is really a power amplifier. If the frequency to be radiated is above about 15 MHz, it is unlikely that the crystal oscillator will be tuned to the final output frequency. The oscillator itself may be tuned to the second harmonic so that the crystal operates in an "overtone" mode, or one of the driver amplifier stages may be used to double or triple the frequency to the final output frequency value.

The final power amplifier is a class C amplifier with self-bias. Provision is made to measure the signal-induced bias current in the input circuit and the direct current supplied to the output circuit. A "keying" circuit is also

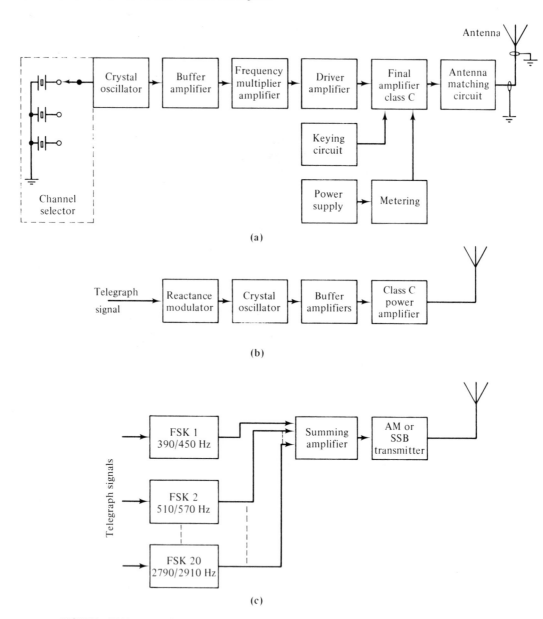

FIGURE 17.19. (a) ICW telegraph transmitter; (b) frequency-shift-keyed transmitter; (c) audio signal transmitter with several telegraph channels frequency-division-multiplexed in the voice channel.

provided which allows the current in the final amplifier to be interrupted, either by a relay contact directly in the current lead, or, as is more likely for high-power transmitters, by biasing the final amplifier into cutoff. In either case, a filter circuit is provided so that the cutoff does not occur suddenly, but rather over many cycles of the carrier. A typical keying time constant would be about 1 ms. This allows the carrier to decay and rise slowly enough so that the sideband frequencies do not spread into adjacent channels where they would give rise to interference heard as clicks or thumps.

Earlier stages of the transmitter might be keyed, but since the output stage relies on a continuous drive signal for most of its bias, heavy amounts of direct current would be drawn during the "key-up" intervals. Thus, the output stage is nearly always keyed.

On–off keying is not the most satisfactory method of transmitting by radio-telegraph, as it may be difficult to detect the off period when background noise is high. Alternative methods are frequency shift keying (FSK) of the carrier, and two-tone modulation. For the frequency shift keying of the RF carrier, a reactance modulator can be provided in which the variable reactance shifts the main crystal oscillator by a small amount. The amount of shift necessary for 60-wpm teletype signals would be about 60 Hz, and a channel bandwidth of about 150 Hz is sufficient to accommodate any sidebands created. Such a transmitter is shown in Fig. 17.19(b). The comments made about keying transients apply equally well here. In order to keep sideband generation to a minimum, the pulse-rise time is limited so that it is perhaps 1/10 of the total pulse length. Higher-speed telegraph systems require more bandwidth, with 100-wpm signals requiring about 170-Hz bandwidth, and 2000-wpm signals needing the full 4-kHz audio channel bandwidth.

Commercial transmission systems more commonly use an audio subcarrier system for transmission. Figure 17.19(c) shows a possible arrangement for such a transmitter. This system uses an ordinary voice-type transmitter, which may be full-AM, SSB, FM, or PM modulated. Up to 20 telegraph channels at 60 wpm can be provided on a single voice channel by frequency-division multiplexing. Figure 17.19(c) illustrates three of the 20 internationally assigned telegraph channels, showing the "mark" and "space" frequencies used for the audio subcarriers which are FSKed. Each telegraph channel is modulated on its own subcarrier, and the subcarriers are then summed to produce the audio input for the transmitter modulator.

17.14 PROBLEMS

1. A polar synchronous transmission takes place at a rate of 64 kb/s. What is the theoretical minimum bandwidth required to transmit this signal? The output spectrum for the system is a raised cosine rolloff function with a rolloff factor of 0.25. Calculate the bandwidth.

2. Plot the normalized output spectrum for the system of Problem 1. What is the -6 dB frequency?

3. On the same set of axes, plot the normalized spectra for raised cosine rolloff functions specified by $f_0 = 100$ kHz and (i) $\rho = 0$, (ii) $\rho = 0.5$, (iii) $\rho = 1$. Compare the bandwidths for these curves.

4. Part of a message being transmitted at a rate of 64 kb/s is ...1010100010.... The output spectrum is a raised cosine rolloff function with $\rho = 1$. Using Fig. 17.4 as a guide, sketch the resulting voltage–time output waveform. Indicate clearly on this the sampling intervals and the positions of the zero crossings on the time axis.

5. The difference equation describing a nonrecursive filter is

$$y_n = x_{n-1} - 0.5x_{n-2} - 0.5x_{n-3}$$

Draw the block diagram for the filter, and determine the output when the input is a unit step function. Assume all $x_i = 0$ for $i < 0$.

6. The input to the filter of Problem 5 is

$$x_n = \sin(n\pi/6)$$

and $x_i = 0$ for $i < 0$. On the same set of axes, plot both the output and input sequences for two cycles of the input sequence.

7. Show that the impulse response for the recursive filter of Example 17.2 is given by

$$y_n = a_0(a_0 b_1)^n$$

Plot the first fifteen values of this output sequence as a function of n.

8. The difference equation describing a recursive filter is

$$y_n = ax_n + by_{n-1}$$

where a and b are constants. Draw the block diagram for this filter. Show that the response of the filter to a unit step input is

$$y_n = \frac{a(1 - b^{n+1})}{1 - b}$$

Plot, as a function of n, the first twenty values of the output response when $a = b = 0.5$. Assume that $x_i = y_i = 0$ for $i < 0$.

9. The input sequence $x_n = \sin(n\pi/6)$ is applied to the filter of Problem 8, the filter coefficients being $a = b = 0.5$. On the same set of axes, plot both the output and input sequences for two cycles of input. Assume that $x_i = 0$ for $i < 0$.

10. The PAM signal of Problem 11.5 is to be transmitted by PCM. (a) Find the number of coding bits required for each sample if the signal-to-quantization-noise ratio is to be at least 30 dB in all channels. (b) Find the minimum channel bandwidth required to transmit the PCM system signal.

11. Describe how a delta-modulation system works, and explain what is meant by slope overload.

12. Explain how adaptive delta modulation improves the system's tolerance to slope overload.

13. A basic 8-bit teletype code is transmitted bit-serially with a 2-bit stop pulse and a 1-bit start pulse delineating each character. The code uses a complement of 165 characters. (a) Find the code efficiency of the basic 8-bit code. (b) Find the overall code efficiency for the 165-character code.

14. A teletype circuit uses the Baudot code and transmits at 200 words/minute. (a) Find the baud rate of the transmission. (b) Find the minimum channel bandwidth required.

15. A block of teletype code contains the following message: "The rate is $2.58." (a) List the ASCII coding for the block in a column with even parity. (b) Generate the parity character for longitudinal odd-block parity for this list.

16. (a) Explain the terms FSK, PSK, and DPSK. (b) Give the advantages and disadvantages of each for transmitting binary information.

17. Draw the block diagram of an ICW telegraph transmitter, and explain the purpose of each block.

18. A PCM system uses 12 bits to encode each sample and handles telephone signals with a volume range of 40 dB (difference between the largest and smallest signals). (a) What is the signal-to-quantization noise ratio for maximum sinusoidal signal level? (b) What is the signal-to-noise ratio for the smallest signal level? (c) If a compander were used which provided a 10-dB compression of the volume range, what would be the new signal-to-noise ratio in (b)?

19. How many bits would have to be used in the system of Problem 18(a) to give an overall improvement in the signal-to-noise ratio of 10 dB?

CHAPTER 18

Facsimile and Television

18.1 INTRODUCTION

In addition to basic signals consisting of speech, music, or telegraph codes, a telecommunications system is often required to transmit signals of visual nature. *Facsimile* means an *exact reproduction*, and in facsimile transmission an exact reproduction of a document or picture is provided at the receiving end. *Television* means *visually at a distance*, and a television system is used to reproduce any scene at the receiving end. It differs from facsimile in that the scene may be "live" (i.e., include movement).

Information is transmitted at a much faster rate in television transmission than it is in facsimile transmission. As a result, television transmission requires a much larger bandwidth, and special wideband circuits are required. The small bandwidth required for facsimile makes it suitable for transmission over normal telephone lines.

18.2 FACSIMILE TRANSMISSION

Some of the uses of facsimile telegraphy are:

1. Transmission of photographs (e.g., for the press).
2. Transmission of documents, weather maps, and so on.
3. Transmission of language texts for which the teleprinter is not suitable (e.g., Chinese). Facsimile also has the advantage that many of the other language difficulties of international operation do not arise.

18.2.1 Facsimile Sender

The message to be scanned may take one of three forms:

1. A single page, which may be a photograph, which is usually wrapped around a cylindrical drum in the sender to permit scanning to take place.
2. Narrow continuous tape.
3. Continuous-sheet paper, which may be thought of as broad tape.

Two methods of scanning are in general use:

1. Optical scanning, in which a light spot traverses the message.
2. Resistance scanning, in which the characters of the message offer varying resistance, and these are brought into circuit by means of a stylus touching and moving over them.

The optical method is the most common and is invariably used for picture transmission. Two scanning systems will be described.

Cylindrical scanning. In this method, the message is first fixed around the drum by means of clips. The drum is then simultaneously rotated about its axis and made to traverse along it under a fixed scanning spot. The light reflected from the scanning area is focused onto a photocell, the electrical output of which represents the signal. The layout of this system is shown in Fig. 18.1. The chopper disk found in older equipment converts the signal into a modulated wave, the carrier frequency being determined by the speed of the disk. The modulated signal is easier to amplify than a direct signal from the photocell. In newer equipment, solid-state amplifiers capable of amplifying the photocell output directly are used, eliminating the need for the "chopper."

In the method shown in Fig. 18.1, a comparatively large area of the message is illuminated, and a mask with a small aperture forms the spot which illuminates the photocell. An alternative method is to focus a very small scanning spot onto the message, thus eliminating the need for the mask and aperture. However, the first method has the advantage of avoiding distortion caused by mechanical irregularities of the drum, and is the most common.

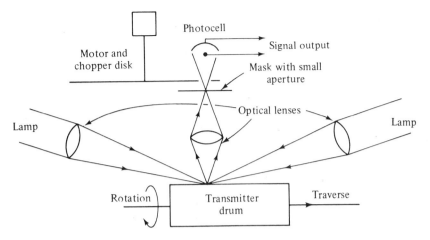

FIGURE 18.1. Facsimile sender system layout.

In the usual scanning arrangement, the spot follows a spiral path around the drum. An alternative, but less common, arrangement is to scan in a series of closed rings, the spot moving from one ring to the next as the fixing clips pass under it. It is also possible to keep the drum fixed and to move the scanning spot instead, but again, this is not common. Typical scanning values are a traversing speed of 1/100 inch per second (3.94 lines per millimeter) and a speed of rotation of 60 revolutions/minute (rpm). This means that there will be 100 scanning lines on each 1-inch width of picture. European systems typically work to four lines per millimeter.

Tape scanning. In this system, the message is taken directly off a printed tape. The scanning beam is arranged to travel across the tape at right angles to the direction of tape travel. One method of achieving this is shown in Fig. 18.2. The light beam will leave the hexagonal prism parallel to the incident beam, but its position will be deflected as the prism rotates. The spot thus travels across the tape as shown and starts a new scan each time a new face of the prism intercepts the incident beam.

The scanning of wide tape has been carried out in practice, but the difficulties encountered have limited its application.

The scanning spot. The shape of the scanning spot is important, as it determines the waveshape of the signal output. Its effect on waveshape is discussed more fully in a later section, but it may be mentioned here that a rectangular-shaped spot is preferred. This is arranged so that there is no overlap or gap at the sides, as shown in Fig. 18.3(a). Less frequently, a trapezoidal spot, as shown in Fig. 18.3(b), is used, and this is arranged such that average length of the top and bottom sides (the unequal sides) is equal to the scanning width, or scanning pitch P. This means that although the corners

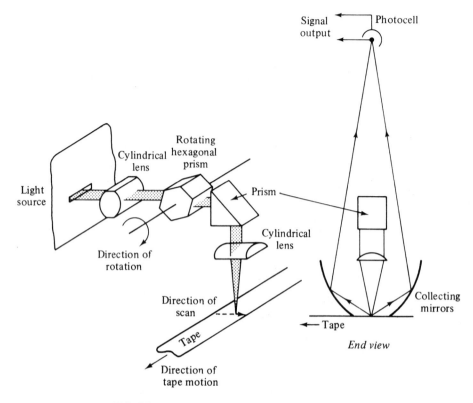

FIGURE 18.2. Tape scanning sender system layout.

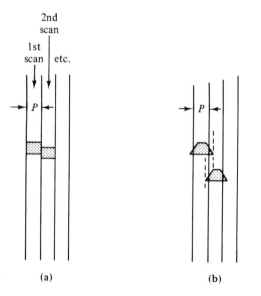

FIGURE 18.3. Scanning spot shapes: (a) rectangular; (b) trapezoidal.

of the spot overlap on successive scans, the average illumination over the area of overlap is equal to the average illumination over the rest of the spot area. It must be remembered that the spot is extremely small and at any given instant can be considered as covering an area of constant illumination on the message. The changes in average illumination convey the picture information.

18.2.2 Facsimile Receiver

The mechanical aspects of scanning in the receiver are similar to those in the sender, and very often identical equipment is used at both ends. Scanning in the receiver must of course produce an optical output from the electrical input signal, the reverse of what happens in the transmitter. This aspect of the receiver is discussed later.

In order for the received signal to have the correct relationship to the transmitted signal, it is necessary for the signals to be synchronized, to be phased correctly, and to have the same height/width ratio.

Synchronization. Where the message is documentary (e.g., a written message), it is sufficient to use synchronous motors for both sender and receiver, and operate off frequency-controlled supply mains. Where picture transmission is involved, a synchronizing signal must be sent, and this by international agreement has a frequency of 1020 Hz. The sender speed bears a known relationship to this, and the receiver speed is adjusted by means of a stroboscope to correspond to the relationship. An accuracy of about 1 in 10^5 can be obtained in this way. For example, if the sender speed is 60 rpm, the receiver speed can be expected to remain within $\pm$ 0.0006 rpm of the sender speed.

With carrier transmission (i.e., when the signal is modulated onto a carrier, as, for example, in frequency multiplexing), it is necessary to send the carrier along with the sideband transmitted (usually the upper sideband). The carrier being present enables the *exact* 1020-Hz synchronizing signal to be recovered. This is an added requirement of facsimile transmission, as with normal telegraph signals it is not necessary to send the carrier, a local oscillator at the receiver being adequate for recovery of the signal. The effect of a constant-speed error is shown in Fig. 18.4(b).

Phase. Correct phasing is necessary to ensure that the image of the clips holding the paper to the drum does not intersect the transmitted picture. Phasing adjustment need only be made once for each picture transmitted and is carried out as follows. The operator at the receiver first adjusts the speed to the correct value by means of the synchronizing signal and then sets the drum in the correct start position. This is held in position by means of a switch. At the sender, a pulsed signal (perhaps the 1020-Hz signal pulsed momentarily) is sent to indicate the start of the transmission, and the pulse releases the switch holding the receiver drum. The effect of incorrect phasing is shown in Fig. 18.4(c).

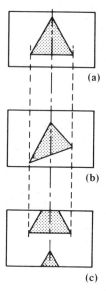

FIGURE 18.4. Facsimile distortions: (a) received image in a properly adjusted system; (b) image distorted by a constant-speed error; (c) image shifted by a phasing error.

Index of cooperation. The height/width ratio must be the same for both the transmitted and received pictures, and this in turn depends on the scanning pitch and the diameters of the drums used in the sender and the receiver.

Referring to Fig. 18.5, let

$$D = \text{diameter of sending drum}$$
$$d = \text{diameter of receiving drum}$$
$$P = \text{scanning pitch of sender}$$
$$p = \text{scanning pitch of receiver}$$
$$n = \text{number of lines scanned}$$

The width of the transmitted picture is nP, and the width of the received picture is np. The height of the transmitted picture is proportional to D, and that of the received picture to d, by the same constant of proportionality. Hence, for the correct height/width ratio to be maintained,

$$\frac{D}{nP} = \frac{d}{np}$$

or

$$\frac{D}{P} = \frac{d}{p} \tag{18.1}$$

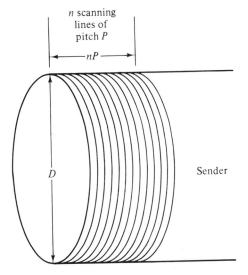
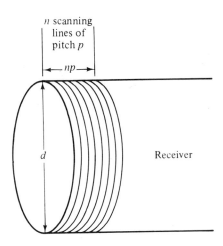

FIGURE 18.5. Facsimile sender and receiver pitch of scan.

Thus, the ratio of the diameter to the scanning pitch is the same for the sender and the receiver. In Europe, the Comité Consultatif International Télégraphique et Téléphonique (CCITT) defines this ratio as the Index of Cooperation (IOC). In the USA, the IEEE defines the IOC as the product of stroke length times scan density. For the drum scanner, the stroke length is πD and the scan density is the lines per unit length $= 1/P$, so that for the drum scanner;

$$IOC\,(IEEE) = \frac{\pi D}{P}$$

$$IOC\,(CCITT) = \frac{D}{P} \tag{18.2}$$

The effect of having different indexes of cooperation at the receiver and transmitter are illustrated in Fig. 18.6.

Example   The drum diameter of a facsimile machine is 70.4 mm, and the scanning pitch is 0.2 mm per scan. Calculate the index of cooperation.

Solution   From Eq. (18.1),

$$IOC\,(IEEE) = \frac{\pi \times 70.4}{0.2} = 1106$$

$$IOC\,(CCITT) = \frac{70.4}{0.2} = 352$$

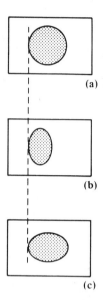

FIGURE 18.6. The effect of different indexes of cooperation: (a) same IOCs; (b) receiver IOC larger than transmitter IOC; (c) transmitter IOC larger than receiver IOC.

When the electrical signal is received, it must be converted to an optical image, and this may be achieved either through photographic reception or through direct recording reception.

Photographic reception. In this method the received signal is used to control the intensity of a light beam, which in turn scans the photographic material. Figure 18.7 shows the essential features of the Duddell mirror oscillograph, which may be used for this. A small coil of fine wire is suspended in a strong magnetic field. Mounted on the coil is a very small mirror, the dimensions which are about 0.03 inch × 0.06 inch. The loop and the mirror, which form the movement of the system, offer negligible inertia. The demodulated signal (i.e., the basic signal representing the message from the sender) is passed through the loop and causes it to deflect, the angle turned through being proportional to the signal current. A beam of light focused onto the mirror is reflected through the various apertures and lenses shown in Fig. 18.7 onto the photographic paper or film on the drum. It can then be arranged that maximum deflection of the mirror corresponds to maximum light through the aperture if photographic paper is to be used (providing positives), and therefore, lesser degrees of deflection put lesser amounts of light onto the photographic paper. Alternatively, if film is to be used to produce negatives, minimum deflection of the mirror can be arranged to give maximum light through the aperture.

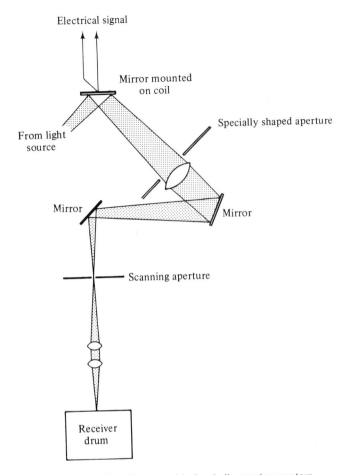

FIGURE 18.7. Photographic facsimile receiver system.

Another method widely used employs the *crater lamp*. This is a lamp which contains neon, argon, or helium, and which glows when a voltage is applied across its terminals. The intensity of the glow is approximately proportional to the applied voltage. The signal voltage is applied to the lamp, and the light from this is focused onto the photographic paper on the drum. This method is not capable of responding to the full tonal range of some pictures, which may be as high as 30 dB (i.e., the differences between maximum light area and maximum dark area on the picture may be 30 dB), which is a voltage ratio of about 32:1. The maximum range of the lamp is about 15 dB, but this is found to be adequate for most purposes.

Photographic positives or negatives may be made from the methods described. Positives are usually made, since these can be processed and finished much faster, and speed is usually the important factor. Negatives have some advantages: more than one copy can be made of the original; they are more sensitive and can therefore be produced with lower lamp power; they allow for retouching, and thus, the quality of the final picture can be improved.

Direct recording reception. In one form of this method, a highly absorbent, chemically treated paper is used, in which the electrolyte held by the paper dissociates when a voltage is applied across it. The signal voltage is applied to the paper via a metal stylus, producing dissociation products, one of which is a metallic salt. This, in turn, reacts with a color chemical in the paper, which produces a mark on the paper. The intensity of the mark depends on the amount of dissociation, which in turn depends on the signal voltage. A steel stylus is often used, as this produces a very intense black coloration.

The paper used is damp and must be kept in sealed containers. It has a lifetime of about one month after opening. It is reasonably cheap, but the tonal range is much less than that obtained with photographic methods. It is usually found to be adequate for low-grade work.

A second form of the method employs a resistance paper known commercially as *Teledeltos paper*. This consists of a metallized backing on which is deposited a substance similar to carbon black, and on top of this, a very thin layer of insulation. A stylus exerts a steady pressure on the paper, and when the signal voltage is applied, burning occurs, which causes blackening of the paper. The tonal range is similar to the previous method, but definition is not so good and the paper is fairly expensive.

18.2.3 Transmission of Facsimile Telegraph Signals

The bandwidth requirements may be found by a similar method to that used for telegraph signals. The worst condition will be assumed, that in which alternate black and white squares occur, each square being the width of the scanning pitch. Each of these squares is a minimum-sized picture element, or *pixel*. It follows that the vertical resolution need be no better than the horizontal resolution (see Fig. 18.8(a)). The ideal output wave would be produced using a scanning *slit*, as shown in Fig. 18.8(b). However, the available output from a slit is very small, and a compromise must be reached between output and desired waveform. A *square* scanning spot would produce a triangular waveform, as shown in Fig. 18.8(c), which would give maximum output. The triangular waveform is not the best, and a compromise is reached using a rectangular scanning spot, shown in Fig. 18.8(b), in which the height is about 0.8 times the width.

As before, let

$$D = \text{diameter of drum}$$
$$P = \text{scanning pitch}$$
$$n = \text{revolutions per second (rps)}$$

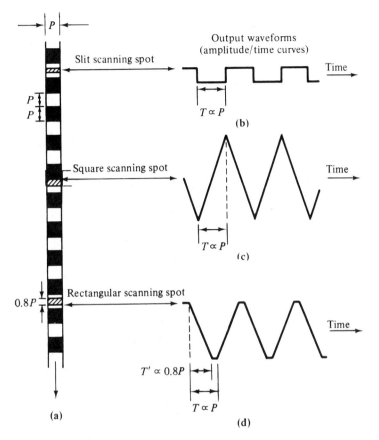

FIGURE 18.8. (a) A one-line scan with equal horizontal and vertical resolution; (b) signal produced by slit scanning; (c) signal produced by square scanning; (d) signal produced by rectangular scanning, midway between (b) and (c).

The number of pixels along one circumference of the drum is

$$\text{pixels per scan} = \frac{\pi D}{P}$$

and therefore this is also the number of pixels scanned in one revolution. The number of pixels scanned in 1 s is

$$\text{pixel rate} = \frac{\pi D n}{P} \tag{18.3}$$

Every two pixels form one cycle of output, and therefore the output frequency is

$$f = \frac{\pi D n}{2P} \quad \text{Hz} \tag{18.4}$$

This only gives the fundamental frequency of the trapezoidal waveform shown in Fig. 18.8(d). The actual wave will also contain higher harmonics, but in practice it is usually found that only the fundamental need be transmitted, bearing in mind that the extreme picture conditions postulated will seldom be encountered. Typical values in a practical system are $D = 66$ mm, $P = 0.1875$ mm (or 3/16 mm), $n = 1.0$ rps, and $f = 553$ Hz.

The bandwidth is determined by the range of frequencies to be transmitted in an actual signal. The equation for f gives the highest frequency, and as the lowest frequency is very close to zero, the bandwidth is approximately f Hz.

Example 18.2 If the drum in the facsimile machine of Example 18.1 rotates at 120 rpm, find the theoretical bandwidth required for transmission of the signal.

Solution From Eq. (18.4),

$$f = \left(\frac{\pi}{2}\right)\left(\frac{D}{P}\right)n \quad (\text{where } n = \text{speed, in rps})$$

$$= \frac{\pi}{2} \times 352 \times 2$$

$$= 1106 \text{ Hz}$$

This is also the bandwidth required.

Line transmission. The basic signal, as obtained from the information scanned, is not suitable for direct transmission, because it is difficult to amplify the low frequencies involved. Modulation (AM or FM) is therefore employed. This also allows frequency-division multiplexing to be used, and two facsimile channels can be fitted in the normal 300- to 3400-Hz telephone channel. The carrier frequencies agreed upon internationally for AM are 1300 Hz and 1900 Hz. All forms of distortion and interference must be kept at a very low level, and a signal-to-noise ratio of at least 35 dB is recommended operating practice. Echo signals must also be avoided, and because of echo signals, long lines are not suitable for phototelegraphy (as distinct from documentary). The gain stability of amplifiers must also be high. Level changes and impulsive-type noise affect AM more than FM, and the latter method, known as subcarrier FM (SCFM) is preferred. However, it must be limited to narrowband FM.

Radio transmission. The main difficulty with radio transmission is the fading which can occur over the radio path, which can completely destroy the picture information. Although special methods are in use to compensate for fading in normal telegraph and telephone radio services (e.g., diversity systems and automatic gain control systems), these are not satisfactory for picture telegraphy. As a result, SCFM is used over some part of the transmission path,

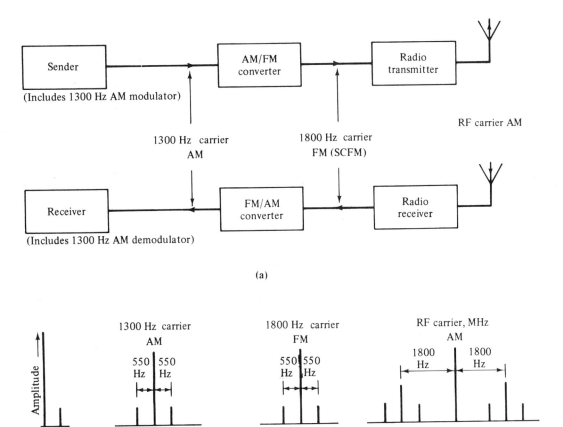

FIGURE 18.9. (a) Subcarrier frequency modulation system for transmitting facsimile signals; (b) frequency spectra of signals at various points in the system.

as it is less affected by amplitude fading. A simplified block diagram is shown in Fig. 18.9(a). The frequency spectra of the various sections in the transmission chain are shown in Fig. 18.9(b). Any fading over the radio path, provided that it is not frequency-selective, will not affect the FM.

Example 18.3   If the signal in Example 18.2 is used to frequency-modulate a subcarrier, and only the first pair of sidebands need be taken into account, calculate the bandwidth of the SCFM system.

$$\text{Bandwidth} = 1106\text{Hz} \times 2 = 2212 \text{ Hz}$$

Solution

In Fig. 18.9, the sender/receiver equipment is shown as producing AM, as would occur, for example, in older equipment utilizing a chopper disk (Fig. 18.1). In this case, AM/FM and FM/AM converters are required as shown. The radio carrier is shown as AM, although FM may be used also.

As quoted previously, a typical traversing speed in facsimile is 1/100 inch per second. Therefore, to transmit a picture 6 inches long requires a scanning time of 10 minutes. In the television systems discussed in Section 18.3, electronic scanning is employed which enables all sections of the transmitted picture to appear almost simultaneously at the receiver. This, however, requires a very wide bandwidth, and although television has been used for the transmission of documentary information, its real development has been for broadcasting (entertainment, news, etc.) to the general public.

Incorporation of modern electronic techniques in facsimile equipment was initially rather slow, with the result that equipment of older design is still to be found in service. However, recent advances include the use of specially designed cathode-ray tubes for electron beam scanning, and use of lasers as light sources. A good account of the history and developments will be found in *Electronic Delivery of Documents and Graphics* by Daniel M. Costigan, published by Van Nostrand Reinhold Co. (1978).

18.3 TELEVISION

Watching television is North America's favorite pastime, so the manufacture of television receiver sets forms a large portion of today's electronics industry. The systems used for television have been thoroughly standardized, so sets vary little from one manufacturer to another. The system presently used in North America is common to all North American countries, Japan, Korea, and a few others. The system common to most European countries differs from the American system mainly in the use of different scanning rates.

The standard frequency assignments for the television channels used in North America are shown in Table 18.1. Channels 2 through 13 are in the VHF band and are the most generally used. Channels 14 through 83 occupy the UHF band and are being used more often as the number of transmitters increases. Earlier receiver sets were built with only a VHF tuner, but all new sets are now required to have a UHF tuner built in. Each channel provides a 6-MHz bandwidth, which is sufficient to transmit the 4.2 MHz of video bandwidth with an additional 1.0 MHz for the vestigial sideband.

Most television broadcasting is now done in color, and although black and white sets may still be obtained, these are usually used for special applications such as security monitoring and microcomputer systems. The color system and all color receivers are made to be compatible with black-and-white transmissions.

TABLE 18.1. Television Channel Frequency Assignments*

Channel Number	Band (MHz)	Channel Number	Band (MHz)	Channel Number	Band (MHz)
2	54–60	29	560–566	57	728–734
3	60–66	30	566–572	58	734–740
4	66–72	31	572–578	59	740–746
5	76–82	32	578–584	60	746–752
6	82–88	33	584–590	61	752–758
7	174–180	34	590–596	62	758–764
8	180–186	35	596–602	63	764–770
9	186–192	36	602–608	64	770–776
10	192–198	37	608–614	65	776–782
11	198–204	38	614–620	66	782–788
12	204–210	39	620–626	67	788–794
13	210–216	40	626–632	68	794–800
14	470–476	41	632–638	69	800–806
15	476–482	42	638–644	70	806–812
16	482–488	43	644–650	71	812–818
17	488–494	44	650–656	72	818–824
18	494–500	45	656–662	73	824–830
19	500–506	46	662–668	74	830–836
20	506–512	47	668–674	75	836–842
21	512–518	48	674–680	76	842–848
22	518–524	49	680–686	77	848–854
23	524–530	50	686–692	78	854–860
24	530–536	51	692–698	79	860–866
25	536–542	52	698–704	80	866–872
26	542–548	53	704–710	81	872–878
27	548–554	54	710–716	82	878–884
28	554–560	55	716–722	83	884–890
		56	722–728		

*Courtesy of Howard W. Sams and Co., Inc.

18.3.1 Television Camera

The TV camera tube is the device which converts the optical image of the program material into an electrical signal, which can be transmitted over cable or by radio to a receiver, where the original image can be reconstructed from the received signals. Several types of tubes using different principles of operation are available, and two of these, the Image Orthicon tube and the Vidicon tube, will be discussed here. Both of these tubes were developed at RCA Ltd., in the late 1940s.

The *Image Orthicon tube* is illustrated in Fig. 18.10. Its action may be explained as follows. The tube consists of an electron-beam gun in one end which emits an electron beam for scanning the target at the other end. Focusing coils mounted around the neck of the tube sharpen the beam so that

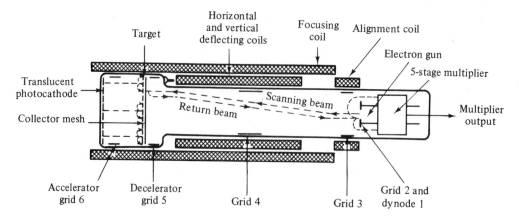

FIGURE 18.10. The structure and operation of the Image Orthicon camera tube. (Courtesy RCA).

it has a very narrow cross section when it reaches the target. Deflection coils also mounted around the neck of the tube cause the beam to be offset from its central axis by a controlled amount so that its point of impact on the surface of the target may be positioned anywhere on the target surface.

In the image section of the tube, the end of the tube is an optical glass plate, the inside surface of which is coated with a photoemissive material. The image to be converted is focused through an optical lens system onto the glass plate and the photoemissive layer. Where light strikes the layer, it emits electrons from its surface in proportion to the light intensity. These electrons are accelerated toward the target by a positively charged screen placed behind the photoemissive layer, so that an electron image of the photo image is transferred toward the target.

The target is a thin layer of semiconducting material which emits several secondary electrons for every accelerated electron that strikes it. The semiconductor material is charged up as it loses the secondary electrons, and since it has a high resistivity, a charge pattern is created on its surface, with the most positive areas corresponding to the lightest regions of the photoimage. The dislodged secondary electrons are collected by a screen or collector mesh through which the original image electrons passed to reach to target and are returned to the photoemissive cathode through an external circuit.

On the scanning side of the target, the electron beam is caused to move from left to right in lines across the target, repeating the lines down the face of the target until the whole target has been scanned. To reach the target, the beam must pass through a repelling or decelerating grid with a negative charge on it, so that the average beam velocity at the target is near zero. Some of the beam electrons will be captured by the positive image charges on the target, while the rest will be turned and accelerated back toward the gun end of the tube by the decelerator grid potential. The larger the positive charge on the target at the point of impact of the beam, the smaller the magnitude of the return beam. Thus, a lighter image causes a reduction in the returning beam current. The returning beam current is collected and passed through a multiple-stage electron multiplier. This is a series of anodes at progressively higher potentials, each one coated with a material that emits secondary electrons in large numbers when struck by high-energy electrons. Thus, the returning beam current is multiplied several hundred times before being passed through a load resistor to develop the video signal voltage.

The Orthicon tube is a very sensitive one which can be used under normal to low lighting conditions, and which responds to light over most of the visible spectrum. This makes it ideal for live television use, and it is widely used for both studio and remote camera applications.

The *Vidicon tube* is much simpler and cheaper to build than the Orthicon, but it is far less sensitive than is the Orthicon because it does not have the built-in electron multiplier stages. This tube finds its greatest application in studios that convert the images from moving picture and still slide projectors, where high light intensity levels are easily maintained. The structure of a Vidicon tube is illustrated in Fig. 18.11.

The Vidicon tube works on the photoconductive principle. The tube is provided with an electron-beam gun in one end, and with focusing and deflection coils on its neck in much the same manner as the Orthicon. The electrons in the beam are not accelerated to very high velocities, and a decelerating screen slows them almost to a stop before impact, so that they do not cause any secondary emission. The target is a layer of photoconductive material on the beam side of an optically transparent conducting film on the inside of the glass window. This conducting film acts as the beam-signal pickup connection. When the beam scans across the target, it charges up the surface of the target with a uniformly large negative charge. Since the photo material normally has a very low conductivity, this charge will remain on the surface for some time after the beam has passed.

The light image is focused through a lens, the window, and the transparent metal film onto the photoconductive layer, and the conductivity of the

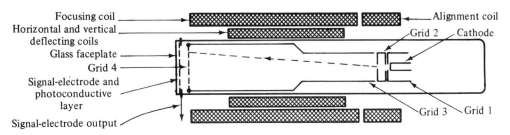

FIGURE 18.11. The structure and operation of the Vidicon camera tube. (Courtesy RCA).

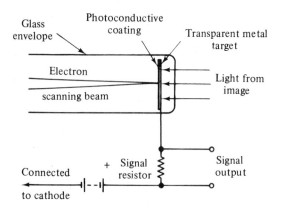

FIGURE 18.12. The structure of the photoconductive camera tube.

layer increases in proportion to the light intensity. This increase in conductivity allows the negative charge to leak off in the more conductive regions, but it is held in the less conductive regions, so that a charge image is created on the target. When the electron beam scans the charge image, the area under the beam is recharged to its negative level, and a current flows in the signal lead which is proportional in magnitude to the amount of decrease of negative charge on the target, or to the intensity of the light that struck the scanned point. The beam current flows through the charged target and its conductive backplate, and through an external load resistor to generate the video signal voltage, as shown in Fig. 18.12.

The video signal voltage is transmitted along with the synchronizing signals to control the receiver scanning functions. At the receiver, the video signal controls the electron beam in a cathode-ray tube in such a way as to reproduce the image on a fluorescent screen. (This is discussed below.) The scanning at the receiver must be tied, or synchronized, with that at the transmitter. The scanning process is very fast, and this, along with the *persistence of vision* of the viewer, makes it appear that all parts of the picture are being reproduced simultaneously.

18.3.2 Cathode-Ray Tube

The *cathode-ray tube* in the receiver performs the opposite function to the camera tube in the transmitter, converting the electrical signal into an optical image. Figure 18.13 shows in outline the action of the cathode-ray tube, where, for the moment, the deflecting and focusing arrangements are omitted. The electrical signal, which represents the picture, is applied to the *control electrode*, which functions in a manner similar to that of the control grid in a tube. The signal causes the potential of the control electrode to vary about a mean level, and as this goes more negative, it reduces the intensity of the electron beam, with a resulting reduction in the brightness of the spot formed on the

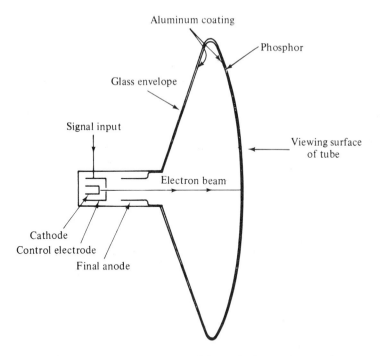

FIGURE 18.13. The structure of the cathode-ray picture tube.

fluorescent screen. Likewise, as it goes less negative, it increases the brightness of the spot on the screen. The fluorescent screen is made out of a *phosphor*, which is a material that emits light when bombarded with an electron beam. In modern tubes, the back of the phospher is coated with a very thin film of aluminum, and this coating is continued down the inside walls of the tube and connected to the final anode. The purpose of the coating is to increase the brightness of the screen (for a given intensity of electron beam) and to ensure that the brightness is even (i.e., no dark patches occur). The coating, being at final anode potential, prevents electrons from the beam from accumulating on the screen, which otherwise would tend to repel the incident beam of electrons, thus reducing brightness. Furthermore, if electrons were allowed to accumulate, they would do so unevenly over the surface of the screen, thus producing dark patches. The aluminum coating improves brightness in another way, by *reflecting* light forward onto the viewing side, light that would otherwise be lost inside the tube. Some of the electron-beam energy is lost in penetrating the aluminum coating, but since this is kept very thin, the energy lost is more than offset by the gain in brightness.

In both the transmitter camera tube and the receiver picture tube, the electron beam must be made to *scan* the screen. The beam usually starts at the top left-hand corner of the screen and moves across to the right in an almost horizontal line. There has to be a small downward deflection of the beam as each line is traversed, so that the complete picture is eventually scanned. When the beam reaches the right-hand side, it is very quickly deflected back to the left-hand side again (this is called line *flyback*), where it starts a new line. The flyback is initiated by a *line synchronizing signal* called the *horizontal sync pulse*. This is a special pulse signal introduced at the transmitter which ensures that the receiver flyback occurs at the same time as the transmitter flyback.

When the scanning beam reaches the bottom right-hand corner of the picture, it must be returned to the top left-hand corner again, to repeat the whole scanning procedure. Thus, a second synchronizing signal must be transmitted at this time to ensure that the receiver recommences complete scanning at the same time as the transmitter. This is the *field synchronizing signal*, called the *vertical sync pulse*.

The deflection of the beam for scanning may be achieved either electrically or magnetically, but magnetic deflection permits a shorter tube to be used, which is an important advantage in television. Hence, magnetic deflection is the most commonly used. It is known that a magnetic field exerts a force on electrons in a beam in just the same way as it does on electrons in a conductor (e.g., in the armature winding of a motor). Figure 18.14(a) illustrates this. In the television tube, the magnetic field must build up steadily so that the spot formed by the beam is deflected at a constant velocity and traces out a line. This is achieved by utilizing an electromagnetic field rather than a permanent magnetic field. The current to the coils can be then be controlled in the desired manner, with the synchronizing pulse causing the beam to return to the start of the next line. Figure 18.14(b) and (c) shows two methods of achieving magnetic deflection. In practice, the coils of Fig. 18.14(b) are saddle-shaped so that they fit closely around the neck of the tube. For practical reasons, it is found best to use the saddle-shaped coils for line deflection and the toroidal-shaped coil of Fig. 18.14(c) for field deflection.

As well as being deflected for scanning, the electron beam must be *focused* in order to form a sharp spot on the screen. Again, focusing may be achieved either magnetically or electrically. Magnetic focusing is usually employed in camera tubes, and either type will be found in receiver picture tubes. Figure 18.15(a) shows how a long coil may be used for focusing, and this is the type used in camera tubes. The coil produces a magnetic field which is directed mainly along the axis of the tube (and hence is known as an *axial field*). Electrons that tend to diverge from the beam are deflected by the magnetic field, and the electron paths become helixes. It is found that the electron paths meet at fixed points along the tube axis as shown in Fig. 18.15(a), and it is only a matter of adjusting the focusing current so that one of these points falls on the target surface. The beam is then focused.

A short magnetic field may also be used and may be obtained either by means of a short coil or a permanent magnet. Either system can be used in television receivers. Figure 18.15(b) illustrates the principle of short magnetic

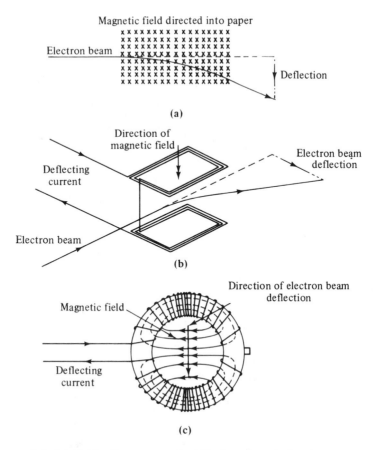

FIGURE 18.14. Electromagnetic deflection of an electron beam.

focusing, where, for clarity, only one divergent electron path is shown. As the electrons enter the magnetic field along this path, they first meet the component of magnetic field directed inward (or the *radial component*). This causes the electrons to be deflected to the right. As they move farther into the field, they encounter the axial component, which results in a downward shift to the particular path shown. As the electrons emerge from the other side of the field, the radial component there is directed outward, which results in the electrons being deflected to the left. The result is that the electron path meets the axis at some point P, and it can be shown that all divergent paths meet at this point. Thus, the strength of the magnetic field is chosen such that point P falls on the screen.

Electric focusing is now becoming widely used in television receiver tubes. Figure 18.16 illustrates the principle. An electric field is set up between cylindrical electrodes, one of which is the final anode in the tube. By definition,

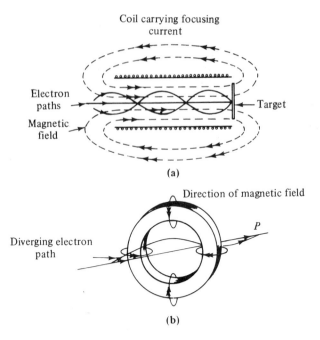

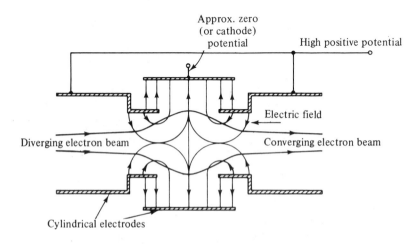

FIGURE 18.15. Electromagnetic beam focusing.

an electric field is a region in which electric charges experience a force, and by convention, the direction of the force is that along which a positive charge would be moved. Thus, the lines of force are directed from the high positive potential electrode to the zero potential electrode. Since the electrons in the beam are negative charges, they will tend to be deflected in the direction opposite to the field. If the beam is divergent on entering the field, it will be

FIGURE 18.16. Electrostatic beam focusing.

convergent on leaving it, thus forming a focal point. The electrodes are shaped as shown in Fig. 18.16 in order to prevent voltage breakdown, as a high potential exists between them.

18.3.3 Interlacing and Picture Scan Repetition Rate

One complete scan of the target area will allow reproduction of one complete picture at the receiver of a television system. A large amount of information must be transmitted during this period, and if the picture is repeated at a high rate, then the bandwidth required for transmission will become excessive, as is shown in Section 18.3.6. In television, if the picture scan rate is made too low, moving scenes will develop a stop-and-go jerky movement in the same manner that slow-motion moving pictures do. Further, the phosphors used in the receiver picture tubes have a relatively low persistence, allowing the picture to fade out between scans, and the scanning will produce a "flicker" at the picture rate. The picture rate must be sufficiently high so that the normal persistence of the viewer's eyes will override the flicker and merge the picture series into smooth motion. This minimum picture rate has been found to be about 35 to 50 pictures per second.

It has also been found that in television systems, if the scanning rate is near, but not exactly equal to, the supply frequency, voltage pickup from the ac power circuits modulate the scanning circuit amplifiers and cause annoying distortion and jitter in the picture. This interference can be minimized by making the picture scanning rate a multiple of the supply frequency. In the American system, the picture or frame scan rate is 30 Hz, which is a submultiple of the supply frequency of 60 Hz.

The 30-Hz picture repetition rate is too low, and if scanning were done in a straightforward sequential manner, the flicker produced because of picture fading between scans would become objectionable. For this reason, the scanning of each picture frame is divided into two separate *fields*, and the entire area is scanned twice during a picture period. A total of 525 lines is included in each picture, so that during the first field $262\frac{1}{2}$ lines are scanned, and during the second field the remaining $262\frac{1}{2}$ lines are scanned.

At the end of the first $262\frac{1}{2}$ lines, a vertical retrace occurs, and the second scan begins at the top center of the picture, as shown in Fig. 18.17. As a result, the second field scan lines fall midway between the positions of the first field lines, so that the two fields merge to form a complete scan frame. The effective flicker rate is now 60 Hz, well above the threshold, and the frame, or picture repetition rate of 30 per second is low enough to minimize video signal frequencies. European supplies are generally operated at 50 Hz, and the picture rate for this system is 25 per second, with a field rate of 50 per second.

Two vertical retraces take place during each complete picture frame period, and for each of these a synchronizing pulse (the vertical sync pulse) is produced in the master camera control unit, which is used to initiate the vertical retracing of both the camera and the receiver scanning beams, so that

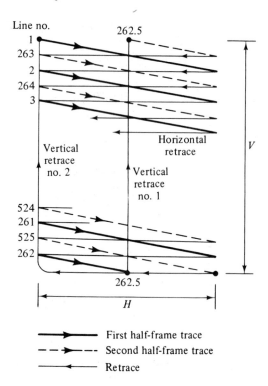

FIGURE 18.17. Television frame scan interlacing pattern.

the two remain in synchronization with each other. The vertical sync frequency, equal to the field frequency, is

$$f_v = 2P \qquad \text{pulses/s} \qquad (18.5)$$

where P is the frame, or picture, repetition rate in pictures/s. The vertical sync pulse is superimposed on top of a longer "blanking" pulse, which is used to turn off the electron beam of the receiver CRT during the retrace period so that the retrace lines cannot be seen. The structure of the combined vertical sync and blanking pulses is shown in Fig. 18.19(d), from which it can be seen that about 20 lines of video information are lost during each vertical retrace period, or about 40 lines out of each complete frame.

18.3.4 Picture Definition

Picture definition is determined by the size of the smallest element in a picture, called a *pixel*. Let N represent the total number of lines in a frame, and let N_s represent the number suppressed during frame synchronization. The number of *active lines* (i.e., lines actually presented to the viewer) is then

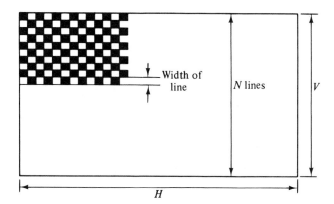

FIGURE 18.18. Checkerboard pattern of alternating black and white minimum elements which yields maximum video signal frequency.

$N - N_s$. If V represents the vertical dimension on the viewer's picture (Fig. 18.18), the active lines will just fill this, and the line width will be

$$W = \frac{V}{(N - N_s)} \tag{18.6}$$

For the same definition in the horizontal direction, H in Fig. 18.18, the number of elements will be

$$
\begin{aligned}
N_{\text{hor}} &= \frac{H}{W} \\
&= \frac{H}{V}(N - N_s) \\
&= a(N - N_s) \tag{18.7}
\end{aligned}
$$

where $a = H/V$ is known as the *aspect ratio*.

Example 18.4 In the American TV system, $a = 4/3$, $N = 525$, and $N_s = 40$. Calculate the number of active lines and the number of smallest horizontal elements.

Solution
$$\text{No. of Active Lines} = 525 - 40$$
$$= 485$$

and from Eq. (18.7),

$$N_{\text{hor}} = \frac{4}{3} \times 485$$
$$= 647$$

18.3.5 Horizontal Sync Frequency

Since two fields go to make up one frame, the total number of lines in a field is $N/2$. The field repetition frequency as given by Eq. (18.5) is $f_v = 2P$, and therefore the number of lines generated per second is $(N/2) \times (2P) = NP$. Since a synchronizing pulse must be sent at the end of each line, the horizontal (or line) synchronization frequency is

$$f_H = NP \quad \text{pulses/s} \tag{18.8}$$

The time available for one electrical line scan (including the *line* synchronization time) is

$$T_H = 1/f_H \tag{18.9}$$

Example 18.5 In the American TV system, $N = 525$ and $P = 30$ frames/s. Calculate the horizontal sync frequency and the total line time.

Solution From Eq. (18.8),

$$f_H = 525 \times 30$$
$$= 15,750 \text{ pulses/s}$$

From Eq. (18.9),

$$T_H = \frac{1}{15,750}$$
$$= 63.5 \mu s$$

The sync-pulse generating circuits in the camera use a very stable oscillator to generate a primary timing pulse train with a frequency of twice the horizontal sync frequency (31,500 pulses per second in the American system). The pulse train is then frequency-divided by 2 using digital circuits to produce the horizontal-sync-pulse train, and divided by 525 to produce the vertical-sync-pulse train.

18.3.6 Video Bandwidth

This is estimated on the basis that two successive pixels of the pattern shown in Fig. 18.18 go to generate one cycle of video output. The time available for the beam to scan the number of horizontal elements in a line N_{hor}, as given by Eq. (18.7), will be T_H less the line suppression time required for line synchronization. Let the line suppression period be represented by T_s; then the highest video frequency is

$$f = \frac{N_{\text{hor}}}{2} \frac{1}{T_H - T_s}$$

$$= \frac{a(N - N_s)}{2(T_H - T_s)} \tag{18.10}$$

Experiments have shown that a reduction in resolution by a factor of 0.7 (known as the Kell factor) can be tolerated, and so the highest video frequency is standardized at

$$f = \frac{0.7a(N - N_s)}{2(T_H - T_s)}$$

$$= 0.35a \frac{(N - N_s)}{(T_H - T_s)} \tag{18.11}$$

The lowest frequency in the video signal will be near zero; therefore, Eq. (18.11) also gives the video bandwidth required.

Example 18.6 Given that $T_s = 10\mu s$ in the American TV system, and the other values are as given in two previous examples, calculate the video bandwidth required.

Solution From Eq. (18.11),

$$f = 0.35 \times \frac{4}{3} \times \frac{(525 - 40)}{(63.5 - 10) \times 10^{-6}}$$

$$= 4.23 \text{ MHz}$$

18.4 THE TELEVISION SIGNAL

The television signal, as transmitted, is a complex one. Four separate components are included in the signal: the sound, the picture brightness or luminance scan information, synchronization information for both vertical and horizontal scans, and color chrominance information. The video information is transmitted as an amplitude-modulated vestigial (partially suppressed) sideband carrier located 1.25 MHz from the lower edge of the channel band. All of the 4.2-MHz-wide upper sideband and 1 MHz of the lower sideband are transmitted along with the carrier, the remainder of the lower sideband being removed. Figure 18.19(a) shows the transmitted video spectrum with the sound

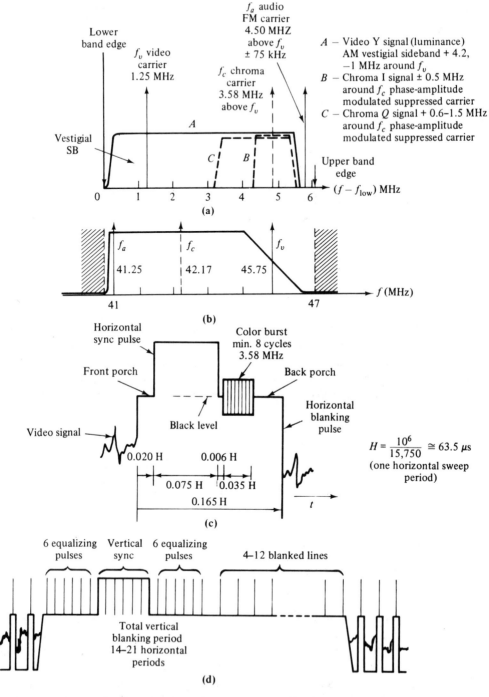

FIGURE 18.19. Television signal characteristics: (a) signal spectrum within the 6-MHz channel assignment; (b) video IF bandpass characteristic; (c) horizontal sync pulse structure; (d) vertical sync pulse structure.

and chrominance carriers indicated. The sideband envelope for the composite video information is shown as well.

In the North American TV system, the sound carrier is located 4.50 MHz above the video carrier and is frequency-modulated, requiring a bandwidth of about 200 kHz centered around the carrier. This is well outside the video band, and can easily be removed with filtering. The chrominance information is carried by a special two-phase phase-modulated suppressed carrier signal which is amplitude-modulated on the video carrier, effectively providing two additional channels on which to transmit the color information without increasing the video bandwidth required. Only the sidebands are transmitted. The main video signal becomes the Y or luminance signal, and the two chrominance channels are the I and Q channels. The Q signal requires a sideband width of ± 0.5 MHz, the I signal a lower sideband of 1.5 MHz and a vestigial upper sideband of 0.6 MHz around the 3.58-MHz subcarrier.

Synchronizing information is carried on the video carrier during retrace periods between lines and between frames. The picture is scanned horizontally 15,750 times per second and vertically 60 times per second in two interlaced frames. Thus, a horizontal retrace occurs every 63.5 μs, and a vertical retrace every $16\frac{2}{3}$ ms. During the horizontal retrace, for which 0.165 of a horizontal period is allowed, the video signal is raised to the black level, to blank out the retrace. A synchronizing pulse 0.075 T_H wide (where T_H is the horizontal scanning period) is provided to trigger the horizontal oscillator of the receiver, and a burst of 8 or more cycles of 3.58-MHz chroma carrier is provided to synchronize the color demodulator circuits. The shape of the video signal as a function of time during a horizontal sync pulse is shown in Fig. 18.19(c).

Vertical retrace takes longer, and a period of from 14 to 21 line periods is blanked out to provide this time. The synchronizing is accomplished by changing the shape of the horizontal sync pulses. As shown in Fig. 18.19(d), 6 regular-sized horizontal sync pulses are provided at twice the rate for equalization, followed by the vertical sync pulse, which is three horizontal periods long and serrated every half-period by a negative-going sync pulse for horizontal stabilization, and then followed by another 6 equalizing pulses. The video signal remains blanked for another 4 to 12 periods to allow sufficient time for vertical retrace to be completed.

18.5 TELEVISION RECEIVERS

Figure 18.20 shows the block diagram of a typical black-and-white television receiver. Starting from the antenna, which is usually of the "rabbit-ear" type, but can be multielement Yagi or a cable-system (CATV) transmission line, the first section is the tuner assembly. The input signal is coupled into the RF-amplifier stage, which is a tuned class A stage, and then to a mixer-oscillator circuit. For the UHF channels, a diode mixer is used with a

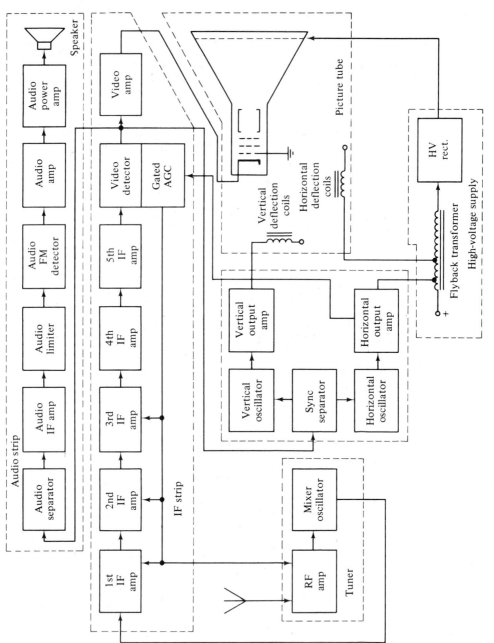

FIGURE 18.20. Black-and-white television receiver block diagram.

separate oscillator circuit operating in the frequency-doubling mode. The second harmonic is used to provide the local oscillator signal to the mixer. Tuning is accomplished by means of a turret switch, which places a different set of coils and capacitors in the circuit for each channel. Means are provided for adjusting the oscillator frequency for each channel internally, as well as a vernier trimmer for fine tuning, which can be controlled from the front panel.

The IF used for all receivers and all channels has been standardized to lie in the band 41 to 47 MHz, and since the sidebands are flipped over in conversion (for oscillator frequency above signal frequency), the video carrier appears at 45.75 MHz, and the sound carrier at 41.25 MHz. The bandpass of the video amplifier strip is designed to pass all of these frequencies, with some sloping because of the vestigial sideband. The typical shape of this bandpass is shown in Fig. 18.19(b). The bandpass characteristics and gain are realized in the IF amplifier string, which can be three to five stages of stagger-tuned amplifiers. AGC is applied to two or more of the IF stages and to the RF-tuner stage. Video detection is accomplished by simple envelope detection at the end of the IF string.

The detected video signal is fed to the input of the audio string, where tuned circuits isolate the audio carrier, which is now at 4.5 MHz. The audio IF amplifier stages are designed to pass and limit a band of frequencies about 200 kHz wide around the 4.5-MHz center frequency. The audio detection circuit is usually a ratio detector, but may be any of the FM detector circuits mentioned in Section 10.5. The detector is followed by the audio amplifier stages, which feed the speaker.

The detected video signal is also passed to the video amplifier, which raises the level of the video signal, removes the sound carrier, and provides the dc bias level for the picture tube cathode. The video signal applied to the picture tube cathode varies the beam intensity, and thus the light intensity produced at the screen.

The detected video signal is also passed to the synchronization circuits, where the video portion is removed by clipping to leave only the sync pulses. The clipped sync pulses are amplified and passed on to trigger the horizontal sweep oscillator. The ramp voltage produced by the oscillator is amplified to drive the horizontal output transformer and the horizontal deflection coil. The large flyback pulse is rectified to provide the high voltage (10 to 20 kV) for the picture tube target and also provides the gating signal to control the AGC circuit.

The color-television receiver is complicated by the necessity to decode the color signals from the video signal before application to the picture tube. The circuitry included by the tuner, the IF strip, the deflection circuits, and the audio strip have exactly the same functions and specifications regardless of whether the set is black and white or color. The video amplifier has different characteristics, the picture tube used has three separate electron-beam guns, and the additional circuits to decode the color information and to control the

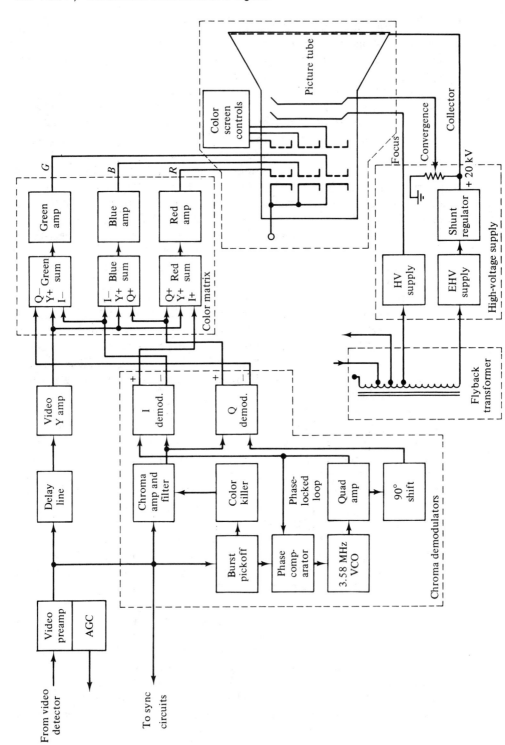

FIGURE 18.21. Color-television receiver block diagram. The diagram shows those parts of the receiver that differ from those of a black-and-white set.

convergence of the three electron beams are the differences. Figure 18.21 shows the segments of circuitry which differ from those in a black-and-white set.

The detected video signal is passed through the video preamp, which raises the level to compensate for the four-way split that follows. This amplifier has a 4.2 MHz bandwidth, with a "trap" (filter) to remove the sound carrier at 4.5 MHz. The video signal is passed to the synchronization separator and deflection circuits, which function in exactly the same manner as those of a black-and-white set. It is also passed through a delay line, which compensates for the delays in the chroma demodulator circuits to the Y video amplifier. This amplifier has the full 4.2-MHz bandwidth and provides the video signal during black-and-white reception periods.

The video signal also passes to the chroma amplifier, which has a bandpass of $+0.6$, -1.5 MHz around the 3.58-MHz carrier frequency, and thence to the I and Q signal demodulators. The video signal is also passed to a special circuit which responds to the short bursts of 3.58-MHz carrier present with each horizontal sync pulse. The output from this burst pickoff circuit provides a gate signal which turns off the chroma amplifier during black-and-white transmission, when no bursts are present (a color killer circuit which prevents random color blooming during black-and-white reception), and also provides the phase synchronization signal for the 3.58-MHz local carrier oscillator for the demodulator. This oscillator is a crystal-controlled voltage-modulated oscillator which is part of a phase-locked loop. Each burst of transmitted carrier causes the phase of the oscillator to shift enough to correct for frequency drift which occurs during the horizontal sweep time period.

The I and Q demodulators decode the two chroma signals from the chroma carrier and present them, along with the Y signal, to the inputs of the color matrix, The signals are added in weighted proportions according to the equations

$$R = +0.62Q + 0.95I + Y$$
$$G = -0.64Q - 0.28I + Y$$
$$B = +1.73Q - 1.11I + Y \tag{18.12}$$

to be compatible with the transmitting conditions given in Eq. (18.13).

The color matrix combines the I, Q, and Y signals to give the three primary color signals, R, G, and B (red, green, and blue), each of which controls a separate electron beam in the picture tube. Special convergence controls steer the electron beams so that they only illuminate the desired phosphors on the screen, which are laid out either in parallel strips or in an alternating dot matrix, and thus recreate the color image.

The picture tube power supplies are complicated by the need for carefully regulated voltages to keep the three beams converged on the correct spots on the screen. The high-voltage target supply of 20 kV must be regulated. It is derived in the same manner as for black-and-white sets, but has a shunt regulator circuit to maintain a constant voltage. This tube is carefully shielded to prevent the radiation of x-rays that may be produced because of the high voltages. A separate rectifier provides the voltage to the focusing anode, and the voltage for the convergence anode is provided by a voltage divider from the target supply.

The design of all television sets is undergoing a major evolutionary step. The ready availability of large-scale integrated circuits in quantity has made it practical to make nearly all of the television set in integrated-circuit form. Sets have been built for several years now using discrete semiconductors for all sections except the picture tube and high-voltage sections. Now whole sections of the receiver are available as integrated circuits. For instance, a single microcircuit is available with all the chroma demodulator and color matrix circuits in one package. Whole IF strips are available as a chip, and whole audio IF strips as well. All the low-level circuits can be built on one small board, which is minuscule in comparison to the picture tube. The next step is the picture tube itself. A semiconductor light-emitting diode display which will produce a coarse color picture has already been produced on an experimental basis by the Japanese. Within a few years such solid-state displays should be on the market, and television sets that are only a few centimeters thick will be available which can be hung on the wall like a picture.

18.6 TELEVISION TRANSMITTERS

18.6.1 Black-and-White Television

The details of the signal structure of a black-and-white television receiver were presented in Section 18.5. The channel is 6-MHz wide, with the video carrier located 1.25 MHz from the bottom edge and the sound carrier 0.25 MHz below the upper edge. The sound is FM, while the video is AM vestigial sideband. Most of the present televisions use the 12 VHF channels, but increasing use is being made of the many UHF channels that have been reserved.

The typical layout of a black-and-white TV transmitter is shown in Fig. 18.22. The sound carrier transmitter, which probably uses an Armstrong modulator system, is typical of those discussed in Section 10.10. The video carrier is generated from a crystal oscillator followed by frequency-multiplier stages up to the final carrier frequency and class C power driver and modulator amplifier stages. The power output to the antenna is typically 50 or 100 kW, so the driver power output may be as high as 1 kW. Amplitude modulation is accomplished from the composite video signal by a plate-modulated class C amplifier. Feedback may be incorporated to reduce modulation distortion. The output from the power amplifier is passed through a filter to suppress the lower

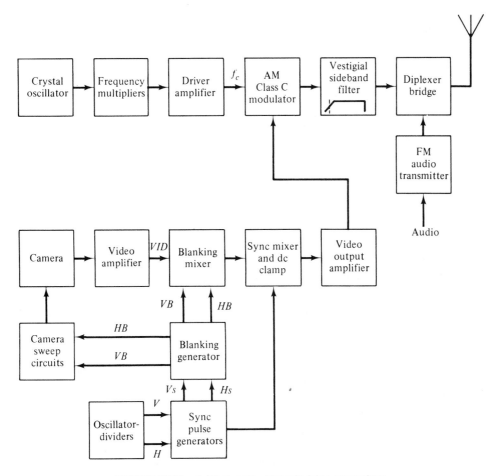

FIGURE 18.22.   A black-and-white television transmitter.

sideband and a portion of the carrier, and then to a diplexer bridge. This diplexer bridge has the function of preventing the video transmitter from loading the sound transmitter, and vice versa, so that all the power from both transmitters is passed through the antenna. It takes the form of a bridge circuit like that shown in Fig. 18.23. The antenna is made up of two phased sections driven from separate drive points, each presenting the same radiation resistance at its input. The two inductances L are equal in value. The video modulated carrier is fed through a transformer T to points Y of the bridge, while the audio modulated carrier is fed to points X and G of the bridge. The voltage between points X and G due to the voltage applied to points Y is zero as long as the two inductances are equal and the two antenna impedances are equal. Similarly, voltage between points Y due to that applied to points X-G is zero. Voltage from the video transmitter is applied across the two antenna

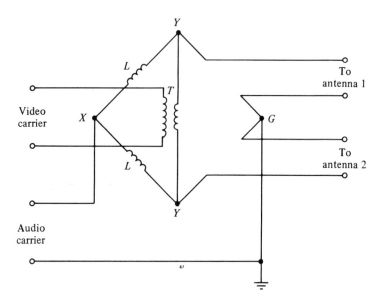

FIGURE 18.23. A diplexer bridge — used for mixing the outputs of two transmitters that use the same antenna system.

sections in series, and that from the audio transmitter is applied in parallel. The two antenna sections are oriented in space quadrature to obtain a circular radiation pattern, so that cancellation in radiation does not take place. The turnstile antenna may be used for this (see Section 15.16.3).

18.6.2 Color Television

The RF portion of a color TV transmitter is identical to that of a black-and-white TV transmitter. The only differences occur in the way in which the composite video signal is made up, since it must also carry information about color. Figure 18.24 shows the block diagram of such a transmitter. The video signal is generated by a special color camera comprised of three separate image tubes of the same type used in black-and-white systems. The color image is optically separated into three components by color filters and converted to red, green, and blue signals by the three image tubes. The three signals, R, G, and B, are combined in the color matrix circuits to produce the luminance signal Y and the two chrominance signals I and Q according to the equations

$$Y = +0.30R + 0.59G + 0.11B$$
$$I = +0.60R - 0.28G - 0.32B$$
$$Q = +0.21R - 0.52G + 0.31B \tag{18.13}$$

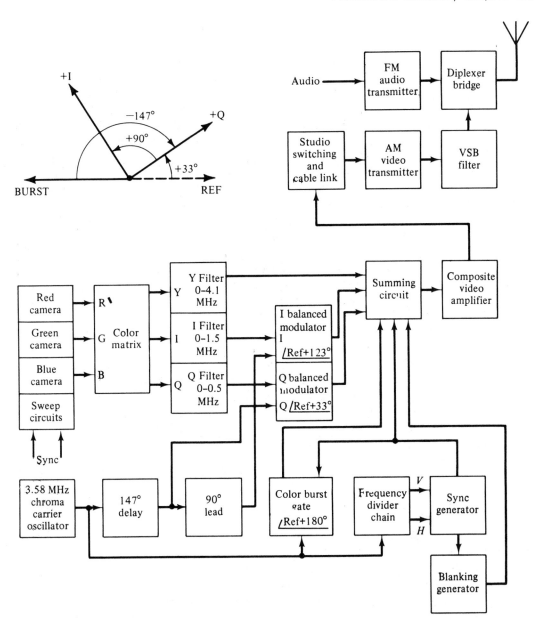

FIGURE 18.24. An NTSC color-television transmitting system.

The luminance, or Y, signal contains all the information required to reconstruct a black-and-white picture from the signal. It is passed through a lowpass filter to limit its output to a bandwidth of 4.2 MHz. The I signal is limited further, with a bandwidth of 1.5 MHz, and the Q signal even further, with 0.5-MHz bandwidth. This limiting is used to reduce interference between the three signals when receiving with a black-and-white set.

A 3.58-MHz crystal oscillator generates the subcarrier for carrying the chroma signals. System reference phase is specified as zero degrees, and the transmitted carrier burst for synchronization lags the reference phase by 180°. The Q-signal carrier leads reference by 33°, and the I-signal carrier leads the Q-signal carrier by 90°, giving a quadrature relationship between the two. These phase relations are shown in the phasor diagram in Fig. 18.24. In the block diagram, the color burst is taken directly from the output of the carrier oscillator. The oscillator output is also passed through a 147° phase delay to give the Q-carrier signal, and this is in turn given a 90° phase lead (or a 270° phase delay) to give the I-carrier signal. The two modulators are balanced modulators and create double-sideband suppressed carrier signals around the 3.58-MHz frequency. A color burst gate allows a short burst of the 3.58-MHz carrier to be transmitted during each horizontal blanking period to allow synchronization of a carrier reinsertion oscillator at each receiver, so that the chroma signals can be demodulated.

The same 3.58-MHz oscillator feeds a divider chain which generates the (approximately) 15,750-Hz and 60-Hz signals from which the horizontal and vertical sync and blanking pulses are generated. These signals are used to control the three color camera systems and are also transmitted to synchronize the receivers.

An adder circuit combines the Y signal, the modulated I and Q signals, the color burst, and the sync and blanking pulses to form the composite video signal. This is then amplified and applied to the video modulator input by a high-power broadband video amplifier. The structure of the composite video signal is shown in Fig. 18.19(b).

18.7 PROBLEMS

1. Discuss and compare television and facsimile telegraphy as methods of transmitting documentary information such as photographs.

2. Discuss one form of facsimile transmitter, and explain in detail how the scanning spot is obtained. Explain also how the basic signal may be modulated during the scanning process.

3. Explain what is meant by synchronization and why this is necessary in picture transmission. Show, with the aid of a sketch, the form of distortion that would occur in a facsimile link if the receiver speed is greater than the transmitter speed.

4. Explain what is meant by index of cooperation and why this is important in facsimile telegraphy. Show also how it affects the bandwidth of the signal. If the index of cooperation of a facsimile transmitter is 352, what must be the index of cooperation of the corresponding receiver?

5. The index of cooperation of a facsimile machine is 352, and its speed of rotation is 60 rpm. The scanning pitch is 5/16 mm. Determine the theoretical bandwidth required.

6. Explain what is meant by subcarrier frequency modulation. A facsimile signal is transmitted on an SCFM system whose carrier frequency is 1800 Hz. The SCFM signal then amplitude-modulates an RF carrier. Assuming that only the first-order sidebands need to be considered in each modulation process, estimate the bandwidth required at RF if the highest frequency in the basic signal is 550 Hz.

7. FCC specifications for a facsimile drum scanner are: line length, 18.85 in.; scan density, 96 lines per inch (LPI). Determine the index of cooperation as defined both by the IEEE and the CCITT. For a scan speed of 90 lines per minute (LPM), determine the bandwidth required.

8. Explain the purpose of the aluminum coating used on the inside of cathode-ray tubes for television. Why is this coating held at final anode potential?

9. Explain how scanning may be achieved in a television system. What is the purpose of the synchronization signals? In the U.S. 525-line, twin-interlace system, how many field synchronizing signals occur during one frame?

10. Describe and compare magnetic and electric methods of focusing an electron beam in a cathode-ray tube.

11. In the British television system, the number of lines is 625, the number of pictures transmitted per second is 25, and the aspect ratio is 4/3. The number of lines suppressed during the field blanking period is 48, and twin interlace is used. The line blanking period is 16% of line duration. Calculate the video bandwidth.

12. Outline the differences between a black-and-white television signal and a color signal. Use a sketch of the waveforms to illustrate, and explain the reason for each component. Also, sketch the signal frequency spectra.

CHAPTER 19

Satellite Communications

19.1 INTRODUCTION

A communications satellite is a spacecraft placed in orbit around the earth which carries on board microwave receiving and transmitting equipment capable of relaying signals from one point on earth to other points. Microwave frequencies must be used to penetrate the ionosphere, since all practical satellite orbits are at heights well above the ionosphere. Also, microwave frequencies are required to handle the wideband signals encountered in present-day communications networks and to make practical the use of high-gain antennas required aboard the spacecraft.

The first commercially operated satellite was launched in August, 1965. Since that time numerous satellites have been launched for communications purposes. Such communications services include point-to-point telecommunications circuits, wide area TV coverage, often referred to as direct broadcasting by satellite (or DBS), and navigational and communications services to ships and aircraft.

Satellite systems may be domestic, regional, or global in character. The service range of a domestic satellite system is confined to the country owning the system, for example, the Canadian Telesat system. Regional systems

TABLE 19.1. Satellite Frequencies

Frequency, GHz	Direction
1.530–1.559	Down
1.6265–1.6605	Up
3.400–4.200	Down
5.850–7.075	Up
7.250–7.750	Down
7.900–8.400	Up
10.70–12.70	Down
12.70–13.25	Up
14.00–14.80	Up
17.30–17.70	Up
17.70–18.10	Both
18.10–20.20	Down
27.00–30.00	Down

involve two or more countries, such as the French–West German Symphonie system. Global systems are best typified by the Intelsat system, which is intercontinental in character. Detailed descriptions of these systems will be found in *Communications Satellite Systems*, published by the IEEE Press (1975), and some of the more general aspects will be described in this chapter.

Coordination of satellite services is carried out by the International Telecommunication Union based in Geneva. Conferences, known as World Administrative Radio Conferences (WARC) and Regional Administrative Radio Conferences (RARC), are held on a regular basis, and recommendations are issued from time to time regarding radiation powers, frequencies, and orbital positions for satellites. Table 19.1 shows present and potential satellite frequencies.

Satellites presently in use are active satellites, that is, the signal received by the satellite is retransmitted rather than being simply reflected back to earth. This means that the satellite has on-board, highly directional transmitting and receiving antennas, and complex interconnecting circuits. Accurate positioning and control mechanisms are required for the satellite. The power requirements for the on-board equipment are usually obtained from arrays of solar cells, with back-up nickel–cadmium batteries for periods of solar eclipse.

19.2 ORBITS

A satellite orbiting the earth stays in position because the centripetal force on the satellite balances the gravitational attractive force of the earth. In addition, atmospheric drag must be negligible, and this requires the satellite to be at a height greater than about 600 km. The choice of orbit is of fundamental importance, as it determines the transmission path loss and delay time, the earth coverage area, and the time period the satellite is visible from any given area. For satellite communication purposes, orbits are conveniently classified as inclined elliptical, polar circular, and geostationary. Each of these is illustrated in Fig. 19.1.

The inclined elliptical orbit is not widely used. Its main advantage is that it provides coverage of the polar regions. For example, it is used for the Russian Molniya/Orbita satellite broadcast system where coverage of the more remote regions of the country is required. The apogee, or highest point of the orbit, is arranged to occur over the region requiring most coverage. This puts the satellite at its greatest height and therefore gives the greatest earth coverage in this region. Also, the transit time is longest at the apogee, making the satellite "visible" for a relatively long period of time over these regions. The inclined elliptical orbit does not permit continuous contact with the satellite from a fixed spot on earth.

The circular polar orbit is not used for communications satellites generally, but it is used for special purposes such as for navigational satellites.

The periodic time is the time taken for one complete orbit, and a synchronous orbit is one for which the periodic time is an integer multiple or submultiple of the earth's rotational period. The geostationary orbit is the synchronous orbit which is most widely used. The rotational period of the earth about its own axis is 23 hours and 56 minutes, and a satellite in geostationary orbit, travelling in the same direction as the earth's rotation, completes one revolution about the earth's axis in this time. The satellite therefore appears stationary to an observer on earth—hence the name *geostationary*. Keeping track of a geostationary satellite is relatively easy, and the satellite is continually visible from within its service area on earth. Another advantage of the geostationary orbit is that the Doppler shift of frequency is

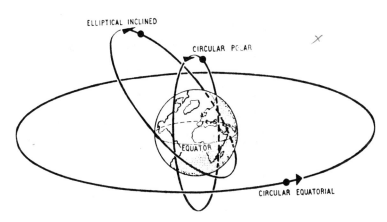

FIGURE 19.1. Typical communications satellite orbits (from *Telecommunications Satellites*, ed. Kenneth W. Gatland, courtesy Prentice Hall, Inc. 1964).

negligible. Doppler shift in frequency results when there is relative movement between source and receiver. If v is the relative velocity between source and receiver, and f is the transmitted frequency, the received frequency is given approximately by

$$f' = f(1 + v/c) \tag{19.1}$$

where c is the speed of light.

The velocity v of a geostationary satellite relative to earth should ideally be zero, although in practice small fluctuations do occur. The height required for geostationary orbit may be deduced from the dynamics of motion. For a circular orbit at height h above ground, the circumferential path is $2\pi(a + h)$, where $a = 6371$ km is the average radius of the earth. Motion in a circle implies that the circumferential speed V is constant, and therefore the time for one orbit is

$$T = 2\pi(a + h)/V \tag{19.2}$$

From the mechanics of the situation, the centripetal force on a satellite of mass M is $MV^2/(a + h)$, and the gravitational force is Mg', where g' is the gravitational acceleration at the satellite height. This in turn is related to the gravitational acceleration $g = 9.81$ m/s at the earth's surface by $g' = g(a/(a + h))^2$. Balancing the centripetal force against the gravitational force gives

$$Mg\left(\frac{a}{a + h}\right)^2 = \frac{MV^2}{a + h}$$

therefore,

$$V = a\sqrt{\frac{g}{a + h}} \tag{19.3}$$

Substituting this expression for V in Eq. (19.2) and solving for h when the numerical values are substituted for a and g gives

$$h = (5075T^{2/3} - 6371) \text{ km} \tag{19.4}$$

Here, T is time in hours.

Substituting for $T = 24$ hours gives $h = 35{,}855$ km as the height of the geostationary orbit.

19.3 STATION KEEPING

Even with the satellite correctly launched in orbit, external forces will act on it to alter its position and orientation with respect to earth. These external forces are variations in the earth's gravity with position, gravitational gradient along the satellite, solar radiation, meteorite bombardment, magnetic field forces, and the gravitational pull of the moon and the sun. The gravitational pull of the sun and moon can cause the orbit of a geostationary satellite to incline, as shown in Fig. 19.2(a). The angle of inclination in degrees is the specified inclination error. This can be kept within $\pm 0.1°$ by means of command controls from earth. Longitudinal drift, also illustrated in Fig. 19.2(a), is a more serious form of error, as satellites are spaced along the geostationary orbit. Again, typical tolerances are $\pm 0.1°$. One degree of longitude represents a distance of about 737 km of arc on the geostationary orbit. Figure 19.2(b) shows the positions of some of the European satellites used for direct satellite broadcasting.

The control routine necessary to keep the satellite in position is referred to as *station keeping*.

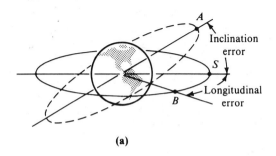

(a)

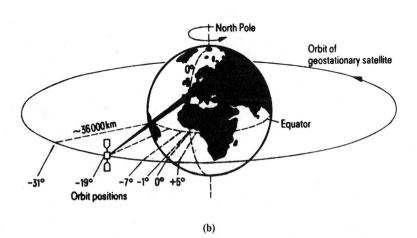

(b)

FIGURE 19.2. (a) Positional errors; (b) broadcasting satellites on geostationary orbit (from Herbert Zwilling, *Direct Broadcasting by Satellite*. Siemens Telecom Report, Vol. 5, No. 1, Mar 1982.)

19.4 SATELLITE ATTITUDE

Satellite attitude refers to the satellite orientation with respect to the earth. Maintaining close control on attitude is required where linear polarized signals are used. For example, in the Anik-B system described later, both vertical and horizontal polarizations defined with respect to the earth's axis are used. Also, where highly directional antennas are used, it is essential that the "spot beams" are directed to the desired earth regions at all times, and solar panels must be oriented for maximum energy pickup from the sun, all of which requires close control of satellite attitude.

The three axes used for attitude are shown in Fig. 19.3(a), and then again in relation to the Canadian Communications Technology Satellite in Fig. 19.3(b). This satellite utilizes a spinning flywheel to provide gyroscopic stiffness in the plane of orbit. Pitch control is obtained by varying the wheel speed, and thrusters are used for yaw and roll control. Pitch and roll error signals are obtained from a fixed infrared sensor pointing towards the earth, and yaw error signals are obtained from sun sensors.

Another method used to obtain gyroscopic stiffness is to impart spin to the satellite itself. This method is used with the Intelsat IV satellite. The satellite is drum shaped, spinning about its drum axis. An antenna platform, which carries a number of antennas, is "despun" so that these antennas always point towards earth.

19.5 TRANSMISSION PATH

The geometry of the situation is shown in Fig. 19.4. It is required to find the one-way distance d. The angle of elevation of the ground station antenna is shown as β. This is the angle the maximum gain direction makes with the horizontal at the antenna site. Applying the cosine rule to the triangle shown gives

$$(a + h)^2 = a^2 + d^2 - 2ad\cos(90 + \beta)$$
$$= a^2 + d^2 + 2ad\sin\beta \tag{19.5}$$

This is a quadratic equation in d which, when solved for positive d, gives

$$d = \sqrt{(a + h)^2 - (a\cos\beta)^2} - a\sin\beta \tag{19.6}$$

The minimum distance is when $\beta = 90°$, giving $d = \sqrt{(a + h)^2} - a = h$, as expected. In theory, the maximum distance occurs when $\beta = 0°$, giving $d = \sqrt{(a + h)^2 - a^2} = 41{,}743$ km for the geostationary orbit. In practice, the

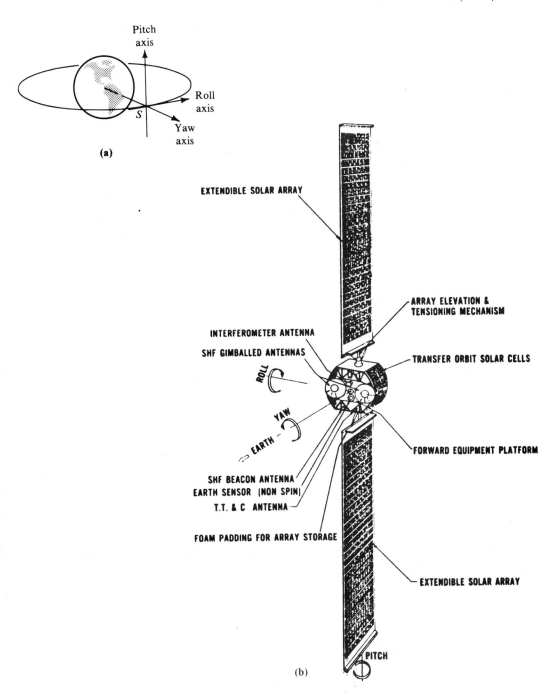

FIGURE 19.3. (a) Axes used in determining satellite attitude; (b) CTS Communications Satellite (courtesy Department of Communications, Canada).

minimum angle of elevation is about $5°$, but this makes negligible difference to the maximum value calculated above.

19.6 PATH LOSS

Most of the path loss occurs through spreading of the signal energy, and, as shown in Section 14.2.1, this loss is given by Eq. (14.8) as

$$L = 32.5 + 20 \log d + 20 \log f \quad \text{dB} \tag{14.8}$$

where d is in km and f is in MHz.

In addition to this loss, absorption and scattering of the signal will occur as it passes through the troposphere and ionosphere, as shown in Fig. 19.5(a). In this case, the loss will be proportional to the path length in the attenuating medium, and this in turn depends on the angle of elevation of the ground antenna. Figure 19.5(b) shows the geometry of the situation. Here, ℓ_v represents the path length for vertical incidence ($\beta = 90°$) through the atmosphere, and ℓ the path length in general. The triangle shown is similar to that shown in Fig. 19.4, and therefore, on replacing d with ℓ and h with ℓ_v, Eq. (19.6) may

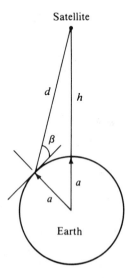

FIGURE 19.4. The ground station antenna angle of elevation, and the transmission path distance d.

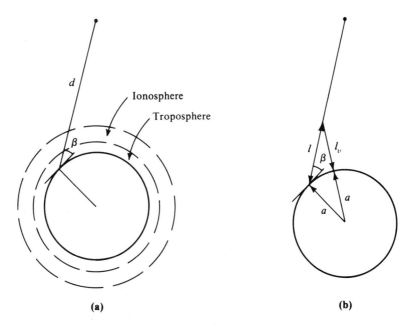

FIGURE 19.5. (a) Transmission path including atmosphere and ionosphere; (b) geometry pertaining to path through atmosphere.

be used to find

$$\ell = \sqrt{(a + \ell_v)^2 - (a \cos \beta)^2} - a \sin \beta \qquad (19.7)$$

For example, with $\ell_v = 100$ km, the minimum path length is simply 100 km, and the maximum path length in the attenuating medium, for the minimum practical value of $\beta = 5°$, is $\ell = 708$ km. The attenuation in the atmosphere varies with frequency as shown in Fig. 19.6. These results are for transmission through a moderately humid atmosphere, measured at sea level. Two absorption peaks are observed. The first, at a frequency of 22.2 GHz results because water vapor molecules go into vibrational resonance at this frequency, and in so doing, absorb energy from the wave. The second peak at 60 GHz is similar, and is caused by resonant absorption of oxygen molecules. The curves show the effect of the angle of elevation on attenuation resulting from the greater path length, given by Eq. (19.7). At 4 GHz, for example, the total atmospheric attenuation for vertical incidence is just over .04 dB, whereas for a 5° angle of elevation, it is about .1 dB. Free-electron absorption which occurs in the ionosphere is negligible at microwave frequencies.

Attenuation will also occur with precipitation, being worse for heavy rain. In system design, a fading margin would be allowed for rainfall, the value depending on the geographical location of the ground station.

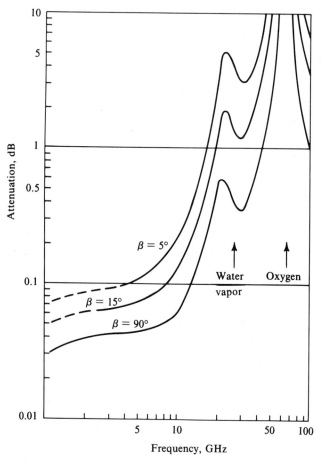

FIGURE 19.6. Composite attenuation curves for the atmosphere, showing both oxygen and water vapor absorption effects. The antenna angle of elevation is β.

19.7 NOISE CONSIDERATIONS

Random electromagnetic radiation occurs from stars, planets, and inter-stellar gas clouds, and is received by an antenna as "noise." It is found that the spectrum density of the general background noise from the sky, usually referred to as galactic or cosmic noise, varies inversely as frequency down to a lower limit set by the particular region of space into which the antenna happens to be pointing. In addition to the general background noise, the earth's atmosphere also introduces noise, since it acts as a loss transmission path, or an attenuator. As shown in Section 4.11.2, an attenuator at a physical temperature T has an equivalent noise temperature $(L - 1)T$. In this case, L is the attenuation of the atmosphere and T is its average temperature. As shown in the previous section, the attenuation increases as a function of frequency, having peaks at the water vapor resonance frequency of 22.2 GHz and the oxygen resonance frequency of 60 GHz. The equivalent noise temperature of the atmosphere, therefore, shows similar peaks. It will be recalled from Section 4.2 that the spectrum density is specified in W/Hz. Let S_n represent the spectrum density of the available noise signal received by the antenna; then, using Eq. 4.2, an equivalent noise temperature for the antenna may be defined as

$$T_{\mathrm{ANT}} = \frac{S_n}{k} \qquad (19.8)$$

where $k = 1.38 \times 10^{-23}$ J/K is Boltzmann's constant.

The equivalent noise temperature for the antenna will consist of a galactic noise contribution, which, as already mentioned, decreases with frequency down to some limit, and the atmospheric attenuation contribution, which increases with frequency. Temperature peaks will occur at the resonant absorption points. Combining these factors, typical curves for T_{ANT} as a function of frequency are shown in Fig. 19.7. These curves are derived under

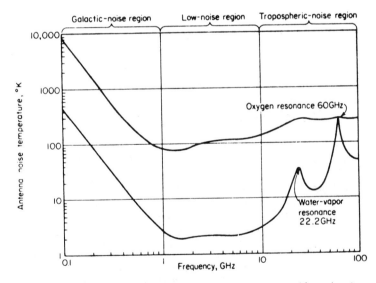

FIGURE 19.7. Irreducible noise temperatures of an ideal, ground-based antenna. The antenna is assumed to have a very narrow main beam without sidelobes or electrical losses. Below 1 GHz, the maximum values are for the beam pointed at the galactic poles. At higher frequencies, the maximum values are for the beam just above the horizon and the minimum values for zenith pointing. The low-noise region between 1 and 10 GHz is most amenable to application of special, low-noise antennas (from Philip F. Panter, Communications Systems Design, McGraw Hill, 1972. With permission).

the assumption that the antenna has a very narrow beam without side lobes, so that no noise is picked up from the earth itself. It is also assumed that the antenna is lossless.

From Fig. 19.7, it can be seen that the antenna noise temperature can range from about 10,000 K to about 2 K, and clearly, this cannot be the physical temperature of the antenna. For example, an antenna situated in the tropics would receive the same noise as an identical antenna situated in the arctic providing they were both pointing to the same region in space and the atmospheric attenuation were the same for each. The physical temperature of the antenna does affect the noise resulting from antenna losses.

As shown in Section 4.11, total receiver noise can be referred to the input by use of an equivalent noise temperature T_e. The total noise temperature referred to the receiver input, taking into account the antenna noise and the receiver noise, is, therefore,

$$T = T_{\text{ANT}} + T_e \qquad (19.9)$$

Thus, the available noise power at the input for a receiver noise bandwidth B_N is

$$P_n = kTB_N \qquad (19.10)$$

This noise power is independent of the antenna gain (if man-made noise received as "signals" is excluded). The signal power received will be directly proportional to the antenna power gain. A figure of merit widely used to characterize a satellite receiving system is the ratio of antenna gain to total input noise temperature, given by

$$M = G/T \qquad (19.11)$$

The ratio is usually specified in decibels:

$$(M)_{\text{dB}} = (G)_{\text{dB}} - 10\log(T) \qquad (19.12)$$

The units for M are often specified as dB/K, or sometimes, dB $\cdot$ K (neither of which is strictly correct).

The higher the value of M, the better the system. Home receivers for satellite TV reception require a value of M of at least 12 dB/K, and the minimum value recommended by the CCIR for class A ground station telephony circuits is 40.7 dB/K. Receivers aboard satellites will have much lower values of M, partly because antenna gain will be lower, but mainly because the antenna points towards a "hot" earth. The resulting T_{ANT} may be about 290 K, compared to perhaps 20–30 K for a good ground station system (see Fig. 19.7). Typically, the value of M for a satellite receiving system is about -8 dB/K.

19.8 THE SATELLITE SYSTEM

A satellite acts as a central or "star" point in a network system. This is illustrated in Fig. 19.8. The satellite network differs from the terrestrial network, which is an "in-line" system, although much of the equipment may be similar. In particular, access to the satellite network requires special schemes, some of which are described in Section 19.10. A central control station is shown by way of illustration, although in practice, various control schemes are in use. Satellite attitude-control (station-keeping) signals may be generated by means of satellite "on-board" equipment. Also, frequency allocation schemes such as are used for multiple access (see Section 19.10) may be assigned through a central control station or by control functions at each ground station, depending on the particular network.

The frequency ranges most commonly used are 4–6 GHz and 12–14 GHz. The first number specified in each range gives the downlink frequency (to the nearest GHz), the second figure the uplink frequency. The Canadian Anik-B satellite system provides a good illustration of satellite technology, and a description of it is given here to bring out some of the techniques used.

The Canadian Anik-B satellite is dual-band, capable of operating in both ranges simultaneously. The 12–14 GHz channeling scheme for the Anik-B satellite is shown in Fig. 19.9. Each RF channel is 72 MHz wide, and there is a guard band of 8 MHz between channels. The satellite transmits in the range 11.7 to 12.2 GHz (downlink), and receives the range 14 to 14.5 GHz (uplink).

As shown in Fig. 19.9, the downlink carriers are horizontally polarized, the uplink carriers vertically polarized. Polarization in this case is with reference to the earth's N–S axis, vertical being parallel to this axis, and horizontal orthogonal to it (see Section 15.5).

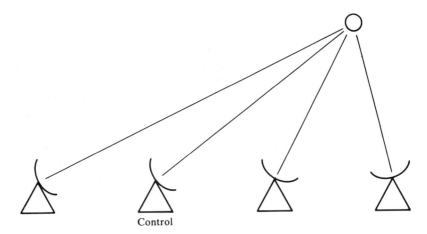

Control

FIGURE 19.8. A satellite network, showing the satellite as the center of the "star" connection.

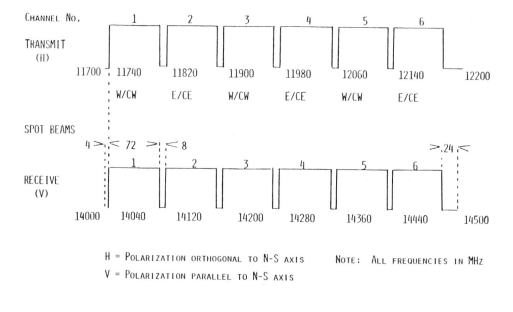

FREQUENCY ALLOCATION AND POLARIZATION PLAN

FIGURE 19.9. ANIK-B channel plan (courtesy Department of Communications, Canada).

The various blocks of equipment aboard the satellite are shown in Fig. 9.10(a) for the 12–14 GHz range. Four antennas are employed which, when transmitting, produce spot beams over Canada in the west (W), central west (CW), central east (CE), and east (E) regions. When receiving, the antennas are coupled together by means of the orthomode couplers, which respond to the received vertical polarization. The combined received signal consists of all six channels, i.e., a wideband signal ranging from about 14 to 14.5 GHz. The received signal passes through a number of units which collectively form what is termed the satellite transponder. In the case of the Anik-B satellite, it can be seen from Fig. 19.10(a) that there are six transponder channels. These share a common front-end receiver section, and the four travelling-wave tube amplifiers (TWTAs) are shared between all six channels on a switched basis. For example, a telephony signal in the channel 4 frequency band received on the W antenna is routed through the common front end, through the input hybrid multiplexer, and, as shown in the figure, is then switched to the E2 TWTA, which then feeds it to the E antenna for retransmission. This channel could be switched to the E1 TWTA and onto the CE antenna.

The received wideband signal is amplified and down-converted to the frequency range 11.7–12.2 GHz in the common front end of the receiver, comprising a low-noise amplifier (LNA), a downconverter, and a local oscillator at 2.3 GHz as shown in Fig. 19.10(b).

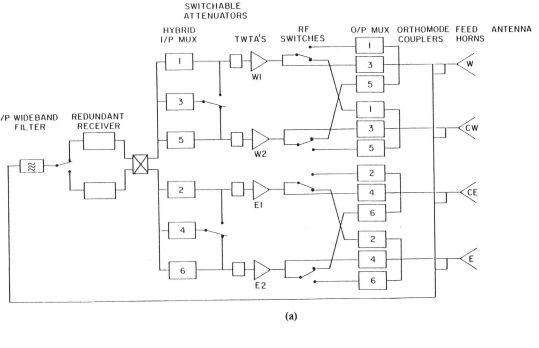

(a)

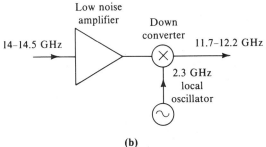

(b)

FIGURE 19.10. (a) ANIK block diagram (courtesy Department of Communications, Canada); (b) satellite single-conversion receiver.

RF switching, controlled from the ground station, enables various combinations of channels to be directed to various antennas. On transmission, RF channels 1, 3, and 5 can be switched between the W and the CW antennas, RF channels 2, 4, and 6 between the E and the CE antennas. Figure 19.10(a) illustrates the situation where channels 1 and 3 are switched to the CW antenna and channel 5 is switched to the W antenna. Channels 2 and 4 are switched to the E antenna, channel 6 to the CE antenna. The "footprints" for these antennas are shown in Fig. 19.11.

On reception, all the antennas are coupled together so that the satellite receives from all regions, as shown in Fig. 19.12.

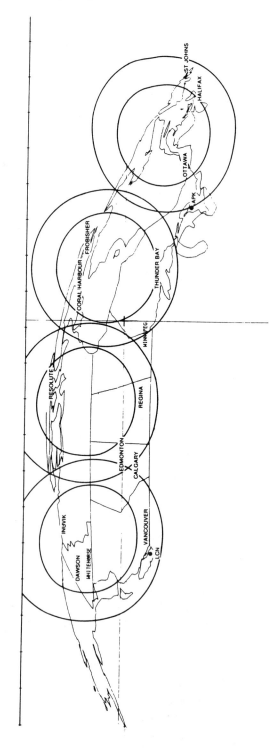

FIGURE 19.11. ANIK-B transmit contours (eirp 46.5 dBw outer contour, 49.5 dBw inner contour). (Courtesy Department of Communications, Canada).

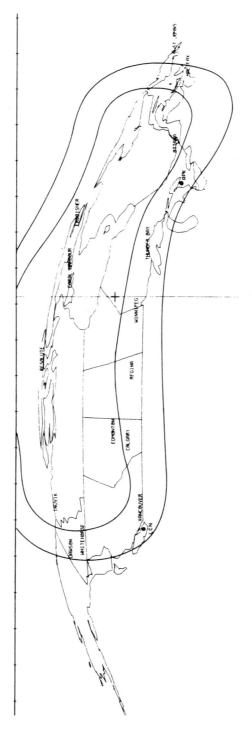

FIGURE 19.12. ANIK-B receive contours (SFD -86dBw / m^2, G / T OdB / K outer contour, SFD -88dBw / m^2, G / T + 2dB / K inner contour). (Courtesy Department of Communications, Canada).

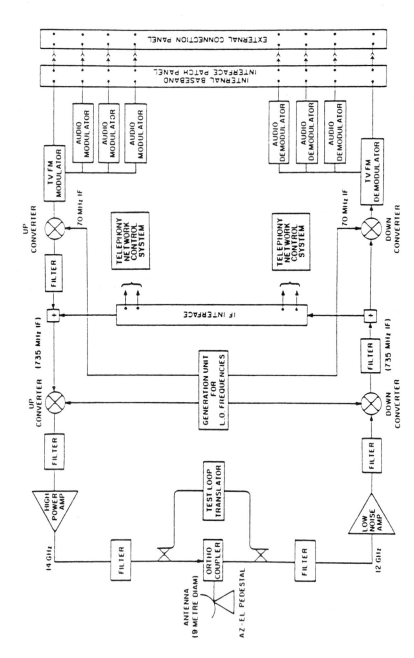

FIGURE 19.13. 9m terminal block diagram (courtesy Department of Communications, Canada).

The letters *eirp* shown on Fig. 19.11 stand for effective isotropic radiated power, which is explained in Section 19.10. Also, the letters *SFD* on Fig. 19.12 stand for saturation flux density, which is explained in Section 19.9.

Various ground terminals are available in the Anik-B system. A main terminal using a 9-m diameter antenna is situated in Ottawa. This terminal acts as a network control station (NCS) for interconnecting telephone traffic between various remote stations. Figure 19.13 shows the block diagram for this ground station. The station can transmit and receive FM video plus three audio channels in any of the six RF channels available through the satellite. In addition, it can transmit telephone messages to remote stations through any of its six channels. Separate telephone carriers, preassigned to the remote stations, are available for this purpose. All telephone messages have to pass through the network control station at Ottawa, and therefore a telephone connection between two remote stations involves two transmission hops, from one station to Ottawa, and from Ottawa to the other station.

The telephone carriers are located in the upper half of the transponder channels, approximately 15 MHz above band center and spaced by 70 kHz. All telephone traffic into the Ottawa station is through transponder channel 4. The telephone carriers assigned to remote stations must of course lie within the transponder channels transmitted by the spot beam for the region. For example, a remote terminal at Whitehorse (see Fig. 19.11) must operate through the W antenna and therefore could be assigned a telephony carrier for reception through channel 5; however, a separate telephony carrier for transmission through channel 4 would have to be assigned.

19.9 SATURATION FLUX DENSITY

The TWTA in each transponder has a power–gain characteristic as sketched in Fig. 19.14. At a certain input value, the output power reaches a maximum value, known as the saturation value. This is an important reference point. The power input is proportional to the power flux density at the receiving antenna (as described by Eq. 15.11), and therefore the saturation point may be specified in terms of the power density at the antenna. This is the saturation flux density (or SFD) in W/m^2. If one watt is used as the reference level, the SFD may be expressed in dBW/m^2. Depending on the nature of the signal being amplified by the TWTA, it may be desirable to operate at saturation, or it may be desirable to operate below saturation, in what is termed a "back-off" condition. The number of decibels of back-off will be specified for both input back-off and output back-off. For example, if the output back-off is 2 dB, this means that the output power will be 2 dB below the power output at saturation level.

Two SFD contour lines are shown in Fig. 19.12, along with the G/T figure of merit for the receiving system (see Eq. (19.11)).

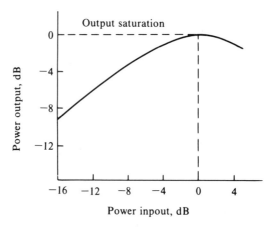

FIGURE 19.14. Travelling Wave Tube (TWT) power transfer characteristic. Both input and output are shown in dB relative to the saturation point.

In link calculations, a quantity of prime interest is the carrier-to-noise density ratio C/N_0, which can be related to the SFD. As shown in Section 4.2, the available noise power density is kT, and if P_R is the available carrier signal power at the receiving antenna, then

$$\frac{C}{N_0} = \frac{P_R}{kT} \tag{19.13}$$

Let ϕ_S represent the saturation flux density. Then $P_R = \phi_S A_{EFF}$, where A_{EFF} is the effective receiving aperture as described in Section 15.8. Hence,

$$P_R = \phi_S \frac{\lambda^2}{4\pi} G_R \tag{19.14}$$

and

$$\frac{C}{N_0} = \phi_S \frac{\lambda^2 G_R}{4\pi kT} \tag{19.15}$$

This equation is often expressed in decibels as

$$\frac{C}{N_0} = \phi_S + \frac{G_R}{T} - G_1 - k$$

$$= \phi_S + M - (G_1) + 228.6 \qquad \text{dB} \tag{19.16}$$

When the equation is written in this fashion, ϕ_S will be in dBW/m², M is the antenna figure of merit as given in decibels by Eq. 19.12, and $k = 10\log(1.38 \times 10^{-23}) = -228.6$ dB is Boltzmann's constant expressed in decibels (or, more strictly, "decilogs"). The gain figure G_1 is the isotropic power gain of an antenna having an effective area of 1 m², which, as seen from Eq. (15.20), is

$$(G_1)_{dB} = 10 \log(4\pi/\lambda^2) \tag{19.17}$$

Example 19.1   Calculate the C/N_0 ratio at the Anik-B satellite for the inner contour line on channel 1.

Solution   The up-link frequency for channel 1 is, from Fig. 19.9, 14.04 GHz; therefore, $\lambda = 0.0214$ m. Hence,

$$G_1 = 10 \log\left(\frac{4\pi}{.0214^2}\right) = 44.4 \text{ dB}$$

For the inner contour, SFD = -88 dBW/m², and $G/T = +2$ dB/K. Therefore,

$$\left(\frac{C}{N_0}\right)_{dB} = -88 + 2 - 44.4 + 228.6$$
$$= 98.2 \text{ dB}$$

19.10   EFFECTIVE ISOTROPIC RADIATED POWER

The *effective isotropic radiated power* (*eirp*) is the actual power radiated by the antenna multiplied by the isotropic power gain of the antenna (see Section 15.7). Let P_T represent the power radiated and G_T the isotropic power gain; then

$$\text{eirp} = P_T G_T$$

or

$$(\text{eirp})_{dBW} = (P_T)_{dBW} + (G_T)_{dB} \tag{19.18}$$

The ratio C/N_0 at the receiving end can be found for a given eirp at the transmitting end provided that the transmission path distance is known. The path loss for free-space propagation is given by Eq. (14.8), and the link equation, derived from Eq. (14.9), is

$$\left(\frac{P_R}{P_T}\right)_{dB} = (G_T)_{dB} + (G_R)_{dB} - (L)_{dB}$$

where L is the path loss as given by Eq. (14.8).

Now, expressing P_R and P_T in decibels relative to one watt enables Eq. (14.9) to be written as

$$(P_R)_{dBW} = (P_T)_{dBW} + (G_T)_{dB} + (G)_{RdB} - (L)_{dB}$$

$$= (\text{eirp})_{dBW} + (G)_{RdB} - (L)_{dB} \qquad (19.19)$$

Hence, from Eq. (19.13), using this expression for P_R and recognizing that $(G_R/kT)_{dB} = M + 228.6$ where M is the receiver figure of merit, gives:

$$\frac{C}{N_0} = \text{eirp} + M - L + 228.6 \qquad \text{dB} \qquad (19.20)$$

As in the previous equations of this nature, decibel or decilog quantities must be used.

Example 19.2   Calculate $(C/N_0)_{dB}$ at the earth receiving station, from a satellite transmitting an eirp of 49.5 dBW on a frequency of 12 GHz. The earth station antenna angle of elevation is $7°$, and the receiving figure of merit is 40.7 dB.

Solution   The path distance is, from Eq. (19.6)

$$d = \sqrt{(6371 + 35855)^2 - (6371 \cos 7°)^2} - 6371 \sin 7°$$

$$= 40{,}973 \text{ km}$$

From Eq. (14.8),

$$L = 32.5 + 20 \log 40{,}973 + 20 \log 12{,}000$$

$$= 206.3 \text{ dB}$$

Therefore,

$$\frac{C}{N_0} = 49.5 + 40.7 - 206.3 + 228.6$$

$$= 112.5 \text{ dB}$$

Since Eq. (19.16) gives the carrier-to-noise density ratio in terms of the saturation flux density ϕ_S, and Eq. (19.20) gives it in terms of the effective isotropic radiated power (eirp), the relationship between eirp and ϕ_S can be found by equating these two expressions:

$$\text{eirp} + M - L + 228.6 = \phi_S + M - G_1 + 228.6$$

$$\therefore \text{eirp} = \phi_S - G_1 + L \qquad (19.21)$$

Equation (19.21) ignores any back-off which might be required to reduce intermodulation distortion in the TWTA as described in Section 19.9. Denoting input back-off by BO_i dB, we see that ϕ_S must be reduced by this amount, and therefore the more general form of Eq. (19.21) is

$$\text{eirp} = \phi_S - G_1 + L - BO_i \qquad (19.22)$$

When the eirp of the ground station is known, the power output of the ground station high-power amplifier (HPA) can be calculated. Again, expressing all quantities in decibels or decilogs, the power transmitted is obtained from Eq. (19.18) as

$$P_T = \text{eirp} - G_T$$

The power output from the HPA has to supply this plus losses. Let L_F represent the loss in the antenna feeder, and L_P the loss resulting from antenna pointing error, both in dB. The power output from the HPA is therefore

$$P_{\text{HPA}} = P_T + L_F + L_P$$
$$= \text{eirp} - G_T + L_F + L_P$$

The HPA may require output back-off. Denoting this by BO_0 dB, we can see that the HPA must be up-rated by this amount in order to achieve the required eirp figure:

$$P_{\text{HPA}} = \text{eirp} - G_T + L_F + L_P + BO_0 \text{ dBW} \qquad (19.23)$$

19.11 MULTIPLE-ACCESS METHODS

As has already been pointed out, a satellite is situated at the star point of a network and therefore must be capable of connecting together earth stations which may be geographically widely separated. If heavy continuous telephone and data traffic is known to exist between any two points, a transponder may be assigned on a fixed basis to relay the multiplexed signal from one station to the other. This is known as fixed assignment. In Fig. 14.1, for example, the terrestrial microwave repeaters (in practice there will be more than the one shown) between the two 4-GHz radio links may be directly replaced by a satellite repeater whose uplink frequency is 4 GHz and whose downlink frequency is 6 GHz.

Such an arrangement does not, however, permit any great degree of flexibility in telecommunications traffic. The problem is that many earth stations will want to have access to the satellite, and because the traffic from any one earth station will vary with the time of day, any given station would like to have access as the demand arises. Thus, most satellites employ what is known as a demand assignment multiple-access (DAMA) system. A number of such systems are in use or are planned, the two most common being frequency-division multiple access (FDMA), and time-division multiple access (TDMA). A system known as SPADE, designed originally for use on the INTELSAT system, provides a good illustration of FDMA and is described in the following section.

19.11.1 SPADE

The word SPADE is a loose acronym for single-channel-per-carrier PCM multiple-access demand-assignment equipment. Each ground station has available a common pool of 800 telephony carriers, and these are assigned in pairs to provide two-way traffic. The ground stations are connected together through the satellite on a high-quality, common signalling channel. When one station wishes to make a telephone connection with another, it selects a channel pair of telephony carriers at random from those that are still available and signals this information through the common signalling channel to the destination station. Assuming that this telephony channel has not been assigned to another request during the signalling period, the two stations are connected and the satellite link tested before connections to the subscribers are made. The round trip signalling time including switching and other kinds of delays is about 600 milliseconds, and it could happen that the channel selected by the originating station would get assigned to another request during this period. The current status of all channels is stored at all stations. The originating station will detect the change of status of its first selection and immediately select a new channel, which it again signals to the destination station.

The telephone signals are pulse-code-modulated (see Section 17.6) and transferred to the telephony carriers by means of phase-shift keying. A four-level form of phase-shift keying, called quaternary phase-shift keying (QPSK), is used. Two-level PSK is described in Section 17.11.2 in which the modulating signal causes the carrier to shift in phase by 0 or 180°. With QPSK, the bits are combined in groups of two (00, 01, 10, and 11), and the difference in each of these levels is represented by a phase shift of 90°. It will be noted that the speech signal is converted to digital form through the PCM, and therefore the system is compatible with data networks.

An important operational point with the SPADE system is that the carriers are "voice-operated"; that is, a carrier is not present unless its modulating signal (speech or data) is also present. In practice this means that out of the total of 800 channels which a transponder channel could be carrying, only about 320 are active at any given time. This reduces the intermodulation distortion which occurs when the TWTA is operated near its saturation level (see Section 19.9). However, it is still necessary to operate the TWTA in a back-off condition, and this reduces the power output and hence the satellite channel capacity. With the INTELSAT IV satellite working into ten class A earth stations, the capacity is reduced from 900 to 450 channels as a result of back-off.

19.11.2 Time-Division Multiple Access (TDMA)

More recently, time-division multiple access (TDMA) has been introduced for satellite operation. This method overcomes the back-off problem and also offers more flexibility in satellite sharing between stations and channels. With TDMA, any given earth station has total access to the satellite transponder for a given time slot. The separate voice and data channels carried by the earth station are time-division-multiplexed (see Section 11.4). Separate voice carriers are not required, and the problem of intermodulation distortion in the transducer does not arise. The transponder can therefore be operated at full power, which is the saturation level.

Each earth station is assigned its own time slot, and since each must access the transponder in the proper time sequence, precise synchronization between earth stations is required to prevent the individual transmissions from overlapping. Some idea of the problem of synchronization can be gained by considering the volume of space about the nominal position in which the satellite can be found. As mentioned in connection with station-keeping, the angular control is typically within $\pm 0.1°$. The orbital arc length subtended by $0.1°$ is $(35,855 + 6371) \times .1/57.3 = 73.69$ km. The distance of 35,855 km itself may vary by $\pm 0.1\%$, or ± 36 km. Thus, the worst case in distance variation is approximately $\sqrt{74^2 + 74^2 + 36^2} = 111$ km, and this results in about 735 μs addition to the round-trip time.

A number of different methods of synchronization are in use or have been proposed. The simplest methods are of the open-loop variety, so-called because the transmission from a given station is not received by that same station, so no feedback correction is possible. A coarse synchronization is possible using predicted orbital parameters to compute delays, and in some tests accuracies of 200 μs have been achieved. Clock-controlled methods are also being tried in open-loop systems in which the earth stations work from a common frequency/time standard. One other open-loop method requires one earth station, known as the reference station, to transmit a reference burst, and all other stations in the network lock their transmissions to this reference.

The other variety of synchronization is known as closed-loop because the station transmitting at any given time also receives its own signal back from the satellite and so can accurately determine the delay time. In the INTELSAT system, a reference station is again used which transmits reference bursts at the frame rate. A frame period contains one time slot from each earth station plus the guard times between these. The reference burst is used by the earth stations to establish the frame synchronization, and this enables any given earth station to access its assigned time slot. However, the position within the time slot may not be correct, and the station then has to transmit its own reference, termed a burst code-word (BCW). This is received back from the satellite, and allows the transmission to be adjusted to the start of the assigned time slot.

A frame period is typically some multiple of 125 μs, which is the Nyquist sampling period for 4-kHz speech. For example, the frame period proposed for the INTELSAT system is 750 μs. The duration of a time slot within the frame assigned to a given earth station can be varied depending on the traffic requirements, and this is one of the versatile features of the TDMA system.

An essential component of the TDMA system is the use of a compression buffer to convert the relatively slow-speed input data to high-speed output data required for the transmission burst. For example, the Bell T1 digital telephone system (see Section 17.6.6) has a transmission rate of 1.544 megabits per second (Mbps). Satellite transmission bursts, however, can have speeds as high as 60 Mbps. The input data are read into the compression buffer at the slower speed, read out at the required burst speed, and, along with the digitized synchronization and control signals, are used to modulate the single RF carrier which is then transmitted to the satellite in the assigned time slot.

Further increase in channel capacity can be achieved by the use of spot beams and frequency reuse. Frequency reuse is when the same frequency can be used in separate beams, achievable by using different polarizations. Thus, different beams may carry the same frequency, but one may be vertically polarized and the other horizontally polarized. Clockwise and counterclockwise circular polarizations may also be used to effect separation. Spot beams are generated by having multiple horn feeds beamed onto a common reflector. The beams are arranged to cover specific areas of a country, as shown, for example, in Fig. 19.11.

More recently, satellite beam switching has been introduced, in which the satellite transponder is switched between antennas in a synchronized and cyclic manner. This is a form of "space multiplexing," and it is called beam switching because, for example, a ground station transmitting on, say, beam A may be interconnected to beam B on the satellite, and then to beam C, and so on, each of these beams going to different ground stations. Satellite switching is used with TDMA, and the combined system is referred to as SS/TDMA. It should be noted that although satellite switching occurs in the Anik-B system illustrated in Fig. 19.10, this is not what is generally meant by SS/TDMA. In this case, the switches are preset by control signals from the ground station, whereas in the SS/TDMA concept, a commutator switch matrix controlled by an accurate clock aboard the satellite achieves the interconnections.

19.12 PROBLEMS

1. The Intelsat IV satellite operates in the bands 5932–6418 MHz and 3707–4193 MHz. The satellite uses 12 transponders, each of bandwidth 36 MHz. An 8-MHz guardband separates the transponder channels into two

groups of six, and the guardband between transponder channels in each group of six is 4 MHz. Draw the frequency channeling scheme, and indicate which are the up-link, and which the down-link, assignments.

2. Calculate the antenna beam angle required by a satellite antenna to give full global coverage from a geostationary orbit. What fraction of the earth's surface is covered by the assumed circular beam?

3. If the beam angle calculated in Problem 2 is the -3 dB beamwidth, calculate the isotropic power gain of the antenna if it is a paraboloidal reflector with an illumination efficiency of 65%.

4. A ground station paraboloidal reflector antenna has an isotropic power gain of 64 dB at an illumination efficiency of 70%. Calculate the -3 dB beamwidth and the spot width of the beam at geostationary height.

5. Show that a positional tolerance of 0.1° on a geostationary orbit is equivalent to an arc distance of 74 km.

6. A ground station is operating to a geostationary satellite at a 5° angle of elevation. Calculate the round-trip time between ground station and satellite.

7. Calculate the difference in round-trip propagation delays for vertical transmission and ground station transmission at 5° elevation to satellites in geostationary orbit.

8. The effective noise temperature of an antenna is 23 K, and the receiver noise temperature is 60 K. The isotropic power gain of the antenna is 46 dB. Calculate the receiver figure of merit M in decibels.

9. A satellite receiving system has a figure of merit of -8 dB/K. The satellite antenna is a paraboloidal reflector type with an illumination efficiency of 70% and a -3 dB beamwidth of 18°. Calculate the total noise temperature at the receiver input.

10. Calculate the transmission path loss at vertical incidence between a geostationary satellite and a ground station operating at a frequency of 4 GHz, allowing .04 dB for atmospheric attenuation and .1 dB for rain attenuation.

11. Repeat the calculation of Problem 10 for an operating frequency of 12 GHz, an atmospheric attenuation of .1 dB, and a rain attenuation of .5 dB.

12. For a satellite system, the up-link frequency is 14.21 GHz, the saturation flux density at the satellite is -81 dBW/m^2, and the satellite figure of merit is 1.9 dB/K. Calculate the satellite carrier-to-noise density ratio in decibels.

13. A satellite has an eirp of 46.5 dBW. The free-space loss is 205.8 dB, and the ground station figure of merit is 30.7 dB/K. Calculate the carrier-to-noise density at the ground station receiver.

14. The transponder bandwidth for the CTS system is 36 MHz, the free-space loss is 207.3 dB, and atmospheric attenuation is .18 dB. Other parameters

are: ground station transmitter power output, 17.86 W; feeder loss, .15 dB; antenna gain, 59.69 dB; satellite antenna gain, 38 dB; satellite system temperature, 1349 K. Calculate (a) the satellite received carrier level in dBW, (b) the satellite receiver noise power, and (c) the carrier-to-noise ratio in dB at the satellite input.

15. The transponder in the CTS satellite has a gain of 120.7 dB and is operated at the power input determined in Problem 14. Down-link parameters are: satellite antenna filter loss, .75 dB; antenna gain, 36.7 dB; ground station antenna gain, 41.2 dB; antenna pointing loss, .2 dB; receiver system temperature, 1020 K. The free-space loss is 205.9 dB, and the atmospheric attenuation is .14 dB. Calculate (a) the satellite eirp, (b) the ground station received power in dBW, (c) the input noise power at the ground station, and (d) the ground station carrier-to-noise ratio in decibels. The system bandwith is 36 MHz.

16. For the Anik-B satellite system, the transponder bandwidth is specified as 72.60 dB–Hz, and Boltzmann's constant as 228.6 dBW/Hz/K. Other system parameters are: uplink—saturation flux density, -83.56 dBW/m^2; spacecraft receiver figure of merit, 2.2 dB/K; $(\lambda^2/4\pi)$ equal to -44.54 dB; downlink—eirp, 50.12 dBW; path loss, 205.8 dB; ground station antenna gain, 40.7 dB; and system noise temperature, 27.61 dB–K. Calculate (a) the bandwidth in MHz, (b) the carrier-to-noise ratio in decibels for the uplink, and (c) the same for the downlink.

17. Satellite parameters are specified as saturation flux density, -81 dBW/m^2, input back-off for the TWTA, 10 dB; and $G_1 = 44.5$ dB. The free-space loss is 207.3 dB. Calculate the eirp of the ground station. Neglect atmospheric attenuation.

18. For the system specified in Problem 17, the ground station antenna gain is 52.3 dB, the waveguide feeder loss is .7 dB, the antenna pointing loss is .8 dB, and the output back-off required is 3 dB. Calculate the power output in watts of the ground station HPA.

Visible Light 0.4 - 0.7 µm.

F0   0.7 - 1.6 µm.

New F0   > 2 µm.

CHAPTER 20

Fiber Optic Communications

20.1 INTRODUCTION

Optical fibers are increasingly replacing wire transmission lines in communications systems. Such optical fiber lines offer several important advantages over wire lines. First, since light is effectively the same as radio frequency radiation, but at a very much higher frequency (about 300 THz, or 3×10^8 GHz), theoretically, the information-carrying capacity of a fiber is much greater than microwave radio systems. Next, the material used in fibers is silica glass, or silicon dioxide, which is one of the most abundant materials on earth, so that eventually the cost of such lines should be much lower than either wire lines or microwave systems. Also, the fibers are not electrically conductive, so that they may be used in areas where electrical isolation and interference are severe problems. And because of the high information capacity, multiple channel routes may be compressed into much smaller cables, thereby reducing congestion in overcrowded cable ducts.

With present technology, fiber optic communications systems are still somewhat more expensive than equivalent wire or radio systems, but this situation is changing rapidly. Fiber optic systems will rapidly become competitive with the other systems in price, and, with their other advantages, will increasingly replace them.

This chapter will describe the principles involved in fiber optic systems and present some of the features of the present state of the art.

20.2 PRINCIPLES OF LIGHT TRANSMISSION IN A FIBER

20.2.1 Propagation within a Fiber

When light enters one end of a glass fiber under the right conditions, most of the light will propagate, or move, down the length of the fiber and exit from the far end. Only a small part of the light will exit through the side walls of the fiber; most of the light will be contained within the fiber and will be "guided" to the far end. Such a fiber is referred to as a light pipe or light guide.

The light stays inside the fiber because it is totally reflected by the inside surface of the fiber. Light entering the end of the fiber at a slight angle to the axis follows a zigzag path through a series of reflections down the length of the fiber. *Total internal reflection* at the fiber wall can occur only if two conditions are met. The first is that the glass inside the fiber core must have a slightly higher index of refraction n_1 than the index of refraction n_2 of the material (cladding) surrounding the fiber. The second is that the light must have an angle of incidence ϕ (between the ray path and the normal to the fiber wall) which is greater than a critical angle ϕ_c, which is defined as the angle of incidence for which

$$n = \frac{c}{v}$$

$$\sin \phi_c = \frac{n_2}{n_1} \tag{20.1}$$

These conditions are illustrated in Fig. 20.1(a).

If a ray of light passes from a zone of low refractive index n_0 into one of higher refractive index n_1 (as it would if it entered the end face of a fiber) at an angle of incidence θ_0 to the normal to the surface which is smaller than the critical angle θ_c, it will enter or be refracted into the zone of higher refractive index at an emerging angle θ_1 to the normal which is less than the incidence angle θ_0 (see Fig. 20.1(b)). If the angle of incidence θ_0 is larger than the critical angle θ_c, then reflection will occur.

Snell's Law says that the incidence angle θ_0 is related to the exit angle θ_1 by the relationship

$$n_0 \sin \theta_0 = n_1 \sin \theta_1 \tag{20.2}$$

Figure 20.2 shows the conditions which exist at the launching end of a typical optic fiber. The core of the fiber has a refractive index n_1 and is surrounded by

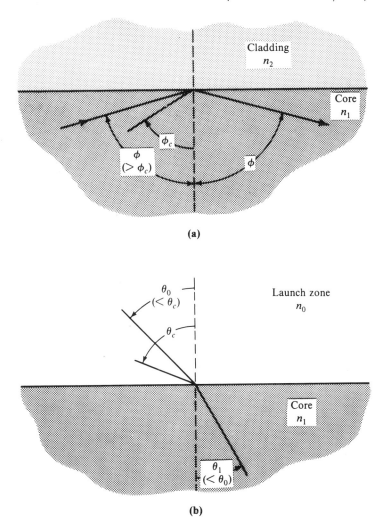

FIGURE 20.1. (a) Reflection at an interface; (b) refraction at an interface.

a cladding of material with a slightly lower refractive index n_2. Light is launched into the end of the fiber, usually from surrounding air with an index $n_0 = 1$. A typical ray of light enters the cable end at an external incident angle θ_0 to the fiber axis (which is usually made normal to the fiber end face). It enters the core at point A at the refracted angle θ_1 to the fiber axis and is then reflected from the core wall at point B at the internal incident angle ϕ. For the right triangle ABC formed by the ray path and the normals,

$$\theta_1 = 90° - \phi \tag{20.3}$$

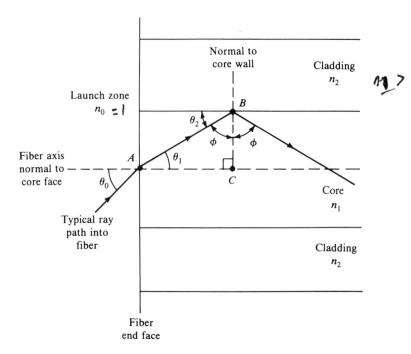

FIGURE 20.2. Path of a typical light ray being launched into an optic fiber.

so that

$$\sin \theta_1 = \sin (90° - \phi) = \cos \phi \qquad (20.4)$$

Substituting this into Eq. (20.2) and rearranging gives the external incident angle θ_0 in terms of the internal incident angle ϕ as

$$\sin \theta_0 = \frac{n_1}{n_0} \cos \phi \qquad (20.5)$$

As long as light enters the fiber at an angle such that the internal incident angle ϕ is not smaller than the critical angle ϕ_c as given as Eq. (20.1), the light will stay in the fiber and propagate to the far end. However, if the critical angle ϕ_c is exceeded, then the light which enters the fiber will immediately be refracted out through the core wall and be lost. As a result, a critical maximum value of the external incident angle θ_0 is defined by substituting for ϕ_c into Eq. (20.5) as follows.

Equation (20.1) is represented graphically in Fig. 20.3 by the right triangle *DEF*, through the definition for the sine of an angle. By Pythagoras'

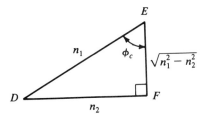

FIGURE 20.3. Pythagoras' theorem relating cos ϕ_c to sin ϕ_c.

theorem and the definition for the cosine, it follows that

$$\cos \phi_c = \frac{\sqrt{n_1^2 - n_2^2}}{n_1} \qquad (20.6)$$

Substituting Eq. (20.6) into Eq. (20.5) gives the maximum value of external incident angle for which light will propagate in the fiber:

Acceptance Angle $\theta_0(\text{max}) = \sin^{-1}\left(\dfrac{\sqrt{n_1^2 - n_2^2}}{n_0} \right) \qquad (20.7)$

This angle is also called *acceptance angle* or the *acceptance cone half-angle*. Rotating the acceptance angle about the fiber axis as shown in Fig. 20.4 describes the *acceptance cone* of the fiber. Any light aimed at the fiber end within this cone will be accepted and propagated to the far end. The larger the acceptance cone is made, the easier launching becomes.

The numerical aperture NA of the fiber, used as a figure of merit for optical fibers, is defined as

$$NA = \sin\theta_0(\text{max}) = \frac{\sqrt{n_1^2 - n_2^2}}{n_0} \qquad (20.8)$$

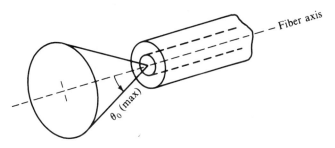

FIGURE 20.4. Acceptance cone obtained by rotating the acceptance angle about the fiber axis.

Usually, with optic fibers launching will take place from air, in which case $n_0 = 1$ and

$$NA \approx \sqrt{n_1^2 - n_2^2} \qquad (20.9)$$

A fractional difference Δ between the core and cladding indexes is defined as

$$\Delta = \frac{n_1 - n_2}{n_1} \qquad (20.10)$$

Substituting equation (20.10) into (20.8) and (20.9), and noting that $\Delta^2 \ll \Delta$ for all practical optic fibers, the numerical aperture becomes

$$NA = \frac{n_1\sqrt{2\Delta}}{n_0} \approx n_1\sqrt{2\Delta} \qquad (20.11)$$

It should be noted that the numerical aperture is effectively dependent only on the refractive indexes of the core and cladding materials and is not a function of the fiber dimensions.

Example 20.1   An optic fiber is made of glass with a refractive index of 1.55 and is clad with another glass with a refractive index of 1.51. Launching takes place from air. (a) What numerical aperture does the fiber have? (b) What is its acceptance angle?

Solution   (a) By equation (20.10), the fractional difference between the indexes is

$$\Delta = \frac{n_1 - n_2}{n_1} = \frac{1.55 - 1.51}{1.55} = 0.0258 \qquad n_1 > n_2$$

By equation (20.11), the numerical aperture is found to be

$$NA \cong n_1\sqrt{2\Delta} = 1.55\sqrt{2 \times 0.0258} = 0.352$$

(b) By equation (20.8), the acceptance angle is

$$\theta_0(\max) = \sin^{-1}NA = \sin^{-1}0.352 = 20.6°$$

20.2.2 Effect of Index Profile on Propagation

The analysis in the previous section is based on what is called a *step-index profile* fiber. This type of fiber is characterized by a core which has a completely uniform distribution of the index of refraction n_1 throughout its bulk, and an abrupt transition at the core boundary to a region of lower refractive index.

Three different situations arise in this surrounding region. The first is the case of the unclad core fiber, in which the medium surrounding the core is simply air, with an index of refraction n_0 of unity. This results in a large fractional difference of index, since the core glass typically has an index of about 1.5, and consequently, large acceptance angles. However, the fiber that typically results is weak mechanically, especially if the core diameter is small. Fibers of this type typically have core diameters in excess of 200 μm.

An *index profile* for a fiber is produced by plotting the index of refraction on the horizontal axis against the radius distance from the core axis on the vertical. The index profile for the unclad core is shown in Fig. 20.5(a).

The second case is that of the glass-clad core as shown in Fig. 20.5(b). In this case the core glass is surrounded by a layer of glass called *cladding* which has a uniform index of refraction n_2 only slightly lower than that of the core. In the simplest case this cladding is surrounded by air, but may have an opaque protective sheath surrounding it. This fiber makes it possible to obtain small core diameters without sacrificing mechanical strength, and low fractional index differences so that single-mode propagation (see Section 20.2.5) becomes possible.

A variation of the clad core gives the so-called *W-profile* fiber. In this case (see Fig. 20.5(c)) the cladding layer n_2 is made only thick enough so that it encloses all of the light energy field ends which extend for a short distance beyond the core interface, and the desired core guiding characteristics are obtained. This first cladding is then surrounded by a second thicker cladding layer with an index n_3 which falls between the core n_1 and the first cladding n_2. This produces the W-shaped profile shown, which, as discussed in Section 20.3.3, also strips out leaky modes from the first cladding. This is the core configuration used for single-mode fibers for long-haul communications. Core diameters as small as 3 μm are possible, while the clad fiber may still have a diameter as large as 300 μm. The second cladding may be surrounded by air or by a protective sheath.

The third type of step index fiber uses a plastic-clad core. The shape of its profile is the same as that in Fig. 20.5(b), but it tends to have higher losses than does its glass-clad counterpart. However, it is much less expensive to manufacture and finds many applications requiring short runs of fiber without repeaters, such as in plant instrumentation.

The profile of a *graded-index* fiber is also shown in Fig. 20.5(c). In this type of fiber, the material in the core is modified so that the index of refraction is at the high value of n_1 at the center and falls off with radial distance from the axis according to a carefully chosen profile. The cladding structure is the same as for a W-profile fiber. The graded-index fiber is used for multimode transmission in which is desirable to minimize the effect of time dispersion of pulses (see Section 20.4.1).

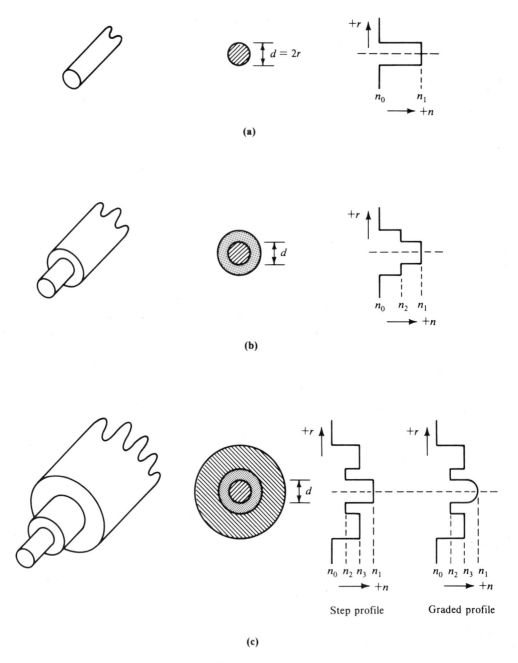

FIGURE 20.5. Core index profiles: (a) An unclad fiber; (b) a simple clad fiber with step index profile; (c) a double-clad fiber with either a step-index profile or a graded profile — the W-profile fiber.

Propagation of light within the core of a step-index fiber is characterized by light rays following a zigzag path of straight line segments. The material within the core of a step-index fiber has a uniformly distributed index of refraction, and light travelling through such a uniform material will continue in a straight line until it encounters a reflecting or refracting surface such as the core/cladding interface. Figure 20.6(a) shows how two rays m_1 and m_2, which are launched into a step-index fiber at less than the critical angle, are propagated down zigzag paths produced when the rays bounce off the core walls. Ray m_3, launched at an angle larger than critical, is refracted out through the cladding and is not propagated to the far end of the fiber.

Figure 20.6(b) illustrates how light propagates in a graded-index fiber. Because the index of refraction within the core is not uniformly distributed, but decreases with distance from the core axis, the light rays are bent by refraction so that they are curved inward toward the axis. Rays which are launched at different angles have different degrees of curvature. If rays are launched at too large an angle (greater than ~~critical~~), they will escape from the core (see m_3 in Fig. 20.6(b)).   *Acceptance.*

A fortunate result of the gradual refraction within the core is that the group velocities of rays of increasing angle can be made almost constant, instead of diverging, as they do for step-index fibers. This creates a periodic focusing effect within the core and greatly reduces the effect of pulse dispersion

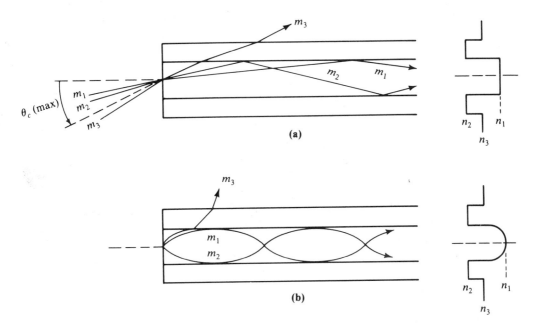

FIGURE 20.6. (a) Light ray propagation in a step index profile fiber; (b) propagation in a graded index profile fiber.

so that much larger data transmission rates may be used than on the step-index fibers.

20.2.3 Modes of Propagation

$T = 10^{12}$

Light propagates as an electromagnetic wave in much the same manner as do microwaves (see Section 13.2 and Appendix B), and when confined to a duct or guide, it propagates in the same manner, but at a much higher frequency. Microwaves occupy the frequency range from 3 GHz to about 100 GHz; that is, they have wavelengths of 10 cm down to about 3 mm. Visible light occupies a narrow range of wavelengths between 0.4 and 0.7 μm, corresponding to frequencies of 750 to 430 THz, six orders of magnitude higher. Most fiber optic communications presently use two bands, between .8 and .9 μm, and between 1.2 and 1.4 μm, where the fibers have low losses.

When a plane electromagnetic wave propagates in free space, it travels as a transverse electromagnetic, or TEM, wave. Under these conditions, the electric field and the magnetic field associated with the wave both lie at right angles to the direction of wave propagation and to each other, as shown by the vectors E_t and H_t in Fig. 20.7(a). The electric field vector E_t is shown in the vertical (y) direction, and the magnetic field vector H_t is shown in the horizontal (x) direction within the field frontal plane. The whole field moves in the z direction at the speed of light c. Under these conditions, neither the electric nor the magnetic fields have vector components in the z direction, i.e., $E_z = 0$ and $H_z = 0$.

When an electromagnetic wave is confined within a guide, it can propagate in one of several types of modes. The first of these is the transverse electric, or TE, mode, which was described for waveguides in Section 13.2. In the TE mode, the electric field E within the guide lies entirely in the transverse plane E_t at right angles to the direction of propagation, but the magnetic field H has a component H_z which lines up with the direction of propagation (see Fig. 20.7(b)), as well as the transverse component H_t. Under these conditions, $E_z = 0$ but H_z is finite. Propagation takes place in the z-direction at the group velocity v_g, while the wave front in the E–H plane moves off at an angle normal to itself in the z–x plane at the speed of light c.

In the transverse magnetic, or TM, mode, the H vector lies entirely in the transverse plane (H_t in Fig. 20.7(c)), while the E vector has both a transverse component E_t and a longitudinal component E_z, so that $H_z = 0$ and E_z is finite. Again, propagation takes place in the z direction at group velocity v_g, while the E–H wavefront travels at an angle normal to itself in the y–z plane at the speed of light c.

In the hybrid (EH or HE) modes, both the E and the H fields have components (E_z and H_z, respectively) lying along the z direction, so that both E_z and H_z are finite. If the transverse component E_t of the E vector is larger than the corresponding transverse component H_t of the H vector (that is, the

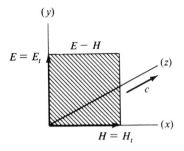

(a) TEM ($E_Z = 0$, $E_H = 0$)

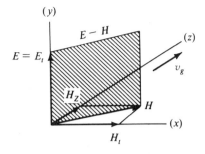

(b) TE ($E_Z = 0$, H_Z finite)

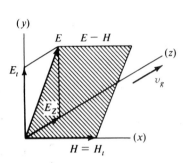

(c) TM (E_Z finite, $H_Z = 0$)

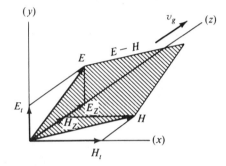

(d) HE (E_Z, H_Z finite, $H_t > E_t$)

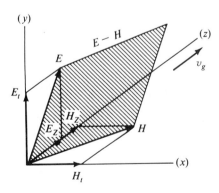

(e) EH (E_Z, H_Z finite, $E_t > H_t$)

FIGURE 20.7. Electromagnetic wave mode classifications: (a) TEM; (b) TE; (c) TM; (d) HE; (e) EH.

E-field contributes more to the transverse field), the mode is called an EH mode, as shown in Fig. 20.7(d). If, however, H_t is larger than E_t, the mode is called an HE mode, as shown in Fig. 20.7(e). In either case, the field moves in the *z*-direction at a group velocity v_g which is only the *z* component of the *E–H* field plane moving normal to itself at the speed of light *c*.

A ray of light is the path which a fixed point in the *E–H* plane would follow as the *E–H* plane moves normal to itself at the speed of light *c*. If that ray encounters a reflecting or refracting surface, it will change direction accordingly. Treating the ray behavior of light, or using *ray optics*, is a convenient way to describe what happens in a guide, since it gives a physical feel for what is happening while still being accurate.

If the rays comprising a mode all pass through the longitudinal, or *z*-axis, of the fiber core, they are called *meridional* rays. Propagation of meridional rays takes place only in the TE or TM modes, which are completely guided.

Rays which do not pass through the core axis, but follow a spiral path of reflecting segments down the fiber core are called *skew* rays. Skew rays propagate entirely in the hybrid EH or HE modes, and some of these modes contribute to losses through leakage and radiation.

20.2.4 Number of Modes a Fiber Will Support

Within each of the classes of modes discussed above, only certain modes will propagate, and each of these will have a particular internal reflection angle φ (and external incidence angle θ_0) associated with it. To illustrate how this can be, consider the propagation of a meridional wave of rays of monochromatic light of wavelength λ_0 propagating between the parallel walls of a glass fiber of diameter *d* as illustrated in Fig. 20.8. The fiber is assumed to have an internal

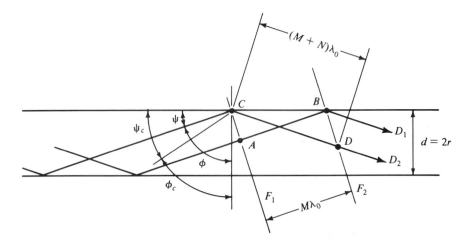

FIGURE 20.8. Two rays in a mode arrive at successive wavefronts in phase with each other.

cutoff angle ϕ_c below which the wave will not reflect, but will refract out through the core cladding.

Rays D_1 and D_2 are two rays which contribute to an equiphase wavefront F_1 which lies at right angles to the ray propagation path. That is, ray D_1 at point A has exactly the same phase as ray D_2 at point C. Ray D_1 continues in the same direction until it meets the core wall at point B, but ray D_2 is reflected at point C and follows the reflection path to point D, which lies on the plane through B which is parallel to front F_2. Both rays follow a zigzag path down the fiber, passing through its axis at each turn. If the guide is to support propagation of front F_1 in a mode, then all rays arriving at F_2 by either a direct path or by a reflected path must arrive with exactly the same phase. Since the distance by a direct path is shorter than one by a reflected path, the *difference* between the direct and reflected path lengths must be an integral number of wavelengths to give the proper phase relationship.

In Fig. 20.8 the direct path AB has a length $M\lambda_0$, where M may be a noninteger number of wavelengths. The reflected path CD has a length $(M + N)\lambda_0$, so that the difference becomes

$$\Delta = CD - AB = (M + N)\lambda_0 - M\lambda_0 = N\lambda_0 \qquad (20.12)$$

After some geometric and trigonometric manipulation based on the figure, it can be seen that the angle which a mode ray makes with the wall of the fiber is given by the expression

$$\sin \psi = \left(\frac{N}{2} \frac{\lambda_0}{d} \right) < 1 \qquad (20.13)$$

where ψ is the complement of the internal reflection angle ϕ. Since parallel wavefronts will form only for integer values of N, the angle ψ can assume only the corresponding values if propagation is to take place. Each value of N gives a corresponding distinct mode.

Equation (20.13) gives a number of discrete angles at which propagation should take place. An upper limit is placed on the value of N since the value of $\sin \psi$ must not exceed unity (corresponding to the point at which the wavefront bounces back and forth between the walls, but does not move along the fiber axis). In general, this means that for propagation, the wavelength λ_0 must be much smaller than the fiber core diameter d. However, this is a theoretical limit, since cutoff will occur before this because with increasing values of N, ψ increases until it exceeds the critical value ψ_c. The number of modes which will propagate will increase if the value of the critical angle ϕ_c is decreased (see Eq. (20.1)) by making the difference between core and cladding indexes larger.

The number of modes that a fiber will support may also be increased by making the ratio d/λ_0 larger. Since the wavelengths of light available are rather limited, this is practically done only by using a fiber with a larger core diameter.

TABLE 20.1.   Cutoff V-Numbers for the lowest order modes that a fiber will support.

Mode	Cutoff V-Number
HE_{11}	0
TE_{01}, TM_{01}	2.405
HE_{21}	2.42
HE_{12}, EH_{11}	3.83
HE_{31}	3.86
EH_{21}	5.14
HE_{41}	5.16
TE_{02}, TM_{02}	5.52
HE_{22}	5.53

The *normalized frequency of cutoff* (*cutoff parameter*, or *V number*) is useful for determining how many modes a given fiber will support. This parameter is found, after a lengthy derivation of the cutoff characteristics which will not be attempted here, to be

$$V = \pi \frac{d}{\lambda_0} \sqrt{n_1^2 - n_2^2} \cong \pi \frac{d}{\lambda_0} NA \qquad (20.14)$$

If the value of the waveguide parameter V found in Eq. (20.14) is considerably larger than unity, then the approximate number of modes which the fiber will support is given by

$$N(\text{modes}) \cong \tfrac{1}{2} V^2 \qquad (20.15)$$

Each different mode has a particular value of the normalized frequency V below which that mode is cut off. The first twelve of these, listed in ascending order, for a fiber guide are given in Table 20.1. There are, of course, many other modes of higher order, but only those which have cutoff frequencies which are less than the V-number for the fiber will propagate.

Example 20.2   The fiber in Example 20.1 has a core diameter of 50 μm and is used at a median light wavelength of 0.80 μm. Find its V-number and how many modes it will support.

Solution   The diameter–wavelength ratio is

$$\frac{d}{\lambda_0} = \frac{50 \ \mu m}{.8 \ \mu m} = 62.5$$

By Eq. (20.14), the V-number is

$$V = \pi \frac{d}{\lambda_0} NA = \pi \times 62.5 \times .343 = 67.3$$

$\pi \times 67.5 \times .352 = 69.1$

$68.69.$

(Recall that NA = 0.343 from Example (20.1).) Only modes with cutoff frequencies below this value will propagate. They number, by Eq. (20.15),

$$N(\text{modes}) = \tfrac{1}{2} V^2 = \tfrac{1}{2} \times 67.3^2 = 267 \text{ modes}$$

This is truly a multimode fiber. $\tfrac{1}{2}(69.1)^2 = 2388$

20.2.5   Single-Mode Propagation   *Mono mode*

Table 20.1 shows the twelve lowest-order modes that a step-index fiber (one with a constant index of refraction throughout the core) will propagate. Each of these modes has a distinct cutoff value given by its V-number. If the V-number of a particular fiber is below the cutoff value for a given mode, that mode will not propagate.

The lowest-order mode which the cylindrical fiber will support is the HE_{11} mode, which has a cutoff V-number of zero. The next higher modes (TE_{01}, TM_{01}) cut off at 2.405, so that if a fiber is made to have a V-number of slightly less than 2.405, all of the modes except the HE_{11} mode will be cut off, and the fiber becomes a single-mode fiber.

Examining Eq. (20.14), the V-number may be reduced by choosing cores with a smaller diameter, or by choosing fibers with a smaller numerical aperture (defined by a smaller fractional difference in indexes). Practically, fibers with cores as small as about 5 μm can be made, and it is found that single-mode propagation has lowest losses for light with a wavelength of 1.3 μm. This gives a practical minimum value for the d/λ_0 ratio of about 4, or a maximum value of numerical aperture at cutoff of about 0.2. The corresponding fractional difference of index between core and cladding is then about 0.01 (with a typical core index of 1.50) as a maximum. Differences of .0001 have been obtained.

Practical problems with the difficulty of lining up connectors and coupling light sources into the very small cores of single-mode fibers have limited their application to date, but these problems are being overcome, and single-mode fibers are finding increasing use in long-distance applications such as submarine cables.

Example 20.3   A single-mode fiber is made with a core diameter of 10 μm and is coupled to a light source with a wavelength of 1.3 μm. Its core glass has a refractive index of 1.55. Determine the maximum value of the fractional difference of index and of the cladding index required for producing single-mode propagation, and find the acceptance angle of the fiber.

Solution
$$d/\lambda_0 = 10/1.3 = 7.69$$

By Eq. (20.14),

$$\text{NA}(\text{cutoff}) = \frac{V(\text{cutoff})}{\pi(d/\lambda_0)} = \frac{2.405}{\pi \times 7.69} = 0.0996$$

By Eq. (20.11),

$$\Delta = \tfrac{1}{2}(NA/n_1)^2 = \tfrac{1}{2}(0.0996/1.55)^2 = 0.00206$$

By Eq. (20.10),

$$n_2 = n_1(1 - \Delta) = 1.55(1 - 0.00206) = 1.547$$

By Eqs. (20.7, 20.8, 20.9),

$$\theta_0(\text{max}) = \sin^{-1}(NA) = \sin^{-1}(0.0996) = 5.716°$$

Obviously, coupling a light transmitter into such a narrow cone requires some special method, such as a carefully positioned lens.

20.3 LOSSES IN FIBERS

20.3.1 Rayleigh Scattering Losses

The glass in optical fibers is an amorphous (noncrystalline) solid which is formed by allowing the glass to cool from its molten state at high temperature until it freezes. While it is still plastic, the glass is drawn out under tension into its long fiber form. During this forming process, submicroscopic variations in the density of the glass and in doping impurities are frozen into the glass and then become reflecting and refracting facets to scatter a small portion of the light passing through the glass. While careful manufacturing techniques can reduce these anomalies to a minimum, they cannot be totally eliminated.

It is also found that the losses induced because of this scattering effect vary inversely with the fourth power of the wavelength ($\propto \lambda_0^{-4}$), so that their effects are reduced to less than about 0.3 dB/km at a wavelength of 1.3 μm. Figure 20.9 shows a plot of the intrinsic minimum Rayleigh scattering losses in silica glass fibers, plotted against wavelength over the usable portion of the spectrum from .7 to 1.6 μm.

20.3.2 Absorption Losses

Three separate mechanisms contribute to absorption losses in glass fibers. These are ultraviolet absorption, infrared absorption, and ion resonance absorption.

Ultraviolet absorption takes place because, for pure fused silica, valence electrons can be ionized into conduction electrons by light with a center wavelength of about 0.14 μm, corresponding to an energy level of about 8.9 eV. This ionization amounts to a loss of energy in the light fields and contributes to transmission loss. The absorption does not only take place at this fixed wavelength, but occurs over a broad band which extends up into the visible portion of the spectrum, with decreased losses at the higher wavelengths. This UV absorption "tail" tails off to give losses of negligible amounts in the 1.2–1.3-μm band.

Doping of the glass with purposely introduced impurities such as germanium dioxide to modify the refractive index causes some increase in the UV absorption tail because of an upward shift in the wavelength of the UV absorption peak. The loss at 1.2 μm for most fibers remains less than 0.1 dB/km however. The UV absorption tails for pure silica glass and for a typical germanium-doped silica are shown in Fig. 20.10.

Infrared absorption takes place because photons of light energy are absorbed by the atoms within the glass molecules and converted to the random mechanical vibrations typical of heating. This IR absorption also exhibits a main spectral peak which, for silica, occurs at 8 μm, with minor peaks at 3.2, 3.8, and 4.4 μm. Again, the peaks are broad, tailing off into the visible portion of the spectrum, to losses of typically less than 0.5 dB/km at 1.5 μm. The typical infrared tail for silica is shown in Fig. 20.10.

Minute quantities of water molecules trapped in the glass contribute OH$^-$ ions to the material, which exhibit absorption peaks at .95, 1.25, and 1.39 μm within the visible spectrum, with the main peak occurring at 1.39 μm. Typically, the water content of the glass has to be kept to below 0.01 parts per million if these peaks are not to spread out and merge to predominate the loss spectrum of the resulting fiber. Figure 20.10 shows the three peaks with typical values for a low-loss fiber.

The presence of some other metals or impurities may also create unacceptable losses within the usable portion of the spectrum. Some of those to be

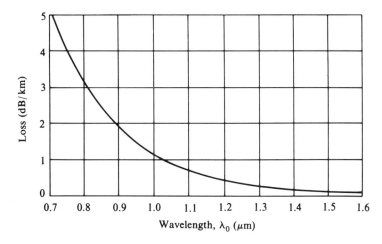

FIGURE 20.9. Rayleigh scattering losses in silica fibers.

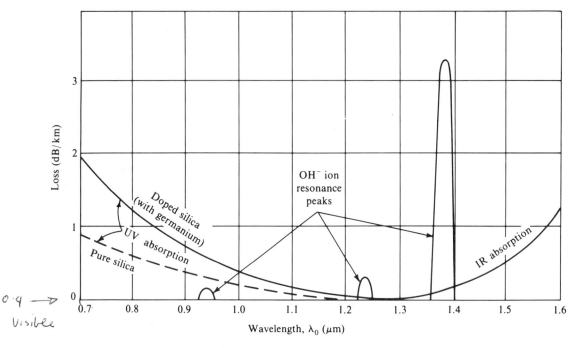

FIGURE 20.10. Absorption loss effects in fused silica glass fibers.

avoided are iron, copper, and chromium. Good refining techniques for purifying the raw materials for the silica are a must if these impurities are to be reduced to acceptable levels. Zone refining of the type used in preparing silicon for integrated circuits is typically used.

20.3.3 Leaky Modes

If at each reflection of a skew mode the angle of incidence is less than critical in *both* the axial and radial directions, the mode will propagate. However, at some of the higher-order modes, the radial component of the incident angle will exceed the critical angle, and part of the radial component of the skew ray will exit from the core by refraction, leaving the remainder to be partially reflected. Successive partial reflections will cause the leaky skew ray to rapidly decay in intensity as it propagates, so that at some distance along the cable only the completely guided modes will remain. These latter are useful for transmission of information.

If the cladding is surrounded by a material with a lower index of refraction, the partial refractions from the skew modes may be trapped within the cladding and propagate along its length to the receiver. In practice, it has been found desirable to remove or "strip" these leaky modes from the core and cladding as rapidly as possible in order to reduce signal dispersion (see Section 20.4). This is accomplished by surrounding the thin cladding layer on the fiber

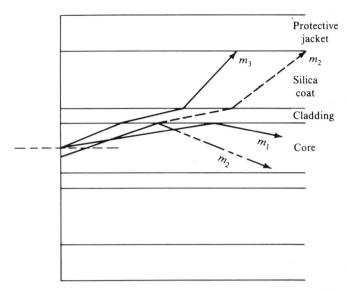

FIGURE 20.11. Leaky mode removal by an additional silica coat over the fiber cladding.

core by a third layer of pure silica which has an index of refraction which is higher than that of the cladding, but lower than that of the core. This silica coating not only adds mechanical strength to the fiber, but also acts to remove the partial refraction rays from the leaky modes and the passed rays from cutoff modes by total refraction, as illustrated in Fig. 20.11. In the figure, m_1 represents the path of a ray in one of the completely guided meridional modes which is completely reflected by the core-cladding interface, and m_2 represents a skew ray which is only partially reflected. Some of the ray energy is refracted through the cladding and then through the silica coat to be absorbed in an opaque protective jacket. m_3 represents a cutoff meridional mode which is completely refracted out of the core and dissipated in the jacket.

The leaky modes introduced at the transmitting end of the fiber usually contain only a few percent of the total guided power, and these are rapidly attenuated near the transmitter. This becomes a fixed loss for the fiber which, if the fiber is complete and unspliced, will occur only once. However, when splices are introduced into the cable, contained modes that propagate in the first section may not be accurately coupled into the next section, with the result that power is transferred to the leaky modes at the splice. This power is lost just after the splice by the stripping effect. Thus, each splice will add to the loss.

20.3.4 Mode-Coupling Losses

Power which has been launched successfully into a propagating mode may be coupled into a leaky or radiating mode at some point farther down the fiber. This coupling effect can take place for several reasons.

Small imperfections in the core glass or in the core-cladding interface, such as small variations in core diameter or cross-sectional shape or bubbles in the glass, can cause energy to be coupled into one of the leaky modes. Losses from this source will be uniformly distributed along the fiber.

Another source of mode coupling is imperfectly formed splices or imperfectly aligned connectors, as mentioned in the previous section. Typically, this loss is held to about 0.5 dB per coupling, or about 0.2 dB per splice. These are discrete losses and can be minimized by reducing the number of splices or connectors required in a given fiber route.

20.3.5 Bending Losses

Two types of bending, microbending and constant-radius bending, contribute to losses in fibers. They arise for different reasons and cause losses by two different mechanisms.

Microbending is a microscopic bending of the core of the fiber that results from slightly different thermal contraction rates between the core material and the cladding material. Microbending can also be introduced if the fibers are subsequently wound into a multifiber cable or wound on spools for transportation.

Losses because of microbending occur because the small bends act as scattering facets which cause mode coupling to occur. Energy from guided modes is cross-coupled into leaky modes and lost through the cladding. Since microbends are randomly distributed over the length of the fiber, losses resulting from them will be uniformly distributed, and a total figure for a fiber can be obtained. Care in manufacturing and handling will minimize these losses.

When optical fibers are installed in cable ducts or on poles for a transmission line, it is necessary to introduce bends in the fiber to negotiate corners. These bends may in some cases be quite sharp, as for instance at a pole hanger point where the cable has some sag between poles, or at right-angle turns in building conduits. These large-radius bends will also introduce some loss of light in the fiber.

In this case, however, the loss occurs because modes which are fully guided in straight sections of fiber are either only partially guided or not guided at all over the curved portion of fiber. Consider a ray of light shown in Fig. 20.12 which is fully guided when it reaches the reflection point A in the straight section of fiber. It is reflected from point A at angle ϕ_A, and, if the fiber were not bent would arrive at point B_1 as the next reflection point. But with the bend in the fiber, the ray encounters the fiber wall early at the outside edge of the bend at point B_2, at an angle of incidence ϕ_B which is smaller than the original angle ϕ_A in the straight section. If ϕ_B is smaller than the critical angle, that ray will escape from the fiber, and some of the energy from the corresponding mode will be lost. If a bare fiber is bent sharply, this lost light can be seen as a bright line along the outer edge of the fiber bend. The sharper the bend is made, the more modes will occur which escape at the bend, so it is

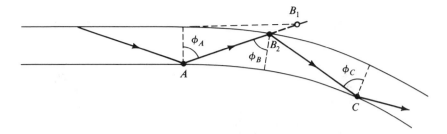

FIGURE 20.12. Ray propagation in a bent fiber.

necessary when installing fibers to ensure that no sharp bends are going to occur. The manufacturer will usually specify a minimum bend radius and give a minimum loss figure corresponding to that.

20.3.6   Combined Losses of Fibers

Four types of losses are inherent to the fiber and must be reduced to a minimum during the fiber manufacturing process. These are the Rayleigh scattering losses, material absorption losses, leaky mode losses, and losses due to mode coupling because of scattering. Of these, the first two, Rayleigh scattering and material absorption losses, are predominant, and when fibers are made, every effort is made to limit losses of these kinds. Figure 20.13 shows the losses in a typical multimode fiber as a function of wavelength, with the

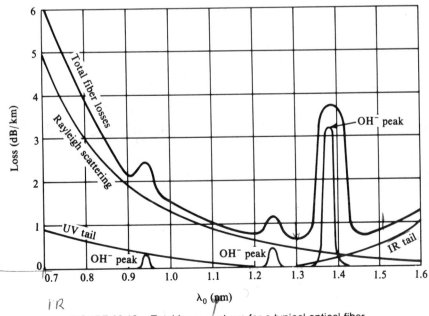

FIGURE 20.13. Total loss spectrum for a typical optical fiber.

Rayleigh and absorption components shown for comparison. This fiber has relatively low losses compared to some fibers, which at 0.8 μm may have losses of as much as 20 dB/km.

All fibers are characterized by a loss spectrum curve of this general shape, although the actual loss values and peaking wavelengths will vary depending on the type of fiber. Figure 20.14 shows actual data for three commercial fibers plotted on the same coordinates as Fig. 20.13. These fibers are all intended for multimode transmission in the 0.8-μm spectrum gap, with information specified for the range 0.5–1.1 μm instead of 0.7–1.6 μm.

As the figure shows, all of the curves have a "window" in the spectrum of losses at about 0.8 μm which coincides with the spectral output from many of the LED light sources presently available for fiber transmission. Within the window, losses are at their lowest, typically between 5 to 10 dB/km. The graded index fiber (a) has the lowest minimum loss of about 5 dB/km at .85 μm, while the plastic-clad fiber (b) has about 8 dB/km minimum loss and the step-index fiber (c) 10 dB/km minimum loss, both at 0.8 μm.

Other spectrum windows occur at 1.2 and at 1.3 μm in the typical spectra for step-index glass clad fibers, and since this is within the gap between the ultraviolet and infrared tails, the lowest losses for any fibers can be obtained in these windows. Typical losses on the order of .5 to 2 dB/km can be obtained

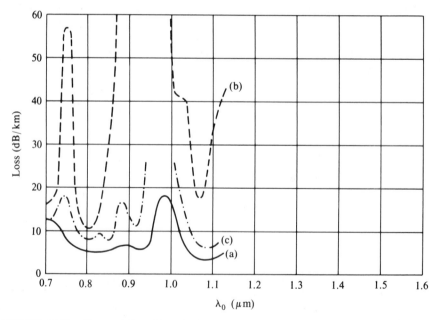

FIGURE 20.14. Comparison of loss spectra for different fibers: (a) graded index multi-mode fiber; (b) plastic clad step index multimode fiber; (c) glass-clad step-index multimode fiber.

in carefully designed fibers. These are the ones which will be used for long-distance monomode transmission systems with widely spaced repeaters.

Those losses already discussed are characteristic of the fiber as it comes from the manufacturer. Other losses, such as mode coupling because of macrobending and microbending and insertion losses because of splices and couplings can be held to a minimum by careful design and installation procedures.

20.4 DISPERSION

20.4.1 The Effect of Dispersion on Pulse Transmission

Theoretically, a pulse of light with a given width and amplitude transmitted into one end of a fiber should arrive at the far end of that fiber with its shape and length unchanged and only its amplitude reduced by losses. If the losses get too large, the pulse amplitude at the receiver will be too small to be detected, and a repeater will have to be included to boost the signal level entering the next section. The length of the pulse (the inverse of pulse repetition rate or bit rate) may be shortened arbitrarily without affecting its detectability if the pulse length and shape remain unchanged during transmission.

Pulses of light transmitted into a fiber encounter several dispersion effects which act to spread the pulse out in the time domain, changing its shape so that it may merge into the previous and succeeding pulses. The pulses can be separated by spacing them out at the transmitter, but this means reducing the maximum bit rate. At some high bit rate, each type of fiber will move from having its maximum length limited by losses to having it limited by dispersion effects.

Three kinds of dispersion, due to three separate mechanisms, exist in a fiber: intermodal dispersion, material or chromatic dispersion, and waveguide dispersion.

20.4.2 Intermodal Dispersion

Each mode that a waveguide will support has a different effective group velocity, even though the phase velocity in each different mode ray may be identical. This arises because the total path followed by a ray in each of the modes is zigzag in nature and has a different total length than each of the other mode rays. Thus, a pulse transmitted into a fiber will propagate over several different paths (corresponding to the excited modes) and emerge at the far end at slightly different times. The received pulse is the summation of these mode pulses, each delayed by a different amount of time.

The shortest total delay will be for a planar wave propagating straight down the fiber core; the longest will be for a wave propagating in the

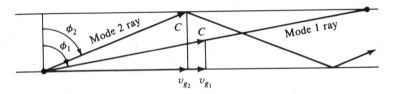

FIGURE 20.15. Group velocities for two modes.

highest-order supported mode with an incident angle just slightly larger than critical. The difference Δ_t in propagation delay time can be readily derived for the case of a step-index fiber as follows.

Let the total fiber length be z. Since it is convenient to state losses and delays in terms of a standard unit length, let z be one kilometer. Referring to Fig. 20.15, which shows two meridional rays of different modes following zigzag paths down the fiber, the total zigzag path length is readily found to be

$$z_t = \frac{z}{\sin \phi} \tag{20.16}$$

where ϕ is the angle of incidence of the mode in question with the normal to fiber wall. In the lowest-order mode, $\phi = \phi(\text{max}) = 90°$, and in the highest-order supported mode, $\phi = \phi(\text{min}) = \phi_c = \sin^{-1}(n_2/n_1)$ by equation (20.1). Now, at $\phi(\text{max})$, the shortest mode path is found to be

$$z_t(\text{min}) = \frac{z}{\sin \phi(\text{max})} = \frac{z}{\sin 90°} = z \tag{20.17}$$

and, at $\phi(\text{min})$, the longest mode path is found to be

$$z_t(\text{max}) = \frac{z}{\sin \phi(\text{min})} = \frac{z}{\sin \phi_c} = z\frac{n_1}{n_2} \tag{20.18}$$

The maximum difference Δ_z in path length because of modal separation is found to be

$$\Delta_z = z_t(\text{max}) - z_t(\text{min}) = z\left(\frac{n_1}{n_2} - 1\right) \tag{20.19}$$

Substituting for the ratio n_1/n_2 from Eq. (20.10) and rearranging gives Δ_z in terms of the fractional index difference Δ:

$$\Delta_z = z\frac{\Delta}{1 - \Delta} \tag{20.20}$$

Since the light rays in the fiber are passing through a dielectric medium with a dielectric constant ε_r which is greater than unity, the rays travel more slowly than they would in free space. The relative permeability μ_r in the dielectric is still about unity. From equation (B.11), the phase velocity in the dielectric along the ray paths is

$$v_p = \frac{1}{\sqrt{\mu\varepsilon}} = \frac{1}{\sqrt{\mu_0\varepsilon_0}\sqrt{\mu_r\varepsilon_r}} \tag{20.21}$$

Substituting for the speed of light c in a vacuum from Eq. 12.5 and for $\mu_r = 1$ in the dielectric gives the phase velocity in the dielectric as

$$v_p(\text{glass}) = \frac{c}{\sqrt{\varepsilon_r}} \tag{20.22}$$

which is the same as Eq. (12.6) for transmission lines.

It can be shown that the dielectric constant is related to the refractive index as

$$\varepsilon = n^2 \tag{20.23}$$

so that in the glass core of refractive index n_1, by substitution in (20.22), it follows that

$$v_p(\text{glass}) = \frac{c}{n_1} \tag{20.24}$$

Finally, the maximum difference in time delay between the lowest- and highest-order modes is found by dividing the difference in path length by the phase velocity, to give

$$\Delta_t = \frac{\Delta_z}{v_p(\text{glass})} = \frac{n_1 z}{c}\frac{\Delta}{1-\Delta} \tag{20.25}$$

$$\Delta = \frac{n_2 - n_1}{n_1}$$

which is usually given in units of nanoseconds per kilometer of length, or ns/km. It should be noted that this type of disperison is characteristic of the fiber and is not affected by the light wavelength. Further, it does not occur in single-mode fibers.

Example 20.4   For the step-index fiber of Examples 20.1 and 20.2, determine the dispersion per kilometer of length and the total dispersion in a 12.5-km length of fiber.

Solution   From Example 20.1, $n_1 = 1.55$ and $\Delta = 0.0258$

(a) for a 1-km length fiber, by Eq. (20.25), the maximum dispersion is found to be

$$\Delta_t = \frac{n_1 z}{c}\frac{\Delta}{1-\Delta} = \frac{1.55 \times 1000 \times 0.0258}{3 \times 10^8 \times (1-0.0258)} = 136.9 \times 10^{-9}$$

$$= 136.9 \text{ ns/km} \quad \text{for A 1km Length .}$$

(b) for a length of 12.5 km, the total dispersion is

$$\Delta_t = \Delta_t(\text{per km}) \times z = .1369 \times 12.5 = 1.71 \ \mu s$$

Multimode graded-index fibers have much lower intermodal dispersion factors than do corresponding step-index fibers, for carefully optimized core index profiles. This comes about primarily because, as a ray passes from the center of the core out to its outer edge on the zigzag path, it passes through a zone of decreasing refractive index (and dielectric constant), with the result that its phase velocity (Eq. 20.22) increases with distance from the core, so that the group velocity at the outer edge is also faster. It then decreases as the ray is refracted back toward the core axis, and then increases outward on the other side. This gives an alternately rising and falling group velocity which has an average higher than that in the lower-order modes which propagate straight down the fiber. The result is a tendency to average out the group velocities of the different modes and produce a much lower total dispersion figure for the graded-index fibers, which have a theoretical minimum value given by

$$\Delta_t = \frac{n_1 z \Delta^2}{8c} \tag{20.26}$$

where n_1 here is the refractive index of the glass at the core center, and Δ is the fractional index difference from core center to cladding as given by Eq. (20.10). While this is a theoretical minimum, practical graded-index fibers have been made with intermodal dispersion factors of much less than 1 ns/km.

Example 20.5   Assume that the fiber in Example 20.4 has been made with an optimally graded core index profile, and determine its maximum dispersion.

Solution   From Example 20.4, $n_1 = 1.55$ at the core center, and $\Delta = 0.0258$. By Eq. (20.26), the dispersion is

$$\Delta_t = \frac{n_1 z \Delta^2}{8c} = \frac{1.55 \times 1000 \times 0.0258^2}{8 \times 300 \times 10^{+6}} = 0.43 \ \text{ns/km}$$

which is very much less than the 137 ns/km for the step-index equivalent. For 12.5 km of fiber, $\Delta t = 0.43 \times 12.5 = 5.38$ ns.

20.4.3 Material Dispersion — *CHromATic Dispersion.*

The index of refraction of the core glass is not the same for lights of different wavelengths, but varies across the spectrum. As a result, if the pulse of light transmitted contains components at several different wavelengths

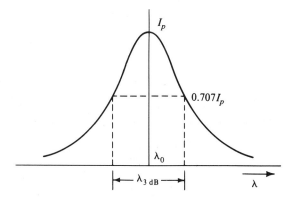

FIGURE 20.16. Defining spectral width of a light source.

centered about a center wavelength λ_0, as shown in Fig. 20.16, the pulse components containing the shorter wavelengths will experience more delay than will those of the longer wavelengths. The result will be an effective time dispersion of the pulse at the received end of the fiber.

The material dispersion has been shown to be proportional to the second derivative of the index of refraction with respect to wavelength, that is, $d^2n/d\lambda^2$, and the resulting dispersion becomes

$$\Delta_t = \left(-\frac{z}{c}\lambda_0\frac{d^2n}{d\lambda^2}\right)\lambda_{3\,dB} \qquad (20.27)$$

where λ_0 is the center wavelength of the light spectrum, and $\lambda_{3\,dB}$ is the spectrum width (analogous to 3-dB bandwidth). This is usually stated in the form

$$D_m = -\frac{z}{c}\lambda_0\frac{d^2n}{d\lambda^2} = \frac{\Delta t}{\lambda_{3\,dB}} \qquad (20.28)$$

in ps/nm · km for a 1-km length of fiber. Typical values for pure and doped silica fibers are shown in Fig. 20.17.

It should be noted that this curve of material dispersion crosses zero near a wavelength of 1.3 μm. As a result, if a fiber is used in single mode at a wavelength near 1.3 μm, the dispersion will nearly cancel out, providing a monochromatic light source (of very narrow spectral width) like a cavity laser is used. Typically, the magnitude of material dispersion in multimode fibers will be much less than that of intermodal dispersion, and will become predominant only in single-mode fibers.

NB! ZERO DISPERSION POINT

MATERIAL DISP << intermodal DISP.

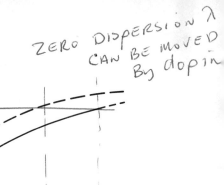

ZERO DISPERSION λ CAN BE MOVED By dopin

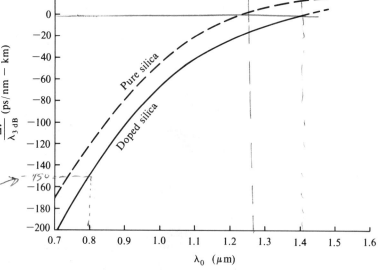

FIGURE 20.17. Material dispersion coefficient as a function of wavelength for silica fibers.

Example 20.6   The fiber used in Examples 20.1, 20.2, and 20.4 is to be used with a 0.8-μm light source with a spectral width of 1.5 nm. What value of material dispersion might be expected?

Solution   From the graph in Fig. 20.17, at $\lambda_0 = 0.8$ μm, the dispersive coefficient $D_m = -0.15$ ns/nm–km for a doped fiber. From the previous examples, $z = 12.5$ km. By equation (20.27), for 1 km of fiber, the dispersion is

$$\Delta_t = D_m z \lambda_{3dB} = -.15 \text{ ns/nm–km} \times 1 \text{ km} \times 1.5 \text{ nm}$$
$$= .225 \text{ ns/km delay.}$$

For a 12.5 km length,

$$\Delta_t = 0.225 \text{ ns/km} \times 12.5 \text{ km} = 2.81 \text{ ns}$$

This figure compares to an intermodal dispersion in a step-index fiber of 1710 ns (Example 20.4) and to 5.38 ns for an optimal graded-index fiber (Example 20.5).

20.4.4   Waveguide Dispersion

If a fiber can be operated so that intermodal dispersion and material dispersion both disappear (as would be the case for single-mode operation near $\lambda_0 = 1.3$ μm), then a third dispersion mechanism will predominate, preventing a totally dispersionless situation from obtaining, except for the case of ideally monochromatic light. Unfortunately, pure monochromatic sources cannot be built, so the light transmitted will contain components at several wavelengths near λ_0 within an envelope like that of Fig. 20.16. The result is that the constant-phase wavefront separation illustrated in Fig. 20.8 will be slightly different for the different light wavelengths, and the corresponding angles of incidence will also be slightly different. This will cause a corresponding shift in the group velocity, which varies in a complicated way with the wavelength, and results in a broadening of the received pulse. This effect is only the result of the guiding properties of the fiber, and hence its name.

It has been shown that for an ideal single-mode step-index fiber, the dispersion coefficient which results from the waveguide dispersion mechanism has a peak value of D_w of about 6.6 ps/nm–km, which places an ideal lower limit on the dispersion at the wavelength where material dispersion disappears. Fortunately, a partial cancellation does take place between material dispersion and waveguide dispersion which has the effect of moving the zero crossover point in Fig. 20.17 slightly higher in wavelength, and by tuning the light source and keeping its bandwidth low, it is possible to obtain practical values of less than the 6.6 ps/nm–km theoretical minimum.

Example 20.7   A 12.5-km single-mode fiber is used with a 1.3-μm light source which has a spectrum width of 6 nm. Find the total expected waveguide dispersion.

Solution
$$D_w = 6.6 \text{ ps/nm–km}$$
$$\Delta_t = D_w z \lambda_{3\,dB} = 6.6 \text{ ps/nm–km} \times 12.5 \text{ km} \times 6 \text{ nm}$$
$$= 495 \text{ ps(or 0.495 ns)}$$

20.4.5 Total Dispersion and Maximum Transmission Rates

Three dispersion mechanisms have been discussed, each of which contributes to pulse broadening during fiber transmission. Each contributes to a greater or lesser degree depending on operating conditions and fiber type. Thus, for multimode step-index fibers, the intermodal dispersion will predominate, while a single-mode fiber operated at the material dispersion null wavelength (about 1.3 nm) will be dominated by waveguide dispersion.

At any wavelength the total dispersion is the root-mean-square combination of these three effects, or

$$\Delta_t(\text{tot}) = \sqrt{\Delta_t^2(\text{imd}) + \Delta_t^2(\text{md}) + \Delta_t^2(\text{wgd})} \qquad (20.29)$$

Of course, each of the values of the dispersions entering Eq. (20.29) is a theoretical value, and because of departures from the ideal in practical fibers,

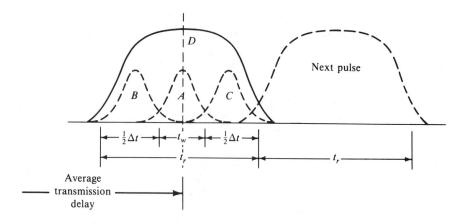

FIGURE 20.18. Illustrating how dispersion causes pulse overlap.

the actual values will be somewhat higher, so that the total value will also be higher than that in Eq. (20.29). However, the result is satisfactory for finding an estimate of the fiber's behavior.

To see how dispersion affects maximum bit rate or information rate, examine Fig. 20.18, which shows how pulses arriving with different delays at the receiver add to form a dispersed or broadened pulse. Pulse A represents a pulse which experiences average delay, pulse B is one which experiences minimum delay, and pulse C is one which experiences maximum delay. The overall received pulse D is the summation of all the pulse components between the limits of B and C.

The pulse width t_w represents the transmitted pulse width, while t_r represents the spread received pulse width, which is larger by the total dispersion time, or

$$t_r = t_w + \Delta_t(\text{tot}) \tag{20.30}$$

If the next pulse is to be detected without ambiguity, then it must not be transmitted until a period t_r after the beginning of the present pulse, to allow time for the spreading to subside. The bit rate corresponding to this ideal situation will be

$$\text{Bit Rate} \rightarrow B = \frac{1}{t_r} = \frac{1}{t_w + \Delta_t} \tag{20.31}$$

Because dispersion factors tend to be larger than ideal in practice, and because the pulse tails tend to spread beyond the average pulse limits, it is found necessary to allow a safety factor of about 5 times on the received pulse width t_r to obtain minimum desired error rates in transmission. Further, if maximum bit rates are to be obtained, the transmitted pulse can be considerably smaller than the total dispersion encountered. Thus, if $t_w \ll \Delta_t$, Eq. (20.31) becomes

$$\text{MAX Bit Rate} \quad B(\text{max}) \cong \frac{1}{5\Delta_t} \tag{20.32}$$

Going in the other direction, if the pulse length is made longer than the dispersion, then the dispersive effect may be ignored if $t_w > 4\Delta_t$, or $B < 1/4\Delta_t$, in Eq. (20.31).

Example 20.8   A single-mode fiber operating at 1.3 μm is found to have a total material dispersion of 2.81 ns and a total waveguide dispersion of .495 ns. Determine the received pulse width and approximate maximum bit rate for the fiber if the transmitted pulse has a width of 0.5 ns. Intermodal dispersion does not occur in a single-mode fiber.

Solution   The total dispersion is found by Eq. (20.29):

$$\Delta_t(\text{tot}) = \sqrt{\Delta_t^2(\text{imd}) + \Delta_t^2(\text{md}) + \Delta_t^2(\text{wgd})}$$

$$= \sqrt{0^2 + 2.81^2 + .495^2} = 2.85 \text{ ns}$$

The received pulse width is approximated by Eq. (20.30):

$$t_r = t_w + \Delta_t = 0.5 + 2.85 = 3.35 \text{ ns} = 0.00335 \text{ μs}$$

And the maximum bit rate is found by Eq. (20.32):

$$B(\text{max}) \approx \frac{1}{5t_r} = \frac{1}{5 \times 0.00335} = 59.6 \text{ Mbits/s}$$

20.5   LIGHT SOURCES FOR FIBER OPTICS

20.5.1   Introduction

Light sources for fiber optics act as light transmitters and, consequently, must meet certain requirements if they are to be acceptable for the purpose.

First, their light must be as nearly monochromatic (single frequency) as possible. Most light sources are not single frequency, but emit light at several frequencies over a band or portion of the spectrum which may be quite broad. A few sources such as gas ionization lamps, light emitting diodes, and lasers emit light over a much narrower portion of the spectrum. But even these are not truly monochromatic and do emit at several frequencies over a narrow band. The emission spectra of some typical light transmitters are compared in Fig. 20.19.

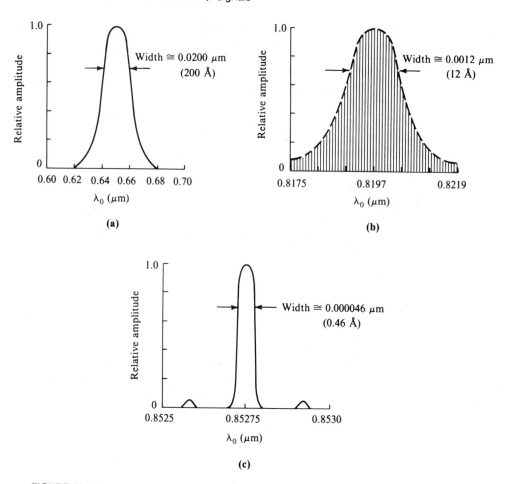

FIGURE 20.19. Light Emission spectra for typical light transmitters: (a) an LED operating at 0.65 nm; (b) a broad-spectrum laser diode at 0.8197 μm; (c) a narrow-spectrum laser diode at .85275 μm.

High Signal Power

Next, they should have a high-intensity light output so that sufficient energy is transmitted to overcome the losses encountered during transmission down the fiber.

The light sources should also be capable of being easily modulated. While most of the devices presently available are capable of being modulated in an analog (amplitude-modulation) manner, the usual mode is binary on–off modulation using PCM because of better noise immunity.

Finally, the devices must be small, compact, and easily coupled to the fibers so that excessive coupling losses do not occur. Also, they must be inexpensive to manufacture.

The light-emitting diode (LED) and the semiconductor laser have both found extensive use in this application. Both of these devices emit light at fixed frequencies because, in their operation, excited electrons which are free to conduct recombine with holes and give off a photon of light each in the process. This photon of light is the result of converting the potential energy of the electron when it becomes trapped by the atom containing the hole.

Each photon contains an amount of energy which is related to the corresponding electromagnetic frequency by the expression

$$E = hf \qquad (20.33)$$

where E is the energy in Joules, h is Planck's constant ($= 6.625 \times 10^{-34}$ Joule-seconds), and f is the frequency in Hertz. The light spectrum is usually stated in terms of wavelength instead of frequency, but the two are related by

$$f = \frac{c}{\lambda_0} \qquad (20.34)$$

where c is the velocity of light in meters/second and λ_0 is the wavelength in meters. Also, the energy is usually stated in terms of electron-volts, so that

$$E = qeV \qquad (20.35)$$

where q is the magnitude of the electronic charge ($= 1.602 \times 10^{-19}$ C/electron) and eV is the energy level in electron volts. Combining Eq. (20.33), (20.34), and (20.35) gives the relationship between wavelength and electron-volt energy level:

$$eV = \frac{hc}{q} \frac{1}{\lambda_0} = \frac{1.24}{\lambda_0} \qquad (20.36)$$

where λ_0 is stated in micrometers (μm).

The energy content of a photon released in a semiconductor is related to the energy band gap of the semiconductor material, which is the amount of energy needed to excite a valence electron in the material to the level where it can leave the parent atom and contribute to a flow of electricity. The width of the energy band gap in a device can be precisely tailored during manufacture by careful choice of the materials used and the doping levels induced.

Example 20.9   Semiconductor diodes are made using materials which produce energy band gaps of 1.9, 1.46, and 0.954 eV. Find the corresponding frequencies and wavelengths of the emitted light.

Solution

By Eq. (20.36), the wavelength is

$$\lambda_0 = \frac{1.24}{eV} = \frac{1.24}{1.9} = 0.653 \ \mu m$$

By Eq. (20.34), the frequency is

$$f = \frac{c}{\lambda_0} = \frac{300 \ (Mm/s)}{0.653 \ \mu m} = 459 \ THz = 459 \times 10^{12}$$

By similar reasoning, for eV = 1.46, $\lambda_0 = 0.85 \ \mu m$, $f = 353$ THz; and for eV = 0.954, $\lambda_0 = 1.3 \ \mu m$, $f = 231$ THz.

20.5.2 Light-emitting diodes

The light-emitting diode (LED) works by the process of spontaneous emission. It is a semiconductor junction diode which emits light when current is passed through it in the forward-biased condition. One side of the diode is p-type semiconductor material containing a very large number of holes (covalent bonds in the crystal structure which have been broken by the removal of one of the pair of electrons forming the bond). The other side of the diode is an n-type semiconductor containing a large number of free conduction electrons. At zero bias, a depletion zone separates the two regions within which all holes and electrons have either recombined or been removed. A barrier potential exists across the depletion zone because the recombined holes and electrons have trapped charge at the impurity atom sites within the depletion zone, as illustrated in Fig. 20.20(a). When enough forward bias voltage is applied to the junction to overcome the junction barrier potential, the depletion zone disappears, and holes are free to move across the junction into the n-region while electrons are free to move into the p-region, where they are minority carriers, as illustrated in Fig. 20.20(b). The minority carriers injected have only a very short lifetime before they meet up with carriers of the opposite type and recombine. When an electron and a hole meet and recombine, they emit one photon of light which has an energy level which corresponds to the amount of energy which was originally needed to free the electron for conduction. The amount or intensity of light emitted depends on the number of minority carriers available for recombination, which in turn depends directly on the forward conduction current in the diode. The frequency of the light emitted is determined by the energy band gap of the materials used to make the junction.

The light-emitting diode is formed by diffusing a very thin layer of p-type donors into the surface of an n-type substrate chip. Light is emitted within the junction zone near the surface of the chip and escapes in random directions through its surface, as shown in Fig. 20.21. The chip is usually arranged so that

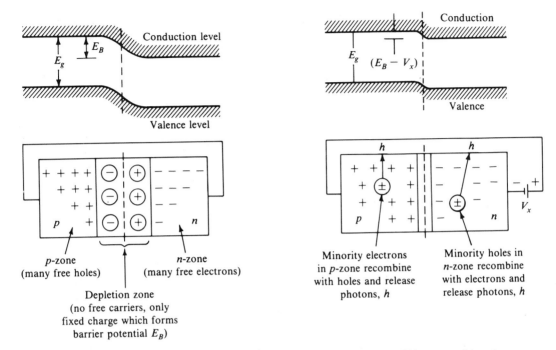

FIGURE 20.20. (a) Energy band diagram and carrier distribution within a zero-biased semiconductor diode; (b) energy band diagram and carrier distribution within a forward-biased semiconductor diode which carrier current.

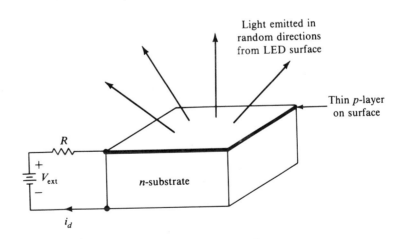

FIGURE 20.21. A forward-biased light-emitting diode (LED) chip.

it is at the bottom of a well or behind a lens, so that the light may be coupled into the transmitting end of a fiber. The light is not concentrated, does not travel in well-defined directions, and has a very broad spectrum, so that the LED is generally useful only for short lengths of large-core-diameter multimode fibers which operate at very low bit rates.

The spectrum of the typical LED is quite broad. LEDs for use with fibers are typically made using a solid solution of gallium arsenide (GaAs) as the semiconductor base, with various doping elements such as phosporus (P), indium (In), and aluminum (Al) used to form the p- and n-regions. Gallium arsenide–phosphide (GaAsP) diodes can be made with band gaps in the range 1.5–2.0 eV (0.62 to 0.83 μm wavelength). Gallium–indium arsenide (GaInAs), indium arsenide–phosphide (InAsP), or aluminum–indium arsenide (AlInAs) diodes cover the range from 0.5 to 1.5 eV, (0.83 to 2.5 μm). Some of these and some others may also be used as semiconductor lasers. The spectrum of a typical LED is shown in Fig. 20.19(a), which is centered on a wavelength of 0.65 μm and has a spectral width of about 0.02 μm (or about 200 Å), typical of the results obtained with GaAsP diodes. The actual diode frequency is chosen for a system so that the system emits in one of the windows in the loss spectrum of the fiber (e.g., that from 0.8–0.9 μm in Fig. 20.14(b)).

20.5.3 Semiconductor Laser Diodes

The term *laser* is an acronym for *l*ight *a*mplification by *s*timulated *e*mission of *r*adiation. Laser action has been obtained using many different types of material, including gases (such as the high-power CO_2 laser used for welding and cutting applications), liquids, and solids (such as the familiar ruby laser used for creating solid images, or holograms). The type of laser used for fiber optic communications is the semiconductor laser.

The semiconductor laser is a special type of solid laser in which the laser action takes place within a semiconductor diode junction of the same type used for LEDs. When current is passed through a diode junction as described in the previous section, light is emitted by spontaneous emission at a frequency or wavelength determined by the energy band gap of the semiconductor material. When a certain critical current level in such a diode is exceeded, the population of minority carriers on either side of the junction and the density of photons reaches such a level that photons begin to collide with already excited minority carriers. The collision causes a slight increase in the ionization energy level which, in effect, makes the carrier unstable, so that it recombines with a carrier of the opposite type at the slightly higher level earlier than it would have if the collision had not taken place. When it does, *two* photons are released, each with the same level of energy.

If the density of excited carriers which may be "stimulated" in this manner is sufficiently high (how high is determined by the diode current level), each photon originally released by spontaneous emission may trigger several other photons by stimulated emission before leaving the junction area. This creates an avalanche effect so that the efficiency of emission rises exponentially with current above the emission threshold value.

The laser action of the semiconductor diode can be enhanced by placing a reflecting surface at each end of the junction zone as shown in Fig. 20.22. One or both of these surfaces can be partially reflecting so that a part of the incident light will transmit through the surface. The two reflecting surfaces are made parallel to each other so that light generated within the junction will bounce back and forth between the two end surfaces several times before escaping, thus increasing the chances of each photon stimulating several more photons into emission. The channel between the reflecting surfaces can be compared to a resonant cavity in a microwave oscillator. It has the further effect that most of the light is emitted in a flat parallel beam from the end of the junction chip, making it easy to couple into a fiber.

While the light beam in a laser such as that in Fig. 20.22 is contained within the junction zone by the optical opaqueness of the p and n layers, the junction may be extremely wide (as much as 2000 μm, the width W of the chip). A variation of the device called a *stripe laser* overcomes this problem by limiting the lasering portion of the junction to a narrow stripe across the chip aligned with the direction of emission, as shown in Fig. 20.23(a). The confining action results from having current introduced into the junction zone only under the narrow deposited stripe contact. As a result, the current density within the junction exceeds the critical value needed to stimulate emission only directly under the contact, so that light emission is also confined to this narrow stripe.

Figure 20.23(b) shows the cross-section of a typical *double heterostructure* stripe laser, so named because of the multiple alloy layers used. In this device, current passes through the copper heat sink through a solder layer and metal

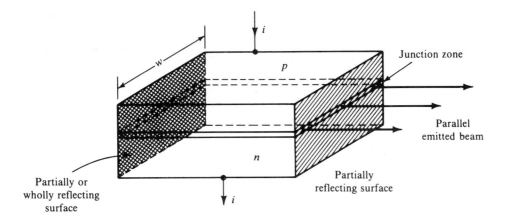

FIGURE 20.22. A broad contact laser diode.

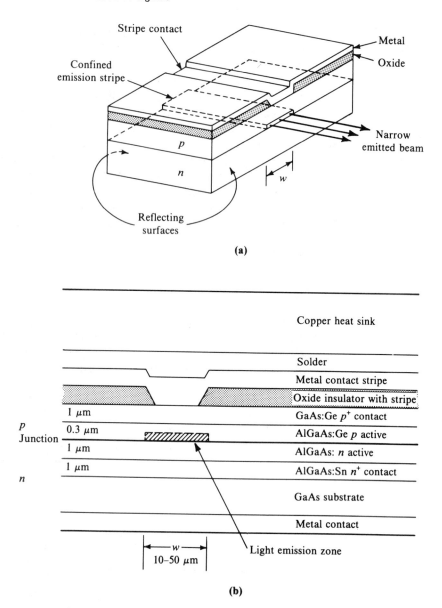

FIGURE 20.23. (a) A simple stripe contact laser diode structure; (b) cross-section of a double heterostructure narrow-stripe laser diode showing the various layers.

contact layer through a narrow stripe gap in an insulating oxide layer into the active layers. The current passes through the first *p*-doped gallium arsenide layer into the second thin active *p*-layer without diffusing to either side, and across the junction to the *n*-layers and on to the metal contact on the other side. The gallium arsenide thick layer acts as a substrate to hold the thin layers before they are soldered to the header.

The active portion of the junction is confined to the thin (0.3 μm) *p*-active layer and between the sides of the contact stripe, which may be as narrow as 3 μm.

Because of the "tuning" action of the reflecting cavity, the semiconductor lasers tend to have very narrow spectral widths. For a laser without a cavity, the laser action will typically produce a spectrum like that shown in Fig. 20.19(b), which is quite broad and contains many "spectral lines"; for a laser with a cavity, the laser action will produce a spectrum like that in Fig. 20.19(c), which may only contain one spectral line with a line width as small as 0.00005 μm (0.5 Å, or 0.05 nm). The latter is about as close to monochromatic light as is possible with present technology.

20.6 PHOTO DETECTORS

20.6.1 Introduction

Several types of photosensitive devices have been tried as detectors for use with fiber optics. These include silicon photodiodes, phototransistors, and photoresistors. Most of them, however, do not have the fast response and extreme sensitivity needed to be useful for communications purposes. The ones that do, include the *p–i–n* diode and the avalanche diode.

20.6.2 The *p–n* Photodiode

An ordinary *p–n* diode may be used as a photodetector, but it tends to have extremely low sensitivity, even though its speed is satisfactory. Figure 20.24(a) shows the structure of a *p–n* photodiode. It has a thin *p*-layer deposited on an *n*-substrate, and light enters through the *p*-layer. This diode has a relatively thin depletion zone around the junction when it is reverse-biased. Photons of light entering the depletion zone ionize hole–electron pairs when they encounter atoms within the crystal structure within the depletion zone. The hole–electron pairs are drawn across the junction by the electric field (see Fig. 20.24(a)) established by reverse bias and contribute to leakage current, so that the leakage current is proportional in magnitude to the incident light intensity.

Unfortunately, in this type of diode, many of the photons entering the thin depletion zone near the junction pass through and on into the *n*-region, where the hole–electron pairs generated are not affected by the junction field and do not contribute to leakage or photo current. Thus, the responsivity or conversion efficiency of the diode is quite low. However since the hole–electron pairs will recombine on moving through the thin depletion zone, their lifetime is quite short, and the diode will respond very rapidly to changes in light intensity.

20.6.3 The *p–i–n* Photodiode

Some improvement in the sensitivity of a diode can be made by including a lightly doped (nearly pure or intrinsic) *n*-layer between the junction and the

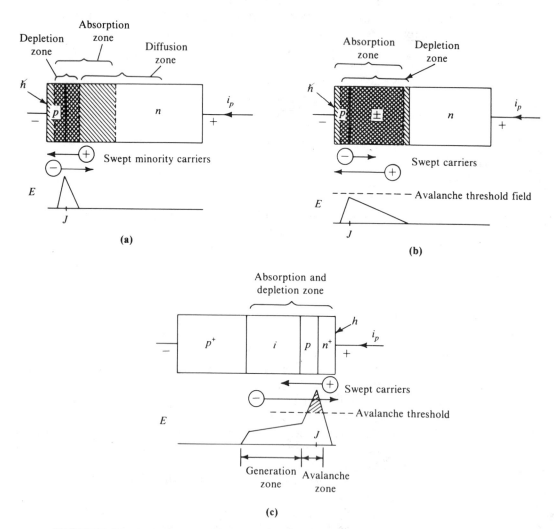

FIGURE 20.24. Photodiode structures with field distribution patterns: (a) pn photodiode without multiplication; (b) pin photodiode without multiplication; (c) pipn avalanche photodiode with multiplication.

more heavily doped *n*-contact region, forming the so-called pin diode. This intrinsic layer, shown in Fig. 20.24(b), is made thick enough so that most of the photons which pass through the junction are absorbed within the layer. The holes created within this region are swept across the junction by the electric field, which now extends deep into the *i*-region, and the electrons are swept out into the *n*-region, thus adding to the total carriers contributing to the photocurrent. This more complete absorption of photons results in a larger photocurrent generation, causing a considerable improvement in the sensitivity of the diode. But since the carriers must travel a greater distance to be removed from the depletion zone, the response time of the pin diode is somewhat slower than that of the *p–n* diode. However, in most applications, this slower speed is not of consequence since the transit time is generally less than a nanosecond, which is shorter than the pulse width of most binary pulse streams used.

20.6.4 The Avalanche Photodiode

If the negative voltage applied to reverse-bias the *p–i–n* diode in Fig. 20.24(b) is increased, a threshold will be reached in which the internal field intensity near the junction zone will become high enough so that the electrons being accelerated through the zone will cause secondary electron–hole pairs to be generated by collisions within the high-field area. The number of carriers generated in this manner multiplies exponentially with the field intensity, causing an "avalanche effect." Quite high "gains" can be obtained in the diode, with the final photocurrent being much greater than it would have been without the avalanche effect.

The *p–i–n* structure, however, is not ideal for avalanche mode operation because, as shown in Fig. 20.24(b), the electric field is distributed throughout a relatively thick depletion zone, which means that the external voltages required to obtain the avalanche threshold value of field intensity are too high for practical operation. To overcome this problem, the *p–i–p–n* structure shown in Fig. 20.24(c) is used. In this structure light enters the diode through a thin *n*-layer which is heavily doped and which forms one side of an abrupt junction with a thin *p*-layer between it and the intrinsic layer. This abrupt junction drops most of the terminal voltage and generates the high field needed for avalanche in its junction region.

As before, the intrinsic layer serves to distribute part of the field deeper into the *n*-region so as to capture and retrieve the deeper photo carriers, sweeping the electrons into the avalanche area. The p^+-i junction acts to sweep out the holes from the *i*-layer, since it is also reverse-biased and drops a portion of the external voltage. Careful adjustment of the doping level in the p^+ layer determines how much of the total voltage it drops, resulting in the field profile shown.

These diodes have the highest sensitivity of any of the presently available photosensors, but do have two drawbacks. First, the carriers have a relatively long transit time before being swept out of the wide depletion zone, so that

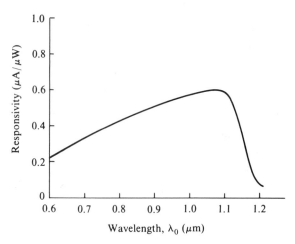

FIGURE 20.25. Spectral responsivity of a typical silicon pin diode photodetector.

their speed of operation is comparable to that of the $p-i-n$ diode. Further, because the avalanche multiplication factor tends to fluctuate randomly, quite a bit of additional "noise" is added to the resulting output signal. Thus, the amount of avalanche gain must be balanced against the amount of additional noise added to the system. In many cases, however, the avalanche diode does give an advantage.

A typical avalanche diode will require a bias voltage of somewhere between 100 and 400 V to produce an avalanche multiplication factor in the region of 100 to 300. This bias voltage must be carefully controlled so that fluctuations in the avalanche multiplication factor are minimized. Bias currents are typically only a few microamperes. Sensitivities of avalanche diodes are such that photo powers received in the order of -70 dBm (at a 1-Mbit/s rate) are usable. (This corresponds to a power level of 0.1 nW.)´

Silicon $p-i-n$ and avalanche diodes typically have a frequency response which extends from about 0.6 μm to about 1.2 μm. Other types of photo diodes are being developed which cover the low-loss windows between wavelengths of 1.0 and 1.4 μm. Figure 20.25 shows the spectral responsitivity (a measure of conversion efficiency, in units of $\mu A/\mu W$) for a typical silicon $p-i-n$ diode. The figure shows a rapid decrease of response for wavelengths above 1.1 μm.

20.6.5 An Optical Receiver Circuit

Figure 20.26 shows the block diagram of a practical optical receiver circuit which makes use of an avalanche photo diode. The light is on–off amplitude-modulated, so that when light is off an extremely small amount of dark or leakage current flows, but when it is on a relatively high value of light current flows (which typically may only be a few microamperes). A transresis-

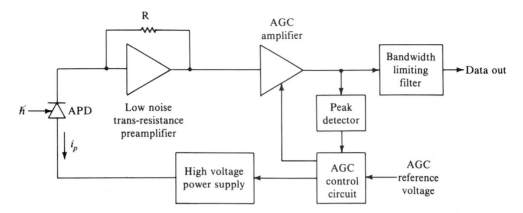

FIGURE 20.26. A practical optical receiver circuit block diagram. (S. D. Personick, M. K. Barnoski, ed., *Fundamentals of Optical Fiber Communications* 2nd ed. (Academic Press, 1981) p. 309. Used by permission.

tance low-noise preamplifier amplifies the photocurrent pulses and converts them to voltage pulses at a usable voltage level, and then an AGC-controlled amplifier acts to maintain an approximately constant peak amplitude of output pulses. The signal is then fed through a bandwidth-limiting amplifier to reduce noise output and is boosted by another amplifier to the final data level.

A signal sample is extracted at the output of the AGC amplifier and passed to a peak amplitude detector circuit to provide the feedback input to the AGC control summing point. The input is compared to a reference level of voltage which establishes the desired output amplitude and puts out control signals which set the gain in the AGC amplifier and which also set the amount of high-voltage bias applied to the avalanche photo diode (APD) to control its multiplication factor. Such a receiver may have a maximum sensitivity of about the −60 dBm optical power level and will accommodate a signal range of up to 37 dB on top of that by way of the AGC control. Its maximum transmission speed is about 10 MHz.

20.7 CONNECTORS AND SPLICES

20.7.1 Factors Affecting Loss in Connectors and Splices

Even in permanently installed fiber transmission lines it is impossible to avoid having at least two connectors or splices, since it is always necessary to terminate a fiber in a transmitter at one end and in a receiver at the other end. The usual process is to provide permanently attached "pigtails" to the transmitter and receiver devices during manufacture which introduce minimum loss, but these are terminated in dismountable connectors so that the transmitter and receiver may easily be taken out for servicing. Further, if the fiber run is at all long, chances are that it will have to be spliced several times along its route. Also, during the lifetime of a fiber link it may encounter several accidental breaks which will have to be repaired. Usually, these are fixed by means of permanent splices rather than connectors.

Several factors affect the amount of loss a connector will add to a cable run. Some of these factors also apply to splices.

Core size mismatch. Poor control of core diameter during fiber manufacture may result in trying to match two cores with different sizes in a connector. If the incoming core is smaller than the outgoing one, no problem is encountered because all the light from one fiber core reaches the other. However, if the outgoing fiber is smaller, then only part of the light from the incoming core will reach it, and part will be lost, adding to the total loss of the fiber. This is illustrated in Fig. 20.27(a), which shows a cone of light escaping

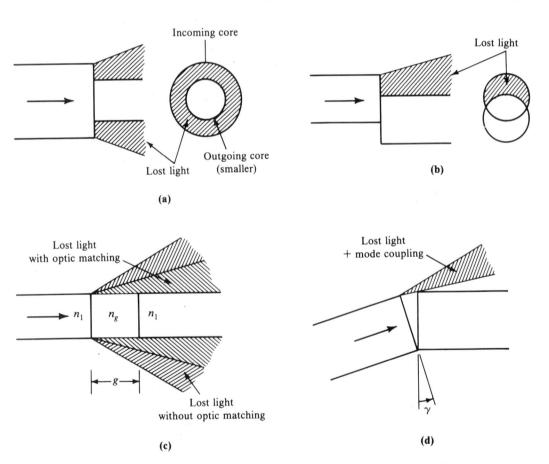

FIGURE 20.27. Connector loss factors: (a) core size mismatch; (b) lateral core misalignment; (c) longitudinal gap and optical mismatch; (d) angular misalignment.

around the smaller outgoing core. The problem may also occur if an attempt is made to couple two different types of fibers.

Lateral core misalignment. If the cores are exactly the same size, but do not line up exactly on each other's axes (lateral displacement), then light will escape from the exposed portion of the incoming core face as shown in Fig. 20.27(b). Such a misalignment may occur because the connector used does not line up the two outside diameters exactly, because the outside cladding diameters are not exactly the same, or because the core of either fiber is not centered in the cladding itself (the cores are not concentric). In any case, the result is more loss. Some types of connectors are arranged so that lateral displacement may be corrected after the connector has been installed, using adjusting screws to center the fibers. This type of alignment problem becomes more serious, of course, as the size of the core is reduced and is extremely critical in the small-core single-mode fibers, where a lateral displacement of 10 μm could mean complete loss of coupling.

Longitudinal gap separation. When light leaves the end of a fiber, it diverges in a cone which is determined by the acceptance angle of the fiber, as discussed in Section 20.2.1. If the mating fiber is not butted right up against the incoming fiber, light will be lost because of this divergence. The amount of light lost increases as the length of the gap g increases (see Fig. 20.27(c)).

Optical gap losses. If the gap between the fibers contains air, light propagating through the gap must pass through two partially reflecting interfaces because of the change of refractive index in going from core n_1 to air n_g and then back to core n_1. This causes what are called Fresnel losses, which are given by

$$\alpha = \left(\frac{n_g - n_1}{n_g + n_1} \right)^2 \tag{20.37}$$

Optical gap, or Fresnel, losses may be almost entirely eliminated by placing an optical matching cement in the gap which has the same refractive index as the core glass. As can be seen in Eq. (20.37), if $n_g = n_1$, the loss α becomes zero. This optical matching has a further advantage in that it decreases the critical acceptance angle until it is the same as the internal reflection maximum angle ϕ_c because there is no longer any refraction as the light passes out of the core. The side losses due to gap separation are also reduced, as indicated in Fig. 20.27(c).

Angular misalignment. As illustrated in Fig. 20.27(d), if the two cores are misaligned so that they meet at a small angle γ, some light will escape through the open gap at one side of the joint. Moreover, light leaving one fiber may be coupled into modes involving loss in the second fiber because of this misalignment and be lost further along the fiber due to leakage. Carefully

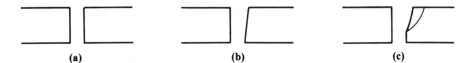

FIGURE 20.28. Fiber cleavage for joining: (a) a perfectly cleaved joint; (b) joint with one fiber cleaved at an angle to the axis; (c) joint with one fiber unevenly cleaved.

designed and installed connectors will minimize the angular misalignment losses.

Improper fiber end preparation. Figure 20.28(a) shows a properly prepared pair of fibers matched up for joining. Both have been cleanly cleaved at right angles to the core axis and have smooth planar mating surfaces. Figure 20.28(b) shows a fiber which was cleaved at an angle to the axis such that when light from the matching fiber strikes the sloping surface, some will be lost by reflection and more by coupling into leaky modes. Figure 20.28(c) shows a fiber which did not cleave evenly due to improper application of the cleaving stress. The end of the fiber has a scalloped irregular surface which scatters the light.

Special jigs have been designed so that when a fiber is clamped into one of them, it automatically scores the glass surface and applies just the right amount of tension in the right direction to achieve good cleavage. Such machines have been used successfully in field splicing in spite of very poor environmental conditions.

Dirt. Any dirt or foreign substance which gets into a connector or splice during or after assembly may increase its losses, or even completely block it. Extreme care has to be taken during installation to prevent any inclusion of dirt. Even so, successful splicing in very dirty circumstances has been accomplished using specially designed equipment and due care.

20.7.2 Connectors

The fiber connectors used at the points where fibers terminate in transmitters and receivers, either at the line ends or at intermediate repeaters, must have dismountable connectors. These connectors must both automatically ensure a minimum loss connection and be easy to disconnect and connect.

In order to ensure low loss, the connector must eliminate the effects of angular and lateral alignment and also ensure that the two fiber ends butt up against each other. Many different schemes have been used to build such connectors, some more successful than others. Two of the more common methods are shown in Fig. 20.29(a) and (b). The first is the V-groove system, where lateral and angular alignment is obtained by holding the fiber ends into the bottom of a V-groove cut in a block. Obviously, the fibers must have the same dimensions, or lateral core misalignment will occur. The fibers may be held in place either by a spring-loading system or by sandwiching them between two V-blocks as shown in Fig. 20.29(a). The V-blocks are made as parts of mating ferrules which press the fiber ends together when they are slid together.

The second alignment method makes use of a precision-bored hole in a jewel bearing similar to the type used in watch making. Figure 20.29(b) shows a self-aligning jewel connector in which one fiber is permanently clamped part-way through the bore of a jewel mount in one ferrule, while the second is held in a flexible rubber mount in a second ferrule. When the two connectors are pushed together, the flexibly mounted fiber is guided into the conical bore of the jewel to butt against the other fiber, while the rubber mount is pushed back and deformed to clamp the fiber in place. In another version of this type of connector, both fibers are held in place by jewels mounted in both ferrules, so that the fiber cleavage faces are flush with the jewel surfaces. When the connector is mated, the holes in the two jewels line up to hold the fibers in the correct position to mate.

Both the V-groove connector and the butt-faced jewel connector are provided with means for adjusting the positions of the fiber mounts in the field.

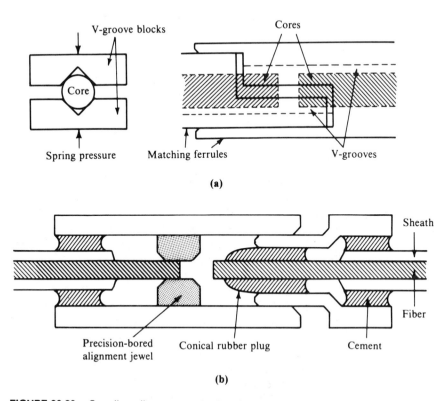

FIGURE 20.29. Coupling alignment methods: (a) V-groove alignment; (b) Precision-bored jewel alignment.

Connection is made, and received signal strength is monitored while adjustment of the connector is made to provide maximum received signal.

Such connectors are rarely attached to the fibers in the field because their assembly is too precise and demanding. Fiber runs are provided with the terminal connectors already installed, so that only in-line splicing need be done in the field.

Losses in connectors of this type, where the fiber gap is not filled with an optical matching material, typically run between 1 and 2.5 dB per connector, with the self-aligning connectors giving figures near the 1-dB level.

Many types of connectors are presently being used; however, the telecommunications industry is planning to accept one or two types as standards in the near future, so that equipment of different manufacturers may easily be interchanged.

20.7.3 Fiber Splices

Fiber-splicing schemes must meet two criteria: splices must be easy to make under the worst of environmental conditions, and they must provide an absolute minimum of introduced loss. Two schemes which meet these requirements and are presently being used are discussed.

The collapsed glass sleeve splice (shown in Fig. 20.30(a)) is convenient because it is easy to assemble in the field without need for sophisticated equipment. The assembly starts by preparing the fibers. The protective sheaths of both fibers are removed a few centimeters from the ends to be spliced. Both fiber ends are then cleaved using a special cleaving tool which may be hand-operated to make the fiber ends smooth and right-angled. Next, a tube of soft glass which melts at a lower temperature than the fiber does, and which has a bore slightly larger than the fiber diameter, is slid over one fiber end and softened with an oxy-hydrogen microtorch until the sleeve collapses around the fiber. Then, after cooling, the other fiber is pressed into the open end of the sleeve with some optical matching cement. When the cement has set, the entire splice is encased in a protective plastic heat-shrink tube and replaced in the cable casing. Obviously, care must be taken to ensure that no dirt enters the splicing ferrule, which would impair the quality of the splice, and only simple hand tools are required for the job. Splice insertion losses of about 0.5 dB per splice are typical of this method.

Currently, fusion splices give the best results from the point of view of losses, but they are more difficult to make in the field. Figure 20.30(b) shows the setup for fusing two prepared fiber ends together. The fibers are stripped and cleaved as described above, or by using a stripping and cleaving tool incorporated in a special splicing machine. The fiber ends are then placed in clamps on a jig which may be aligned using vernier screws. The fiber ends are butted and aligned while being observed through a binocular microscope (not shown), a process which requires considerable operator skill, and then are heated with either a microtorch or an electric arc until the two ends fuse together. This is also critical since too much heat will cause the molten glass in

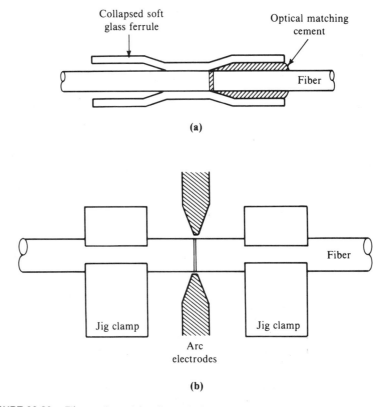

FIGURE 20.30. Fiber splices: (a) collapsed glass sleeve splice; (b) making a fusion splice.

the splice to flow and deform the fiber. Again, operator skill is required. Finally, a protective plastic sleeve is placed on the fiber and it is returned to the cable casing.

In spite of the difficulty involved in making fusion splices, the effort is worth it, and successful field splicing has yielded splices with insertion losses of less than 0.1 dB per splice.

20.8 FIBER OPTIC COMMUNICATION SYSTEMS

At low information rates, the length of a fiber optic transmission line is limited almost entirely by the losses introduced along its length. If the transmission bit rate is increased, then at some high bit rate the length of line will become limited by the amount of pulse dispersion which takes place along the line. If the line is to be longer than that allowed by loss or dispersion limiting, then repeaters must be included in the line at intervals to amplify, reshape, retime, and retransmit pulses into the next line section.

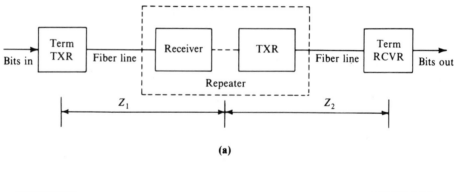

(a)

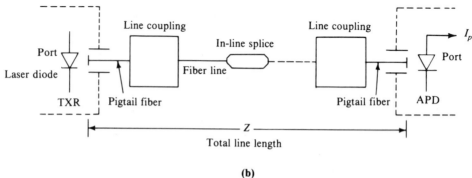

(b)

FIGURE 20.31. (a) A fiber optic line with two links and one repeater; (b) elements in a typical fiber link.

Figure 20.31(a) shows the major components of a two-link fiber optic line. At the sending end, an optical transmitter such as an LED or a laser diode couples light into the fiber, turning it on and off according to the bit stream to be sent. At the repeater, an attenuated and dispersed version of the transmitted light pulses is detected by an avalanche photodiode or a $p–i–n$ diode and is amplified, reshaped, retimed, and sent out on the second link. The receiver at the second terminal converts the light pulses to electrical pulses for distribution at the receiving end, again, after reshaping and amplifying them.

In a loss-limited fiber link, the total losses introduced by the fiber and all of the connectors and splices in between must not exceed the difference between the transmitted optical power P_t and the lowest allowable received optical power P_r (both expressed in dB) for a given type of detector and transmission error rate. Figure 20.31(b) shows a single fiber link. A total light flux or power P_t is emitted from the optical source device (a laser diode is shown here). Because of inefficiency in coupling the diode to the fiber end, which protrudes into an optical port around the diode, only part of the light from the source actually gets into the fiber, and a port loss L_{pt} which may be as high as 6 or 8 dB is introduced. A short piece of fiber (1 or 2 m) of negligible loss is permanently attached to the optical port, with a quick-disconnect connector permanently attached to its end. This mates with another connector permanently fastened to the end of the main fiber line. The connection so made may introduce a typical connector loss L_c of about 1 dB into the line.

Light passing down the fiber encounters losses of L_f (dB/km) to give total fiber losses of $L_f z$ (dB). It will also encounter N_s splices which each contribute a loss L_s, for a total splice loss of $L_s N_s$ (dB).

At the receiving end (which may be a repeater) another pigtail, line connector, and port feed the light from the fiber to the photo detector diode, introducing a second connector loss L_c and a second port loss L_{pr}. A 5- or 10-dB loss margin M is usually included to account for any unsuspected losses which may occur, such as aging of the transmitter diode or receiver diode, bending losses in the fiber, or extra splices due to accidental breaks.

This summation of losses is stated mathematically as

$$P_t \, (\text{dB}) - P_r \, (\text{dB}) = M \, (\text{dB}) + L_{pt} + L_{pr} + N_c L_c + N_s L_s + z L_f$$

$$(20.38)$$

from which the total allowed fiber length in the link can be found.

If the bit rate for the link is to be higher, then the spacing may become dispersion-limited. In this case, the total length may be determined by an extension of Eq. (20.32) in Section 20.4.5, where the total dispersion Δ_t is the dispersion in μs/km multiplied by the length z in km, and B is the desired maximum bit rate in Mbits/s. That is,

$$B = \frac{1}{5z \, \Delta_t/\text{km}} \tag{20.39}$$

The shorter length z obtained from Eq. (20.38) or from Eq. (20.39) is the longest fiber length which may be used in the link.

Example 20.10 A fiber link is to be installed with the following characteristics:

Transmitter-laser diode with $P_t = 0$ dBm (or 1 mW)
Receiver—APD with minimum $P_r = -57$ dBm for an error rate of 1 in 10^9

Tx/Rx port losses	$L_{pt}, L_{pr} = 6$ dB each
Two connectors at	$L_c = 1$ dB each
Five splices at	$L_s = 0.5$ dB each
Fiber total loss	$L_f = 2$ dB/km at the
Fiber total dispersion	$\Delta_t(/\text{km}) = 0.505$ ns/km
Maximum desired bit rate	$B(\text{max}) = 35$ Mbits/s

(a) Determine the loss-limited line length for a loss margin of 5 dB. (b) Determine the maximum bit rate that the link of part (a) will support. (c) Determine the dispersion-limited length for the bit rate given.

Solution

(a) By Eq. (20.38),

$$P_t - P_r = M + N_c L_c + N_s L_s + L_{pt} + L_{pr} + zL_f$$
$$0 - (-57) = (5 + (2 \times 1) + (5 \times 0.5) + 6 + 6) + 2z$$
$$z = 17.8 \text{ km}$$

(b) By Eq. (20.39),

$$B(\text{max}) = \frac{1}{5\Delta_t z} = \frac{1}{5 \times 0.000505 \ \mu\text{s/km} \times 17.8 \text{ km}} = 22.2 \text{ Mbit/s}$$

which is less than desired. The link is dispersion-limited.
(c) By Eq. (20.39), the dispersion-limited length is

$$z = \frac{1}{5\Delta_t B(\text{max})} = \frac{1}{5 \times 0.000505 \times 35} = 11.3 \text{ km}$$

Thus, repeater spacing may not exceed 11 km for this link.

20.9 PROBLEMS

1. A glass-clad fiber is made with core glass of refractive index 1.500, and the cladding is doped to give a fractional index difference of 0.0005. Find (a) the cladding index, (b) the critical internal reflection angle, (c) the external critical acceptance angle, and (d) the numerical aperture.

2. Core glass of index 1.6200 is to be used to make a step-index fiber with an acceptance cone half-angle of $5°$. (a) What will the internal critical reflection angle be? (b) What should the cladding index be? (c) What fractional index difference does this give?

3. (a) Explain the difference between a step-index fiber and a graded-index fiber. (b) What is a W-index fiber?

4. (a) What is the advantage of using a graded-index core in a fiber? (b) Explain how that advantage is obtained.

5. The fiber in Problem (1) has a core diameter of 50 μm and operates with a light source of wavelength 0.843 μm. (a) What is the V-number of the fiber? (b) How many modes will it support?

6. A step-index fiber is made with a core of index 1.52, a diameter of 29 μm, and a fractional difference index of 0.0007. It is operated at a wavelength

of 1.3 μm. Find (a) the fiber V-number, (b) the number of modes the fiber will support, and (c) the names of each of those modes.

7. A step-index fiber has a core index of 1.5000, a fractional difference of 0.0005, and operates at a wavelength of 1.2 μm. (a) What is the largest diameter the core may have if it is to operate as a single-mode fiber? (b) Which mode does a single-mode fiber propagate?

8. Explain the differences between TE mode, HE mode, TM mode, and EH mode.

9. A germanium-doped silica fiber is to be used at a wavelength of 0.85 μm. Find approximate loss factors caused by (a) scattering, and (b) absorption. (c) Assuming maximum mode-coupling loss of 0.5 dB/km, what will be the total loss factor for the fiber? (d) If maximum loss in the fiber is to be less than 30 dB, how long can the fiber be?

10. Explain how energy is lost from a "leaky mode."

11. (a) Explain how mode coupling can occur. (b) Explain how mode coupling causes loss.

12. Explain how energy is lost from a fiber at a sharp bend.

13. (a) At what wavelength is a glass fiber likely to have its lowest loss? (b) What would be a typical value for that lowest loss, and what would be its main cause? (c) Why would a fiber not likely be used at a wavelength of 0.94 μm or 1.38 μm?

14. Explain the term *dispersion* as applied to optical fibers.

15. A step-index multimode fiber has a core index of 1.5000 and a cladding index of 1.49800. Find (a) the intermodal dispersion factor for the fiber, (b) the total dispersion in an 18-km length, and (c) the maximum bit rate allowed, assuming dispersion limiting.

16. If the fiber in Problem 15 is operated at 1.1 μm with a light source of 1.5 nm bandwidth, what is the total material dispersion?

17. A step-index fiber operated at 1.3 μm is found to have dispersion factors of 10 ns/km intermodal, 0.5 ns/km material, and 0.8 ns/km waveguide. The fiber is 15 km long and is restricted so that it only propagates one mode. (a) What is the total effective dispersion? (b) What maximum bit rate will it handle?

18. If a dispersion-limited fiber with a total dispersion factor of 3.3 ns/km is to be used at 1000 Mbits/sec, how long can it be without repeaters?

19. A light-emitting diode is made with gallium-arsenide doped with aluminum and has an energy band gap of 1.55 eV. What is its dominant emission wavelength?

20. A laser diode has an energy band gap of 1.1 eV, but experiences a 20% increase in emitted wavelength when operated as a laser. What wavelength does it emit?

21. Explain how a light-emitting diode generates light.
22. Explain the term *stimulated emission* as applied to laser diodes.
23. What advantages do semiconductor laser diodes have over light-emitting diodes when used for fiber transmission.
24. What is a stripe laser? What advantage does it have?
25. (a) Explain how a $p-i-n$ photo diode works. (b) How is this diode an improvement over the $p-n$ diode?
26. Can a $p-i-n$ diode be used as an avalanche diode? (b) If so, why would it not be so used?
27. (a) Explain how an avalanche diode works. (b) How does the avalanche diode compare to the $p-i-n$ diode for sensitivity? (c) What limits the ultimate sensitivity of an avalanche diode?
28. (a) Describe the spectral response of a typical silicon $p-i-n$ diode. (b) What type of diode might be used for a 1.3-μm receiver?
29. (a) List the factors which are most likely to affect the quality of a fiber connector. (b) List those which may affect the quality of a splice.
30. (a) What precautions should be taken when installing or repairing fiber equipment, connectors, or splices? (b) Explain why these precautions are so important in terms of impairment of transmission.
31. (a) Which splice technique gives the lowest insertion losses? (b) Which is the easiest done in the field? Why? (c) List typical insertion losses for the two types of splices.
32. List the steps involved in making a collapsed sleeve splice in a manhole full of water.
33. A fiber link is to be installed using the following items:
    (1) A laser diode with power output of 250 μW
    (2) A $p-i-n$ diode detector with a sensitivity of -45 dBm
    (3) A transmitter port loss of 5 dB
    (4) A receiver port loss of 0.5 dB
    (5) Two terminal connectors with 1.5 dB loss each
    (6) Ten splices with losses of 0.15 dB each
    (7) Fiber losses of 3 dB/km at 0.9 μm
    (8) Fiber dispersion of 0.9 ns/km
    (9) System loss margin of 5 dB
    (10) Maximum bit rate of 15 Mbits/s
        (a) Determine the maximum repeater spacing.
        (b) Is spacing loss-limited or dispersion-limited?

APPENDIX A

Logarithmic Units

A.1 THE DECIBEL

If two powers P_1 and P_2 differ, their difference can be expressed in decibels as

$$D = 10 \log_{10}\left(\frac{P_1}{P_2}\right) \qquad \text{(A.1)}$$

A positive number of decibels indicates that P_1 is greater than P_2, a negative number, that P_1 is less than P_2. For example, 50 W is greater than 10 W by $D = 10 \log_{10}(50/10) = 7$ decibels (dB), whereas 10 W is less than 50 W by $D = 10 \log_{10}(10/50) = -7$ dB. The decibel is one-tenth of a larger unit, the bel, defined as $\log_{10}(P_1/P_2)$; in practice, the bel is inconveniently large, and the decibel is the preferred unit.

Although the decibel is based on power ratios, it can also be used to express voltage and current ratios. Let the power P_1 be developed in a resistor R_1 across which the voltage is V_1 and the current through which is I_1; let the corresponding quantities for P_2 be R_2, V_2, and I_2. Then,

$$D = 10\log_{10}\left(\frac{V_1^2/R_1}{V_2^2/R_2}\right)$$

$$= 20\log_{10}\left(\frac{V_1}{V_2}\right) + 10\log_{10}\left(\frac{R_2}{R_1}\right) \tag{A.2}$$

By similar reasoning, it is easily shown that D can be expressed as

$$D = 20\log_{10}\left(\frac{I_1}{I_2}\right) + 10\log_{10}\left(\frac{R_1}{R_2}\right) \tag{A.3}$$

Therefore, it can be seen that a knowledge of the resistance values is needed, in addition to that of voltage or current values, in order to determine the decibel value. Because of the widespread use of the decibel, its meaning has been extended to cover voltage and current ratios. By definition, if two voltages V_1 and V_2 differ, their difference D_v in decibels is given by

$$D_v = 20\log_{10}\left(\frac{V_1}{V_2}\right) \tag{A.4}$$

Similarly, if two currents I_1 and I_2 differ, their difference D_i in decibels is

$$D_i = 20\log_{10}\left(\frac{I_1}{I_2}\right) \tag{A.5}$$

As already shown, D_i and D_v can only be related to D when the ratio R_1/R_2 is known.

It is meaningless to state absolute values of voltage, current, or power in decibels. The decibel represents a ratio, and therefore a reference level must also be stated if absolute values are required. For example, the phrase "a power of 20 dB" is meaningless, but "20 dB relative to 1 W" means 100 W. Very often the reference level is implicitly understood. A common example of this is where noise levels are quoted in decibels, the reference level being the threshold of hearing (see Fig. 3.5). Another common example, in telecommunications engineering, is the selectivity curve of a tuned circuit, where the resonance value is used as reference (see Fig. 1.22).

Sometimes the reference level is indicated in abbreviated form, the best example of this being the dBmW, meaning "decibels relative to 1 milliwatt." Other examples are dBμV and dBV, these being decibels relative to 1 microvolt and 1 volt, respectively.

A.2 THE NEPER

The neper is a logarithmic unit originally introduced to express the attenuation of current along a transmission line, using natural (or naperian, hence the name "neper") logarithms rather than common logarithms. If two currents I_1 and I_2 differ, their difference N in nepers is

$$N = \ln\left(\frac{I_1}{I_2}\right) \quad \left(= \log_e\left(\frac{I_1}{I_2}\right)\right) \qquad (A.6)$$

The relationship between decibels and nepers is easily established. Equation (A.5) gives

$$D_i = 20\log_{10}\left(\frac{I_1}{I_2}\right)$$

and on changing the logarithmic base,

$$D_i = (20\log_{10}e)\left(\ln\frac{I_1}{I_2}\right)$$

$$= 8.686\,N \qquad (A.7)$$

Example A.1 Express (a) 3 nepers in decibels; (b) 3 decibels in nepers.

Solution (a) $\qquad\qquad D = 8.686 \times 3$

$\qquad\qquad\qquad\qquad = 26.06\ \text{dB}$

(b) $\qquad\qquad N = \dfrac{3}{8.686}$

$\qquad\qquad\qquad\qquad = 0.345\ \text{N}$

A.3 LOGARITHMIC SCALES

By graduating graph axes in lengths proportional to the logarithms of numbers rather than to the numbers themselves, a much wider range of values can be accommodated on a given scale. Scales based on common logarithms are usually employed. Figure A.1 illustrates graph paper that has one axis graduated logarithmically and the other linearly. This is known as "semilog" graph paper, and because the logarithmic scale pattern repeats itself four times, it is referred to as "four-cycle." The logarithmic scale of Figure A.1 can

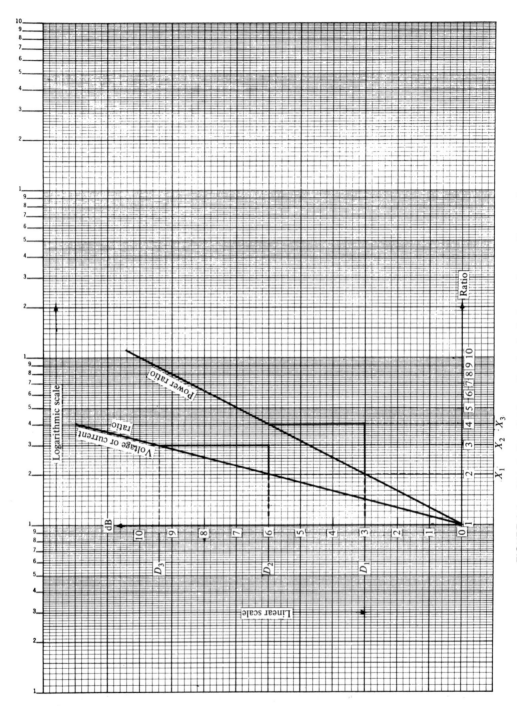

FIGURE A.1. Conversion chart — ratios to decibels — using semilog graph paper.

accommodate values ranging over four orders of magnitude (e.g., 1 to 10^4, 10 to 10^5, 0.01 to 10^2, etc.). It will be apparent that a zero origin cannot be shown on a logarithmic scale, and care must be taken when interpreting graphs plotted on logarithmic scales. Also, the slope of the graph must be carefully interpreted, as it involves the logarithms of numbers. Figure A.1 shows the decibel equations (D, D_i, and D_v) plotted against a ratio scale, which is logarithmic. The slope of the *voltage or current ratio line* should be 20, and this is obtained from

$$\text{slope} = \frac{D_3 - D_2}{\log_{10} X_2 - \log_{10} X_1} \qquad \left(\text{not } \frac{D_3 - D_2}{X_2 - X_1}\right)$$

$$= \frac{9.5 - 6}{0.4771 - 0.3010}$$

$$= 20 \tag{A.8}$$

Similarly, the slope of the *power ratio line* is

$$\text{slope} = \frac{D_2 - D_1}{\log_{10} X_3 - \log_{10} X_1} \qquad \left(\text{not } \frac{D_2 - D_1}{X_3 - X_1}\right)$$

$$= 10 \tag{A.9}$$

The Transverse Electromagnetic Wave

Electromagnetic energy is propagated through space and is guided along transmission lines in the form of a *transverse electromagnetic wave* (TEM wave). For this type of wave, the electric field E (V/m), the magnetic field H (A/m), and the direction of propagation x, along which the wave travels with phase velocity v_p (m/s), are mutually at right angles, as shown in Fig. B.1(a). If the TEM wave is reversed in direction, then either the E or the H field (Fig. B.1(b)) must reverse. This is similar to the condition required when reversing direction of rotation of a dc generator while maintaining same polarity of induced EMF.

To illustrate some of the basic properties of the TEM wave, a sinusoidal variation will be assumed; the field equations are then

$$e = E_{max}\sin(\omega t - \beta x) \quad \text{V/m} \quad (B.1)$$

$$h = H_{max}\sin(\omega t - \beta x) \quad \text{A/m} \quad (B.2)$$

where $\omega = 2\pi f = 2\pi/T$, and $\beta = 2\pi/\lambda$. The periodic time T and the wavelength λ are shown in Fig. B.1(c) and (d).

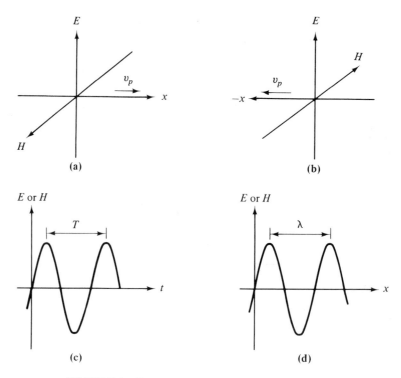

FIGURE B.1. Transverse electromagnetic (TEM) waves.

Clearly, since one cycle occurs in T seconds, the number of cycles in one second, which is the frequency f in Hertz, is

$$f = 1/T \tag{B.3}$$

Also, since the wave generates f cycles in one second and covers a distance v_p meters in one second, the wavelength λ, which is the distance spanned by one cycle, is

$$\lambda = \frac{v_p}{f}$$

or

$$\lambda f = v_p \tag{B.4}$$

In the following equations, rms values, indicated by E, H, B, and D, will be used for simplicity.

The electric field E will be accompanied by an electric flux density D (C/m^2) given by

$$D = \varepsilon E \tag{B.5}$$

The magnetic field H will be accompanied by a magnetic flux density B (teslas), given by

$$B = \mu H \tag{B.6}$$

Here, ε is the absolute permittivity, and μ the absolute permeability, of the medium in which the wave travels.

Two further experimental facts are: (1) an electric field density D moving at velocity v_p will induce a magnetic field H given by

$$H = Dv_p \tag{B.7}$$

and (2) a magnetic field density B moving at velocity v_p will induce an electric field E given by

$$E = Bv_p \tag{B.8}$$

The vector directions in each case are as shown in Fig. B.1(a). The fact that each type of field, when moving, generates the other in the proper direction gives rise to a self-sustaining electromagnetic wave, once the initial disturbance of either type of field is made.

The relationships in Eqs. (B.7) and (B.8) may not be familiar, but a common application of the latter is in the electromechanical generation of an EMF by rotating armature conductors to cut a magnetic field at right angles.

From the four field equations (B.5), (B.6), (B.7), and (B.8), the following important relationships are easily obtained.

Wave impedance:

$$Z_0 = \frac{E}{H} \ \Omega \tag{B.9}$$

$$= \sqrt{\frac{\mu}{\varepsilon}} \tag{B.10}$$

(This should be compared with Eqs. (12.15) and (12.16).)

Phase velocity:

$$v_p = \frac{1}{\sqrt{\mu\varepsilon}} \quad \text{m/s} \tag{B.11}$$

Average power density in wave:

$$P_D = EH \quad \text{W/m}^2$$

$$= \frac{E^2}{Z_0}$$

$$= H^2 Z_0 \tag{B.12}$$

(These relationships should be compared with the circuit equations relating V, I, R, and power P.)

APPENDIX C

Answers to Odd Numbered Problems

SECTION 1.16   1. $I = 200$ A; $R = .01$ Ω
   3. .2
   5. 5.5 dB
   7. $R = 285.7$ Ω; $R_3 = 487.1$ Ω
   9. $Z = 50 + j100$ kΩ; $V_R = 50$ V; $V_L = 100$ V; $V = 112$ V (peak voltages)
   11. $R_p = 158$ kΩ (frequency dependent); $L = 2$ μH (frequency independent)
   13. $C_{MIN} = 3333$ pF; $C_{MAX} = .03$ μF
   15. $C = 5$ pF; $|V_L| = |V_C| = 314$ V (rms)
   17.   50 MHz
   19. 188; 150
   21. 3.46 : 1 step-up
   23. see Section 1.12.2
   25. 38.46 kΩ; 5 pF
   27. $R = 50$ Ω, $C = 79.6$ pF; $P_{AV} = .045$ pW
   29. $.4\underline{/40°}$
   31. (a) 1 V; (b) .0397; (c) 0 dB and -27.97 dB
   33. (a) .992; (b) .970; (c) .625

SECTION 2.4   1. Fig. 2.7 (g), $V_p = \pm 1$ V
   3. 18.19 mV each
   5. .0157

SECTION 3.13 13. 3 kHz, 1.585 octaves

23. 13.9 pF, 2.78 nC, 1.6 mVrms

25. 0.1 V, No.

SECTION 4.13 1. 4.41×10^{-17} watts; 1.05×10^{-17} watts

3. (a) 8.94 μV; (b) 15.5 μV; (c) 5.16 μV

5. 500 Hz

7. 1.38 μV

9. 0.31 μV

11. 1829 : 1 or 32.6 dB

13. 8.2 μV

15. 39.2 dB; 34 dB

17. 0.92

19. 1.09

SECTION 5.9 1. 30.53 pF, 11.8, 2.12 MHz

3. 30 pF

5. 0.127 μH

7. 125 μS

13. 1.77 MHz, 162 μH, 2.95 MHz

15. Fig. 5.18 (b), 0.239 μH, 955 pF, 1.50

17. 816 pF, (ignore C_o, r_o, R_L) 85.8, $j250$, 51.4 kΩ, 7.5 kHz.

SECTION 6.10 1. 50 kΩ, 130 pF, 2.92 kΩ

3. 2.60 MHz, 3

5. 11.25 MHz, 12 dB

7. 22.4 μS, 7.12 MHz

SECTION 7.11 1. 43.525–70.525 MHz, 84.050–111.050 MHz, low-pass filter, cutoff between 40.525–43.525 MHz

3. 7.6–13.6 MHz, −29.9 dB

5. 483 pF, 3.88 μH

7. 7.49 pF, 319.9 pF

SECTION 8.5 1. 28 V

3. $E_c = 10$ V; $m = .5$

5. 0.8

7. 0.43

9. 0.95; 0.3

11. .99

13. (a) .66; (b) $P_c = 112.5$ W; (c) $P_T = 137.5$ W; (d) 9.2 W; (e) 93.75%

SECTION 9.8   3. 0.5 kHz, 1.8 kHz, 910.5 kHz, 911.8 kHz.
   5. 1 kHz at 3.535 V_p, 499 kHz at 1.105 V_p, 501 kHz at 1.105 V_p
   7. Interchange carrier signals to modulators 1, 2
   9. Signals at 252, 248.5 kHz, L.S.B. and 252.5 to 247.5 kHz
  11. 0.5 and 2 kHz give 2.5 and 1 kHz out
  13. Channels 5.5 kHz wide starting at zero. Within each, modulation 0–4 kHz, Pilot at 4.5 kHz. BW 76 kHz

SECTION 10.8   1. 15 kHz, 5
   3. Spectral lines 4.8 V at 90.00 MHz, 0.9 V at 89.985, 90.015 MHz, 0.1 V at 89.970, 90.03 MHz
   5. 0.1 Rad
  11. 6 kHz, 3 kHz, 36 kHz, 45.7 dB, 25.23 dB
  13. 0.01745 kHz, 1719 x, 3 triplers to 27 MHz, down conversion to 1.875 MHz ($f_o = 28.875$ MHz), 6 doublers to 120 MHz

SECTION 11.6   1. (a) 8 kHz, (b) 40 kHz, (c) 9 MHz
   5. (b) 80 kHz; (c) 127.3 MHz

SECTION 12.7   3. 5 mN/m; $.2\pi$ rad/m
   7. (a) 3.33 ns; (b) 3×10^8 m/s
   9. $.632\big/\underline{108°}$
  11. (a) 2 : 1; (b) 3.65 : 1
  13. (a) 1/8; (b) 3/8
  15. $5.3 - j2.8$ mS; $R_p = 188.7$ Ω; $X_p = 357.1$ Ω (inductive)
  17. $24 + j\ 31$ Ω
  19. $.2632\ \lambda$; $.072\lambda$ shorted stub
  21. (b) and (d) can be matched
  23. 33.3 Ω
  25. $R = 1.984 \times 10^{-3}$ Ω/m; $G = 2.5 \times 10^{-7}$ S/m; $L = 1.189$ μH/m; $C = 30.89$ pF/m
  27. $\lambda/4$ section, 29.44 Ω; open circuit stub, 25.97 Ω

SECTION 13.4   1. $\lambda_c = 3.16$ cm; $\lambda_g = 9.55$ cm; $\lambda = 3$ cm
      $v_p = 9.55 \times 10^8$ m/sec
      $v_g = .942 \times 10^8$ m/sec
   3. see section 13.2.2
   5. 1.36
   9. Highest frequency for TE$_{10}$ mode is 6.30 GHz. At $f = 6$ GHz only the TE$_{10}$ mode is excited

SECTION 14.10 1. (a) 150.6 dB; (b) 0.871 μW

                 3. 1.93 m

                 5. see section 14.3.6

                 7. see section 14.4.1

                 9. see section 14.4.7

                11. see section 14.4.9

                13. see section 14.5

SECTION 15.20 1. (a) 50 $\underline{/-11°}$ ohms; (b) 61.09 $\underline{/1°}$ ohms

                5. $d = 11.36$, $\phi = 26.56°$

                7. (a) 8.66 μV; (b) 5 μV; (c) 0

                9. (a) .796 W/m; 17.3 V/m

                    (b) 7.96 mW/m; 1.73 V/m

                    (c) .796 μW/m; 0.173 V/m

                    (d) 7.96 nW/m; .00173 V/m

               11. $\theta_{-3\text{ dB}} = 65.54°$; $\theta_{-10\text{ dB}} = 111.6°$

               13. 8 m

               15. .667 mV/m

               17. 2.1 μA

               21. .317 λ

               23. Point A 2764 m; Point B 1916 m

               27. Lengths in metres, working from reflector to front director: 2.17, 2.07, 1.96, 1.86, 1.77. Spacing = .413 m

               29. (a) 180°, (b) 106.3°, (c) 90.5°

SECTION 16.3 1. 1121 Ω, 5.94 dB

                5. 37 dB

                9. 1.74 dB, 3140 km

               11. 4.57 dB, Yes

SECTION 17.14 1. 32 kHz; 40 kHz

                3. Bandwidths are 100, 150 and 200 kHz

                5. $y_0 = 0$, $y_1 = 1$, $y_2 = .5$, $y_3 = .5$, $y_{n>3} = 0$

               13. (a) .922; (b). 67

               15. (a) See Fig 17.13 (d). (b) 01000100 parity bit first

               19. 14 bits

SECTION 18.7   5. Assume IOC(CCITT), 553 Hz

7. 1809, 576, 1358 Hz

9. 2

11. 4.59 MHz

SECTION 19.12   1. Uplink carrier frequencies in MHz; 5950, 5990, 6030, 6070, 6110, 6150, 6200, 6240, 6280, 6320, 6360, 6400.
Downlink carrier frequencies in MHz: 3725, 3765, 3805, 3845, 3885, 3925, 3975, 4015, 4055, 4095, 4135, 4175

3. 20.37 dB

5. 74 km

7. 35.7 ms

9. 659 K

11. 145.8 dB

13. 100 dB

15. (a) 59.23 dBW, (b) -105.81 dBW, (c) 5.067×10^{-13} W, (d) 17.14 dB

17. 71.8 dBW

CHAPTER 20.9   1. 1.4993, 88.2°, 2.7°, 0.0474

5. 8.83, 39

7. 19.4 μm, HE_{11}

9. 2.3 dB/km, 1.0 dB/km, 3.8 dB/km, 7.9 km

13. 1.3 μm, 0.7 dB/km, absorption by OH^- radicals

15. 6.68 ns/km, 120 ns, 1.66 Mbits/s

17. 14.2 ns, 14 Mbits/s

19. 0.80 μm

33. 8 km loss limited

Index